VIRUSES

Biology Applications Control

VIRUSES

Biology Applications Control

David R. Harper

Garland Science
Taylor & Francis Group
NEW YORK AND LONDON

Garland Science
Vice President: Denise Schanck
Editor: Elizabeth Owen
Assistant Editor: David Borrowdale
Production Editor: Ioana Moldovan
Illustrator: Xavier Studio
Cover Design: Andrew Magee
Copyeditor: Jo Clayton
Typesetting: EJ Publishing Services
Proofreader: Susan Wood
Indexer: Medical Indexing Limited

David Harper is the Chief Scientific Officer of AmpliPhi Biosciences. After completing a BSc in Microbiology and Virology at the University of Warwick and a PhD at the University of Newcastle-upon-Tyne studying viral genetics, he carried out postdoctoral work at St. Bartholomew's Medical School in London and at the University of Iowa in the USA. He joined the faculty at St. Bartholomew's in 1991 as Lecturer in Molecular Virology and was leader of the herpesvirus research group in the Department of Virology. In 1997 he founded Biocontrol Limited, which combined with Targeted Genetics in the USA to form AmpliPhi Biosciences in 2011.

ISBN 978-0-8153-4150-5

Library of Congress Cataloging-in-Publication Data

Harper, D. R. (David R.)
Viruses : biology, applications, and control / David R. Harper.
p. ; cm.
Expanded version of Molecular virology / David Harper. 2nd ed. 1998.
Includes bibliographical references.
ISBN 978-0-8153-4150-5 (pbk. : alk. paper)
1. Molecular virology. 2. Virus diseases--Molecular aspects. I. Harper, D. R. (David R.). Molecular virology. II. Title.
[DNLM: 1. Viruses. 2. Antiviral Agents. 3. Immunity. QW 160]
QR389.H38 2012
616.9'1061--dc23

2011015551

Published by Garland Science, Taylor & Francis Group, LLC, an informa business, 711 Third Avenue, New York, NY 10017, USA, and 2 Park Square, Milton Park, Abingdon, OX14 4RN, UK.

Printed in the United States of America

15 14 13 12 11 10 9 8 7 6 5 4 3 2 1

Visit our website at http://www.garlandscience.com

Cover images courtesy of the Centers for Disease Control and Prevention. **Top row**: Marburg virus (*Filoviridae*); measles virus (*Paramyxoviridae*); influenza virus (*Orthomyxoviridae*); vesicular stomatitis virus (*Rhabdoviridae*); Ebola virus (*Filoviridae*); Tacaribe complex virus (*Arenaviridae*). **Middle row**: Marburg virus (*Filoviridae*); smallpox virus (*Poxviridae*); norovirus (*Caliciviridae*); an Old World arenavirus (*Arenaviridae*); Orf virus (*Poxviridae*); influenza virus (*Orthomyxoviridae*). **Bottom row**: an adenovirus (*Adenoviridae*); infectious bronchitis virus (*Coronaviridae*); herpes simplex virus (*Herpesviridae*); La Crosse encephalitis virus (*Bunyaviridae*); norovirus (*Caliciviridae*); smallpox virus (*Poxviridae*).

Preface

Viruses: Biology, Applications, and Control is a much-expanded version of the second edition of Molecular Virology. It provides a solid grounding in biomedical virology and also broad coverage of the ways in which viruses relate to the modern world. It is never possible for a book of this size to be as comprehensive as the author might wish, but by focusing on key mechanisms and developments the book presents an overall picture along with sufficient detail for the reader to develop a useful understanding of the field as a whole.

Many new or expanded sections have been included to cover recent developments in this rapidly expanding field including virus evolution and extinction, emerging infections, gene therapy, bacteriophage therapy, and diagnostics. A comprehensive listing of antiviral drugs underpins coverage of ways in which virus infections can be controlled, and also included are the interactions between viruses and the immune system. The main text gives an overview of virus replication strategies with detail on specific viruses being given in a special appendix. By providing such extensive coverage in a single volume, this book allows readers at all levels from novice to expert to dip in and retrieve information on a wide variety of topics.

ACKNOWLEDGMENTS

Many others have provided material that has been used to illustrate the points made in this book. The author is particularly grateful to the CDC Image Library, and to Sally Adams, Ian Chrystie, Linda Ebbs, A.M. Field, Charles Grose, C.R. Madely, Keith Nye, Jackie Parkin, and Didier Raoult, and also to the authors of the *Sourcebook of Medical Illustration*.

Thanks are also due to the staff at Taylor & Francis for their patience and support through the lengthy gestation of this work.

In writing this book the author and publisher have benefited greatly from the advice of many virologists. We would like to thank the following for their suggestions in preparing this edition.

John L. Casey (Georgetown University, USA); Andrew D. Davidson (University of Bristol, UK); David D. Dunigan (University of Nebraska, USA); Sheila Graham (University of Glasgow, UK); Stephen Rice (University of Minnesota, USA); Alexander Voevodin (Vir&Gen, Toronto, Canada).

DEDICATION

This book is dedicated to my daughters Sophia and Evie, and to their favorite cats, Mao Nong Chai and Kipling.

Contents

CHAPTER 1
Virus Structure and Infection

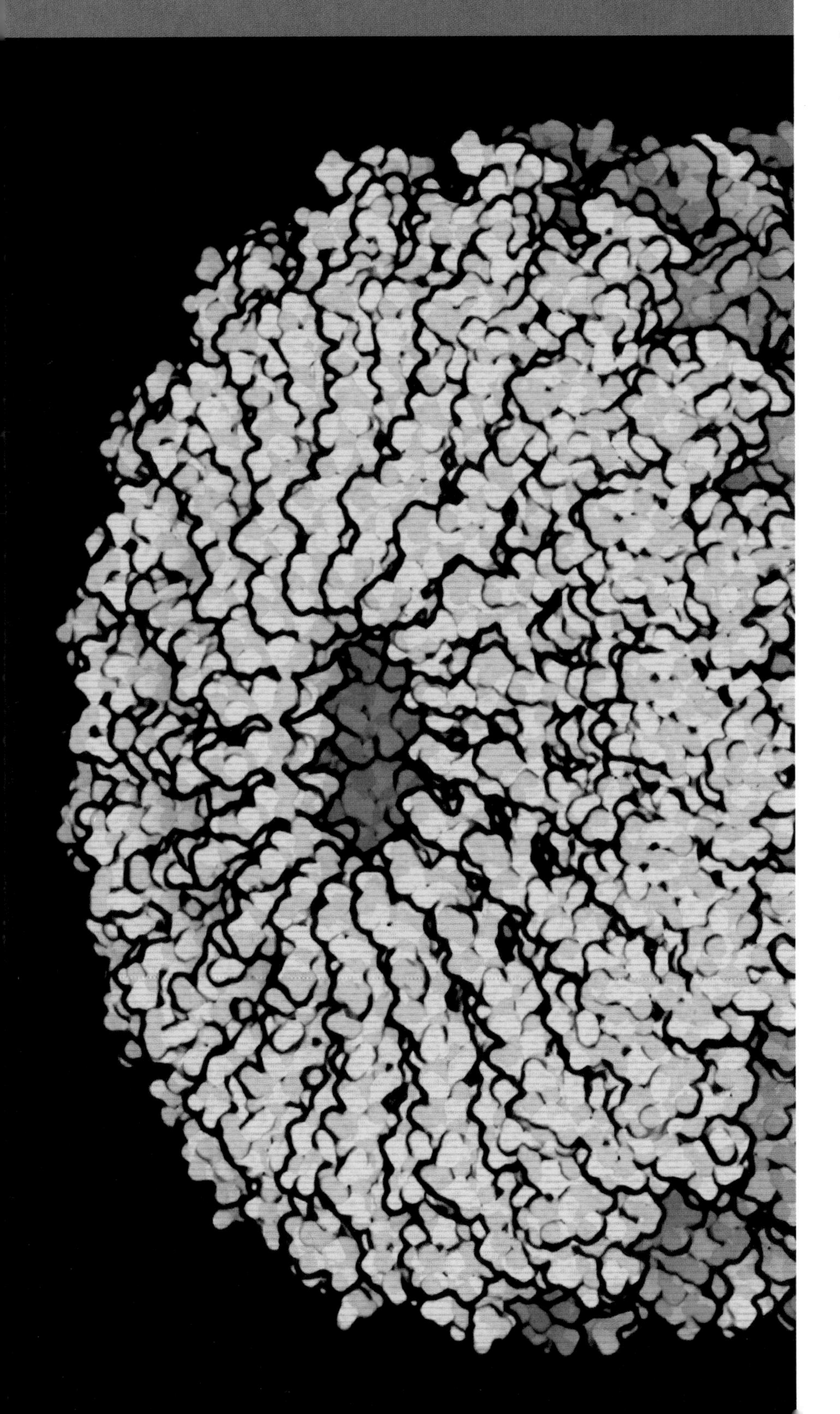

INTRODUCTION

Viruses are the most numerous and the most ubiquitous form of life on the planet Earth. By many estimates there are more than ten million trillion trillion (10^{31}) viruses on the planet—though the accuracy of this figure is of course impossible to verify. Viruses infect every species that has been examined, from bacteria to humans. Even some viruses have parasitic viruses.

About the chapter opener image
Tobacco mosaic virus
(Courtesy of the Research Collaboratory for Structural Bioinformatics Protein Data Bank and David S. Goodsell, The Scripps Research Institute, USA.)

While the common perception of the effect of viruses on their hosts is of damaging and often lethal infections, this is usually inaccurate. The majority of viruses live far more quietly. In fact, many have adapted to produce no apparent effects at all, and foremost among these are those virus remnants that live permanently within the genomes of their hosts. More than eight percent of the human genome appears to be made up of such remnants. Some may even be beneficial. The ability of the female mammal to nurture a fetus without rejecting the half-foreign tissue that is growing within the mother's body appears to rely on the effects of such "endogenous" viruses.

It seems self-evident that a life form that requires cells to live must have evolved after these cells were available to support its lifestyle, but more recent theories suggest that viruses may have existed in a pre-cellular world and that, billions of years ago, they may have played a key role in generating more complex forms of life. Even today, it is becoming clear that viruses have a major role in the transfer of genes between both related and unrelated species, and that viral effects exert control at the largest scale in every ecosystem that has been studied.

Thus, while the spectacularly lethal effects of Ebola, human immunodeficiency virus (HIV), or bird flu draw attention to the damage that viruses can cause this is very much the tip of the iceberg. Viruses are involved at all levels from the start of life billions of years ago in the primordial soup to the beginning of new life in the womb today.

Virology, the study of viruses, and of their nature and effects, is a relatively young discipline. The early light microscopes of van Leeuwenhoek showed the details of cellular structures and even the "little animalcules" of protozoa (in 1674) and bacteria (in 1676). However, light microscopes can only resolve down to approximately half the wavelength of the light used (approximately 200 nm) and most viruses are smaller than this. The direct observation of viruses had to wait for the development of the electron microscope in the 1930s. Up until these first observations, there was still a strong body of opinion that held that viruses were no more than chemical toxins (**Box 1.1**). We now know rather more.

1.1 WHAT IS A VIRUS?

A virus is a subcellular organism with a parasitic intracellular life cycle. It requires the replication machinery of a cell to replicate. It has no metabolic activity outside the host cell. Rather, it has the potential for life, in the same way that a disk containing the code for a computer program is only a potential program until it is put into the host computer. Viruses are not (as is often proposed) the simplest form of life, since their life cycle involves not only their own metabolism, but also that of the cell whose replicative machinery they use.

The effects of viruses on the infected cell and on the organism as a whole can be highly variable (see Section 1.4 and Chapter 4). Cell killing by the virus as the cause of acute disease is actually unusual. Some infections may lack any obvious signs, and where disease does occur in most cases the symptoms mainly reflect the immune response to infection rather than the infection itself. In some cases such immune-mediated symptoms may occur, and even kill, after the infection is resolved. In other cases, the infection is much longer term, with virus remaining within the body at low levels or in specific locations. In some cases infection may even lead to oncogenesis, the induction of cancers.

As experimental systems, viruses provide us with relatively straightforward approaches to studying biochemical events, and have helped to explain many aspects of biology. A generalized outline of virus replication is shown in **Figure 1.1**. Viruses bind to specific receptors and then enter the cell by a variety of mechanisms. Once within the cell, the virus begins making messenger RNAs and then proteins, usually in staged groups. The early proteins are the machinery of replication, while the later proteins are used to form the next generation of viruses, along with new copies of the viral genome.

Box 1.1 A brief history of virology

The origins of virology are usually dated to 1892. In that year, Dimitri Ivanovsky, a Russian biologist working in the Ukraine (**Figure 1**), reported to the Academy of Sciences of St. Petersburg that the infectious component of mosaic disease of tobacco was capable of passing through an unglazed porcelain filter that would hold back bacteria and fungi. These filters were then a recent development, and had been designed to remove infectious agents from liquids, in particular for the purification of water.

In fact, it was Adolf Meyer in 1879 who reported that tobacco mosaic disease was capable of transmission between plants. Meyer tried to isolate a bacterium or fungus, the only known infectious agents at that time, but was unsuccessful. Ivanovsky's contribution was to show that the cause was apparently much smaller than a bacterium or fungus. The term "filter-passing" was to remain an important part of the definition of a virus for many years.

Despite his status as the father of virology, Ivanovsky did not actually isolate the cause of tobacco mosaic disease. It was left to Martinus Beijerinck, a collaborator of Meyer's, to show in 1898 that the agent could amplify itself in tobacco plants, indicating that it was not simply a chemical toxin. Beijerinck also used the term *contagium vivum fluidum* (infectious living liquid) to describe his finding. Other filter-passing viruses were identified; the first virus of animals (foot and mouth) in 1898, and the first virus of humans (yellow fever) in 1901.

However, the nature of the *contagium vivum fluidum* remained controversial. Even after Felix d'Herelle established the basic techniques of modern virology from 1917 onward, other workers continued to support the idea of a chemical toxin. Through the 1930s, the chemical nature of viruses was clarified, and it became clear that they were composed mainly of protein, with a smaller amount of nucleic acid.

Figure 1
Dimitri Iosifovich Ivanovsky. The Russian biologist who is usually credited with the first report of viruses in 1892.

Figure 2
A replica of Ernst Ruska's 1933 electron microscope, the first to have a resolving power greater than a light microscope, constructed in 1980 by Ernst Ruska. Courtesy of J Brew under Creative Commons Attribution-Share Alike 3.0 Unported.

It was not until the development of the electron microscope in the 1930s (**Figure 2**) that the physical nature of viruses was firmly established. Almost all viruses are too small to be seen in the light microscope, which can only resolve down to approximately half the wavelength of the light used (approximately 200 nm). It took even longer to show the functions of the various components and the processes involved in virus infections, but eventually the nature of viruses was confirmed.

What's in a name?

The first recorded use of the term "virus" dates from 1392, when it was used to define a venomous substance. Its use to describe an infectious agent in general dates from 1728. By the time of Ivanovsky's work, the generally accepted meaning was of a poisonous or toxic substance, and the term was often used to describe any infectious agent. Thus, the original term "filter-passing virus" referred to a toxic agent rather than a virus in the modern sense. It was only as further investigations clarified the nature of these new, smaller agents of infection that the term began to refer specifically to the infectious nucleic acid–protein complex that today defines a virus. And of course the language continues to evolve. From 1972, the self-copying programs that can "infect" computers have also been known as viruses.

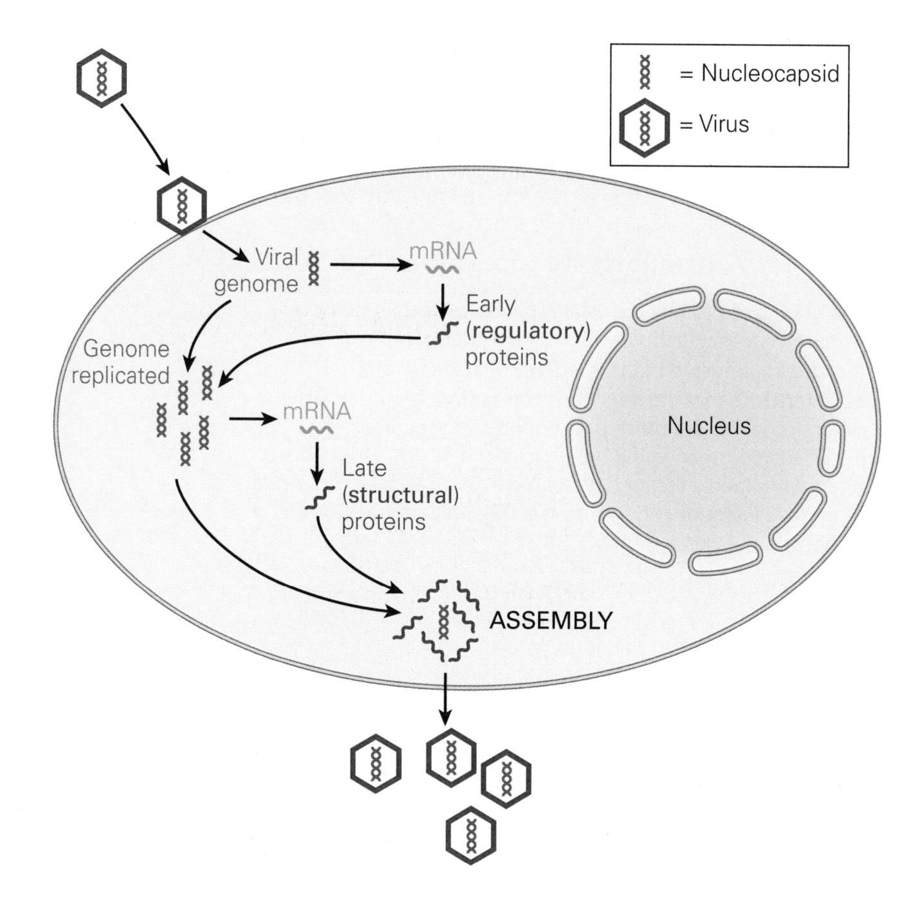

Figure 1.1
Schematic diagram of virus replication. Virus particles (virions) attach to target cells and gain entry. The viral genome (purple) is transcribed into mRNA (green) and is also replicated. The early proteins (red) translated from viral mRNA often have a regulatory function, while those produced later have a structural role. Viral proteins and nucleic acids are assembled into progeny viral particles, which are released from the cell.

Once formed and filled with the viral genome, progeny viruses can leave by a range of routes ranging from budding through the cell membrane to destruction of the infected cell.

The discovery of extremely complex viruses and extremely simple cells with intracellular replication is blurring the boundaries of what makes a virus, and it is likely that future discoveries will show this to be a continuum rather than a border, with some forms of life exhibiting the properties of both.

What makes a virus?

At its most basic, a virus particle (a **virion**) consists of a (DNA or RNA) nucleic acid *genome* surrounded by a shell of protein (the *capsid*), which often also contains lipids and sugars. The basic function of a virion is to stabilize the nucleic acid that carries the information needed to make the next generation of viruses, and to deliver it into a cell where it can replicate. This requires:

- Structures to contain and protect the nucleic acid genome together with any associated proteins which are required for its replication. The core complex of nucleic acid packaged with structural and replicative proteins is called the **nucleocapsid**.
- Specific receptors/effectors on the virion surface that allow the virus to bind to and enter the target cell. These are often complexes of protein and sugars (**glycoproteins**) and may be contained in an **envelope** derived from host cell lipid.

These requirements are fulfilled in very different ways by different viruses. The complexity of a virus is a direct reflection of the size of the viral genome. The more information a virus can encode, the more proteins it can make.

Viruses infecting humans produce anything from one to more than a hundred proteins. Viruses come in a range of shapes and sizes (morphologies).

The viral genome can be either DNA (as with all cellular life) or (uniquely to viruses) RNA. Virions do not contain both types of nucleic acid. An RNA genome, while relatively common among viruses, has many implications for the virus, discussed in Chapter 3.

1.2 VIRUS STRUCTURE AND MORPHOLOGY

Capsids

The **capsid** is the protein structure surrounding the viral genome and is formed of repeating protein subunits (**capsomers**) assembled around the nucleic acid to form the nucleocapsid. The use of small protein subunits reduces the amount of genetic coding capacity that has to be dedicated to producing the capsid proteins. These subunits are also often capable of self-assembly around the viral genome, a property first demonstrated with tobacco mosaic virus in the 1930s. Saving on coding capacity is important since even the largest viral genomes are small by cellular standards (**Figure 1.2**). However, many cellular genomes have very high levels of apparently noncoding DNA. In viruses, use of coding capacity is very efficient by comparison. In the case of the *Hepadnaviridae*, overlapping genes enable the genome to encode RNAs that are greater in total than the size of the genomic DNA (see below). Even in the largest virus, 90% of the DNA is thought to code for proteins. This is an extremely high figure by cellular standards; for humans the figure is generally agreed to be less than 5%.

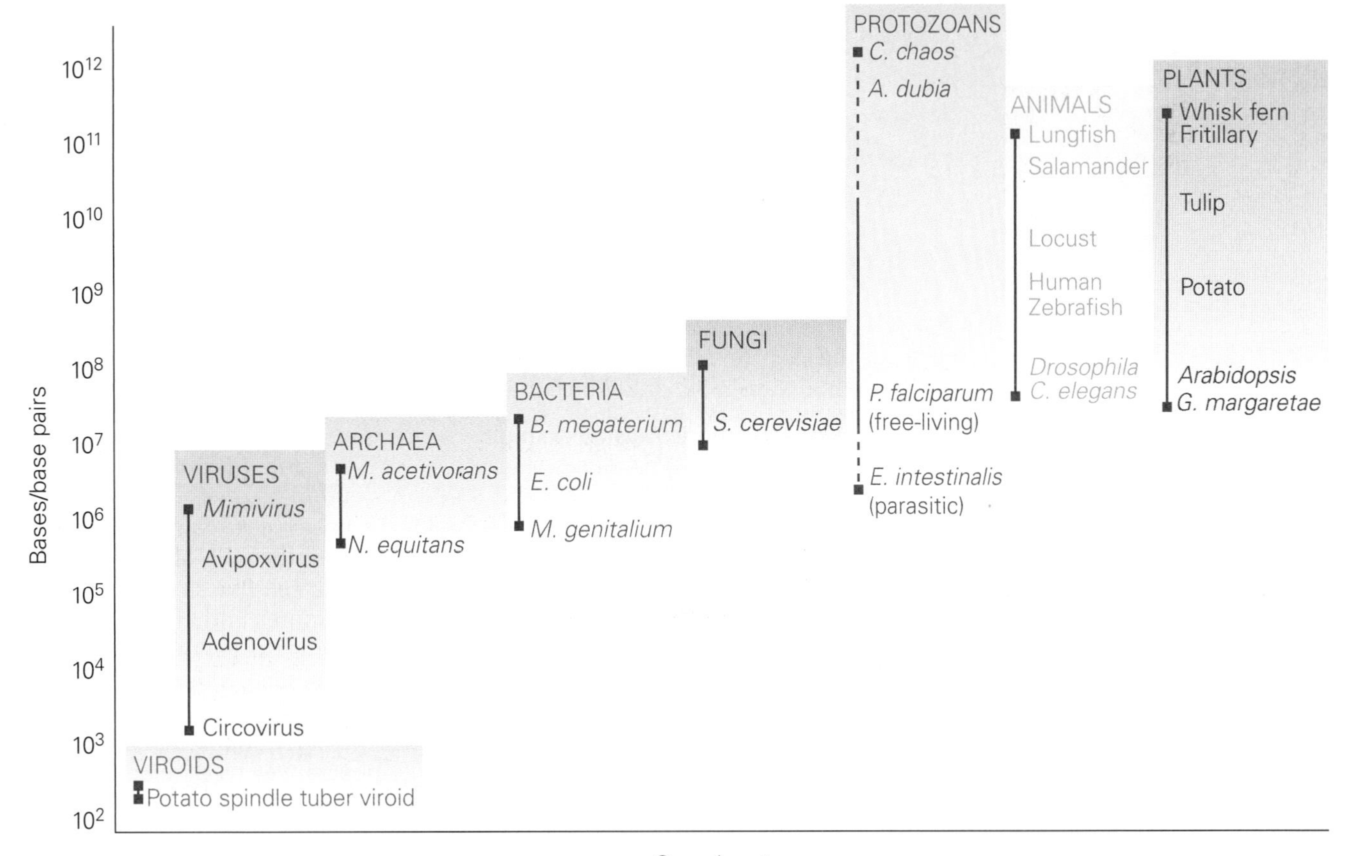

Figure 1.2
Sizes of genomes of representative organisms. Although viral genomes are much smaller than those of most cellular organisms, most viral DNAs are highly functional, in some cases with overlapping genes that increase coding capacity still further within their small genome. The largest protozoan genomes may simply reflect chromosome duplication (polyploidy), possibly to support the nuclear structure of a very large cell. Despite the large size of its genome, the whisk fern is a very simple plant, the genome of which appears to contain very high levels of repetitive DNA.

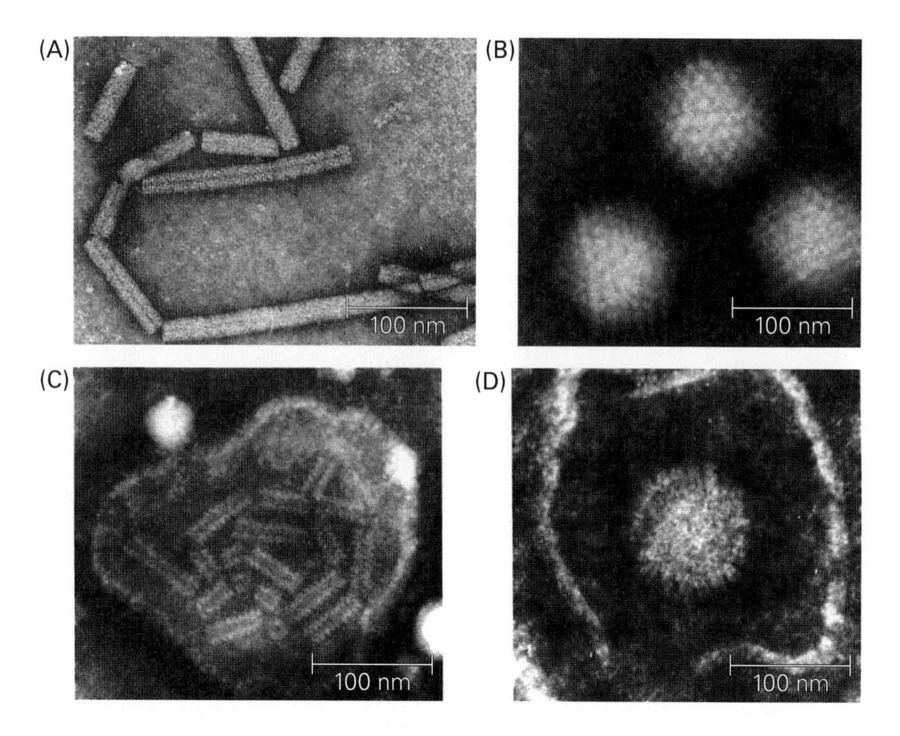

Figure 1.3
Examples of viruses from the main morphological groups. (A) Unenveloped, helical (tobacco mosaic virus). (B) Unenveloped, icosahedral (adenovirus). (C) Enveloped, helical (paramyxovirus). (D) Enveloped, icosahedral (herpesvirus). (A) Courtesy of CR Madeley, Royal Victoria Infirmary, Newcastle-upon-Tyne. From Madeley CR & Field AM (1988) Virus Morphology. With permission from Elsevier. (B) to (D) Courtesy of Ian Chrystie, St Thomas' Hospital, London.

Viruses representing the standard morphologies are shown in **Figure 1.3**. These are:

- Non-enveloped/helical capsid (exemplified by tobacco mosaic virus, as no such viruses are known to infect humans)
- Non-enveloped/icosahedral capsid [exemplified by adenovirus (*Adenoviridae*)]
- Enveloped/helical capsid [exemplified by measles virus (*Paramyxoviridae*)]
- Enveloped/icosahedral capsid [exemplified by herpes simplex virus (human herpesvirus 1, *Herpesviridae*)]

Most viruses fall into these four groups. However, there are many exceptions, and some viruses have morphologies that differ from that of the main groups. In addition, most viruses can produce variant forms which may be totally unlike their 'classical' appearance and the common descriptions of virus morphology may be very different from what is actually seen under the electron microscope. Influenza (*Orthomyxoviridae*) is particularly variable when grown in tissue culture, as shown in **Figure 1.4**.

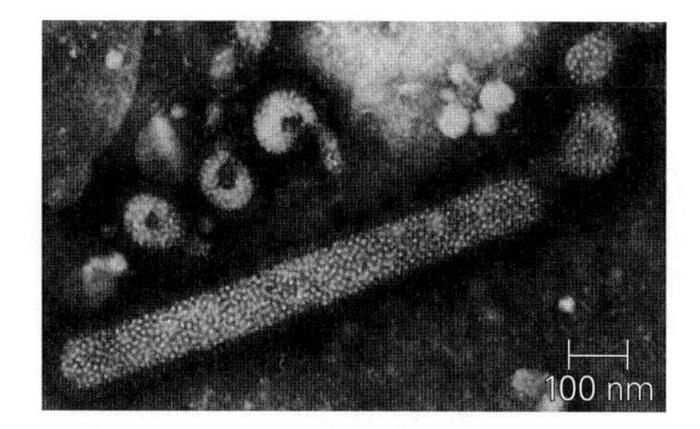

Figure 1.4
Variant forms of influenza virus. Courtesy of Ian Chrystie, St Thomas' Hospital, London.

Two of the more unusual examples of virus morphology are the poxviruses and filoviruses. Poxviruses (including smallpox and vaccinia) have very large genomes (for viruses) and are highly complex, producing over 100 proteins. Poxviruses are just about visible by light microscopy, and can be almost 0.5 μm long. A poxvirus is often described as brick-shaped, and is in some ways reminiscent of the complexity of a cell-like structure. It has an outer coat containing lipid and protein structures and complex internal structures including a core structure containing the genome. The outer lipid-protein shell is not the envelope commonly seen with other viruses, but rather is made up of lipoprotein subunits which form the virion surface, although this may be enclosed by an additional membrane. Historically, in the electron microscope the two typical appearances were referred to as the C (capsule) and M (mulberry) forms: the former smooth and relatively featureless, representing the hydrated form, while in the latter (dried) form the repeating subunits could be seen (**Figure 1.5A**). Inside this a complex arrangement of proteins, with both core and lateral structures, contains the many viral enzymes that a poxvirus uses to replicate. The M form, while often cited in early work, is the result of drying during preparation for electron microscopy, which is now avoided by the use of non-drying (cryogenic) procedures.

Filoviruses (Marburg and Ebola hemorrhagic fever viruses) have an approximately helical nucleocapsid in an extremely long enveloped particle which, although only 80 nm wide, can be up to 14,000 nm long, making them easily the largest viruses. However, these are very simple viruses (with a genome about one-twentieth the size of that of the poxviruses), and their massive length appears to be the result of poorly controlled elongation rather than a genuinely complex structure. In support of this, maximum infectivity is actually associated with the shorter (but still very large) 800–900 nm forms (**Figure 1.5B**).

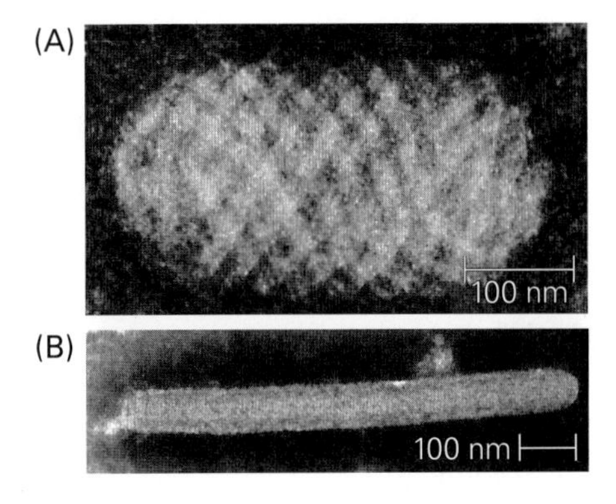

Figure 1.5
Forms of (A) poxvirus and (B) filovirus. Courtesy of Ian Chrystie, St Thomas' Hospital, London, and A.M. Field, HPA Colindale, London.

In a further addition to complexity, many types of viruses may contain additional elements. Examples include the thick proteinaceous **tegument** underlying the herpesvirus envelope or the cellular ribosome, which are incorporated into arenaviruses.

While it is usually relatively simple to identify the type of a virus by electron microscopy, the limitations of using virus structure are apparent with the small non-enveloped icosahedral viruses. The actual shape of these is usually unclear, and they are generically referred to as small round viruses, which is entirely reasonable since all that can be said from their morphology in the electron microscope is that they are small, round, and (probably) viruses.

Capsid symmetry

Proteins will naturally adopt the most energetically favorable state, and so certain structures are favored. These often have helical or icosahedral symmetry. In a capsid with helical symmetry, proteins are aligned in a helix around the nucleic acid and are rod-like in appearance (**Figure 1.6**). In a capsid with icosahedral symmetry, the proteins form an outer shell around the nucleic acids in the center.

An icosahedron is a 20-sided solid with faces formed of identical equilateral triangles. Icosahedral symmetry requires 60 units to form the faces of the capsid. Some of these (e.g. adenovirus, Figure 1.3B) are visibly of icosahedral shape, but almost appear just roughly spherical. This is because of one key

Figure 1.6
Helical and icosahedral capsid structures. Tobacco mosaic virus and adenovirus.

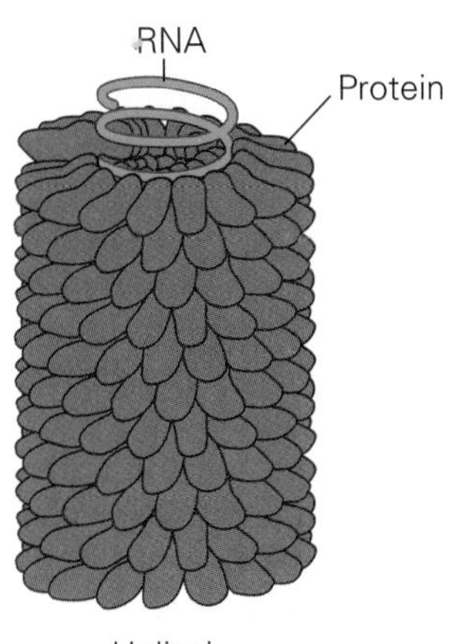

Helical
(part of tobacco mosaic virus)

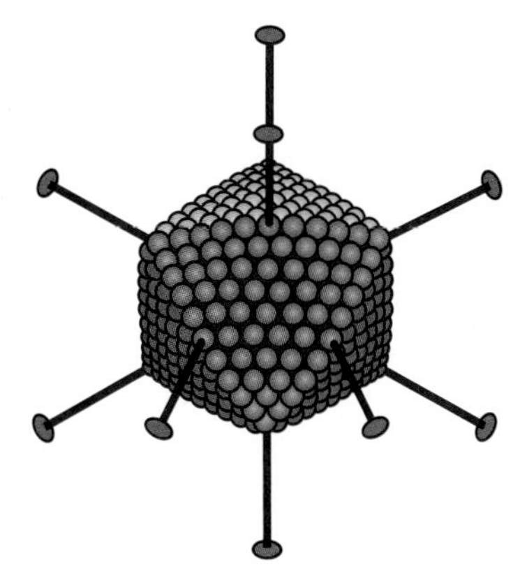

Icosahedral
(adenovirus)

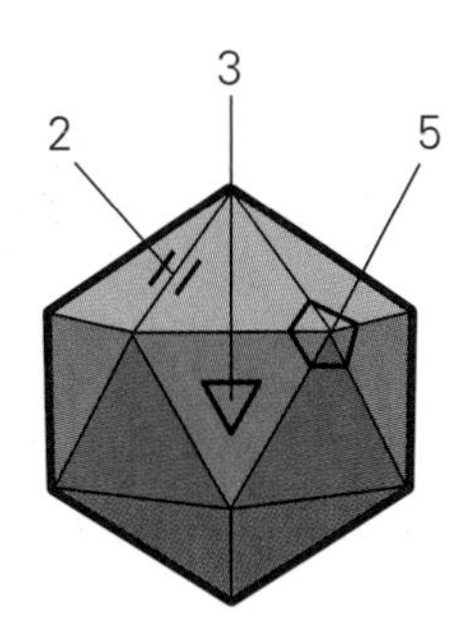

Icosahedral symmetry
showing axes of rotation

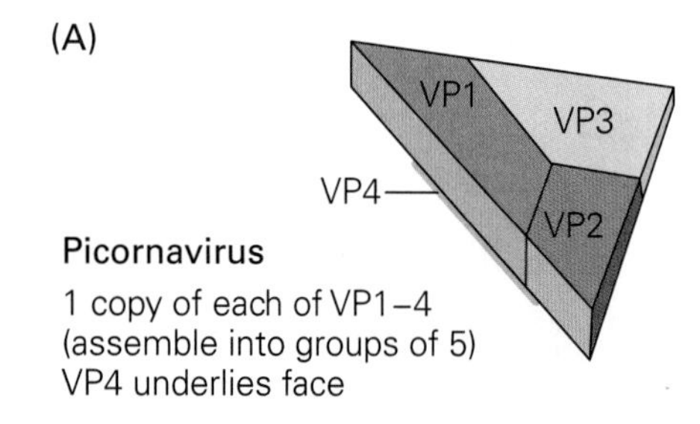

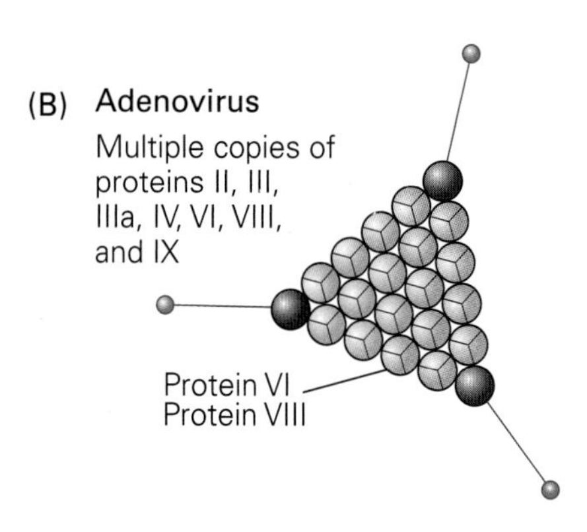

Protein II (hexon subunit), 3 copies
Protein III (penton base), 5 copies
Protein IV (fiber), 3 copies
Protein IIIa joins penton bases to hexons
Protein VI links face to nucleoprotein core
Proteins VIII and IX stabilize hexon interactions

Figure 1.7
The proteins of the capsid face for viruses with icosahedral symmetry.
(A) Picornavirus, (B) adenovirus.

point: symmetry is not the same as shape. Many capsids with icosahedral symmetry are generally spherical, while the capsid of the picornaviruses is predominantly dodecahedral in shape, with twelve pentagonal sides (containing the minimum sixty units needed for icosahedral symmetry).

In fact, the definition of symmetry is based on the presence of the correct axes of symmetry: icosahedral symmetry is defined by twofold axes of rotational symmetry around the face edge, threefold axes around the face center, and fivefold axes around the face vertex (Figure 1.6).

The adenovirus genome is a double-stranded DNA (dsDNA) of 36 kilobase pairs (kbp); the picornavirus genome comprises 7–8.5 kb single-stranded RNA (ssRNA). Both virions are non-enveloped with icosahedral capsid symmetry. However, the fourfold difference in genome size allows the adenovirus particle to have a far more complex structure.

The picornavirus particle is made up of only four main viral proteins (VPs 1–4; in one genus, *Parechovirus*, VP2 and VP4 remaining joined together as VP0) plus one copy of a very small protein (VPg) on the end of the genomic RNA. All are cleaved from a single precursor protein. VP1, VP2, and VP3 form the outer face of the capsid, underlaid by the VP4 protein. Each face of the capsid is made up of five triangular subunits, each containing one copy of each protein (**Figure 1.7**). These pentamers then form the 12 faces of a dodecahedron. Nevertheless, this structure has the two-, three-, and fivefold axes of icosahedral symmetry.

By contrast, adenovirus capsids are formed of "hexons," each of which contains three copies of protein II. The twenty hexons on each face of the icosahedron are stabilized by protein VIII internally and protein IX externally, and are attached to the DNA-protein core by protein VI. Penton (vertex) faces contain five copies of protein III to which protruding fibers made of three copies of protein IV are attached (Figure 1.7). Penton bases are held to the hexons by protein IIIa. The hexons and pentons together make up the triangular faces of a regular icosahedron (Figure 1.6). The adenovirus particle as a whole contains 11 proteins.

Despite these differences, the basic structure achieved has icosahedral symmetry in both cases. Readers interested in the complex rules of geometry which govern the precise formation of capsids should consult the further reading at the end of this chapter.

Virus envelopes

Another important structural element present in many viruses is the **envelope**. This is derived from the membranes of the host cell, although precisely which cellular membrane varies between viruses. Envelopment is not simply a passive process of picking up some membrane from the cell, since specific changes do occur to the membrane before it envelops the virus. Changes to membrane fluidity resulting from the preferential incorporation of specific lipids may be important. However, the most apparent change is the presence of viral proteins (seen as "spikes" or a "fringe" when the virion is viewed by electron microscopy, see Figure 1.3) projecting through the envelope. Clearly, viral proteins must be present on the outside of the envelope membrane in order to perform specifically viral functions such as binding to the host cell. These proteins are usually glycoproteins, with sugar groups attached to the polypeptide. The sugars make the protein locally hydrophilic, and are usually essential for function. All enveloped viruses have such proteins and, due to their nature and their position on the surface of the virion, they are usually highly immunogenic.

Lipid is also found inside many of the nucleocytoplasmic large DNA viruses (NCLDVs), where its function is less certain.

1.3 THE VIRAL GENOME

Viral genomes may be double or single stranded and may consist of DNA or RNA, the latter being unique to viruses and requiring specialized biochemical adaptations. The sizes of viral genomes range from the 1759 bases of single-stranded DNA that make up the genome of porcine circovirus (*Circoviridae*) to the 1,181,404 bp double-stranded DNA genome of *Mimivirus*, which is larger than some bacterial genomes.

Genome size and virus structure

As stated above, capsid assembly involves the building of the capsid from small, repeating protein subunits. This has the effect of minimizing the need for dedicated genome space to produce the structural proteins, an important consideration when a typical virus has a genome a hundred times smaller than that of a typical bacterium, as shown in Figure 1.2.

A large viral genome, while still very small by cellular standards, means that the virus can produce many proteins, allowing a complex structure. Viruses with small genomes are more restricted. In general, the bigger the viral genome, the more complex the structure can be.

An excellent example is provided by comparing adenoviruses with picornaviruses as detailed in Section 1.2. The larger adenovirus genome permits the adenovirus particle to have a far more complex structure despite the basic icosahedral symmetry in both virus types.

Even simpler than picornaviruses are the parvoviruses, which have nearly the smallest genomes of any virus infecting humans (only some of the *Picobirnaviridae* are smaller). The parvovirus virion is 18–22 nm in diameter with icosahedral symmetry (though this is very difficult to see in the fuzzy blob that is shown in the electron microscope), and contains as few as two (more usually three) different types of protein. This contrasts with the poxviruses, which have the largest genome of any type of virus infecting humans (about 50 times the size of the parvovirus genome). Poxviruses have highly complex virions containing more than 100 different proteins in multiple structures with an intricately structured appearance (**Figure 1.8**).

With small virus genomes, a single protein frequently has multiple functions, since there is simply not enough coding capacity to produce one protein for each task. The capsid proteins of the picornaviruses (7–8.5 kb ssRNA) or papillomaviruses (5.3–8 kbp, dsDNA), for example, can assemble themselves into capsids. This ability to self-assemble is common to many viruses with small genomes. By contrast, the proteins of the herpesvirus (120–260 kbp, dsDNA) and adenovirus (36 kbp, dsDNA) capsids require the help of other (nonstructural) proteins which do not form part of the final virion.

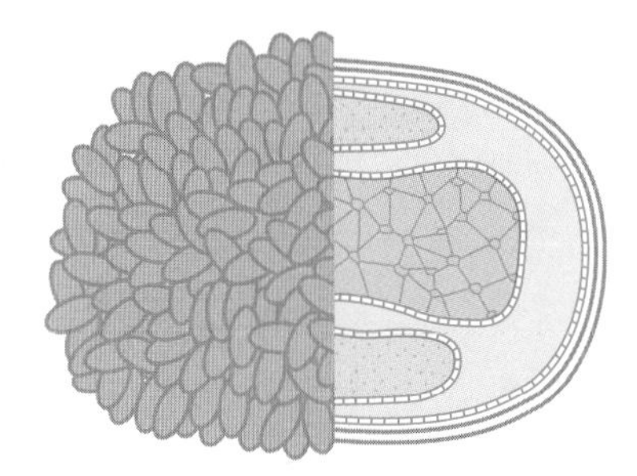

Figure 1.8
The structures of parvovirus and poxvirus, showing different levels of complexity. Poxvirus adapted from, Cull P (ed) (1990) The Sourcebook of Medical Illustration. With permission from DA Information Services.

Genome size and infection

Viruses with larger genomes are also able to provide more of the functions that they need to replicate, and therefore they rely less on the production of specific enzymes by the cell.

Viruses with RNA genomes do not generally use the cellular machinery for DNA synthesis. With the exception of the *Retroviridae* and other reverse-transcribing viruses, they have to rely on their own synthetic machinery since their host cells do not make RNA from RNA. This means that they generally function independently of and have little effect on the DNA synthesis machinery of the host cell.

However, among viruses with DNA genomes, the very small viruses, such as the *Parvoviridae*, rely on the cell to provide the machinery to copy the

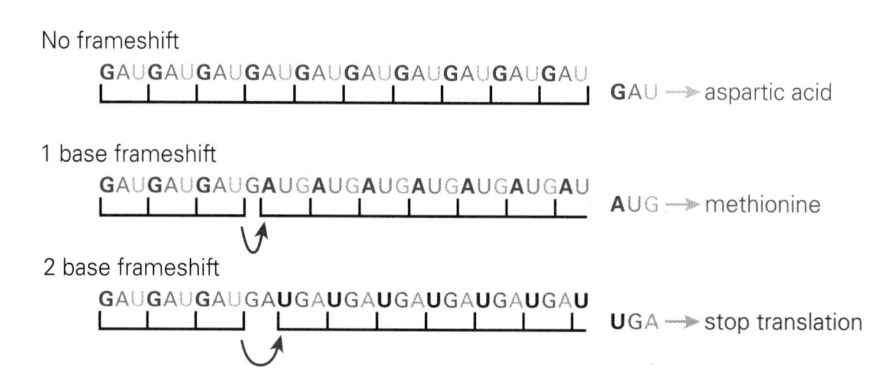

Figure 1.9
Effects of frameshifting and the use of different reading frames. A one-base frameshift alters translation from reading frame 1 (GAU → aspartic acid) to reading frame 2 (AUG → methionine); a two-base frameshift alters translation from reading frame 1 to reading frame 3 (UGA → stop translation); a three-base frameshift resumes translation in the original reading frame, with the deletion of one GAU codon.

viral DNA. This limits their ability to infect cells quite severely, as they are only able to infect actively dividing cells (those making DNA) and most cells within the body are not dividing at any one time.

Above these, in terms of complexity, is a range of small DNA viruses. They do not have the genetic capacity to produce what they will need to replicate in a nondividing cell, but they are able to avoid the limitations of the very smallest viruses. They often do this by coding for proteins that force the infected cell into active division. Examples are the T antigens of the *Polyomaviridae* and the E6 and E7 early proteins of the *Papillomaviridae.* If the infection proceeds and the host cell is killed, the damage is limited to that cell. However, if the infection is (as is often the case) abortive, and is unable to complete replication of the virus, then the infected cell is forced into division but not killed. While other mechanisms are used (notably by the larger DNA viruses), this is one important mechanism of virus-induced cellular transformation (see Section 1.5).

Larger viruses are able to use the machinery of the cell to make more viruses regardless of whether the cell is dividing, since they have the genetic capacity to produce essential enzymes, such as the many enzymes mediating for nucleotide metabolism in the *Herpesviridae* (for example, ribonucleotide reductase, thymidine kinase, and thymidylate synthetase).

The very largest viruses, the nucleocytoplasmic large DNA viruses (NCLDVs; *Asfarviridae, Iridoviridae, Mimivirus, Poxviridae,* and *Phycodnaviridae*), actually set up a synthetic center in the cytoplasm of the cell, which is sometimes referred to as a "second nucleus."

Making the most of a limited genome

Due to the small sizes of their genomes, many viruses use methods of making one genome sequence produce multiple proteins.

Frameshifting is one common method. Since each amino acid is encoded by three bases (a codon), the protein produced by reading the mRNA can be produced in one of three 'reading frames' (**Figure 1.9**). This allows very different proteins to be produced from 'overlapping genes' in one nucleic acid sequence, but requires that the sequence produce a useful protein when read in the different reading frames. This is potentially a very demanding requirement. An illustration of the difficulties involved might be to try thinking of a single sentence that makes sense in English, French, and German.

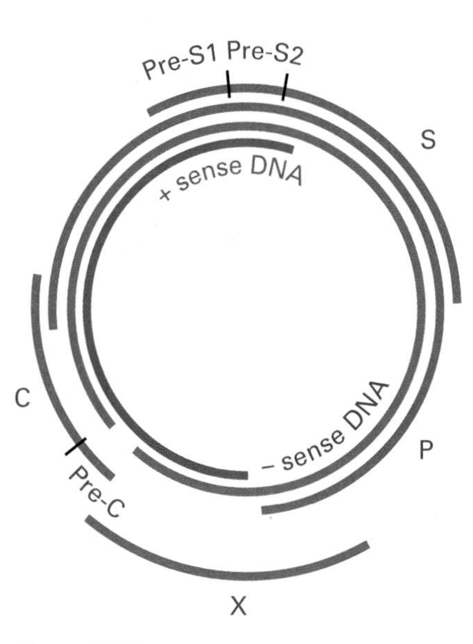

Figure 1.10
Use of overlapping genes in the hepatitis B genome. Note the partially double-stranded genome. Proteins shown are: C, the nucleocapsid (core) protein; S, the surface protein, produced in three forms of differing length, Pre-S1 or L (for large), Pre-S2 or M (for medium), Surface or S (for small); P, the viral polymerase; X, a protein of indeterminate function.

The hepatitis B virus (*Hepadnaviridae*) (**Figure 1.10**) has a very small genome (3.2 kb) along with a very complex replication process, meaning that it needs to make efficient use of its coding capacity. Much of the genome codes for proteins in two different reading frames (Figure 1.9), and extensive use is made of alternative promoters and translational start sites. Additionally, the entire genome codes for proteins, with the regulatory sequences for individual genes lying within the coding regions of others.

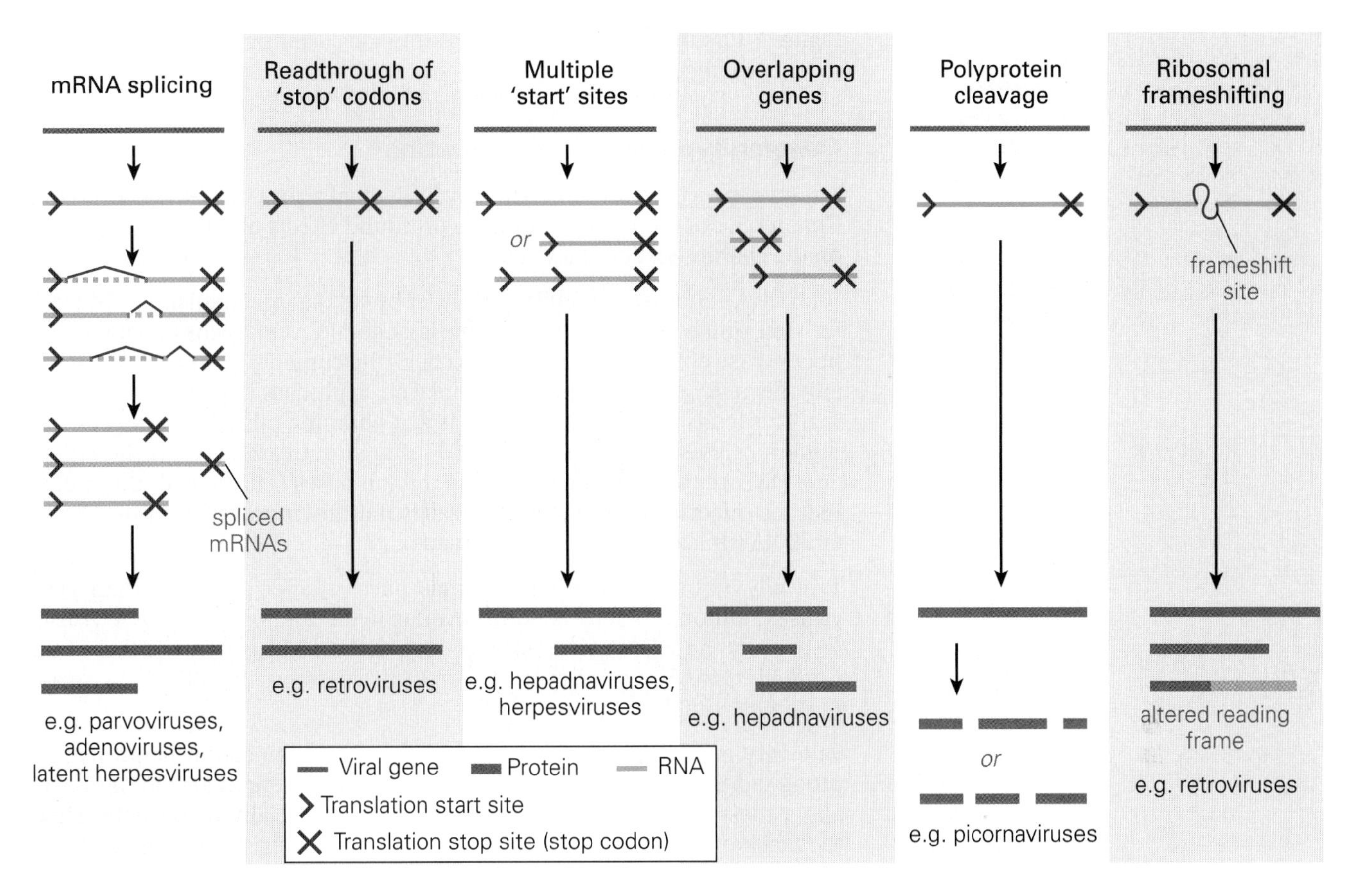

Figure 1.11
Strategies used by viruses for producing multiple proteins (red) from a single gene (purple). These include the use of multiple translation start sites (arrowheads) or readthrough of stop codons (crosses), alternative mRNA splicing, overlapping genes, and polyprotein cleavage. Frameshifting (see Figure 1.9) can also result in the production of alternative proteins downstream from the frameshift site.

The use of such mechanisms is not restricted to viruses with small genomes. Adenoviruses cut and join mRNAs differently to produce different proteins from one original RNA, a process known as **RNA splicing**. Some of the RNAs produced by latent Epstein-Barr or herpes simplex herpesviruses from genomes of greater than 100 kbp are also spliced. Herpesviruses also use multiple start sites for translation of mRNAs to produce some proteins, while retroviruses use inefficient stop sites to produce low levels of proteins from the mRNA after such sites.

An analogous system is used at a transcriptional level by paramyxoviruses, with inefficient transcription stop sites between genes, allowing RNA synthesis to "run on" in some cases.

The use of different reading frames can also result from 'slippage' of the translating ribosome at a frameshift site. In retroviruses, this site is a combination of a sequence at the gene junction and a fold in the mRNA downstream of this site, as shown in **Figure 1.11**.

Polyprotein cleavage can produce different proteins by cutting the polyprotein at different sites.

These methods are used by viruses, but not exclusively. For example, most eukaryotic genes appear to contain many introns (regions of RNA removed by splicing and not present in the mature mRNA), so it appears that size alone is not the reason for such strategies.

Figure 1.11 summarizes the various strategies used by viruses to make the most of their genomes.

Genome type and virus replication

The strategies used for replication by individual virus families are covered in Chapter 3, but it is worth noting the profound effects of the type of genome on the characteristics of a virus.

Viral DNA genomes are replicated either by cellular mechanisms or by similar viral processes. DNA replication is generally very accurate since DNA polymerase enzymes generally also check the copied sequence and remove any mismatches, which are then replaced; a process known as proofreading. This leads to a high level of accuracy, with one mistake made for every million or even billion bases copied, after proofreading. With the largest viral DNA genomes being just over a million bases (*Mimivirus*), this means that the majority of DNA genomes will be faithful reproductions of the parent DNA (unless other factors intervene).

Typically, viruses with RNA genomes have smaller genomes than DNA viruses, although there is some overlap with the smallest DNA viruses (especially the ssDNA viruses). The largest RNA virus genome is that of the *Coronaviridae*. However, even at over 30 kb this is less than 3% of the size of the largest viral DNA genome.

One very important reason for this is that RNA replication is not usually proofread by the cell. To cells, RNA is not the genetic material. It is usually a messenger molecule, short-lived and essentially disposable. Thus, the mechanisms for proofreading do not generally exist for RNA synthesis, resulting in a huge increase in the mutation rate, which ranges from one in a thousand bases to one in a hundred thousand bases. Thus, even the smallest RNA viral genomes, such as the 3569 bases of the bacterial virus MS2, may contain mutations.

For a complex organism that produces small numbers of offspring, mutation generally is to be avoided, since most mutations are harmful to the complex organization that is needed for that organism to function. However, for a virus, the deleterious mutations will be less damaging since some of the vast numbers of progeny virus will be viruses with no or insignificant mutations which can reproduce as well as the parent virus. Of the remainder, many mutations will be harmful, but the occurrence of beneficial mutations, while rare, will also be favored by the production of large numbers of progeny virus.

Mutation is an important mechanism in helping viruses to evade immune surveillance (see Chapter 4). It is worth emphasizing again that the mutation rate of RNA genomes is so high (one error in every 1000–100,000 bases copied) that it is unlikely that any copy of a viral RNA genome is exactly the same as the template from which it is copied.

The presence of a segmented genome in many RNA viruses also reflects the low fidelity of RNA replication, since one long molecule may be more likely to contain deleterious mutations. Segmented genomes also aid in the reassortment of RNA genes; while RNA can recombine (exchange sequences with a similar nucleic acid), most of the mechanisms to allow this are cellular, and are targeted at DNA. Segmentation provides an alternative route by which exchange of genetic information can occur. Influenza virus (*Orthomyxoviridae*) has seven (influenza C) or eight (influenzas A and B) segments of RNA making up its genome and is the best known and most studied of the viruses with a segmented genome. Other viruses with segmented genomes have from two to twelve segments. Influenza has a very high rate of inherent, mutational genetic variation (referred to as **antigenic drift**), but also has the ability to appear with new and very different genes (**antigenic shift**), apparently by exchanging RNA segments with influenza viruses from or in influenza-infected "animal" reservoirs (**Figures 1.12 and 1.13**). In particular, the pig is thought to be susceptible to both avian and human influenza viruses, and to act as a source of novel variants produced by antigenic shift resulting from such mixed infections. It is these novel influenza strains that are responsible for the worldwide pandemics of influenza (see Box 8.2).

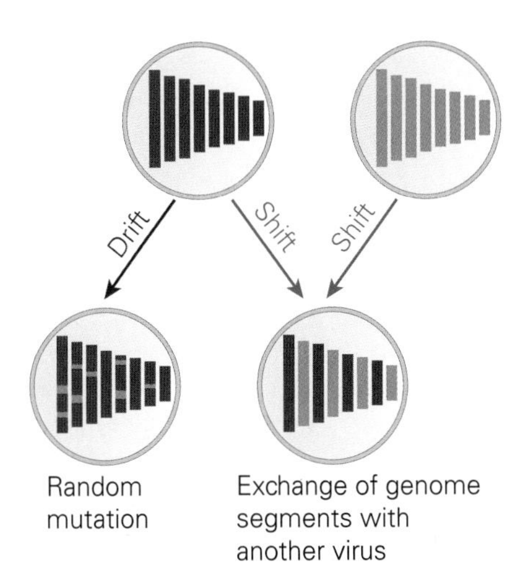

Figure 1.12
Principles of antigenic drift and shift. Colored bars and segments represent novel RNA sequences.

Mutation in the HIV genome

Even RNA viruses with unsegmented genomes may vary rapidly. HIV is one of the most rapidly changing viruses, despite the fact that HIV has an unsegmented genome and also has two copies of the genome in each virion, which could moderate the effect of mutations. The use of the specialized polymerase enzyme **reverse transcriptase** that copies the RNA genome into DNA seems to favor mutations of HIV, and the replication strategy used by HIV to allow full-length copies to be produced (copying a section of the genome from each genomic RNA molecule to produce a full-length copy) may also increase variation.

This enables it to evade the immune system by changing the targets for the immune response. However, since the virus is diploid, mutations should be moderated. Four hypotheses exist to explain the high rate of retrovirus mutation:

1. Reverse transcriptase is a very low-fidelity polymerase. Combined with the high rate of mutation inherent in RNA genomes, this would ensure that almost all copies of the HIV genome were different in some way from their parent molecule.

2. "Jumping" of the reverse transcriptase between genome copies, if it occurs, could introduce errors, and could also mean that the transcriptase is prone to jumping elsewhere, particularly at specific sequences.

3. Where multiple retroviruses infect a cell, the genomes might be able to reassort, giving one copy of the genome of each "parent." It is thought that the close interaction between the two genomes within the virion requires that they be very similar, so while this might occur with viruses of the same 'species,' it would probably not occur between widely different viruses.

4. Since retroviruses produce the specific enzymes required for integration and excision of their DNA, these may give a far higher rate of recombination than normally occurs. Since cells appear to contain a significant number of endogenous retroviruses (see Section 1.5), it is possible that recombination with these could occur, increasing the resultant variation.

The precise roles of the above mechanisms are uncertain, but they could singly or collectively help to account for the high mutation rate of retroviruses. While several human retroviruses have now been identified, it is clear that many retrovirus-related elements are present in the genomes of

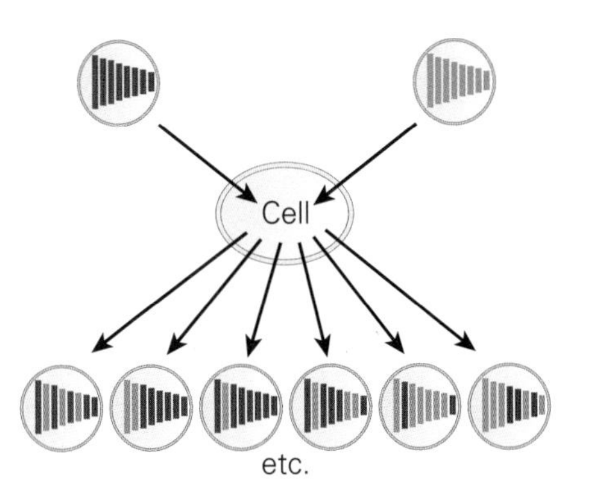

Figure 1.13
Products of genetic mixing of RNA segments (antigenic shift). Two viruses infecting the same host cell will produce a range of progeny, examples of which are shown here. Colored bars and segments represent novel RNA sequences.

most higher organisms, and the role of these elements is not yet known. Some appear to be only fragments, while others can move around inside the cell. There is a lot still to be learned about the nature and effects of these elements.

Some DNA viruses can also mutate very rapidly, notably the small *Parvoviridae* and the *Geminiviridae* of plants. The *Hepadnaviridae* (including hepatitis B virus) have even inserted an RNA stage into the replication of a DNA genome, the role of which is likely to include aiding mutational evolution. The general processes of replication of viruses will be covered in more detail in Chapter 3.

1.4 THE EFFECTS OF VIRUS INFECTION

The general perception of the result of virus infection is sickness for the host and, at the cellular level, the killing of the host cell. However, in many cases the infection itself may be nonsymptomatic (subclinical). It is often the case that severe symptoms result from infections with viruses that have not yet had time to adapt to life in a new host. This is seen in the severity of many **zoonotic** infections (transferred from animals to humans) such as the filovirus diseases Marburg and Ebola (probably transferred from bats), and with Hantavirus pulmonary syndrome transferred from the deer mouse.

When symptoms are produced, these commonly result from the responses of the immune system to the infecting virus, as detailed in Chapter 4. Such symptoms include inflammation, rashes, and fever, some of which are actually defensive responses that help to limit the infection.

At the cellular level, there are only a limited number of infections where the killing of cells (cytopathic effect) by the virus is responsible for disease. Even where there is a high level of cell killing, as in infection of the cells of the intestinal lining by rotavirus (*Reoviridae*) or the destruction of the anterior horn cells of the spinal cord by poliovirus (*Picornaviridae*) or West Nile virus (*Flaviviridae*), other factors are involved and these are often immune mediated.

Viral infections are often studied in cultured cells rather than an intact host organism. Culturing cells chosen for susceptibility to virus infection in the absence of any immune protection is a very unnatural system, and the acute, lytic infections often seen in cultured cells may be a poor reflection of events occurring during viral infection of the whole organism. However, even in cultured cells, rapid killing is by no means universal.

It is not in the interest of the virus to kill its host too rapidly, since the production of the next generation of virus may be compromised. This is an important factor to consider in studying the processes of virus replication. However, this does not mean that viruses are harmless, since they all require the machinery of the host cell to be redirected to the production of the next generation of viruses. To increase the efficiency of virus production many viruses produce specific effects on the host cell that are mediated by viral functions. For example, the virion host shutoff (Vhs) protein produced by herpes simplex viruses is carried within the virus particle (in the tegument) and, when introduced into the host cell, it then degrades cellular mRNAs.

Many other viruses also decrease cellular synthesis. This can involve simple competition for the cellular protein synthesis machinery by the production of massive amounts of viral mRNA, as seen with vesicular stomatitis virus (VSV), the best studied of the *Rhabdoviridae*. However, VSV also appears to interfere with mRNA transport from the nucleus to the cytoplasm, further indicating that biological systems rarely rely on a single pathway.

Other viruses use different mechanisms, from the acquisition of caps from cellular mRNAs by influenza virus (*Orthomyxoviridae*) to the redirection of the cellular protein synthetic machinery to use uncapped mRNAs by the *Reoviridae*.

1.5 TYPES OF VIRUS INFECTION

In cultured cells, some viruses are only capable of an active (acute) infection, where the host cell is rapidly killed as a result. However, many others can infect a cell and actively produce virus without immediate cell death by using less destructive methods of exit, such as budding from the cell surface. Alternatively, the virus may become **latent** within the cell, as mentioned below, either integrating with the chromosome or remaining as a self-replicating extrachromosomal nucleic acid (an **episome**).

Viral infections at the cellular level can be subdivided into four types (**Figure 1.14**):

- Acute or lytic: rapid cell killing and production of progeny virus
- Persistent or chronic: long-term infections with low levels of virus production
- Latent or proviral: viral genome is maintained in an inactive state within the cell
- Transforming: producing altered cell growth (usually only part of virus genome present)

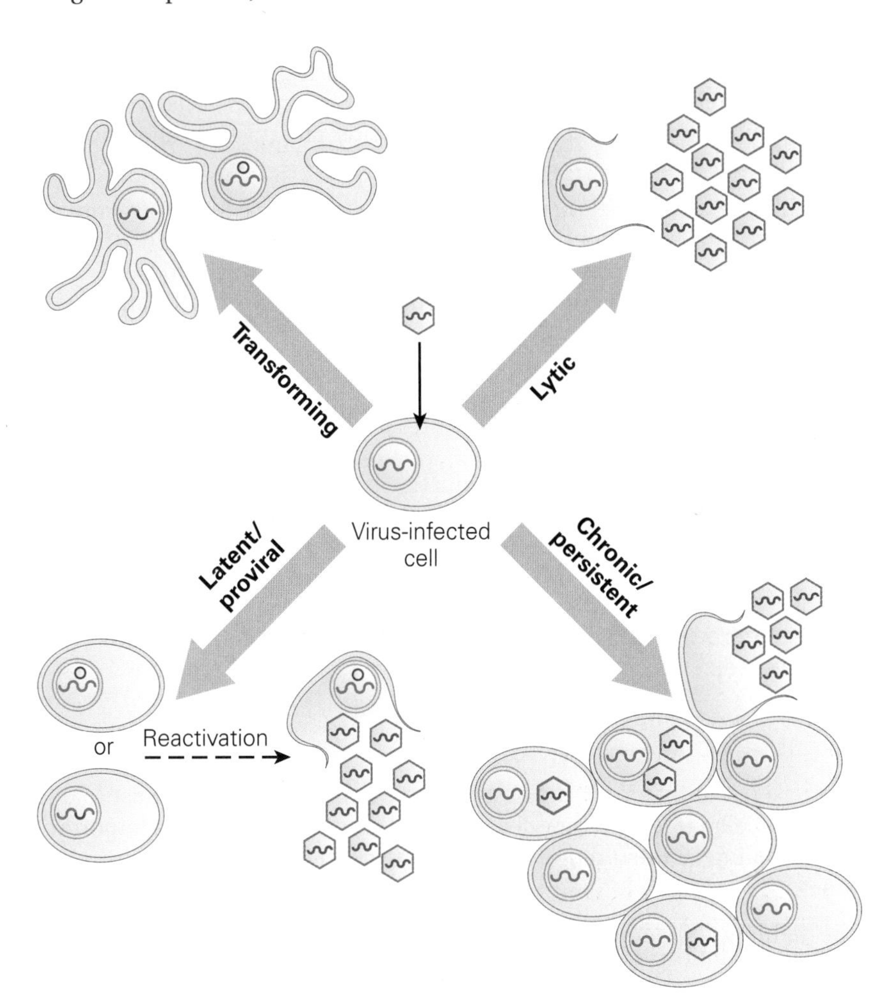

Figure 1.14
Differing effects of virus infection. While viruses may kill infected cells rapidly in a lytic infection, many other types of infection exist. In a chronic or persistent infection, cells may remain infected for a long time and produce virus at low levels; most cells are not killed and not all cells are infected. Latently infected cells may maintain the whole viral genome either within the cellular DNA or as an extrachromosomal element (an episome) until reactivation induces virus production. Transformed cells usually contain only part of the viral genome, exhibiting altered growth properties that may underlie oncogenesis.

Lytic infections

An acute or lytic infection results in rapid cell death, and is commonly seen in virus infection of cultured cells. Despite this, it appears that such rapidly **cytotoxic** infections may be less common *in vivo*, where interactions with the immune system can occur. **Lysis** of infected cells may be an active process involving viral enzymes, or simply a reflection of indirect damage to the cell resulting from virus infection. Lysis of the cell has the effect of releasing virus, which can then initiate further infections.

Low-level infections that kill only a small proportion of cells have been observed in cell culture for several types of virus, although an external limitation on infection such as antiviral drugs or cytokines is often necessary to prevent a lytic infection. Examples include the use of interferon or antiviral drugs such as aciclovir (for herpesviruses). Infections of this nature appear to be relatively common *in vivo* following infection with a range of viruses, since many elements not present in cell culture (notably the effects of the immune system) act to limit virus replication.

Persistent or chronic infections

The terms chronic and persistent are not distinct. In such infections, virus replication is at low levels, and the immune response may be moderated by the development of tolerance to viral antigens, allowing a long-term, low-key infection to become established. The type of cell infected may also play a major role, since some cell types will not permit full virus replication.

A commonly quoted example of a persistent infection in humans is the rare sub-acute sclerosing pan-encephalitis (SSPE) which can follow infection with measles virus, typically 1 in 10,000 to 1 in 100,000 cases. In SSPE, virus replicates at low levels without producing infectious virus, with mutations in and altered production of viral proteins, and possible alterations in immune responses.

Probably due to measles virus immunization programs, SSPE is becoming even more rare, but it does represent a good example of a virus continuing to replicate at a very low level over a long period of time. This and other persistent infections may also be moderated by defective interfering viruses (see Chapter 2).

More common chronic infections include the long-term liver infections caused by both hepatitis B (*Hepadnaviridae*) and hepatitis C (*Flaviviridae*) viruses.

An alternative example of prolonged virus infection is that of **slow viruses**, where the virus is present and replicating at low levels for a long time before producing apparent disease. Examples in humans may include JC virus (*Polyomaviridae*) and the transmissible spongiform encephalopathies (see Chapter 2). Infections of animals by viruses of the genus *Lentivirus* (*Retroviridae*) provide additional examples, in particular that of maedi/visna virus of sheep.

Latent infections

In a true latent infection, the viral genome is present but is almost totally inactive. Cellular factors as well as viral factors are believed to be important in controlling latency, acting to block functional expression of most virus genes.

The defining characteristics of a latent infection are that only a limited part of the genome is active, and no virus replication occurs. A frequently quoted example of such an infection is that of herpes simplex virus (HSV) where, following the acute infection, the virus becomes latent in ganglionic nervous tissue. Under these conditions, only a very small region of the viral genome is active, producing three RNA latency-associated transcripts (LAT transcripts) which appear to be involved in reactivation from the latent state. The full-length 8.3 kb transcript is present at low levels, while two smaller transcripts (2.0 kb and 1.45 kb) are more highly expressed.

Recently a group of *micro-RNAs* (see Box 3.3 and Section 6.5) produced from the LAT transcripts have been identified, which appear to play a role in controlling latency by inhibiting the expression of multiple proteins that are required for productive infection.

During latency, the HSV genome appears to be maintained as an episome, expressing only the LAT transcripts, with no firm evidence of protein expression. In contrast, during acute infection, more than 70 proteins (and corresponding mRNAs) are produced by HSV. The contrast with the latent state is striking. Clearly, while latent, the virus is effectively silent, presenting no targets for the immune system and producing very limited (if any) effects on the host organism.

Other viruses, including some herpesviruses, may have a more active latent state that is more accurately described as a chronic, low-level infection. In such situations, immune responses to viral antigens may become problematic and cause disease symptoms.

An alternative latency is that of **proviral integration**, seen with retroviruses, where the genome is copied into DNA and integrated into the cellular chromosomal DNA. The formation of a reservoir of latent, inactive viral genomes which are nevertheless capable of reactivation is important for HIV (*Retroviridae*), and contributes to the extreme difficulty of eradicating this infection. Latency has parallels in the lysogeny of certain viruses of bacteria (*bacteriophages*) such as bacteriophage lambda, which can integrate their DNA into the bacterial chromosome until reactivation is triggered by external events (see Chapter 7).

Integrated DNA is replicated along with the host cell DNA, and may (as noted above) account for a significant percentage of cellular DNA. It is believed that the human genome contains almost 100,000 integrated retroviral genomes, accounting for around 8% of the genome. Many of these are incomplete or damaged and are incapable of reactivation – they are effectively "trapped" in the genome of the cell.

The role of endogenous retroviruses (if any) is uncertain, but there have been numerous suggestions including involvement in the pathogenesis of transmissible spongiform encephalopathies (see Chapter 2) and the (possibly damaging) modulation of the immune response to virus infections.

In contrast, a potential beneficial effect that has been proposed is the production of localized immunosuppression to prevent rejection of genetically foreign placental tissue during pregnancy in mammals. It has even been suggested that it is the action of such endogenous retroviral elements that has permitted the evolution of mammalian viviparous reproduction, permitting the fetus to mature inside the mother's body without being rejected by the immune system.

Cellular transformation

Transformation of cells results in altered growth patterns and can in some cases lead to **oncogenesis**, the formation of cancers, as a result of uncontrolled cell growth. Transformation occurs due to genetic changes in the cell and can be the result of a number of factors including the effects of radiation, mutagenic chemicals, or virus infection.

Transformed cells exhibit a number of altered properties. These include:

- Loss of regulation of cell growth (cell density inhibition, need for growth factors, need for solid substrate)
- Changes to cell appearance and structure (cytoskeleton, expression of unusual cell surface proteins, reduced adhesion, proteinase secretion)
- Abnormal chromosome numbers (aneuploidy)
- Altered patterns of transcription and the production of altered patterns of growth factors
- Loss of limitation of number of divisions before onset of failure to grow in culture (immortalization) for cells from some species (including human and mouse, but not chicken)
- Where cells are transformed by viruses, the viral genome is present (although suggestions of other mechanisms exist), usually integrated into the cellular genome. Specific viral **oncogenes** (genes associated with cancer) may be expressed and cellular functions altered

In the case of virus infection, transformation may result from an abortive infection. Some small viruses such as the *Papillomaviridae* or the *Polyomaviridae* drive their host cells to replicate in order to gain access to the enzymes needed to replicate their DNA (which are not present in resting cells). If the infection proceeds and cell death results, this prevents further growth. However, if the infection is abortive and fails to kill the cell, uncontrolled growth can result, often with fragments of the viral genome present within the cellular DNA which then alter the expression of cellular genes (covered in more detail in Chapter 4). Viral DNA can also be introduced into the cellular DNA as part of the normal viral life cycle, as in the case of retroviruses (*Retroviridae*) (Chapter 2). Expression of integrated DNA or effects resulting from insertion into cellular genes can then lead to transformation.

A wide variety of viral functions are linked to transformation:

- Viral nucleic acid integrating in cellular tumor suppressor genes, inducing transformation
- Expression of viral oncogenes, often as partial or fusion proteins, freed from normal controls
- Virus-induced alterations of cellular gene expression, either nearby (*cis*-acting) or via factors affecting distant genes (*trans*-acting)
- Translocation within host chromosomes altering expression of cellular genes

Due to their altered growth properties, cells immortalized by transformation are often used in medical research and in virus culture, since they permit the establishment of a cell line with the potential for unlimited growth. The first of these to be established was the cervical adenocarcinoma HeLa cell line arising from cervical tumor tissue taken from Henrietta Lacks in 1951. HeLa cells have a greatly increased number of chromosomes and multiple insertion sites containing genes from human papillomavirus type 18 (*Papillomaviridae*), a known oncogenic virus. This work was followed by direct demonstration of the ability of Rous sarcoma virus (*Retroviridae*), a virus known to cause cancers in chickens, to transform cells in culture due to presence of *v-src*, a viral homolog of a cellular tyrosine kinase gene.

An increasing number of viruses are being linked to oncogenesis in both animals and humans, the most recent being Merkel cell polyomavirus (*Polyomaviridae*) (see Chapter 4).

Key Concepts

- Outside the host cell, a virus is biologically inert. It requires access to the machinery of its host cell in order to replicate. Thus, viruses are not the simplest form of life since they need not only themselves but also the machinery of a host cell to complete their life cycle.
- Viruses consist of RNA or DNA wrapped in a protein coat (a capsid) formed from small, repeating protein subunits that stabilizes the genome. The capsid may take many forms, though helical or broadly icosahedral forms are most common.
- Viruses may also have an outer envelope of host-derived lipid through which viral proteins project. The external proteins of the virus mediate binding to the host cell.
- Viruses are unique in that their genome may be RNA or DNA, though never both at the same time. RNA genomes appear to be more common and mutate extremely fast, providing a source of variation that helps the virus evade host defenses.
- Most viral genomes are smaller than those of cellular organisms. Thus, the use of viral genomic coding capacity has to be very efficient compared to that of cells.
- Viruses with larger genomes rely less on the host cell, making more of their own enzymes.
- Rapid killing of the host is often the result of a poorly adapted virus. Viruses that have had time to adapt to their ecological niche tend to evolve ways to coexist with their host, reducing the symptoms of disease and often leading to long-term infections.
- Viral infections can result in rapid killing (lysis) of infected cells, but others result in a persistent or chronic infection, where only a proportion of infected cells are killed. Some viruses are latent; the viral genome persists in a dormant form which can be reactivated at a later stage. Other viruses integrate with the cellular genome and alter cellular function (transformation), which can lead to cancer (oncogenesis).

DEPTH OF UNDERSTANDING QUESTIONS

Hints to the answers are given at http://www.garlandscience.com/viruses

Question 1.1: Why do viruses use small, repeating subunits to form the virus particle?

Question 1.2: Why do most viruses have RNA genomes?

Question 1.3: How do viruses adapt to life in a new host?

FURTHER READING

Arrand JA & Harper DR (1998) Viruses and Human Cancer. BIOS Scientific Publishers, Oxford.

Boccardo E & Villa LL (2007) Viral origins of human cancer. *Curr. Med. Chem.* 14, 2526–2539.

Butler PJ & Klug A (1978) The assembly of a virus. *Sci. Am.* 239, 62–69.

Campbell K (2010) Infectious Causes of Cancer. John Wiley & Sons, Chichester.

Claverie JM, Ogata H, Audic S et al. (2006) Mimivirus and the emerging concept of "giant" virus. *Virus Res.* 117, 133–144.

Daròs JA, Elena SF & Flores R (2006) Viroids: an Ariadne's thread into the RNA labyrinth. *EMBO Rep.* 7, 593–598.

Dayaram T & Marriott SJ (2008) Effect of transforming viruses on molecular mechanisms associated with cancer. *J. Cell. Physiol.* 216, 309–314.

Fields BN & Howley PM (2007) Virology, 5th ed. Lippincott Williams & Wilkins, Philadelphia.

Javier RT & Butel JS (2008) The history of tumor virology. *Cancer Res.* 68, 7693–7706.

Macville M, Schröck E & Padilla-Nash H (1999) Comprehensive and definitive molecular cytogenetic characterization of HeLa cells by spectral karyotyping. *Cancer Res.* 59, 141–150.

Skloot R (2010) The Immortal Life of Henrietta Lacks. Crown Publishing, New York.

Zuckerman AJ, Banatvala JE, Griffiths P et al. (2008) Principles and Practice of Clinical Virology, 6th ed. John Wiley & Sons, Chichester.

INTERNET RESOURCES

Much information on the internet is of variable quality. For validated information, PubMed (http://www.ncbi.nlm.nih.gov/pubmed/) is extremely useful.

Please note that URL addresses may change.

All the Virology on the WWW. http://www.virology.net (general virology site)

The Big Picture Book of Viruses. http://www.virology.net/Big_Virology/BVHomePage.html

CHAPTER 2
Virus Classification and Evolution

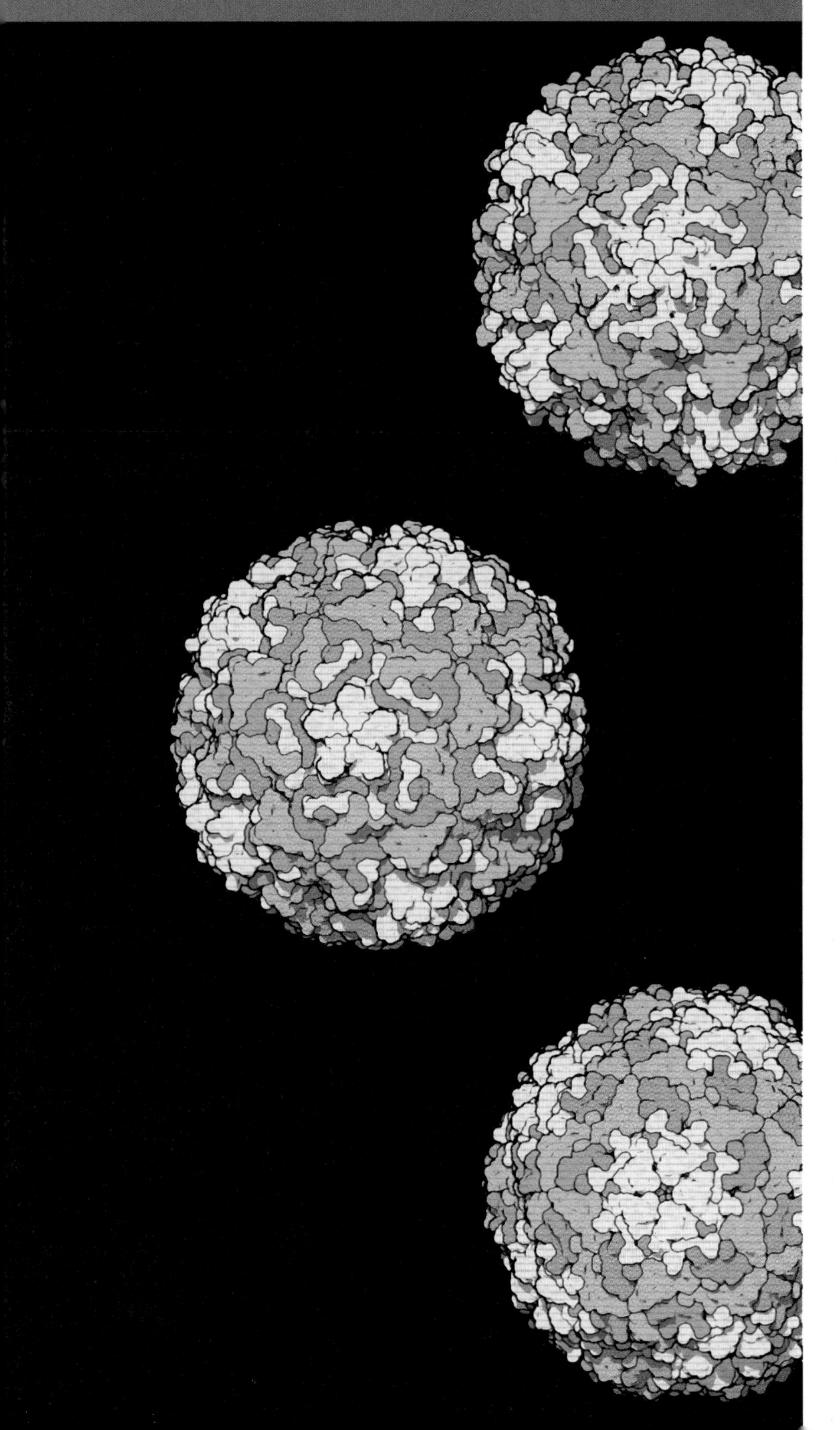

INTRODUCTION

The earliest classification systems used for viruses date from before their nature was understood. These systems were based entirely on the clinical signs and symptoms produced by infection. We can still see their effects today, for example in the hepatitis viruses, named for their effects on the liver. Hepatitis A, B, C, and E are all very different types of virus with RNA (hepatitis A, C, E) or DNA (hepatitis B) genomes, while hepatitis D is not properly a virus at all. The poxviruses are a recognized group, with smallpox, orf virus, molluscum contagiosum virus, tanapox, and sometimes cowpox and monkeypox infecting humans. But chickenpox (from *tzuiken* pox, or itchy pox) is caused by a herpesvirus, and the great pox is a name formerly used for the bacterial disease syphilis. Thus, a more defined system was needed.

One classification was established based on the observed characteristics of viruses (size, morphology).

About the chapter opener image
Three Picornaviruses
(Courtesy of the Research Collaboratory for Structural Bioinformatics Protein Data Bank and David S. Goodsell, The Scripps Research Institute, USA.)

2.1 VIRUS CLASSIFICATION

The formal taxonomical system for the classification of viruses is administered by the **International Committee for the Taxonomy of Viruses (ICTV)**, and is based on a broad range of characteristics including morphology and genome type (**Box 2.1**).

Viruses are classified into **families** (ending *-viridae*) and **genera** (ending *-virus*) on the basis of these criteria and, increasingly, on genetic relatedness. Below this is the virus **species**, the lowest formal level of virus classification (**Box 2.2**).

The classical taxonomical definition of a species as a group of organisms capable of interbreeding and producing fertile offspring is essentially meaningless for viruses, which do not reproduce by this means. Although the term is used by the ICTV, it is used in a much more flexible sense than is the case for more complex organisms (Box 2.2).

Generally, viruses are identified by clinical, immunological, structural, and molecular means and are then grouped into the categories defined above. Such groupings may change as more information about the viruses

Box 2.1 Criteria for classifying viruses (from Fauquet 2006)

I. Virion properties

A. Morphology properties of virions
1. Virion size
2. Virion shape
3. Presence or absence of an envelope and peplomers
4. Capsomeric symmetry and structure

B. Physical properties of virions
1. Molecular mass of virions
2. Buoyant density of virions
3. Sedimentation coefficient
4. pH stability
5. Thermal stability
6. Cation (Mg^{2+}, Mn^{2+}) stability
7. Solvent stability
8. Detergent stability
9. Radiation stability

C. Properties of genome
1. Type of nucleic acid—DNA or RNA
2. Strandedness—single stranded or double stranded
3. Linear or circular
4. Sense—positive, negative, or ambisense
5. Number of segments
6. Size of genome or genome segments
7. Presence or absence and type of 5′-terminal cap
8. Presence or absence of 5′-terminal covalently linked polypeptide
9. Presence or absence of 3′-terminal poly(A) tract (or other specific tract)
10. Nucleotide sequence comparisons

D. Properties of proteins
1. Number of proteins
2. Size of proteins
3. Functional activities of proteins (especially virion transcriptase, virion reverse transcriptase, virion haemagglutinin, virion neuraminidase, virion fusion protein)
4. Amino acid sequence comparisons

E. Lipids
1. Presence or absence of lipids
2. Nature of lipids

F. Carbohydrates
1. Presence or absence of carbohydrates
2. Nature of carbohydrates

II. Genome organization and replication
1. Genome organization
2. Strategy of replication of nucleic acid
3. Characteristics of transcription
4. Characteristics of translation and post-translational processing
5. Site of accumulation of virion proteins, site of assembly, site of maturation and release
6. Cytopathology, inclusion body formation

III. Antigenic properties
1. Serological relationships
2. Mapping epitopes

IV. Biological properties
1. Host range, natural and experimental
2. Pathogenicity, association with disease
3. Tissue tropisms, pathology, histopathology
4. Mode of transmission in nature
5. Vector relationships
6. Geographic distribution

is obtained. Two other levels of classification are recognized by the ICTV; **orders** (ending *-virales*, containing several virus families), and **subfamilies** (ending *-virinae*, between family and genus). No higher level of classification (kingdom, phylum, or class) is recognized for viruses (**Figure 2.1**).

For a long time only two orders containing viruses infecting humans had been defined; the orders *Mononegavirales* (families *Filoviridae, Paramyxoviridae,* and *Rhabdoviridae*) and *Nidovirales* (*Coronaviridae* and *Arteriviridae*). The recently defined family *Bornaviridae* has also been assigned to the *Mononegavirales,* but infection of humans is still controversial. Two additional orders have now been confirmed; the *Herpesvirales* (*Herpesviridae* and the new families *Alloherpesviridae* and *Malacoherpesviridae*) and the *Picornavirales* (*Dicistroviridae, Iflaviridae, Marnaviridae, Picornaviridae,* and *Sequiviridae*).

Genetic relationships determined from nucleic acid sequence data are now routinely used to determine or refine relationships between viruses. On some occasions these have produced surprises where viruses with similar morphology are much less closely related at a genetic level.

One example is the Ictalurid herpes-like viruses, exemplified by channel catfish virus (ictalurid herpesvirus 1). While this is morphologically similar to the herpesviruses, analysis showed that the nucleotide and predicted amino acid sequences of this virus, (according to the ICTV) are "only tenuously related to those of other herpesviruses and identify a distinct lineage."

In consequence, based on genetic data, it was assigned to a new family, the *Alloherpesviridae,* within the newly created order *Herpesvirales.*

Taxonomic rearrangements also have to take into account **metagenomic analysis**. This is nucleic acid sequence analysis that identifies multiple genomes present (often in very large numbers) within nucleic acids recovered directly from the environment or from patients. Such analysis shows the presence of a much larger number of viral sequences in the environment than can be cultured, many of them entirely novel. All of these have to be taken into account (see Chapter 9 for more details).

Since the discovery of the giant *Mimivirus* (see Section 2.4), it has been proposed that viruses should be allocated a new taxonomical division: the superdomain *Acytota* or *Aphanobionta*—non-cellular life (sometimes also referred to as the Virosphere). All cellular forms of life (*Archaea, Bacteria,* and *Eukarya*) would fall into the other superdomain, *Cytota*—cellular life. There are a number of problems with this classification, notably

Box 2.2 A viral species

The term "species" was accepted in 1991 by the International Committee on the Taxonomy of Viruses (ICTV; www.ictvonline.org) as the lowest formal taxonomical level for classifying viruses, but has a very different and more flexible meaning than for most other uses. The ICTV definition is that "a virus species is a polythetic class of viruses that constitute a replicating lineage and occupy a particular ecological niche." A polythetic class is defined by ICTV as "one whose members have several properties in common, although they do not necessarily all share a single common defining property. In other words, the members of a virus species are defined collectively by a consensus group of properties." This somewhat looser classification is necessary at this level. The higher level classifications used for viruses are different, and are "universal," defined by properties that are necessary for membership. ICTV is working to standardize the species definition, but this remains a work in progress.

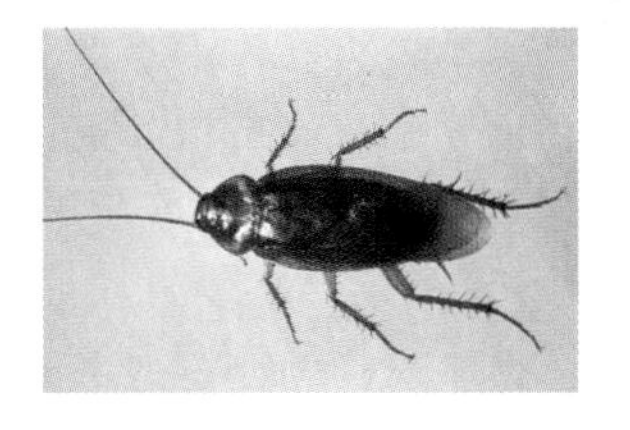

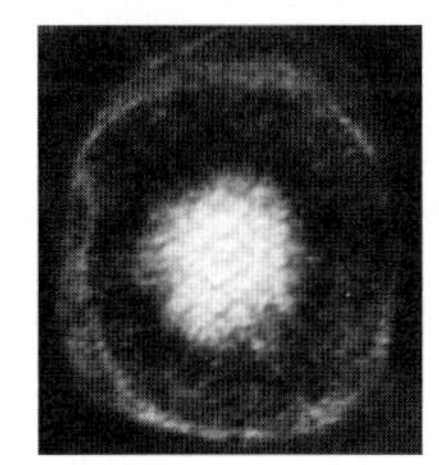

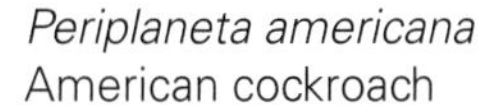

Periplaneta americana
American cockroach

Human herpesvirus 1
Herpes simplex virus type 1

	Example	
	Non-viral	Viral
Kingdom	*Animalia*	Not used
Phylum	*Arthropoda*	Not used
Class	*Insectae*	Not used
Order	*Blattaria*	*Herpesvirales*
Family	*Blattidae*	*Herpesviridae*
Subfamily	*Blattinae*	*Alphaherpesvirinae*
Genus	*Periplaneta*	*Simplexvirus*
Species	*americana*	*Human herpesvirus1*

Figure 2.1
Taxonomical structure. Comparing a virus with a eukaryotic organism. From, CDC Public Health Image Library (http://phil.cdc.gov/).

Box 2.3 The virocell concept

In the virocell concept, viruses are regarded as the equivalent of spores or eggs, with the true metabolically active state being the infected cell—in particular, the cytoplasmic virus factory that is formed by some large viruses. Under such a scheme, it would be inappropriate to classify viruses as non-cellular, since their metabolically active state is that of host cell plus replicating virus. An equivalent would be to look at a sperm cell and classify it as a haploid (single gene set) motile protozoan and wonder why it has so many apparently irrelevant genes.

However, the concept remains controversial. Since some viruses can infect a wide variety of cells from many different host organisms, which is the true virocell? And if the concept is accepted, what of the many forms of life that rely on others? Is a flea on a dog actually a combination organism—a dogflea? And what of the malaria parasite *Plasmodium*, which actually spends part of its life inside human red blood cells?

While the virocell concept is an interesting interpretation, it is far from being universally accepted.

the need for the *Acytota* to use the *Cytota* as an essential part of their life cycle. Can they then truly be classified separately? The question forms the basis of the **virocell concept** (**Box 2.3**). There is also the problem of how to classify nonviral obligate intracellular parasites, including sub-viral forms (see **Figure 2.2** and Section 2.4) along with degenerate cells ranging from

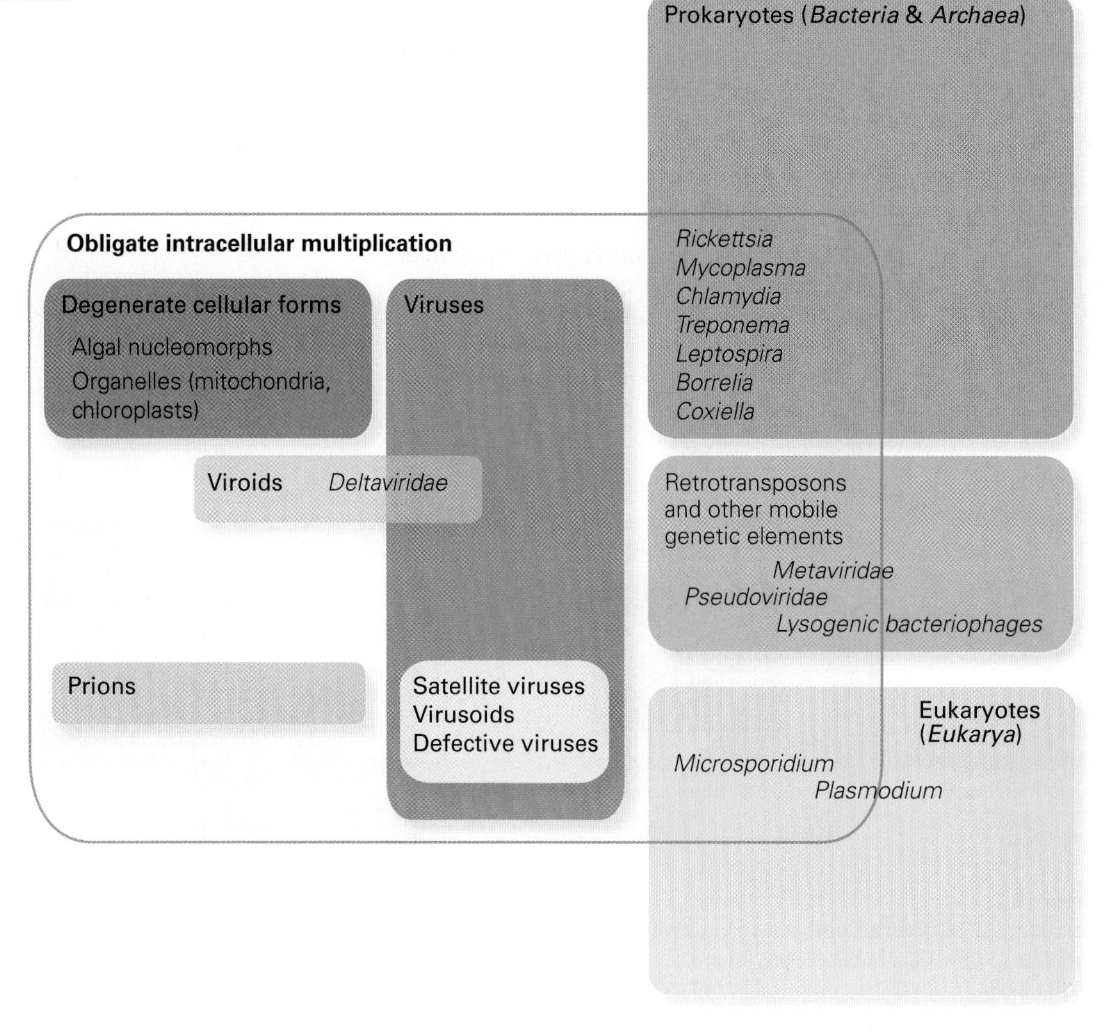

Figure 2.2
Viruses as obligate intracellular parasites. Viruses rely on a host cell to provide the machinery for their replication, as do the smaller viroids (including hepatitis D, the only known viroid-like agent infecting humans, which acquires a protein coat from co-infecting hepatitis B virus) and the protein-based prions. Mobile genetic elements are also obligate intracellular parasites. Among these the *Metaviridae* and the *Pseudoviridae* of plants, insects, and fungi are predominantly intracellular but are able to produce viral proteins and leave the host cell, while lysogenic (latency-capable) bacteriophages may sometimes adopt an intracellular existence. Degenerate cellular forms cover a wide range from recent endosymbionts to established intracellular organelles such as mitochondria. Among bacteria, there is a wide range of bacteria that require an intracellular environment to supply their needs. Even among eukaryotic organisms, the malarial parasites (*Plasmodium*) have an obligate intracellular stage in their replication cycle, growing within the liver and red blood cells of their non-insect hosts.

nucleomorphs (small, reduced eukaryotic nuclei found in some plant organelles) down to the mitochondrion which retains an independent, if small (16.6 kb in humans) genome which provides only a few of its proteins. Add to that the non-nucleic acid prions (see Section 2.4) and it is clear that this level of taxonomy would be as confused as any other.

An alternative exists to the morphology-based classification system, based on genome type and replicative strategy. It is known as the *Baltimore system* (after its creator, the Nobel Laureate Professor David Baltimore) and is widely used alongside the ICTV system. The system divides viruses into seven classes and is covered in more detail below. However, it also has problems coping with the complexity of biological systems.

The existence of multiple systems of classification illustrates the problems inherent in any formal taxonomy for viruses (and even in some cases for bacteria and other microorganisms, where species boundaries can get very blurred).

It is always tempting to try to classify everything using the same methods, but the basic concept of a species as a group of organisms capable of interbreeding and producing fertile offspring is of course not applicable to viruses, which do not interbreed in the conventional sense. While gene transfer between viruses is common, this does not use sexual reproduction and if anything complicates the taxonomical picture.

Below the species level, matters get even more complex. How to classify the multiple subtypes, strains, and variants that together make up a species? Even within an individual patient there are a large number of variants, often termed **quasispecies**. The question does not yet have an adequate answer, though genomics may in time provide one.

As an example of this, **Figure 2.3** shows a phylogenetic tree for one group of viruses. What is not shown is the mass of related variants for each named virus type, which can on occasion blur into those of an adjacent virus type. Effectively, the situation for viruses is not amenable to division into

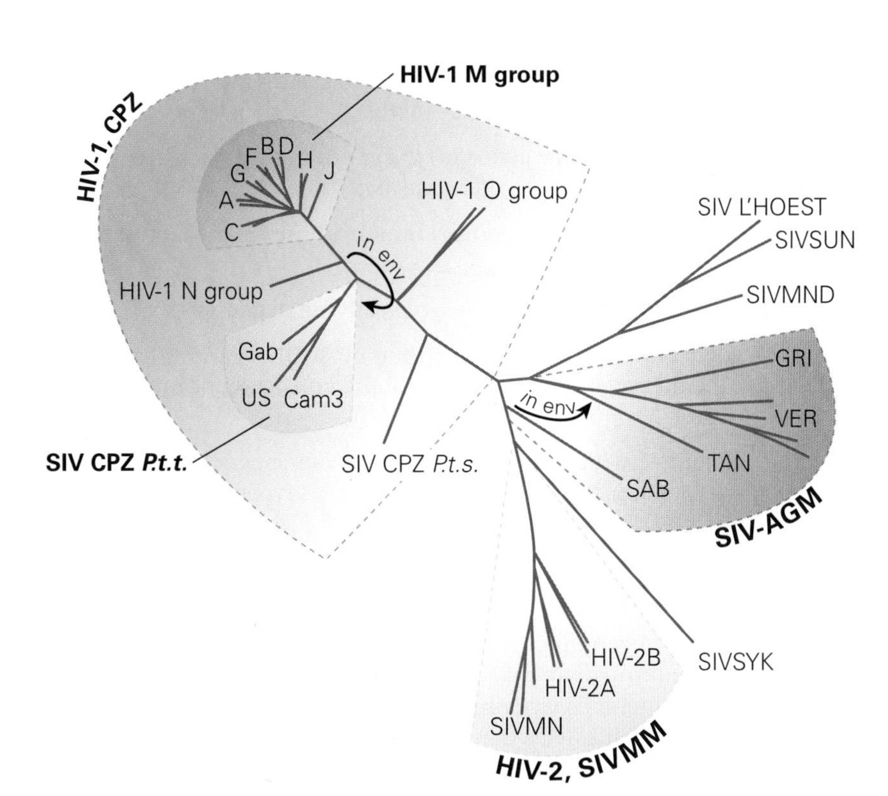

Figure 2.3
Phylogenetic tree of the simian immunodeficiency virus (SIV) and HIV viruses showing genetic relatedness.

Figure 2.4
A "Lichtenberg figure" fractal caused by high-voltage dielectric breakdown within a block of plexiglass. The branching discharges are thought to extend down to the molecular level. Courtesy of Stoneridge Engineering (http://www.capturedlightning.com).

separate individual types. Rather, it is a simplified way of describing a spectrum of related organisms that becomes more complex the closer one looks—a fractal structure (**Figure 2.4**).

Viral taxonomy is a tool with distinct limitations, rather than a solid definition, and is always subject to review. As in so many situations, the complexity of a biological system is far beyond our efforts to define it. That does not, however, stop attempts being made to do so.

2.2 VIRUS MORPHOLOGY

Morphology is an important element of virus classification, and is usually determined using the electron microscope. Most viruses can be readily classified by these criteria, and they provide a straightforward basis for the taxonomical groupings used. The morphologies of the families of viruses infecting humans are shown in **Figure 2.5** (and also in Figure 1.3). The basic classifications are:

- Non-enveloped virus with a helical capsid
- Non-enveloped virus with an icosahedral capsid
- Enveloped virus with a helical capsid
- Enveloped virus with an icosahedral capsid
- Complex (a "catch-all" group for those which do not fit the above)

While most viruses fall into the first four groups, there are many exceptions. In addition, most viruses can produce forms which are totally unlike their 'classical' appearance and the common descriptions of virus morphology may be very different from what is actually seen under the electron microscope (see Figure 1.4).

2.3 VIRAL CLASSIFICATION BY GENOME TYPE

The replication strategy of a virus is heavily influenced by the nature of the viral genome, and this is commonly used when classifying viruses. Indeed, the **Baltimore system** is based entirely on this.

Viruses may be subdivided by genome type as outlined for viruses infecting humans in **Table 2.1**. The types of genome are:

- Double-stranded (ds) DNA genomes; among these are the largest of viral genomes
- Single-stranded (ss) DNA genomes; typically small genomes
- dsRNA genomes tend to be smaller than most DNA genomes; most viruses with dsRNA genomes have a genome consisting of multiple segments
- ssRNA genomes, which are transcribed as messenger RNAs (positive sense); small genomes, but many different virus types fall into this class
- ssRNA genomes which need to be transcribed into complementary strands to produce mRNAs (negative sense); small genomes, some have segmented genomes
- Viruses with RNA genomes that use reverse transcriptase activity to produce a DNA intermediate that is used to make the RNA genome; small genomes; also includes virus-like agents with predominantly or exclusively intracellular lifestyles that use a reverse transcriptase stage

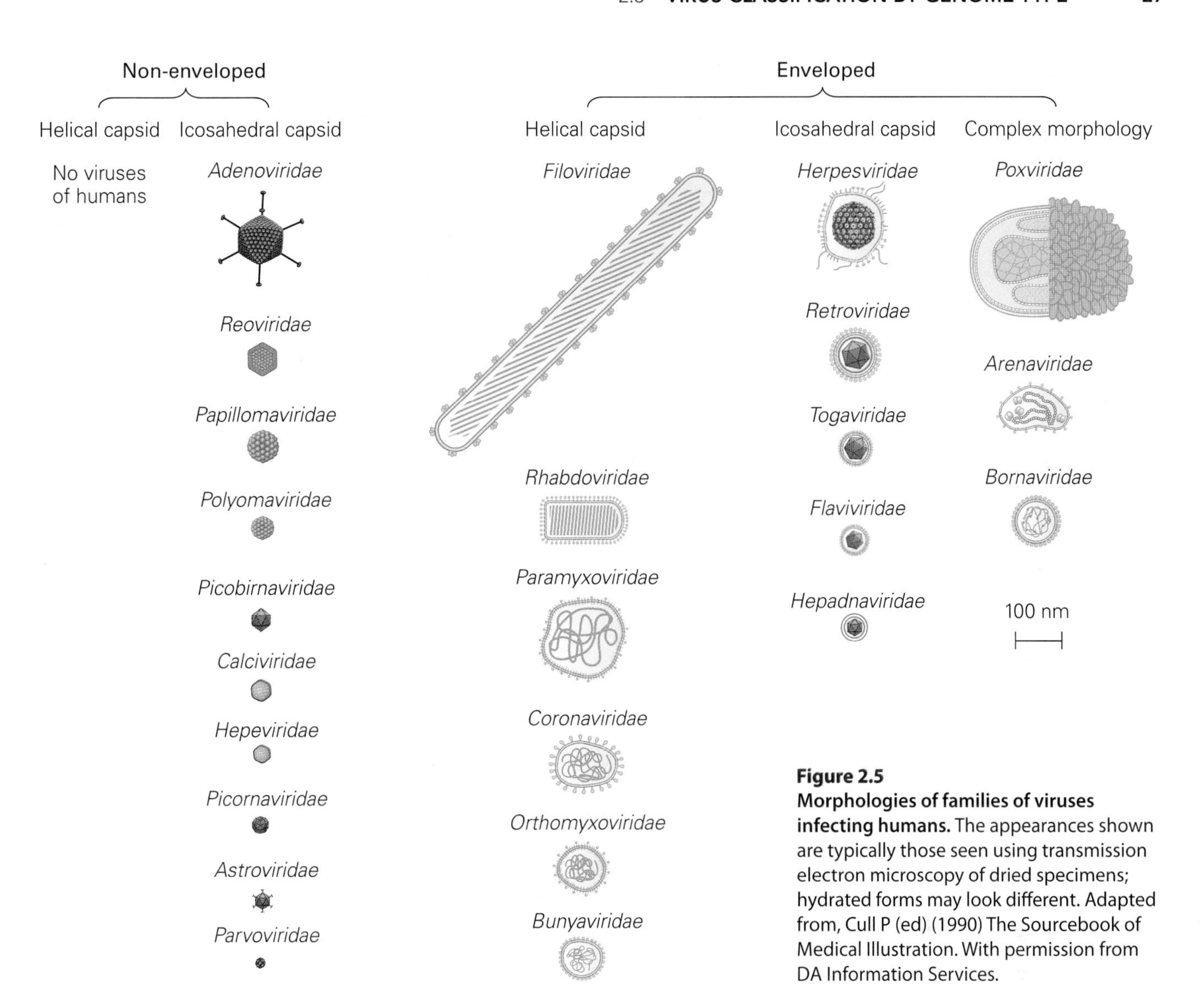

Figure 2.5
Morphologies of families of viruses infecting humans. The appearances shown are typically those seen using transmission electron microscopy of dried specimens; hydrated forms may look different. Adapted from, Cull P (ed) (1990) The Sourcebook of Medical Illustration. With permission from DA Information Services.

- Viruses with DNA genomes that use an RNA intermediate stage to produce the DNA genome via a reverse transcriptase activity; small genomes, often with overlapping genes

These seven categories as listed above form classes 1 to 7 of the Baltimore system (see Figure 3.5). Unfortunately, as with the more formal classification used by the ICTV, there are many issues with attempting to classify a complex biological system into a few fixed categories. For example, class 5 (negative-sense ssRNA) was subdivided into those with non-segmented genomes and those with segmented genomes. However, no such division has been attempted in other classes where there are both segmented and non-segmented genomes, such as the dsRNA viruses of class 3.

Other viruses fall between the defined classes. For example, both the *Bunyaviridae* and the *Arenaviridae* have genomes that are predominantly negative sense, but can also contain positive-sense regions (though admittedly these do not appear to be directly translated). This would make them class 4.5. Added to that, they also have segmented genomes, suggesting the possible subdivision even of class 4.5.

Additionally, an eighth category that may need to be included is the *viroids*, self-replicating RNAs without any protein stage in their replication that use the cellular machinery to copy themselves. One human agent

Table 2.1 Virus families affecting humans

		Members		Genome size (kb/ kbp)	Envel- ope	Capsid	Size (nm)	Virion proteins	Example of human disease	Main host types
		All*	Human							
dsDNA										
	Adenoviridae	31	6	35.8–36.2	No	Icosahedral	80–110	10	Common cold	Vertebrates: mammals (primates, humans), birds, reptiles, fish
	Herpesviridae	66	8	120–260	Yes	Icosahedral	120–200	24–71	Cold sores, chickenpox	Mammals (humans), birds, fish
	Papillomaviridae	44	26	8	No	Icosahedral	52–55	2 + cellular histones	Warts, cervical cancer	Mammals (humans)
	Polyomaviridae	13	2	5	No	Icosahedral	40–45	3 + cellular histones	Progressive multifocal leukoencephalopathy	Mammals (humans), birds
	Poxviridae	62	6	130–375	Yes	Complex	140–260 × 220–450	75+	Smallpox	Mammals (humans), birds, insects
ssDNA (–)										
	Parvoviridae	37	6	5	No	Icosahedral	18–26	2–3	Anemia	Mammals (humans), birds, insects
Segmented										
	Picobirnaviridae	2	1	4	No	Icosahedral	35–40	4?	Gastrointestinal illness	Mammals (humans)
dsRNA										
	Reoviridae	79	13+	18.2–30.5	No	Icosahedral	60–80	10–12	Diarrhea	Mammals (humans), birds, reptiles, fish, molluscs, arthropods, insects, plants, fungi
ssRNA (+)										
	Astroviridae	9	1	6.8–7.9	No	Icosahedral	27–30	3	Gastroenteritis	Mammals (many, inc. primates, humans), birds
	Caliciviridae	6	2	7.4–8.3	No	Icosahedral	35–39	1–2	Gastrointestinal illness	Mammals (primates, humans), birds, reptiles, fish
Nidovirales	*Coronaviridae*	20	7	25–33	Yes	Elongated helical	120–160	5	Common cold	Mammals (humans), birds

		Members		Genome size (kb/kbp)	Envel-ope	Capsid	Size (nm)	Virion proteins	Example of human disease	Main host types
		All*	Human							
	Flaviviridae	58	Variable	9.5–12.5	Yes	Polyhedral	40–60	3–4	Yellow fever, hepatitis	Mammals (humans), birds, arthropods (arboviruses)
	Hepeviridae	1	1	7.2	No	Icosahedral	27–34	1	Hepatitis E	Humans
	Picornaviridae	22	11–13	7–8.5	No	Icosahedral	27–30	5	Polio, meningitis	Mammals (humans), birds
	Togaviridae	40	13+	9.7–11.8	Yes	Spherical/ pleomorphic	70	5–7	Rubella	Mammals (humans), birds, arthropods (arboviruses)
ssRNA (ambisense, segmented)										
	Bunyaviridae	104	Variable	6.3–12	Yes	Helical (circular) × 3	80–120	4	Hemorrhagic disease	Mammals (humans), insects, plants (arboviruses)
	Arenaviridae	22	Variable (7+)	11	yes	Filamentous × 2	110–130	5	Meningitis, hemorrhagic disease	Mammals (humans), arthropods (arboviruses)
ssRNA (–)										
Mononegavirales	*Bornaviridae*	1	1?	8.9	Yes	Crescentlike	80–100	5–7	Neurological disease?	Mammals (many, inc. primates, humans), birds
	Deltaviridae	1	1	1.7	Yes	Polyhedral	36	2+3	Hepatitis	Humans
Mononegavirales	*Filoviridae*	5	5	18.9–19	Yes	Elongated helical	790–1400 × 80	5	Hemorrhagic fever	Primates (humans)
Mononegavirales	*Paramyxoviridae*	34	10	15.2–15.9	Yes	Elongated helical	150–200	6–7	Measles, mumps	Mammals (humans), birds
Mononegavirales	*Rhabdoviridae*	38	4+	11–15	Yes	Bullet-shaped	45–100 × 100–430	5–11	Rabies	Mammals (humans), fish, plants
Segmented										
	Orthomyxoviridae	6	3 (+ 2)	10–14.6	Yes	Helical	80–120	7	Influenza (encephalitis)	Mammals (humans), birds, arthropods, fish
RNA/DNA (ssRNA, diploid)										
	Retroviridae	53	4+	7–11	Yes	Spherical/ pleomorphic	80–100	3–9	AIDS	Mammals (humans), birds, reptiles
DNA/RNA (partially dsDNA)										
	Hepadnaviridae	6	1	3.0–3.3	Yes	Icosahedral	40–48	4	Hepatitis	Mammals (humans), birds

* Many subtypes exist, and additional viruses are provisionally assigned to individual families, but the above shows confirmed members according to reference (b).

Main sources: (a) International Committee on the Taxonomy of Viruses—ICTVdb Descriptions 2002 (http://www.ncbi.nlm.nih.gov/ICTVdb/ICTVdB/index.htm); (b) International Committee on the Taxonomy of Viruses—ICTVdb Master Species List 2005 (http://phene.cpmc.columbia.edu/Ictv/ICTV8thReport%20Master%20species%20list.htm); (c) Knipe DM & Howley PM (eds) (2006) Fields Virology. Lippincott Williams & Wilkins.

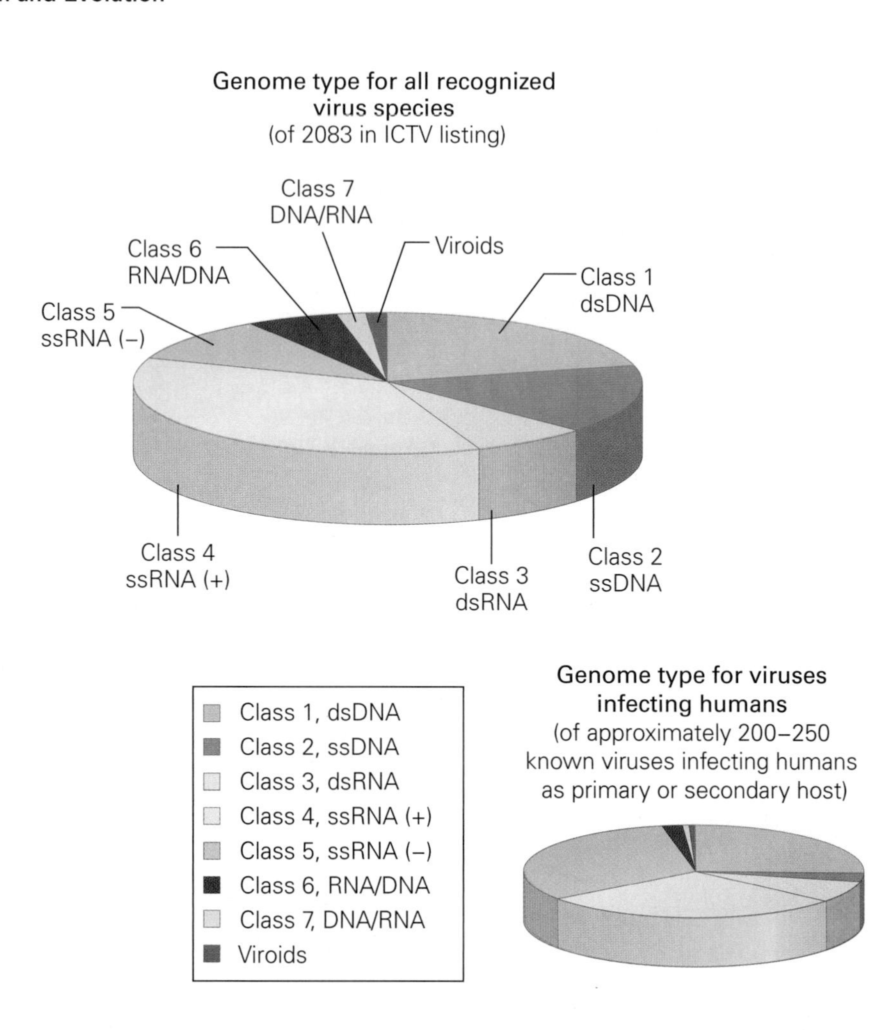

Figure 2.6
Number of viruses by genome type. Because a number of viruses infect humans only rarely (as secondary or occasional hosts or as dead-end zoonotic infections) it is impossible to be precise about the number of viruses that can infect humans. The number 200–250 is a useful guide, however. It must be noted that the upper pie chart represents only virus species recognized by the ICTV. There are vastly more species in existence that have been identified, as is revealed by metagenomic analysis of environmental samples (see Chapter 9). Current estimates suggest that there are more than ten million trillion trillion (10^{31}) viruses on Earth. They are unlikely to fall into only 2000 (2×10^3) species.

(hepatitis delta) has extensive similarities to viroids (and is often classified with them), although it produces two small proteins and is classified by the ICTV as the sole member of the genus *Deltavirus* rather than with the viroids. Further details are given in Section 2.4.

The number of viruses within each of the Baltimore classes varies. Positive-sense ssRNA genomes (class 4) appear to be the most common among those viruses infecting humans, while DNA genomes that replicate via an RNA intermediate (class 7) appear to be the least common, unless one counts the even rarer viroids. The frequency of individual genome types is detailed in **Figure 2.6**. It should be noted that this only reflects formally recognized virus species. For example, the huge number of dsDNA bacteriophages that have yet to be characterized are not included in this figure.

2.4 SUPERVIRUSES AND SUB-VIRUSES

With increasingly precise knowledge of exactly what constitutes a virus at the molecular level, it has become clear that there are various infectious agents that stretch the classification.

Until recently, the nucleocytoplasmic large DNA viruses (NCLDVs) were considered the largest known viruses and were still smaller than cellular forms of life.

Supervirus

In 1992, that all changed. *Mycoplasma genitalium* (580,073 bp) was succeeded as the smallest true cellular organism by *Nanoarchaeum equitans* at 490,885 bp. At the time, the largest virus known was *Bacillus* bacteriophage G at 497,513 bp—very slightly larger than the genome of *Nanoarchaeum equitans*.

This small difference in the size of the largest viruses and the smallest cells was shattered when a sample of water from a Bradford, United Kingdom (UK) cooling tower was found to contain a new organism. At first the sheer size of the new find led to its tentative classification as a Gram-positive bacterium, *Bradfordcoccus*. But it wasn't a bacterium. After eleven years of investigation, the new organism was identified as the first **supervirus**, *Mimivirus*. Named for its apparent mimicry of a bacterium, the new virus was huge; 400 nm in diameter, with filaments projecting a further 100 nm on all sides, it was clearly larger than other viruses (**Figure 2.7**). But it was the genome that was to be the real surprise.

At 1,181,404 bp it was more than twice the size of the smallest cellular organism. It codes for at least 911 genes, many of which are homologs of genes that were previously only seen in eukaryotic organisms, such as four different aminoacyl-tRNA synthetases and a range of DNA repair enzymes. However, about half of *Mimivirus* genes have no known homologs. Among those genes with identified relationships, *Mimivirus* contains relatively few genes that appear related to other NCLDVs, but contains genes with homologs in all three domains of life: *Archaea*, *Bacteria*, and *Eukarya*. This has led to suggestions that it is an ancestral remnant from before the divergence of these forms of life that may have played a role in the development of the cellular nucleus that distinguishes eukaryotes. Other proposals were that *Mimivirus* represents an intermediate stage between viruses and cellular life (the term "girus," from Giant vIRUS, was proposed), and even that *Mimivirus* represents a fourth domain of life. Later work has suggested that many of the *Mimivirus* genes have been acquired by the virus, and thus that it is a product rather than a progenitor.

To resolve these questions, more examples of these giant viruses are needed for study. However, at present, this first *Mimivirus* is the only recognized species; *Acanthamoeba polyphaga mimivirus*. *Mimivirus*-like nucleic acid sequences have been identified from a range of aquatic environments, notably the Sargasso Sea, and it is likely that many more exist in nature. Once these are characterized, more insights are likely.

One intriguing possibility is a suggested association between *Mimivirus* and human pneumonia. This remains to be proven, although reports from experimental infections of mice appear to support the suggestion.

(A)

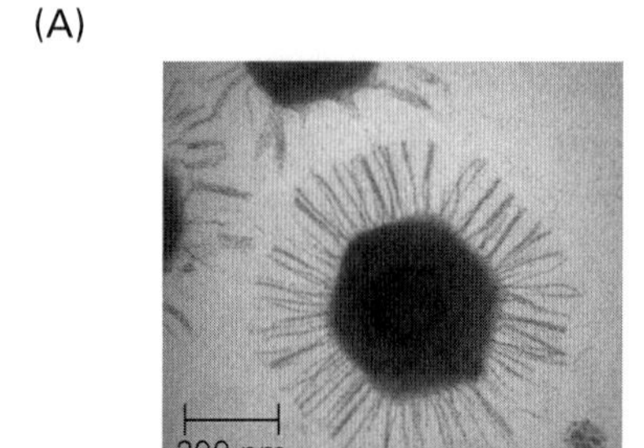

(B)

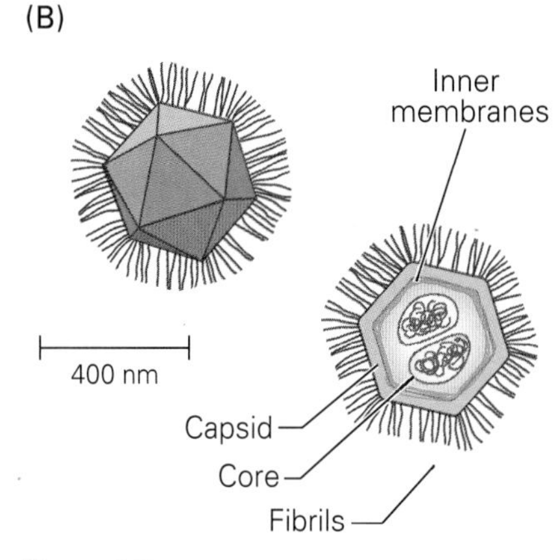

Figure 2.7
Mimivirus. (A) Electron micrograph of *Acanthamoeba polyphaga mimivirus*. Courtesy of Didier Raoult, Rickettsia Laboratory, La Timone, Marseille, France. (B) Schematic showing the structure of *Mimivirus*. Courtesy of Xanthine under the Creative Commons Attribution 2.5 Generic license.

Sub-viral infections

Contrasting with the complexity of the largest viruses, there is a range of agents that do not show the full spectrum of characteristics associated with most viruses (as shown in Figure 2.2). These range from replication-deficient satellite viruses which need to use one or more elements provided by a replication-competent (helper) virus, down to naked RNA (viroids) and even infectious proteins (prions). There is also a range of defective viruses as well as mobile genetic elements seen in every organism from bacteria to humans that bear at least some resemblance to viruses.

Satellite viruses

A **satellite virus** can resemble a replication-competent virus in many ways and produces virus-specific structural proteins, but has adapted to require co-infection of the cell with a different but specific helper virus, which allows the satellite virus to complete its replicative cycle. This is a parasitic relationship, which usually has a negative effect on the replication of the helper virus.

The only known satellite virus of humans is adeno-associated virus (AAV), a member of the *Parvoviridae*, which is closely related to the replication-competent parvoviruses. Despite this, it is unable to replicate without the presence in the same cell of a helper adenovirus (or, under some circumstances, the herpesvirus HSV or physical factors such as heat, radiation, or chemical toxins). The helper function appears to involve a range of functions associated with DNA replication. While cell extracts may support replication, this is at low efficiency, supporting the need for the helper virus.

If no helper virus is available, AAV is capable of integrating into the cellular genome until helper functions become available. It has even been suggested that latency is the preferred state for AAV, and that the helper virus is actually disrupting this. AAV does not appear to be associated with any known human disease, but can interfere with some functions of the helper virus.

Interestingly, the largest known virus (*Acanthamoeba polyphaga mimivirus*) has recently been shown to have what appears to be a satellite virus; "Sputnik." This is an as yet unclassified 50 nm icosahedral virus with an 18.3 kb dsDNA genome, which also appears to interfere with infection by its helper.

Virusoids

Virusoids are the next step down from satellite viruses although the terms are often combined. Virusoids are associated with viruses of plants, and no human examples are known. They have very small genomes (200–400 bases of circular ssRNA), and do not produce their own structural proteins. Rather, they acquire these entirely from the helper virus. Virusoids replicate in the cytoplasm using helper virus functions.

Viroids

Viroids infect plants and produce a range of diseases. There are 29 recognized species (defined on the basis of RNA homology), falling into two families; the *Avsunviroidae* and the *Pospiviroidae*. Viroids are very small (220–375 bases) circular RNA molecules, with very pronounced secondary structure. Although they are nominally single stranded, the vast majority of the viroid RNA is paired with other regions, forming a tightly coiled structure (**Figure 2.8**). This convoluted structure helps to stabilize them, since they must survive without a protective coat of protein in a world full of nuclease enzymes that would normally destroy a single-stranded RNA very rapidly indeed. They are the smallest self-replicating infectious agents to have been characterized, but despite this (and unlike satellite viruses or virusoids) they can replicate without any requirement for a helper virus.

Viroids do not actually code for any proteins. The viroid genome is replicated in the cellular nucleus entirely by cellular enzymes, and there is no DNA or protein stage in their life cycle. The viroid RNA is transcribed by the

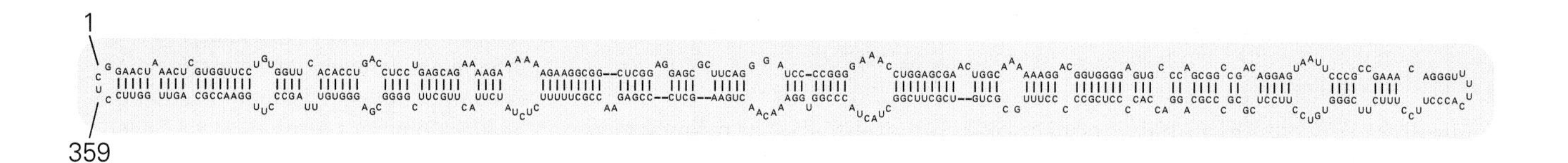

Figure 2.8
Base pairing of potato spindle tuber viroid (PSTV) RNA. Courtesy of Jakub Friedl under the Creative Commons Attribution-Share Alike 3.0 Unported license.

cellular DNA-dependent RNA polymerase II; the unusual secondary structure of viroid RNAs means that they are used as templates at high efficiency by this enzyme, which normally makes messenger RNA from the cellular DNA.

Since the viroid RNA is produced as **concatamers** (polymers of the genome) and no viroid proteins are produced, there was a very significant question regarding how the genomes were cut to unit length. Surprisingly, this is not by a cellular protein, but rather is mediated by the viroid RNA itself, functioning as a **ribozyme**, a word derived by combining ribonucleic acid with enzyme. This ability of RNA to catalyze a reaction without the involvement of a protein was very surprising when it was first discovered in 1981, and has been studied intensively using viroids as a model system, although similar cleavage events occur in virusoids, satellite viruses, and other systems. The mechanisms of ribozyme cleavage and their potential uses are discussed further in Chapter 6. It has also been proposed that this may be a fossil of an RNA-only world, before the development of proteins or DNA.

Deltavirus

Despite the fact that all known viroids only infect plants, they have very marked similarities to the hepatitis delta virus (the delta agent or HDV). Co-infection with HDV is associated with more severe forms of hepatitis B infection. Although it is classified as the sole member of the genus *Deltavirus*, HDV has some characteristics of viroids, virusoids, and satellite viruses.

HDV codes for two proteins, the small and large delta antigens, which are produced from a single open reading frame; 195 amino acids are translated for the small delta antigen, and 214 for the large delta antigen, which is thus identical to the small delta antigen for over 90% of its length.

HDV is dependent upon a helper virus (in this case, hepatitis B) to form HDV particles. The HDV genome can replicate independently, using RNA polymerase II as is seen with viroids. It then uses hepatitis B structural proteins to form the HDV particle, along with the large delta antigen.

The genome of HDV is larger than a viroid, a 1680-base circular ssRNA. However, it has a viroid-like secondary structure in which 70% of bases are paired, and contains base sequences related to those seen in viroids but distinct from those seen in virusoids or satellite viruses. In addition, its replication appears to involve a viroid-like self-cleavage of the RNA. From the case of HDV, it is clear that the arbitrary classification of satellite/virusoid/viroid actually covers a spectrum of related sub-viral agents.

Replication-defective viruses

In any discussion of sub-viral infectious agents, the various replication-defective viral forms generated by replicating viruses must be considered. In the case of influenza, as discussed in Chapter 3, virions that do not contain the full set of genome segments could replicate if co-infecting viruses provided the missing functions. Similarly, retroviruses that have replaced part of their own genome with cellular oncogenes require a co-infecting competent helper virus (Section 4.8).

Many viruses, particularly those with RNA genomes, produce **defective interfering (DI) particles**, particularly at high multiplicities of infection. These contain only part of the viral nucleic acid which, being smaller, can replicate more rapidly than whole viruses, provided that a replication-competent virus is present in the same cell to supply the missing functions. DI particles appear to be important in establishing persistent virus infections, since they can alter the course of infection by interfering with viral replication. All of these viruses are very similar to their 'parent,' and (necessarily) exist as mixed populations. In the case of DI viruses and of oncogenic retroviruses where an oncogene replaces essential viral functions, they can have profound effects on the course of infection.

Mobile genetic elements

Mobile genetic elements were first studied in bacteria, where it became clear that some self-replicating genetic elements, or plasmids, could transfer from cell to cell. Some complex plasmids can actually initiate this transfer, coding for transfer-associated structures on the cell surface. Plasmid transfer between bacteria, taking with it the genes coding for resistance to antibiotics, has caused significant clinical problems.

Eukaryotic cells have direct equivalents of plasmids, similar to those seen in some virus infections. As with prokaryotes, there are other, mobile genetic elements. The mobile genetic elements known as transposons contain a subgroup (**retrotransposons**) which use an RNA stage and a reverse transcriptase function to move themselves about. A limited number of these also code for proteins and are capable of assembling into virus-like particles. These have now been grouped into the *Metaviridae* (infecting fungi, insects, and nematodes) and the *Pseudoviridae* (infecting fungi and insects). These two families use reverse transcriptase and retrotransposon-coded proteins to relocate themselves within the genome of their host cells. Thus, they further extend the range of known sub-viral agents.

Endogenous viruses

Both prokaryotes and eukaryotes contain virus-derived sequences of varying degrees of completeness, many derived from lysogenic bacteriophages and integrated retroviral sequences, respectively. In humans, 98,000 such genetic elements have been identified, accounting for around 8% of the total genome. Levels of transposons may be considerably higher.

Endogenous retroviruses and retroviral elements have the potential to interact with both the host organism and with other viruses, as described in Section 1.5.

Clearly, the spectrum of sub-viral agents extends all the way down to a few bases of virus-derived genetic material integrated into the cellular genome which may have some (even if very limited) potential to produce effects on the cell.

Prions

Prions appear to be quite different from the spectrum of sub-viral agents described above. They exist both in animals and fungi, but the origins of the field came from work during the 1960s that showed that the agents causing spongiform encephalopathies in humans and animals were transmissible.

The best-studied spongiform encephalopathy at the time was scrapie, a disease of sheep, which was used as a model system for the other spongiform encephalopathies. All are long-incubation diseases causing severe localized neurological damage, producing holes in the brain which give it a sponge-like appearance. Human diseases of this type are detailed in **Table 2.2**.

Scrapie may be transmitted experimentally to other animals, and has been observed to transmit to captive animals fed with sheep-derived materials, notably deer and mink. Despite extensive work, no transmission of scrapie to humans has been demonstrated.

Table 2.2 Prion diseases of humans[1]

Disease	Host	Mode of transmission	Cases in UK, 1990–2008
Sporadic Creutzfeldt-Jakob disease (CJD)	Any human	Somatic mutation of PrP gene	998
Iatrogenic Creutzfeldt-Jakob disease (CJD)	Hospital patients	Infection via contaminated tissue (dura mater, corneal tissue, human growth hormone[2]) or instruments	59
Familial Creutzfeldt-Jakob disease (CJD)	Humans with specific inherited mutations in PrP gene	Inherited mutations conferring susceptibility to disease	68
Variant Creutzfeldt-Jakob disease (vCJD)	Any human	Dietary intake of prion-contaminated bovine material or (more rarely) blood transfusions	164
Gerstmann-Sträussler syndrome (GSS)	Humans with specific inherited mutations in PrP gene	Inherited mutations conferring susceptibility to disease	35
Fatal familial insomnia (FFI; sporadic form also exists)	Humans with specific inherited mutations in PrP gene	Inherited mutations conferring susceptibility to disease	None[3]
Kuru	Members of the Fore tribe in New Guinea	Ritual consumption of human brain tissue (now ceased)	None, disease eliminated

[1]Spongiform encephalopathies of animals include scrapie (sheep), BSE (cows), transmissible mink encephalopathy (TME, mink), chronic wasting disease (deer, elk), feline spongiform encephalopathy (cats), and exotic ungulate encephalopathy (kudu, nyala, oryx). Downer cow syndrome has some similarities but is not as yet confirmed as a spongiform encephalopathy. [2]Contamination was from cadaver-sourced human growth hormone (HGH), which has now been replaced with recombinant versions. [3]Approximately 60 cases reported worldwide over this period.

The prion hypothesis

Despite much work in the 1950s and 1960s to identify a slow virus (see Chapter 1) causing spongiform encephalopathies, none was identified. In fact, the agent had many properties suggesting that it was a protein with no accompanying nucleic acid, most notably its remarkable resistance to inactivation. Infectivity appeared to be linked to a modified form of the cellular protein PrPC. The modified form is known as PrPSc, and was present in fibrils within the brain associated with areas of neuronal damage. The term prion was selected to describe such an agent, although this was not the first use of the word; the term prion has actually been used for far longer for a type of Antarctic seabird. However, in virology, a prion was defined by Stanley Prusiner in 1998 as "proteinaceous infectious particles that lack nucleic acid." Studies showed that they resist inactivation by agents which destroy nucleic acid and contain an essential modified isoform of a cellular protein.

The prion hypothesis was very much against the prevailing orthodoxy at the time, where the use of nucleic acids to encode and copy the genetic information was not only an essential property of living organisms, but actually formed the definition of life, and an alternative virino hypothesis was proposed suggesting that scrapie infectivity involves a small (undetected) nucleic acid associated with the PrP protein. There was some evidence to support the virino hypothesis, including the (now discounted) finding that appropriate nucleic acids may co-purify with infectivity, and an observation of very small (10–12 nm) virus-like particles associated with fibrils in hamsters. However, this was countered by evidence that highly purified protein can be infectious and, since even an extremely small 10–12 nm "virus" could be expected to have a molecular weight of about 750 kD, such infectivity would not be expected to co-purify with a 30 kD protein.

A biocatalytic model which could explain how a protein causes disease was then proposed. In the original such model, PrPSc (the disease-associated form) converts the normal (PrPC) form into PrPSc, causing disease by the loss of normal PrPC function. Unfortunately for this idea, transgenic mice completely lacking the PrPC gene appear broadly normal. Though a number of mild effects have been reported, including some loss of Purkinje (motor control) cells in the brain with aging and (possibly) some differences in transmission of neural impulses, these effects are much less severe than mouse scrapie.

It has now been shown that conversion of PrPC to PrPSc can be mediated *in vitro* by PrPSc, albeit at low rates. This provides support for the core hypothesis of the biocatalytic model. Potential methods for refolded PrPSc to exert a direct pathogenic effect have been proposed, possibly by exposing internal regions of the protein which have been linked to neuronal apoptosis.

Extensive work has been undertaken to demonstrate the nature of the infectious agent, and the absence of any significant amount of nucleic acid now seems to have been proven. The balance of evidence at present strongly favors the prion hypothesis of a biocatalytic, protein-only infectious agent.

From cows to man

Interest in prion disease was increased greatly as a result of the appearance of bovine spongiform encephalopathy (BSE), or mad cow disease. Since its first appearance in the mid-1980s in the UK, almost two hundred thousand cases have been reported in cows. Of these, almost 184,000 have been in the UK, with less than 5000 in the rest of the world.

The disease is characterized by loss of motor control and a classic spongiform encephalopathy in the afflicted cow. At first it was thought that the disease had arisen by transfer of scrapie in sheep-derived cattle feed, but, more recently, it was suggested that the disease arose in the UK beef herd and was transmitted by cow-derived cattle feed. While this could decrease concern about a super-scrapie that has acquired the ability to jump between species, it leaves the concerns surrounding an entirely novel disease (see below).

The feeding of animal protein to ruminant animals, described in the journal *Nature* in 2000 as "feeding cows to cows," was widespread but has now been modified in the light of the events surrounding BSE. It should be noted that while some countries are free of clinically confirmed BSE, related conditions exist, notably "downer cow syndrome" in the United States (which has had only three confirmed cases of BSE), although this is a much less well-defined syndrome. Downer cows may lack spongiform pathology, but material derived from them has been reported to induce such disease in mink.

The human degenerative brain disorder Creutzfeldt-Jakob disease (CJD) is also a spongiform encephalopathy. There is no epidemiological link between scrapie and CJD, but a novel form of CJD was identified in humans from 1994 onward. This was initially referred to as new variant CJD (nvCJD), but is now more generally referred to simply as variant CJD (vCJD). It is characterized by onset at an earlier age than "classical" CJD (a mean of 26 years, far lower than for other forms of spongiform encephalopathy) and by a significantly different and more aggressive pathology.

It is now generally accepted that vCJD represents a human form of BSE, transmitted to humans by consumption of tissue from the estimated 400,000 BSE-contaminated cows that entered the food chain in the UK, mainly during the 1980s. Even though BSE prions differ in only seven or eight amino acids from scrapie prions, that appears to be enough to allow a jump to man, something that has never been shown for scrapie.

However, despite all the concern and the huge economic costs resulting from BSE (with over 4 million British cows slaughtered and export bans in place for many years), the relative numbers of vCJD are limited (**Figure 2.9**), with a total of 164 confirmed cases in the UK by mid-2008. Reported cases elsewhere number approximately 30. Of these, many are linked to previous residency in the UK.

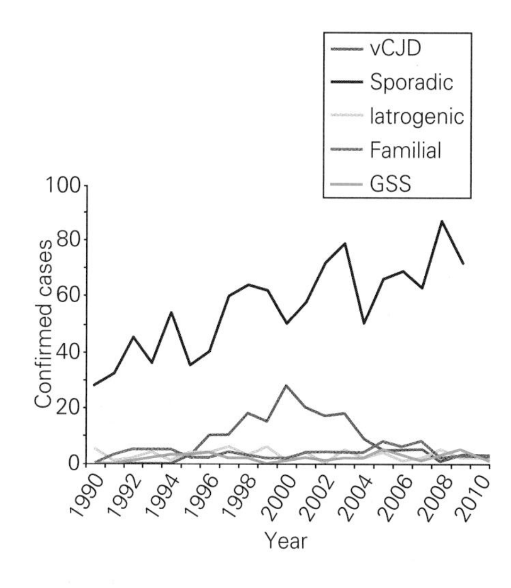

Figure 2.9
Number of cases of spongiform encephalopathies in the United Kingdom by year. See Table 2.2 for descriptions of individual diseases.

The concern initially was that the observed cases were only the first few cases of the bell curve of a major epidemic, but this is now thought unlikely. From the first fatality in 1995, numbers rose to a maximum of 28 deaths in 2000, but fell to 3 per year in 2009 and 2010. It now appears that transmission to humans may be rare, despite apparently high levels of exposure.

However, it is known that genetic factors influence susceptibility to vCJD. All but one case identified to date have been homozygous for methionine at residue 129 of the PrP protein. This particular genotype is only present in 39% of the UK population. The one exceptional case was heterozygous at this residue, expressing both valine and methionine. However, this case did not show infection of the brain (prions were observed in lymphoid tissue), and appears to have been infected by the relatively unusual route of a blood transfusion. The patient died five years after transfusion, from a non-neurological disease, so it is difficult to determine what (if any) the course of vCJD might have been in this case.

Many different approaches for diagnosing vCJD infection have been evaluated, including testing of blood and urine, and it is known that prions may be detected in lymphoid tissue in some cases. However, confirmation of vCJD is typically made after death by examination of brain tissue.

A range of therapeutic options are under investigation, but none is yet established as effective.

Clearly, while methionine homozygosity at residue 129 is associated with the development of vCJD, far less than 39% of the UK population has developed the disease (the actual figure is less than 0.0003%). Thus, it is likely that other (unknown) genetic factors are also required for the disease to become established, restricting the susceptible population.

While the possibility exists of a second, coming wave of the disease, this now seems far less likely. It may be, as with the dosing of millions of polio vaccine recipients with the potentially oncogenic SV40 virus between 1955 and 1963 (see Section 5.10), that we have dodged another bullet. However, this will not be certain for many years.

2.5 WHERE DO VIRUSES COME FROM?

There are a variety of different theories as to where and how viruses originated, and it is highly likely that several of them are correct. All of the available evidence indicates that viruses have multiple origins (they are polyphyletic).

Studies at the molecular level indicate some surprising relationships, such as structural similarities in the outer proteins of herpesviruses infecting humans and viruses of bacteria (bacteriophages). Recent findings with the very largest viruses have suggested links between all three domains of life (see Section 2.4).

However, despite some high-profile similarities there are many differences, which suggest a variety of mechanisms for the generation of viruses. Clearly viruses could not be the earliest form of life, since they need host cells to grow, but some of the molecular mechanisms used by viruses could predate cellular life and indeed could have contributed to its evolution.

Proposed origins for viruses include:

- Fossils from a pre-DNA, pre-protein world in which catalytic RNAs were the effector mechanisms
- Remnants from a pre-cellular world
- Degenerate cells, which have lost the functions needed for independent existence
- Escaped genes, mobilized from within the genomes of cellular organisms
- Escaped nuclei, mobilized from within eukaryotic cells

The potential fossil origin of viruses is best shown by the self-replicating RNAs known as viroids, while the role of DNA viruses in the creation of cellular life is being earnestly debated.

Escaped genes may be most like the reverse-transcribing (RNA making DNA) *Retroviridae*, and their predominantly intracellular cousins, the *Metaviridae* and *Pseudoviridae*. Degenerate cells and escaped nuclei are a probable origin for many of the highly complex viruses, some of which use only the most basic machinery of the cell, setting up what is sometimes referred to as a "second nucleus" for their replication.

It is of course impossible to determine the origins of viruses, for several reasons. First, they leave a very poor fossil record due to their small size. Second, any ancestral virus will by now have collected so many additions and alterations as to be unrecognizable except at the most basic level. It is known that viruses can acquire genes from each other and from their host cells—many of the proteins identified in viruses, such as the oncogenes of the genus *Oncornavirus* (family *Retroviridae*), show clear signs of cellular origins. It is highly likely that the vast majority of viruses contain genetic elements from a variety of origins, selected and adapted by the unforgiving processes of evolution, to produce the most efficient possible replication within their particular ecological niche.

Molecular studies of both proteins and nucleic acid sequences are the best means that we have available to look for clues to the origins of viruses, and for indicators of evolutionary relationships. With the development of high-speed sequencing systems, spelling out the full sequence of a viral genome takes hours instead of years. This is permitting broad studies of both known and unknown viruses, and is providing huge amounts of data to support studies of virus evolution. The challenge now is to make the most of this flood of data, which is so great as to be capable of producing congestion even in the most powerful systems.

We can also see signs of evolution in how viruses infect their host. Some viruses kill both cell and host rapidly, though this is rarely an efficient way to spread the virus itself and these are often viruses that have only recently moved into a new host. Other viruses have evolved a more stable relationship with the host, some for brief periods, and others (such as herpes simplex virus) for the entire life of the host organism.

Viruses are highly adaptable, and have evolved to fill a wide variety of ecological niches. Some have developed ways to remain stable for many years, such as the baculoviruses of insects with their thick protein coats. Others are relatively unstable, but have developed highly efficient methods of transfer.

By most accounts, assuming that one concedes they are indeed a form of life, viruses are the most numerous type of life on Earth, with over ten million trillion trillion (10^{31}) viruses on the planet. Wherever they came from, they have clearly been highly successful.

Key Concepts

- Many different taxonomical methods have been used to classify viruses, but none of those developed to date appear to be capable of reflecting the true complexity of the systems involved.
- The International Committee on the Taxonomy of Viruses uses a classification based on a broad range of properties of the virus: morphology and physical characteristics, the nature of the viral genome and its replication strategy, properties of the viral macromolecules, antigenic and biological properties.
- The Baltimore system categorizes viruses into seven groups (with numerous subdivisions) based on the replication strategy of the viral genome.
- A wide range of sub-viral infections have been identified, from the protein-only prions through the naked RNA of viroids to satellite "viruses of viruses" that parasitize the largest virus known.
- Prions are now generally accepted to be a form of infectious protein—the only known form of life that does not use nucleic acid.
- Viruses are unlikely to have come from a single origin. Of the multiple theories that exist (escaped genes, escaped nuclei, degenerate cells, or fossils from a pre-cellular or even a pre-DNA world) different origins are likely to apply to different viruses.

DEPTH OF UNDERSTANDING QUESTIONS

Hints to the answers are given at http://www.garlandscience.com/viruses

Question 2.1: What are the problems with existing attempts to classify viruses?

Question 2.2: How do the viruses we know about reflect the true diversity of viruses?

Question 2.3: Viruses may be remnants of a pre-cellular or even pre-protein world. How might such a world have looked?

Question 2.4: How does a protein-only infectious agent propagate itself?

FURTHER READING

Claverie JM, Ogata H, Audic S et al. (2006) Mimivirus and the emerging concept of "giant" virus. *Virus Res.* 117, 133–144.

Daròs JA, Elena SF & Flores R (2006) Viroids: an Ariadne's thread into the RNA labyrinth. *EMBO Rep.* 7, 593–598.

Fauquet CM (2006) Virus classification and nomenclature. In Encyclopedia of Life Sciences. John Wiley & Sons, Chichester. http://www.els.net/.

Fields BN & Howley PM (2007) Virology, 5th ed. Lippincott Williams & Wilkins, Philadelphia.

Forterre P (2006) Three RNA cells for ribosomal lineages and three DNA viruses to replicate their genomes: a hypothesis for the origin of cellular domain. *Proc. Natl Acad. Sci. USA* 103, 3669–3674.

Forterre P (2010) Giant viruses: conflicts in revisiting the virus concept. *Intervirology* 53, 362–378.

Iyer LM, Balaji S, Koonin EV & Aravind L (2006) Evolutionary genomics of nucleo-cytoplasmic large DNA viruses. *Virus Res.* 117, 156–184.

Koonin EV, Senkevich TG & Dolja VV (2006) The ancient virus world and the evolution of cells. *Biol. Direct* 1, 29–55.

La Scola B, Desnues C, Pagnier I et al. (2008) The virophage as a unique parasite of the giant mimivirus. *Nature* 455, 100–104.

Prusiner S (1998) Prions. *Proc. Natl Acad. Sci. USA* 95, 13363–13383.

Safar JG, Kellings K, Serban A et al. (2005) Search for a prion-specific nucleic acid. *J. Virol.* 79, 10796–10806.

Somerville RA, Bendheim PE & Bolton DC (1991) Debate: the transmissible agent containing scrapie must contain more than protein. *Rev. Med. Virol.* 1, 131–144.

Tsagris EM, Martínez de Alba AE, Gozmanova M & Kalantidis K (2008) Viroids. *Cell. Microbiol.* 10, 2168–2179.

Zuckerman AJ, Banatvala JE, Griffiths P et al. (2008) Principles and Practice of Clinical Virology, 6th ed. John Wiley & Sons, Chichester.

INTERNET RESOURCES

Much information on the internet is of variable quality. For validated information, PubMed (http://www.ncbi.nlm.nih.gov/pubmed/) is extremely useful.

Please note that URL addresses may change.

GiantVirus.org. http://www.giantvirus.org (information on *Mimivirus* and other large DNA viruses)

ICTVdb. http://www.ncbi.nlm.nih.gov/ICTVdb/ (the universal virus database of the International Committee on Taxonomy of Viruses)

ICTV Master Species List 2008 (downloadable). http://talk.ictvonline.org/files/folders/documents/entry272.aspx

UK National Creutzfeldt-Jakob Disease Surveillance Unit (NCJDSU). http://www.cjd.ed.ac.uk

CHAPTER 3
Virus Replication

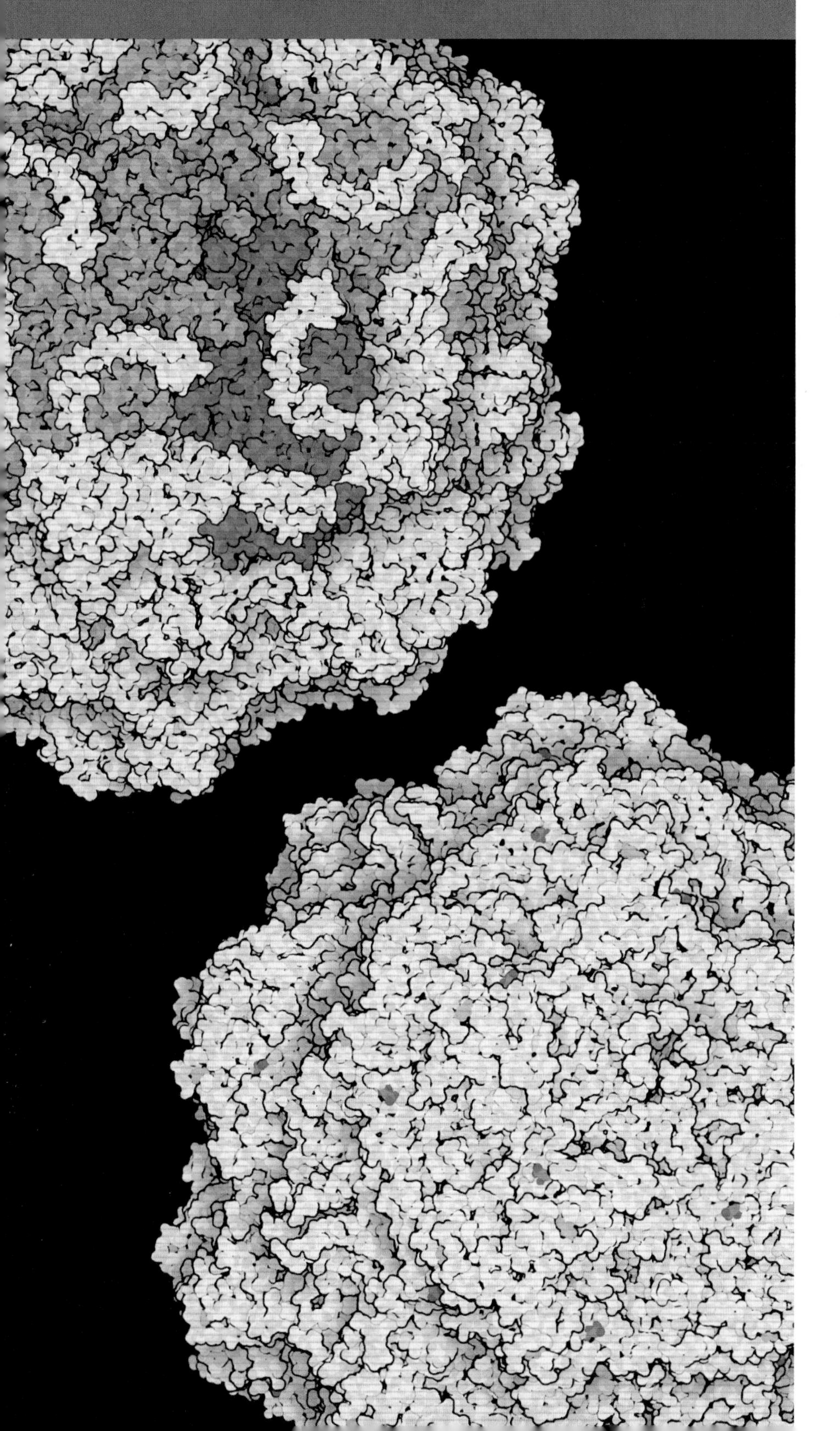

INTRODUCTION

As noted in Chapter 1, viruses are not the simplest form of life. They are actually more complex than their host cell, since they add their own metabolism to that of the cell in order to replicate. To do this, they need to be able to identify their target cell, bind to it, and enter it. This then allows them to begin the process of converting the cell into a virus factory, after which they have to assemble the new generation of viruses and leave the cell.

About the chapter opener image
Parvovirus
(Courtesy of the Research Collaboratory for Structural Bioinformatics Protein Data Bank and David S. Goodsell, The Scripps Research Institute, USA.)

It would be wrong to suggest that things always proceed smoothly for the infecting virus. Viruses cannot move themselves and so they rely on passive processes to encounter their host. If the cell is of the correct type, proteins (often glycosylated) on the outside of the virus will bind to specific receptors on the cell surface. In some cases, viruses can cut themselves free if the binding is inappropriate, as with binding of influenza virus (*Orthomyxoviridae*) to receptor-like structures present in the respiratory mucus.

Once bound, a virus must get inside the cell. This can involve a wide range of processes, including virus-mediated activities such as fusion with the outer membrane and cell-mediated uptake through a variety of pathways.

Within the cell, the virus needs to move to the correct cellular compartment. For some viruses this is the nucleus and for others it is in the cytoplasm. Once it has reached the correct location, the virus begins the process of diverting the activities of the cell to the production of the new generation of viruses. This is a complex process. The simplest viruses need almost everything necessary for replication to be provided by the cell, and are thus very restricted as to the type of cell that can support a productive infection. More complex viruses produce many of their own enzymes, just using the basic machinery of the cell. Once again, matters do not always proceed smoothly, and infection does not always proceed to the production of new viruses. Some types of cells will not support the full process of replication, resulting in an abortive infection. In such cases, some virus function may cause the cell to modify its activities, potentially leading to uncontrolled growth and even to tumor formation. Alongside this, the production of virus proteins within the cell provides signals that the cell uses to call in an immune response, a process with which viruses interfere at many levels to prevent the killing of their host cell before they are ready to leave it.

A range of virus-specific macromolecules are produced, including genomic nucleic acids and messenger RNA, along with proteins and glycoproteins. These must then be assembled into new virus particles (virions). Lipids, where required, are harvested from specific cellular structures, often after modification to suit the needs of the virus.

Finally, the assembled virus must leave the cell. A range of processes are used, from simple killing of the cell to budding through the cell membrane or transport in cellular vacuoles. Then, of course, the virus must find new host cells, while evading the effects of the immune system, to begin the whole process once again.

Owing to the limitations of space in this book, only a general overview of the replication mechanisms and strategies that are used by viruses will be presented here. Examples for major families of viruses infecting humans are presented in the Appendix. Readers wishing to obtain more detailed information are referred to the references and citations at the end of this chapter.

3.1 ATTACHMENT AND ENTRY

In order to replicate, a virus needs to gain access to a host cell. A virus particle (a *virion*) both protects the viral genetic material, and delivers it into a host cell able to support its replication. To achieve delivery to a suitable cell, the virion surface must have proteins (often glycosylated) that can identify and bind to surface **receptors** present on the required type of target cell. This specificity for a certain type of host cell is referred to as **cell tropism**, but also involves a number of other factors such as the ability of the virus to replicate within the infected cell.

For some viruses, a wide variety of cells can support replication, while for others the binding process is highly specific, reflecting a need for a very

Box 3.1 Different species, different receptors

With the influenza viruses (*Orthomyxoviridae*), those infecting humans prefer receptors containing *N*-acetyl neuraminic acid linked by an α2,6 linkage (typically found in the human respiratory tract) while those infecting the aquatic birds that seem to be the natural reservoir of influenza prefer an α2,3 linkage, commonly seen in the avian gut. While not absolute, this difference indicates the role of receptors in species specificity. This is an area of considerable concern with H5N1 influenza since this different receptor requirement is a key factor in limiting the spread of infection from birds to humans. However, some species are able to become infected with both types of virus and thus act to generate new strains of influenza by mixing the genes of the two types. Traditionally the pig has been seen as the "mixing vessel," but there is now evidence that other species, including the chicken, may be able to do this, and even that avian-type linkages may be present in humans.

specialized type of host cell. Even the smallest of viruses exhibit varying cell tropism. Both parvovirus B19 and adeno-associated virus (AAV) are members of the *Parvoviridae*: B19 virus infects only cells expressing a very specific antigen found on specific cells of the erythrocyte lineage, while AAV can use a wide variety of receptors, enabling it to infect a wide variety of cells.

The binding is mediated on one side by specific viral proteins, and on the other side by structures (often, but not always proteins) on the cell surface.

Cellular receptors for viruses

The range of cellular receptors that is known to be used by viruses is steadily increasing, and some examples are shown in **Table 3.1**. These include both proteins and sugars. Some viruses appear to use a single receptor, while others can use a wide range, often on different types of cell. A good example is provided by members of the *Herpesviridae*. The initial receptor for Epstein-Barr virus (EBV; human herpesvirus 4) is complement receptor 2 (CD21), a surface protein specific to B lymphocytes. In contrast, herpes simplex virus type 1 (HSV-1; human herpesvirus 1) can use multiple proteins of different types and is able to infect a wider variety of cells.

Sugars are often involved in early, nonspecific binding, as with HIV (*Retroviridae*), but they may also be the main element of the receptor, as with influenza viruses (*Orthomyxoviridae*). Since such sugars are common in the mucus of the respiratory tract, influenza viruses also carry a surface protein that can detach the virus from non-cellular sugars (neuraminidase). Inhibition of this activity can limit virus infection. Different influenza species also bind to different receptors (see **Box 3.1**).

Even where a single receptor has been identified for a particular virus, this may simply reflect a limited understanding of the processes involved. Viral attachment to cells is, in many cases, a multi-stage process. HIV provides a good example of this (**Figure 3.1**). Initial, nonspecific attachment to

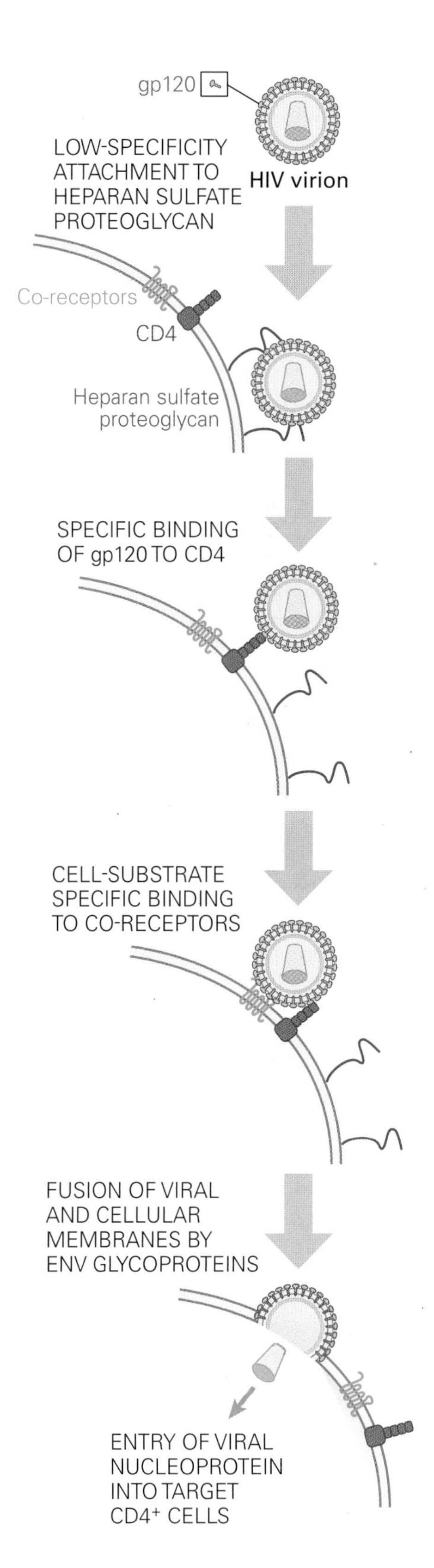

Figure 3.1
Attachment of HIV to CD4$^+$ cells. The first stage in viral attachment is a low-specificity interaction between glycoprotein 120 (gp120) on the HIV virion and heparan sulfate on the cell surface. This contact is then reinforced by the specific binding of gp120 to CD4 proteins on the cell surface. This binding induces structural rearrangements in gp120, unmasking other binding regions that can then interact with cell-specific co-receptors including CCR3, CCR2b, CCR8, though the main co-receptors are CXCR4 (for the so-called X4 viruses) and CCR5 (for the so-called R5 viruses). Successful and specific binding is followed by fusion and viral entry.

Table 3.1 Examples of receptors for viruses infecting humans

Family	Virus	Cellular receptor type	Cellular receptor
Adenoviridae	Adenovirus subgroups A, C, D, E, F	Immunoglobulin superfamily protein	Coxsackie adenovirus receptor
	Adenovirus subgroup B	Inhibitory complement receptor	CD46
Arenaviridae	Lassa fever virus and others	Dystrophin-associated glycoprotein	α-dystroglycan
Bunyaviridae	Hantaan virus	Integrins	α3 integrins
Coronaviridae	Human coronavirus 229E	Metalloproteinase inhibitor	Aminopeptidase N
	Human coronavirus OC43	Sugar structure on cell surface	*N*-acetyl-9-*O*-acetylneuraminic acid
Flaviviridae	Dengue, Japanese encephalitis	Polysaccharides plus others	Heparan sulfate and others
Herpesviridae	Herpes simplex virus	Mucopolysaccharides on cell surface then cell adhesion molecules/cell surface polysaccharide/TNF receptor family protein	Glycosaminoglycans then nectins/ heparan sulfate/HVEa
	Cytomegalovirus	Polysaccharide and integrins	Heparan sulfate and various integrins
	Epstein-Barr virus	Complement receptor	CD21
	Human herpesvirus 7	T-cell surface marker	CD4
Orthomyxoviridae	Influenza virus types A and B	Sugar structure on cell surface	*N*-acetylneuraminic acid (α2,6-linked[1])
	Influenza virus type C	Sugar structure on cell surface	*N*-acetyl-9-*O*-acetylneuraminic acid
Paramyxoviridae	Measles virus	Lymphocyte activation molecule	CD150
Parvoviridae	Parvovirus B19	Glycosphingolipid	Erythrocyte P antigen
	Adeno-associated virus	Glycoproteins on cell surface and human fibroblast growth factor receptor or integrin	Heparan sulfate proteoglycan and FGFR1 or $\alpha V\beta 5$
Picornaviridae	Cocksackievirus A	Complement inhibitor and intracellular adhesion molecule/integrin	CD55 and ICAM-1/$\alpha V\beta 3$ integrin
	Cocksackievirus B	Complement inhibitor/immunoglobulin superfamily protein/others	CD55/coxsackie adenovirus receptor/others
	Echoviruses 1 and 8	Complement inhibitor/integrin	CD55/$\alpha 2\beta 1$
	Foot-and-mouth disease virus	Integrins	Various integrins
	Hepatitis A virus	T-cell immunoglobulin mucin (TIM) family protein	HAVCR1/TIM1
	Poliovirus	Epithelial adhesion molecule	CD155
	Rhinoviruses	Intercellular adhesion molecule (some use lipoprotein receptors)	ICAM-1 (some use LDLR/α2MR/LRP)
Reoviridae	Rotavirus	Integrins	Various integrins
	Reovirus serotype 3	Sugar structure and adhesion protein	*N*-acetyl-neuraminic acid and JAM1
Retroviridae	Human immunodeficiency virus	Polysaccharides, immune cell surface marker, and chemokine receptor	Heparan sulfate/galactosyl ceramide, CD4, and CXCR4/CCR5/ others[2]

[1] Human influenza viruses; avian influenza viruses use α2,3-linked neuraminic acid. [2] Cell tropism is determined by co-receptor type; HIV that uses CCR5 is selective for $CD4^+$ T lymphocytes, while HIV that uses CXCR4 is selective for $CD4^+$ macrophages. Other co-receptors may be used in cultured cells. TNF, tumor necrosis factor.

Box 3.2 Clathrin-coated pits and entry to cells

Clathrin is a protein that is involved in transport of material into the cell. By binding to a range of receptor proteins on the inner face of the cell membrane, it curves the cell membrane, forming pits that permit the import into the cell of extracellular material (**Figure 1**). This curvature is based on the ability of clathrin itself to form curved structures composed of hexagons and pentagons in a way that is strikingly similar to icosahedral virus structures (see Chapter 2). Clathrin-coated pits deepen to form vesicles, which then pinch off and migrate within the cell, frequently merging with and becoming acidified by digestive endosomes. The clathrin itself is rapidly recycled to the membrane, allowing some types of cell to import very large quantities of material, since a new vesicle can form in less than a minute.

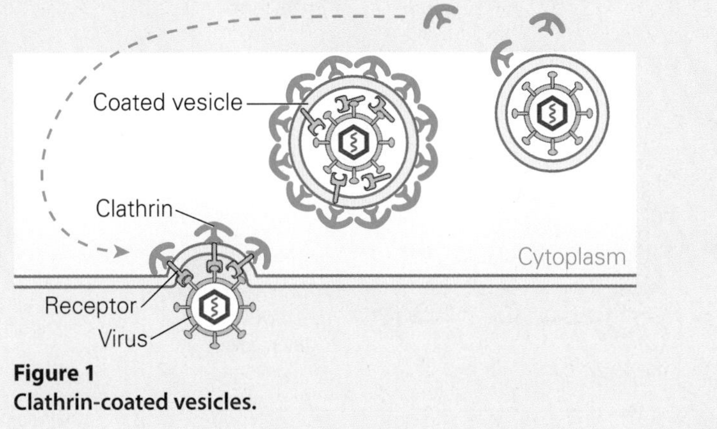

Figure 1
Clathrin-coated vesicles.

polysaccharides is followed by sequential, highly specific interactions with cell surface proteins, positioning the viral Env glycoprotein to mediate fusion of the virion envelope with the cell membrane. The precise use of different second receptors is associated with different growth characteristics, both *in vitro* and *in vivo*.

Virus entry

Once a virus is bound to the cell, the next step is for it to gain entry. Many non-enveloped viruses are internalized by the cell itself, often by binding to areas of the membrane referred to as **clathrin-coated pits** (see **Box 3.2**). Others are internalized within vacuoles and then released within the cell by various mechanisms. Some viruses seem to rely on a single mechanism, while others can use multiple routes to enter the cell. The range of methods of viral entry is shown in **Figure 3.2**.

Where viruses enter the cell in vacuoles, these may be acidified by fusion with digestive **endosomes** (vacuoles containing acids and digestive enzymes) within the cell, in an effort to digest their contents. However, many viruses, both enveloped and non-enveloped, have adapted to use this by having proteins which, when their conformation is altered by the low pH, become able to mediate the exit of the viral genome (and other necessary components) into the cell itself.

Other vacuolar systems may also be used by viruses, with or without the direction to digestive endosomes that is associated with the clathrin pathway. These include the small **caveolae** (50–100 nm pits in the cell membrane that are associated with the protein caveolin) and the much larger structures of **macropinocytosis**.

A good example is provided by influenza virus, which can enter vacuoles by a variety of methods. In all known cases, these then become acidified, and a conformational change in the HA protein exposes regions that mediate fusion of the viral and vacuolar membranes. At the same time, the viral M2 protein permits acidification of the interior of the virion itself, dissociating the ribonucleoprotein from the virion structure and allowing it to move out into the cytoplasm through the opening made by the membrane fusion. These multiple functions illustrate clearly the central role of acidification in the entry of influenza virus to cells.

The end result of viral entry is that the viral nucleic acid–protein complex is within the cell, ready to begin making the components of the next generation of viruses.

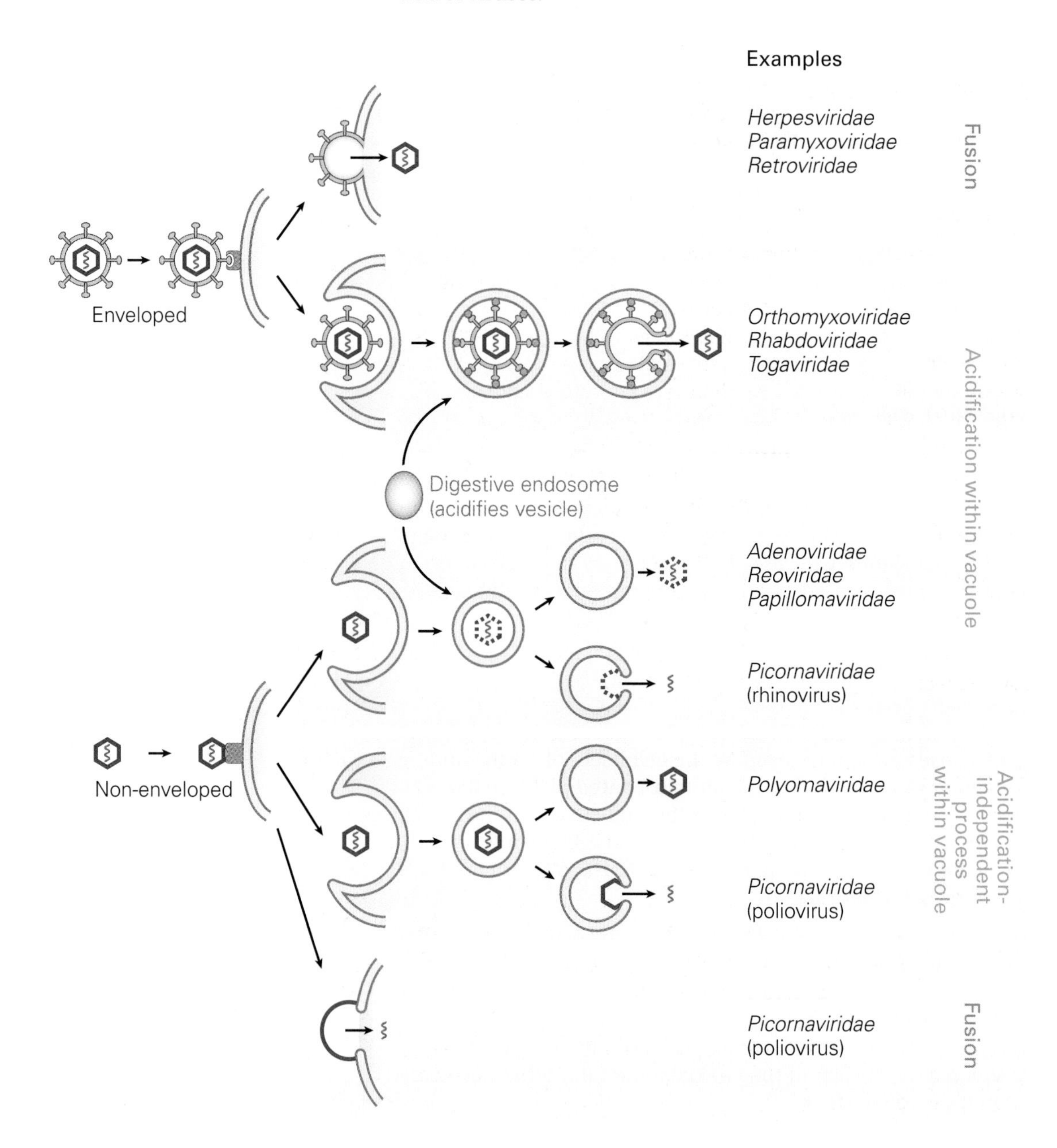

Figure 3.2
Methods of viral entry into cells. Once bound to the cell surface (left), some viruses can fuse with the cell's plasma membrane and release their contents into the cytosol (upper and lower panels). Other viruses are internalized within vacuoles of varying types and sizes (center panels). Release from vacuoles created by the clathrin pathway is triggered by acidification of the vacuole following fusion with digestive endosomes. For enveloped viruses, the envelope may fuse with the acidified vacuole and release the intact virion, while for non-enveloped viruses, conformational changes in the virion can cause either the altered virion or just the genetic material to be released. Other acidification-independent pathways allow release of some non-enveloped viruses.

3.2 REPLICATION

Where virus entry into the cell is followed by an active infection, a 'one-step' growth curve is followed (**Figure 3.3**). In the first 'eclipse phase,' the virus fragments in order to begin replication and has not yet made and assembled the components of progeny virus. Infectious virus cannot be recovered. In a eukaryotic cell, this lasts for anything from a few hours to a few days. However, it is much shorter in viruses of bacteria (bacteriophages), where it was first studied.

In the next logarithmic phase, progeny virus begins to appear. The progeny virus genomes assemble into new viruses, which are also used to make yet more viruses, until the cell is totally dedicated to the production of viruses. This amplification results in a logarithmic increase in the amount of virus present within the cell, and is referred to as secondary transcription. In an unrestrained infection, cell death then follows.

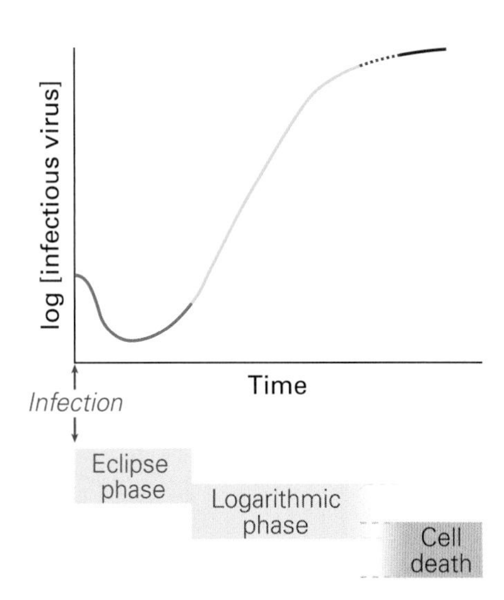

Figure 3.3
The one-step growth curve of an active (acute) viral infection.

Sites of virus replication

The actual site at which replication occurs varies between viruses. In eukaryotic cells, some viruses replicate in the cytoplasm while others do so in the nucleus (**Figure 3.4**). In prokaryote cells, there is only a single cellular compartment. Even within the cytoplasm there are a number of distinct compartments which are associated with different stages of virus replication. For example, cellular membrane structures can be involved at different stages. Some nucleic acid synthesis is associated with membranes close to the nucleus, while acquisition of membranes to form virus envelopes can occur at a variety of sites. Specific cellular functions needed by the virus are also located at specific sites, with protein synthesis occurring at ribosomes on the rough endoplasmic reticulum, while the attachment of sugars to form glycoproteins occurs in the Golgi apparatus.

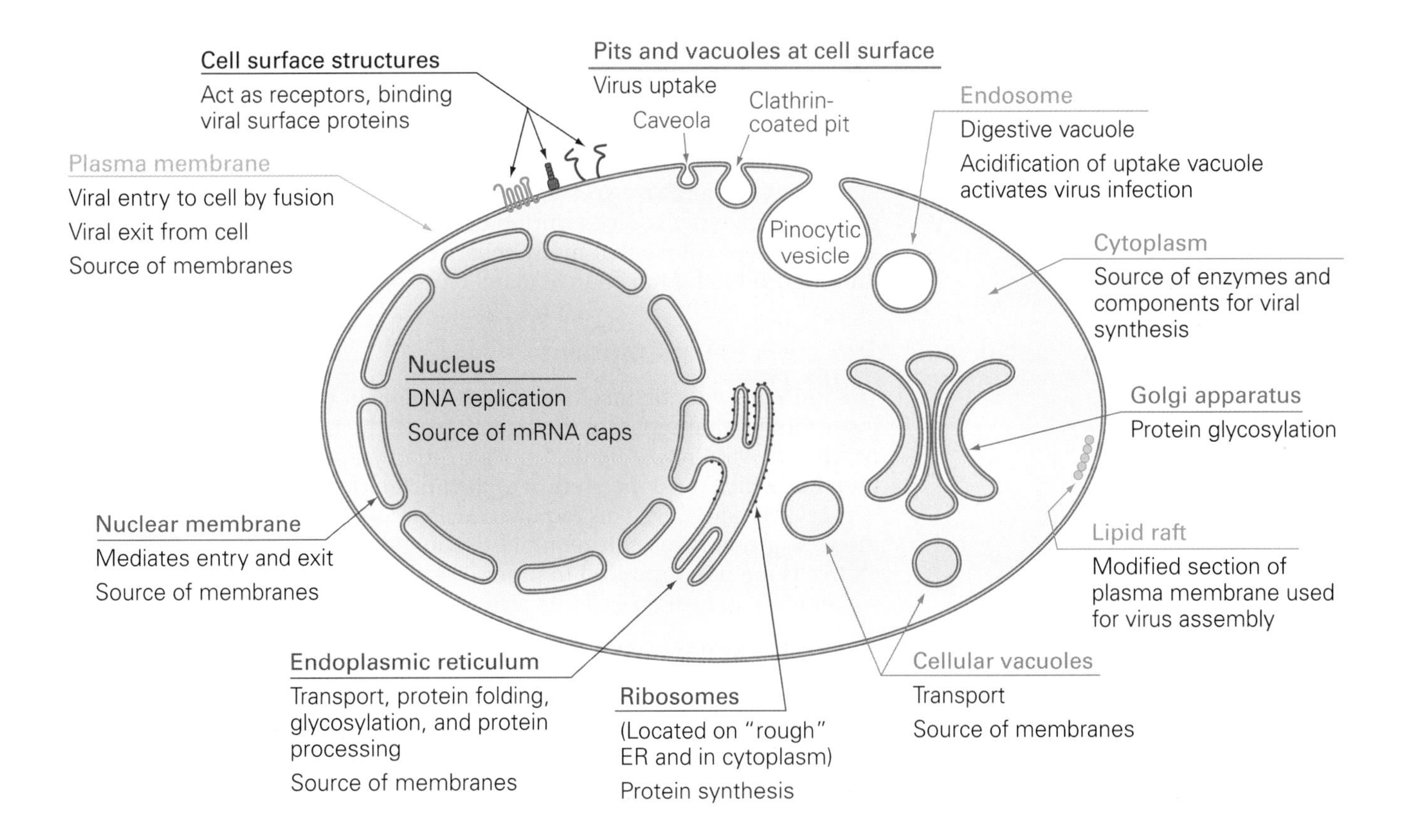

Figure 3.4
Major cellular structures involved in virus replication.

Most viruses with DNA genomes (DNA viruses) and some with RNA genomes (RNA viruses) replicate in the nucleus, where the cellular machinery for the synthesis of nucleic acids is located. Cells do not routinely contain enzymes capable of copying RNA into RNA and so most RNA viruses replicate in the cytoplasm. However, some of the largest DNA viruses set up what has been called a virus-coded "second nucleus" in the cytoplasm, where replication occurs. These DNA viruses, known as nucleocytoplasmic large DNA viruses (NCLDV), have large genomes and the capacity to code for the many functions needed for cytoplasmic replication. The NCLDV include the families *Poxviridae, Asfarviridae, Iridoviridae, Phycodnaviridae,* and the as yet unassigned genus *Mimivirus*. Of these, only the *Poxviridae* are known routinely to infect humans.

Such an approach is not universal—while other large and complex DNA viruses of eukaryotes such as the *Herpesviridae* and the *Polydnaviridae* have large genomes and may produce proteins with functions similar to cellular enzymes, their replication occurs in the nucleus.

Entry into the nucleus

Those viral genomes that replicate in the cell nucleus must cross the nuclear membrane, which surrounds the eukaryotic cell nucleus. Nuclear localization signals (NLS) on viral proteins (which are similar to those present on cellular proteins) target viruses to the nuclear membrane. Specific receptor proteins then mediate binding to nuclear pores, which are large protein complexes that cross the nuclear membrane, permitting movement of material between the cytoplasm and the nucleus. There are approximately 2000 nuclear pores in the nuclear envelope of a typical vertebrate cell.

Nuclear pores limit the size of what can enter the nucleus to approximately 25–40 nm, blocking all but the smallest viruses such as the *Parvoviridae* and the *Hepadnaviridae*. These viruses have had to make many adaptations to achieve such small size, and it is tempting to speculate that the effects of this particular exclusion limit may not be coincidental.

To pass through a nuclear pore, larger viruses may break up into subunits that are small enough to enter, as with the individual entry of the encapsidated genome segments of the *Orthomyxoviridae*. Alternatively, the nucleic acid genome may enter alone or without the majority of the nucleocapsid proteins as seen with the *Herpesviridae*. Another approach, as seen with many of the *Retroviridae*, is for the virus to enter the nucleus when the membrane is dissolved during cell division. A limitation of the latter approach is of course that many cells are not actively dividing, which can prevent productive infection of many types of cell.

Synthesis of viral proteins

The protein synthetic machinery of the cell is in the cytoplasm. Thus, for those viruses that replicate their genomes in the nucleus, such as the herpesviruses, viral mRNAs must move to the cytoplasm to be translated to proteins, which must then return to the nucleus to form capsids around the herpesvirus genomes. This requires nuclear localization signals that target the viral proteins to their required destination. It should be remembered that cells are not optimized to produce viruses, and significant alterations are required to turn the cell into a virus factory.

Many viruses express different genes at different stages of their replication, but the large DNA viruses have particularly marked phases in their replication. They produce different groups of proteins at particular and appropriate times. Such timing is often controlled by the use of different promoters (regions of the genome which promote transcription) for each group of proteins, which are activated by different but specific stimuli.

Table 3.2 Examples of herpes simplex virus protein classes

Class	Gene	Protein	Function
Pre-α	ORF-P	ORF-P	Gene repressor
Immediate early / α	α0	ICP0	Transactivator
Early / β_1	U_L29	ICP8	DNA unwinding protein
Early / β_2	U_L23	TK	Thymidine kinase
Late / γ_1 (leaky late)	U_L19	ICP5	Major capsid protein
Late / γ_2 (true late)	U_L44	Glycoprotein C	Virion surface glycoprotein

One of the best-studied examples is HSV-1 (*Herpesviridae*), in which there is a separation into three main groups: immediate early (IE, or α), early (E, divided into β_1 and β_2), and late (L, divided into γ_1 and γ_2) (**Table 3.2**). In reality, these classes are not exact, with most proteins falling somewhere along a continuum. There are also some proteins (such as the ORF-O and ORF-P proteins) which are classified as "pre-α" and appear to repress (and be repressed by) α protein expression. These may have a role in establishing viral latency, though the role and expression of proteins in this state has yet to be confirmed.

The cascade of viral protein expression for herpes simplex virus is both complex and tightly regulated. After the virus enters the cell, the proteins contained in the herpesvirus *tegument* (a specialized region lying between capsid and envelope, see Figure 1.3) and capsid alter cellular functions and allow the synthesis of the first batch of viral mRNAs and proteins. These are referred to as immediate early or α proteins, and are mainly regulatory proteins. These, in turn, are necessary to allow the synthesis of the next set of proteins, early or β proteins, with β_1 synthesis occurring very shortly after α protein synthesis and before β_2 synthesis. Most β proteins are concerned with synthesizing the viral DNA and also with preparing the cell for the manufacture of virus. Early proteins, together with the newly synthesized viral DNA, allow the synthesis of the late or γ proteins. These fall broadly into two classes: γ_1 or "leaky late" and γ_2 or "true late" which appear to have an absolute requirement for prior viral DNA synthesis. The γ proteins include the structural components of the new virus along with proteins that guide assembly. Most of the proteins that entered the cell with the virus were late proteins (with some important exceptions). This completes the cycle by starting off the synthesis of the immediate early proteins. This cycle is illustrated in the Appendix.

Typically, parvoviruses make mRNAs for two to three structural proteins slightly later than mRNAs for up to four nonstructural proteins. This means that there must be some rather limited form of temporal control, but the methods of this control are not clear.

The much smaller circular dsDNA of the *Polyomaviridae* have two broad classes of gene expression; early and late, with the former being regulatory proteins and the latter the structural proteins present in the virus particle (VP1–VP3) as well as the nonstructural agnoprotein. Replication of these viruses occurs in the nucleus.

Some RNA viruses use an even simpler strategy. For poliovirus (*Picornaviridae*), all of the viral proteins are produced at once as a single large "polyprotein" that is then cleaved to produce all of the regulatory and structural proteins. This is covered in more detail below.

3.3 SYNTHESIS OF VIRAL GENOMES

As well as the viral mRNAs and proteins, a virus needs to make copies of its genome for inclusion into the next generation of viruses. For viruses with single-stranded genomes, this involves the synthesis of an opposite sense "antigenome" from which new genomes are transcribed, while viruses with double-stranded DNA genomes can copy both strands at once as with cellular DNA synthesis. There are many different types of virus genome using a variety of replication strategies (**Figure 3.5**). However, the requirement to make full-length copies of their genome applies to all forms of life, and viruses are no exception.

DNA synthesis

DNA synthesis, catalyzed by **DNA polymerase** enzymes, is required for the replication of all DNA viruses (and even for some RNA viruses). DNA synthesis needs a **primer** from which to start. Usually this is a short RNA copied

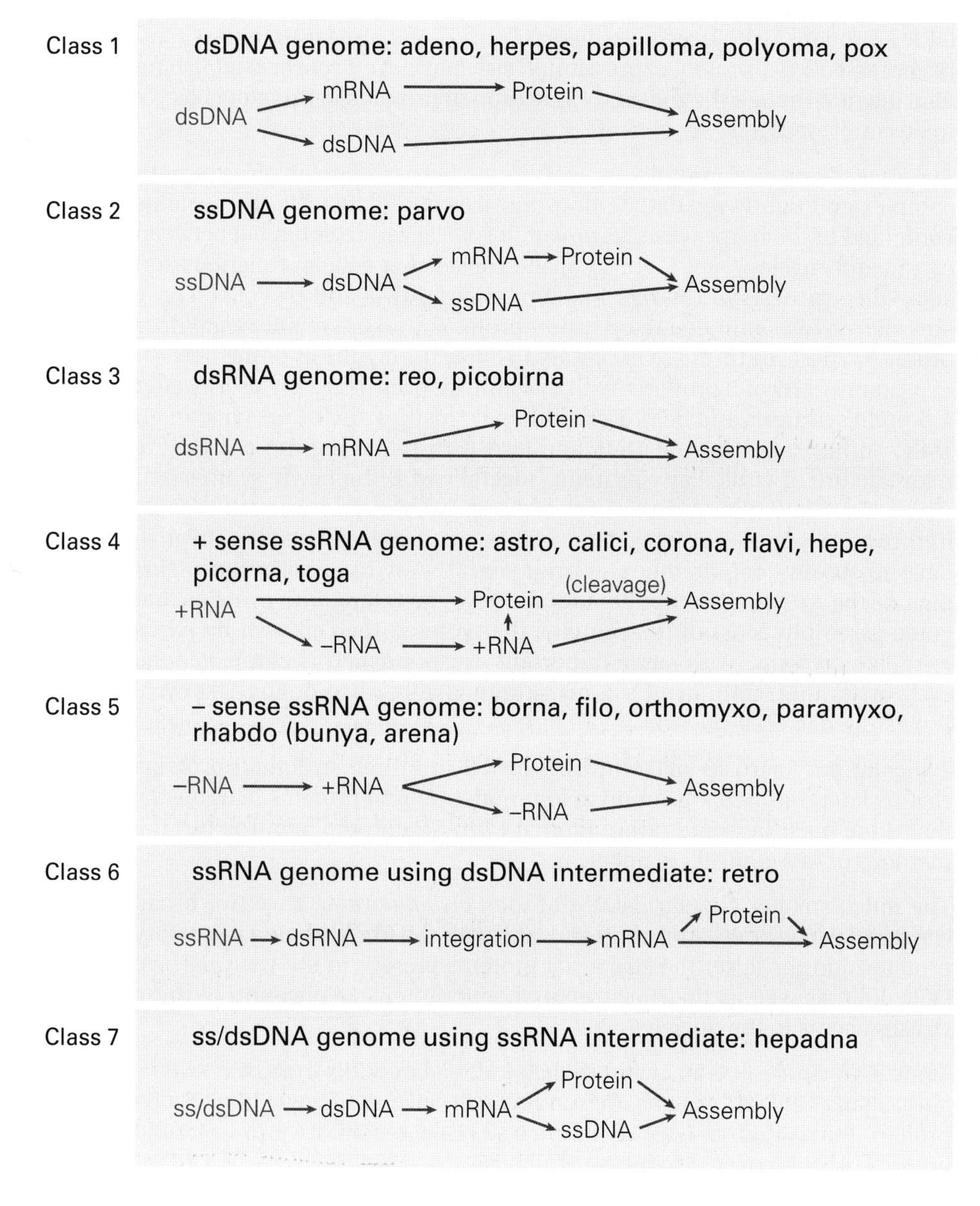

Figure 3.5
General methods of virus replication by genome type. These methods relate to the Baltimore classification of viruses described in Chapter 2. Class 1 dsDNA genomes: adeno-, herpes-, papilloma-, polyoma-, and poxviruses; Class 2 ssDNA genomes: parvoviruses; Class 3 dsRNA genomes: reo- and picobirnaviruses; Class 4 positive-sense ssRNA genomes: astro-, calici-, corona-, flavi-, hepe-, picorna-, and togaviruses; Class 5 negative-sense ssRNA genomes: borna-, filo-, orthomyxo-, paramyxo-, and rhabdoviruses; Class 6 ssRNA genomes using dsDNA intermediate: retroviruses; Class 7 ss/dsDNA genomes using ssRNA intermediate: hepadnaviruses.

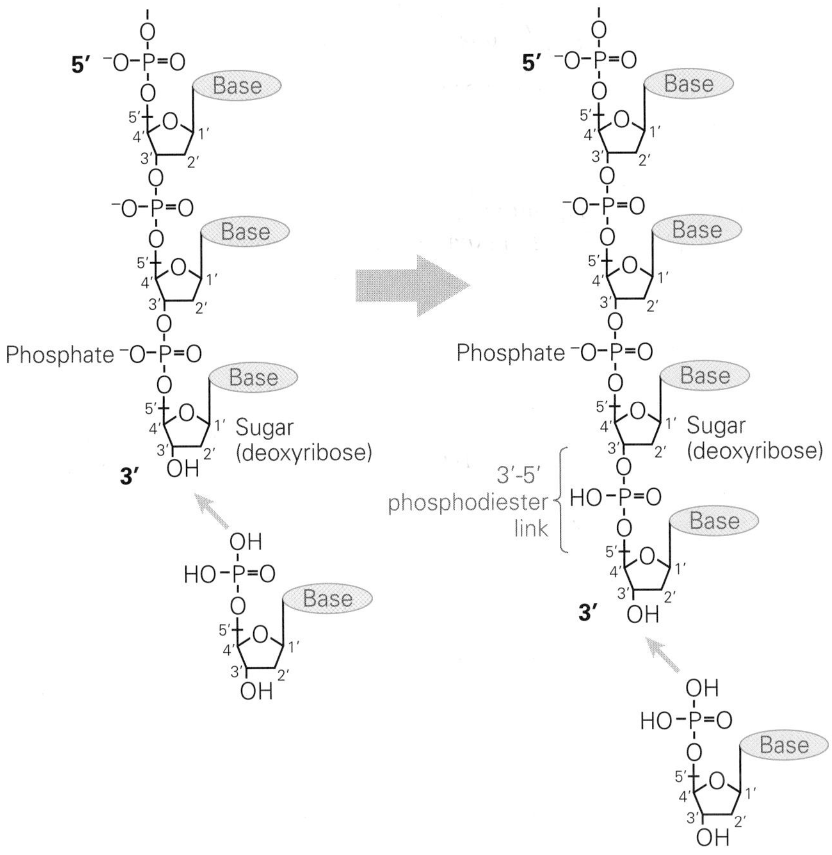

Figure 3.6
DNA replication requires an RNA primer and proceeds in the 5′ to 3′ direction.

from a specific region of the template DNA, which is later degraded by a specific ribonuclease (RNase) activity. Due to its different structure, RNA can be produced without a primer.

The formation of a new DNA chain takes place from the end of the forming DNA chain 5′ toward the 3′ end, as shown in **Figure 3.6**. The carbon at the 5′ position on the five-member sugar ring is attached via a phosphodiester link to the carbon at the 3′ position of the most recently attached nucleotide.

The use of RNA primers creates a problem, since these are copied and bound to a section of the template DNA. How is this region copied into the new DNA? DNA cannot be synthesized back up the chain toward the 5′ end of the molecule, so that the primer binding region cannot be produced in this way. While only a few bases might be lost if this region is not replicated, over many generations the whole DNA genome would be lost. The mechanisms by which full-length copies of any viral genome are made without missing out these, together with other regions at the ends of the molecule(s), are highly varied and often involve complex folding of the genome during transcription.

Viruses with double-stranded (ds) DNA genomes

Viruses with double-stranded DNA genomes often have large genomes (up to 1200 kb for *Mimivirus*), allowing them to have complex virions and to produce a range of virus-specific enzymes, unlike more limited viruses, which depend to a very great extent on cellular enzymes and processes. For example, the enzyme thymidine kinase, which supplies components of DNA synthesis and is produced by *Herpesviridae* and *Poxviridae*, is only present in actively dividing cells. By producing viral versions of cellular enzymes, the virus ensures that they are available when and where the virus wants them.

Smaller dsDNA viruses (such as the *Polyomaviridae*), which cannot encode so many proteins, use more cellular functions and appear to have less complicated controls of replication. These small dsDNA viruses use an alternative approach to obtaining enzymes only produced by dividing cells. They produce a protein known as the large T antigen which, by interaction with the pRb gene (which is important in cell cycle control), drives the host cell into the S phase, a pre-division state in which DNA synthesis occurs. In this way, the virus obtains the enzymes required for DNA replication.

For viruses with a linear dsDNA genome such as the *Herpesviridae*, the basic mechanism of replication involves circularization of the linear genome. DNA synthesis occurs in one direction, with the result that viral dsDNA genomes are rolled off as very long **concatamers** of the genome, which are cut to unit length during viral assembly (**Figure 3.7A**). The production of concatameric copies of the viral genome is very common, and is one (relatively simple) system to allow full-length genome copies to be produced, since the polymerase "completing the circle" will displace the primers that initiated nucleic acid synthesis.

For the *Polyomaviridae*, which has a circular dsDNA genome, extension from the RNA primers proceeds in both directions. Since DNA synthesis can only progress from 5′ to 3′, synthesis on the "lagging" strand (where DNA formation has to proceed toward the 5′ end) takes place by the formation of a series of short 5′–3′ DNA molecules that are then joined to form the new DNA strand. The process produces interlinked full-length copies of the circular viral genome (**Figure 3.7B**) which are then released by cleavage, a process known as **decatenation**.

Viruses with single-stranded (ss) DNA genomes

All viruses with single-stranded (ss) DNA genomes have very small genomes, the largest being a plant virus genome of about 11,000 nucleotides. The smallest is that of porcine circovirus, which at 1759 nucleotides is also the smallest genome of a virus known to infect vertebrates. DNA from this virus has recently been identified as a contaminant of rotavirus vaccines, which has raised concerns over human exposure.

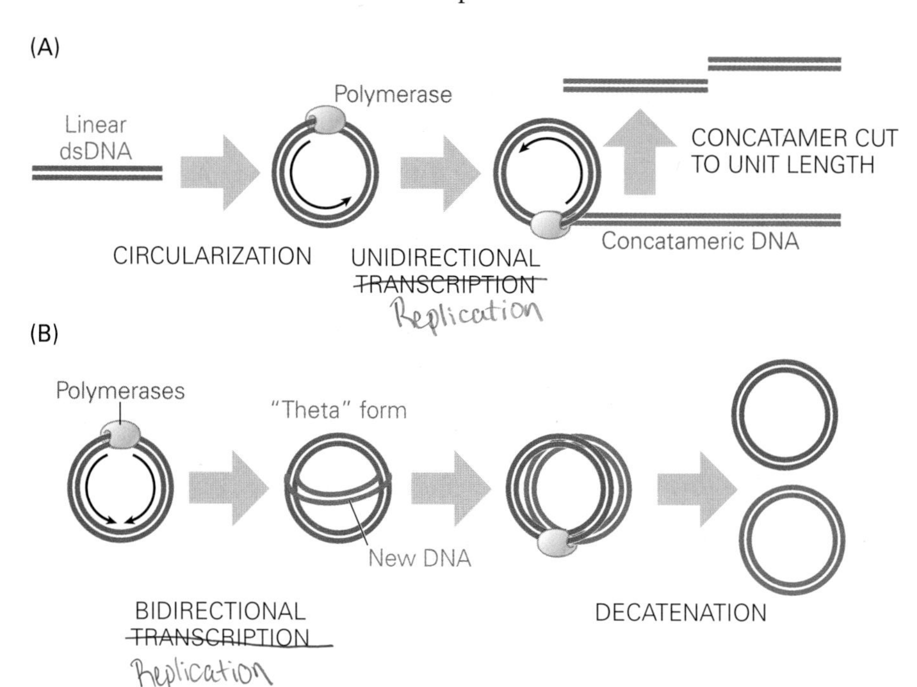

Figure 3.7
Genome replication strategies for dsDNA viruses. (A) Linear dsDNA (*Herpesviridae*). (B) Circular dsDNA (*Polyomaviridae*). Newly synthesized DNA is shown in pink, and the origin of DNA synthesis in yellow.

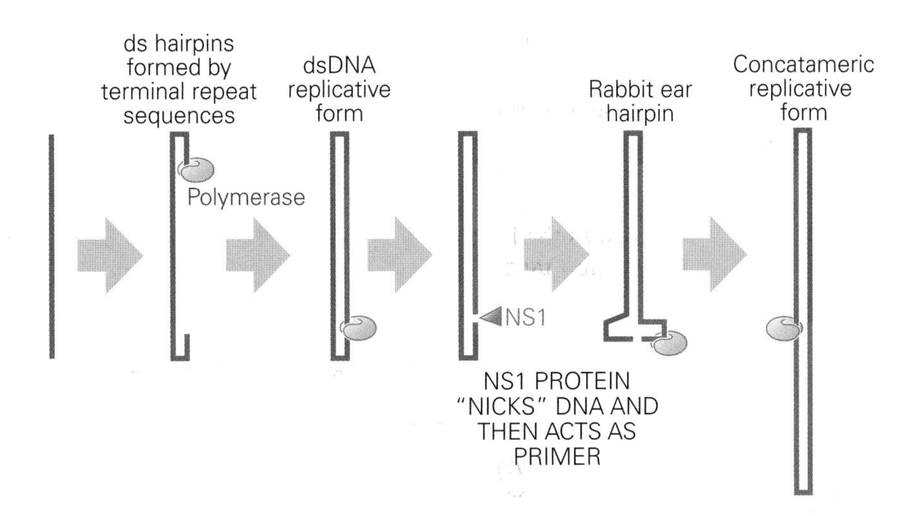

Figure 3.8
Genome replication strategy for linear ssDNA viruses. Viruses that use this strategy include the *Parvoviridae*. The concatameric replicative form may be extended further, or nicked and used to produce unit-length genomic DNAs which are then encapsidated. Newly synthesized DNA is shown in pink, and the origin of DNA synthesis in yellow.

The *Parvoviridae* family provides the best-studied examples of ssDNA viruses that infect vertebrates. Their genome is a typically small ssDNA viral genome of 4100–6200 nucleotides. Replication occurs in the nucleus, where the ssDNA genome is copied by a complex process involving the formation of double-stranded "hairpin" structures at the genome termini and the formation of a dsDNA replicative form (RF). This is then used to produce mRNA or is further extended to a double-stranded concatameric form (**Figure 3.8**).

The *Parvoviridae* are among the most limited of viruses due to the small size of their genome, and do not have the ability to turn on cellular DNA synthesis. Instead, they can only productively infect a cell that is actively making DNA. This restricts the types of cell that will support their replication to actively dividing cells, such as erythrocyte precursor cells.

DNA viruses that use an RNA intermediate

The only virus infecting humans that replicates via an RNA intermediate is hepatitis B virus, a member of the *Hepadnaviridae*, although a similar strategy is used by the *Caulimoviridae* that infect plants.

In the virion, the hepatitis B genome is usually a partially double-stranded DNA. The genome is approximately 3200 nucleotides long and is double stranded for 50–85% of its length, with a full-length negative-sense DNA strand and a shorter positive-sense strand of variable length. RNA primers and the viral **reverse transcriptase** (polymerase) enzymes are covalently attached to the genomic DNA. A minority of virions (5–20%) contain a fully dsDNA, but this appears to be an aberrant form.

Once within the target cell, the partly double-stranded genome is converted to a fully double-stranded, covalently closed circular form (CCC DNA) by unknown enzymes and the uncoated nucleocapsid enters the nucleus (**Figure 3.9**). Following this, a full-length pre-genomic (pg) mRNA is synthesized along with smaller, subgenomic mRNAs. The pgRNA acts both as an mRNA for the viral core and reverse transcriptase (polymerase) proteins, and as a template for the viral reverse transcriptase to produce a full-length negative-sense DNA strand. Binding of the viral reverse transcriptase (polymerase) to the negative-sense DNA strand triggers encapsidation within a core of the viral C (core) protein. The 1700–2800-nucleotide, variable length, positive-sense DNA strand is then formed within the viral capsid. Transcription ceases before a full-length copy is produced, possibly due to exhaustion of the available nucleotides within the capsid; this might explain the variable length of the positive strand.

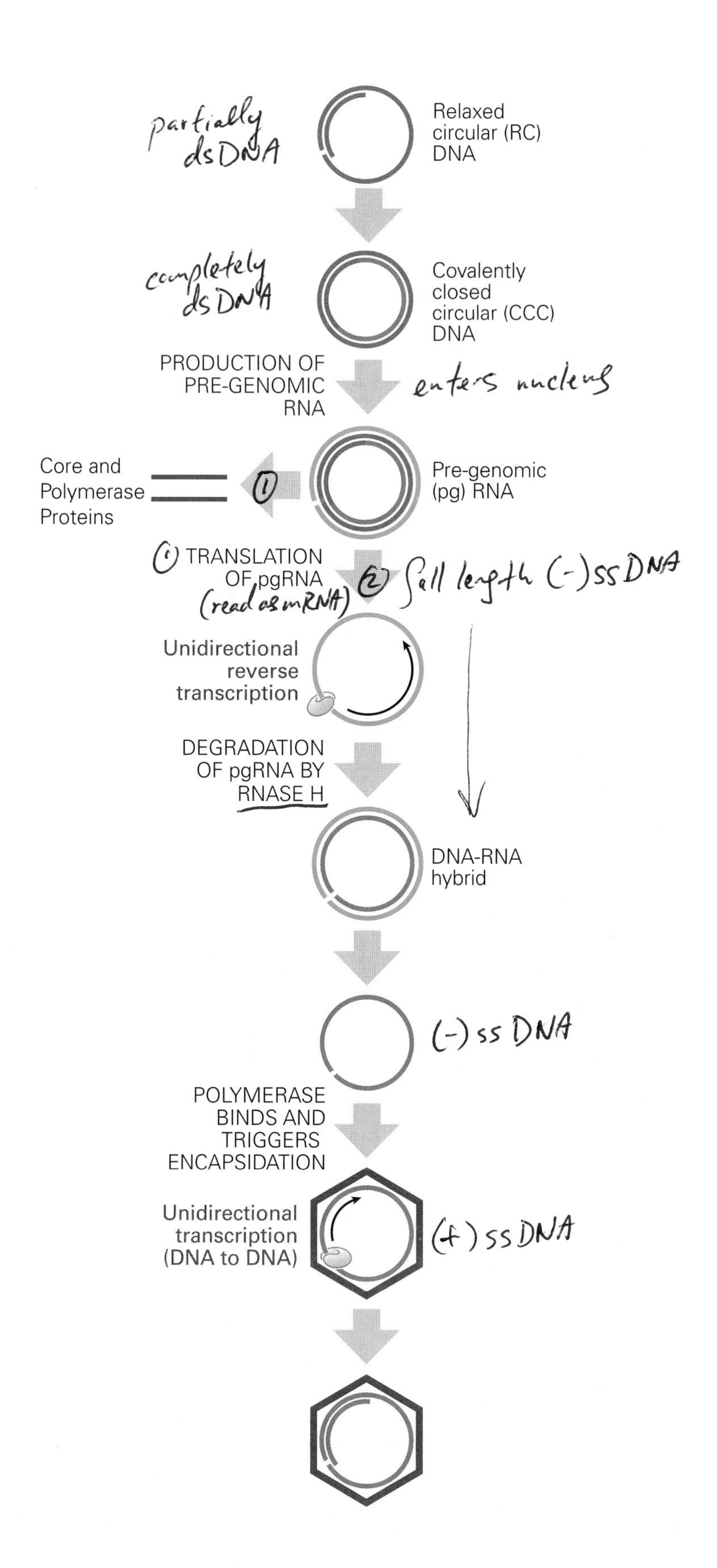

Figure 3.9
DNA genome replication strategy involving an RNA intermediate. Hepatitis B (*Hepadnaviridae*) replicates using this strategy. Newly synthesized RNA is shown in green, new DNA in pink, and proteins in red.

Quite why a virus with such a small genome has such a complex replication cycle is not clear. The viral reverse transcriptase also functions as a DNA polymerase and uses over half of the coding capacity on the viral genome. A viral or even a cellular DNA-dependent DNA polymerase is not used.

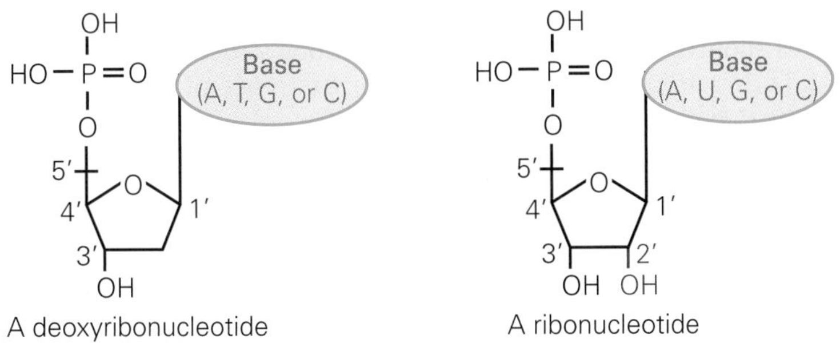

Figure 3.10
Ribonucleotides differ from deoxyribonucleotides by the presence of a 2′-hydroxyl group.

However, the low fidelity of RNA transcription (see below) may provide an evolutionary advantage through the generation of mutants able to survive changing evolutionary pressures.

RNA synthesis

All cellular organisms synthesize RNA but these are generally relatively short-lived, functional molecules such as **messenger**, **transfer**, and **ribosomal RNA** (mRNA, tRNA, and rRNA, respectively). However, in many viruses the genome itself may be RNA. In fact, among viruses known to infect humans, RNA genomes appear to be substantially more common than DNA genomes (see Figure 2.6).

As with DNA, RNA is referred to as **positive sense** (containing the mRNA sequences, and usually functioning as an mRNA) or **negative sense** (complementary to the mRNA sequences). Some viral genomes contain regions of both positive and negative senses, and are referred to as **ambisense genomes**.

As with DNA, the formation of a new RNA chain proceeds from the 5′ end toward the 3′ end. Most RNA synthesis does not require any primers, as does DNA synthesis, and so the process of copying RNA genomes does not usually require the complex arrangements required to ensure complete replication of viral DNA molecules as described above. However, some RNA viruses do use primers, such as the VPg protein primer bound to the end of the positive-sense ssRNA genome of the *Picornaviridae*.

RNA is more reactive (and thus less stable) than DNA, due to the presence of an extra hydroxyl group at the 2′ position on the sugar ring (**Figure 3.10**). Degradation of the ends of viral RNA genomes can be a problem given the relatively unstable nature of RNA, and many RNA viruses control this by having protective structures present at the genome termini. In negative-sense RNA genomes these may involve complex folding of the RNA, while in positive-sense RNA genomes these are often a 5′ **cap** (a 5′-5′-linked methylguanylate group, required for mRNA translation) and 3′ **polyadenylation** [the poly(A) tail] seen on mRNAs (**Figure 3.11**).

Viruses with RNA genomes exhibit a very wide variety of structures (see Table 2.1). The genomes can be double stranded or single stranded (positive, negative, or ambisense). They can be single molecules or broken into anything from two to twelve segments. To make matters even more complex, some viruses with RNA genomes replicate using a DNA intermediate.

The largest RNA genomes (33,000 nucleotides for the *Coronaviridae*) are far smaller than those of DNA viruses (approximately 40 times smaller than the largest DNA virus), reflecting the low fidelity limits of RNA replication (see Section 1.3).

The presence of a **segmented genome** in many RNA viruses also appears to be an adaptation to the low fidelity of RNA replication, since one long

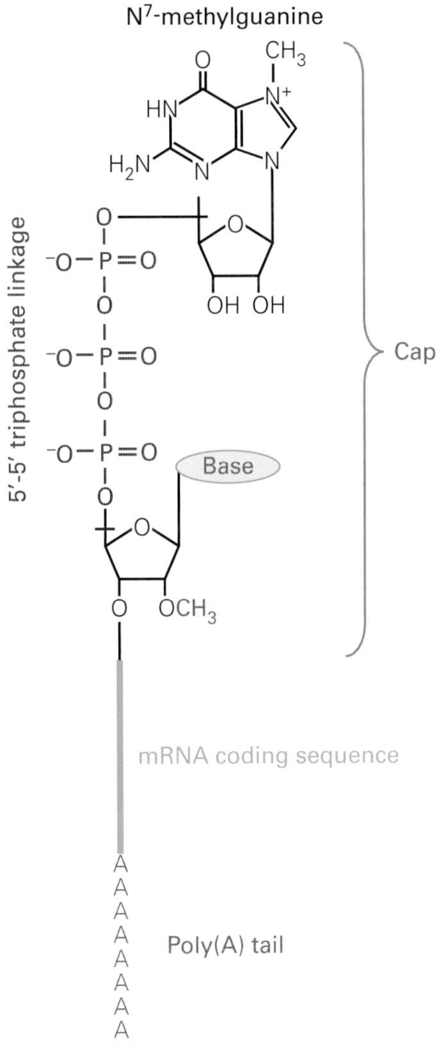

Figure 3.11
Messenger RNA (mRNA) structure. The mRNA coding sequence is of variable length, with a 5′ cap—a 5′-5′-linked methylguanylate structure—and a 3′ polyadenylate tail of variable length. 2′-Methyl groups (shown on the first base here) may be present on a variable number of nucleotides.

molecule may be more likely to contain deleterious mutations. Segmented genomes allow the reassortment of genes among viruses infecting the same cell—it is this process that supports the appearance of the new antigenic variations that cause pandemic influenza.

Influenza virus (*Orthomyxoviridae*) has seven (influenza C) or eight (influenzas A and B) segments of RNA making up its genome and is the best known and most studied of the viruses with a segmented genome. Other viruses with segmented genomes have from two (*Arenaviridae*, *Picobirnaviridae*) to twelve (genus *Coltivirus*, family *Reoviridae*) segments.

Replication of an RNA genome poses a particular problem for the virus. While most DNA viruses make use of cellular polymerase enzymes (a few do produce their own polymerase enzymes), very few viruses with RNA genomes can use cellular enzymes, since cells do not contain useful levels of RNA-dependent RNA polymerase. This means that even very small RNA viruses need to produce a polymerase enzyme or find some way to use a cellular enzyme not originally intended for that purpose.

Viruses with double-stranded (ds) RNA genomes

Reoviruses have genomes made of double-stranded RNA, a molecule not found at significant levels in normal cells. They provide a good model for dsRNA viruses in general. Their genome is divided into 10–12 blunt-ended segments (that is, with no single-stranded regions at the termini), most of which code for a single protein. The genomic positive RNA strand is an mRNA, complete with a 5′ cap modification (produced by viral enzymes during maturation). However, it lacks the 3′ poly(A) tail seen on most eukaryotic mRNAs.

The transcriptionally active form of a reovirus is a partially digested viral nucleoprotein core. The first round of mRNAs are transcribed from negative-sense RNA, capped within the core, and released into the cellular cytoplasm; the parental RNA stays inside the core. These mRNAs appear to be produced in two phases, with synthesis of the second group coding for the late proteins possibly requiring inactivation of a cellular control.

Both the amount and activity of mRNA is regulated, controlling the amount of each viral protein made. The released mRNAs are used both as mRNAs for protein synthesis and as virion components that are encapsidated into virion precursors. How all of the segments are correctly inserted is not fully understood, but sorting does appear to occur. Once within the nascent virus particle, a single round of negative-sense RNA copies are made from the mRNAs, and these are then used to produce yet more mRNA. There is evidence that this secondary transcription produces uncapped mRNAs that are then preferentially translated late in infection. Since all cellular mRNAs are capped, this effectively turns off cellular protein synthesis.

Viruses with positive-strand ssRNA genomes

A positive-sense RNA genome is infectious in itself, since it can function as an mRNA for the production of viral proteins. Of the many viruses with this type of genome, the best-studied example is poliovirus (a member of the *Picornaviridae*). As the genome can act as an mRNA, there is no need for the virus particle to contain an RNA-dependent RNA polymerase. The viral genome can be translated to produce the polymerase and other necessary enzymes after entry to the cell. This simple replication strategy is clearly efficient, since this type of virus is very common.

The poliovirus genomic RNA is translated to produce a single protein of approximately 220 kD. It lacks the normal translational initiation structure

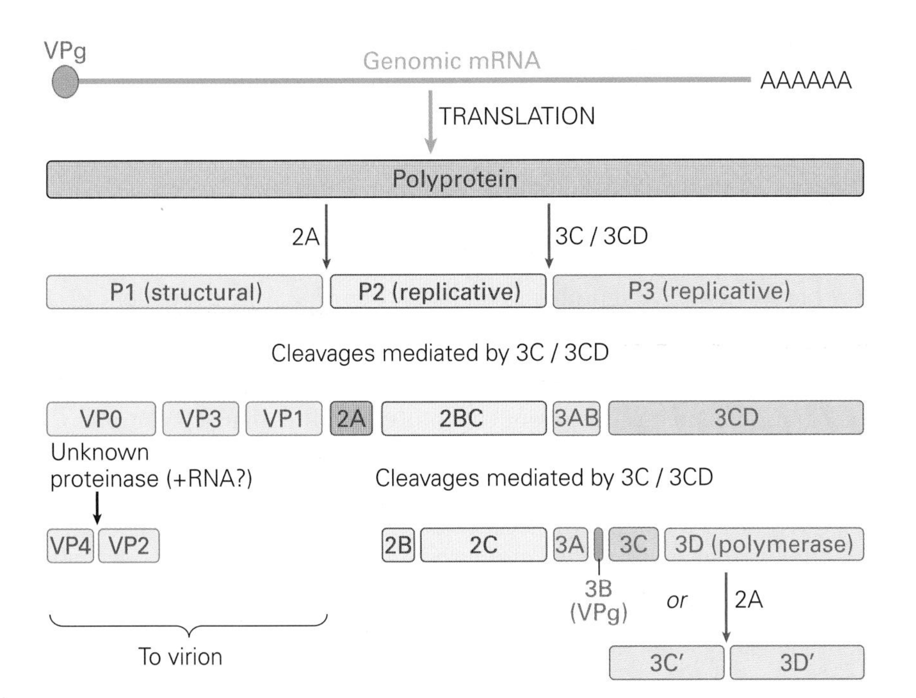

Figure 3.12
Cleavage of the poliovirus polyprotein. The action of proteinases 2A and 3C/3CD is shown. The proteinase responsible for cleavage of VP0 is unknown.

of a 5′ cap, but a specialized region of the RNA, the internal ribosomal entry site (IRES), binds directly to the ribosome. Poliovirus suppresses cellular protein synthesis by degrading multiple components of eukaryotic initiation factor 4 (eIF-4), a complex essential for translation of capped mRNAs.

The large polyprotein is then repeatedly cleaved by proteinases to produce both the structural proteins that make up the virion and all of the enzymes necessary for replication, including the polymerase (**Figure 3.12**).

One problem with such a simple scheme is that the virus produces equal amounts of every protein; enzymes such as the polymerase are not needed in the same amounts as the structural subunits of the capsid. However, by control of the cleavage process, some regulation does occur.

Once released, the viral 3D protein polymerase then produces full-length negative-sense RNAs which, in turn, are copied into positive-sense RNAs.

The process of RNA synthesis by poliovirus results in 30–70 times more positive strands than negative strands. Thus, most positive strands are found free in cells while all negative strands are either bound to positive strands in double-stranded replicative forms (RF), or more commonly in a partially dsRNA called a replicative intermediate, or RI (**Figure 3.13**). This consists of one full-length negative-sense RNA on which multiple copies of the positive strand are being synthesized. The matching strands stay together until they are displaced by the next polymerase.

The RI is part of a replication complex associated with the smooth endoplasmic reticulum. Some of the positive-sense RNAs are translated, while others are packaged as viral genomes by the newly synthesized coat proteins, which are still being cleaved even as assembly occurs.

One copy of a small viral protein (3B or VPg) is attached at the 5′ end of the genomic form of the positive RNA, where it acts as a primer. VPg is also associated with controlling the use of individual viral RNA molecules, with VPg forming part of a complex that favors a genomic role rather than function as an mRNA.

Other viruses with positive-sense ssRNA genomes may produce more than one mRNA, allowing greater control of the production of individual

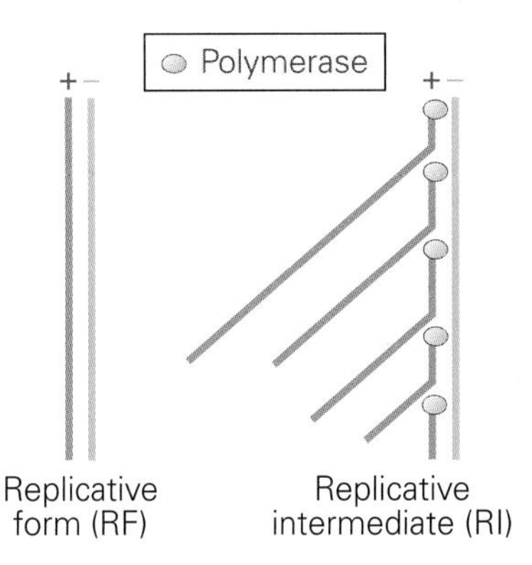

Figure 3.13
Replicative RNA structures. Positive RNA is shown in red, and polymerases in yellow.

proteins, and some exhibit limited temporal control, with early (replication) and late (structural) proteins being produced at different times in the replicative cycle.

Viruses with negative-strand ssRNA genomes

Single-stranded RNA genomes complementary to mRNA cannot function as mRNAs. Thus, an RNA polymerase is needed to produce mRNAs before any production of viral proteins can occur. The viral RNA-dependent RNA polymerase is carried in the nucleocapsid. Once within the cell, both mRNAs and genomic copies are made. During this process, a variety of dsRNA intermediates are seen (RFs and RIs; Figure 3.13), and a range of viral mRNAs are each translated to make different viral proteins, rather than a polyprotein.

Given the lack of complexity inherent in their small size, most RNA viruses do not have complex control mechanisms. However, all of the viral proteins are not produced to the same level, and a number of controls are used. For example, in replication of the *Paramyxoviridae* and the *Rhabdoviridae* (both members of the order *Mononegavirales*), initiation of transcription occurs at a single promoter at the 3′ end of the genome (**Figure 3.14**). Sequences between each gene produce the poly(A) tail of the mRNA followed by the cap of the new RNA. The efficiency of initiating the new mRNA is less than 100% (70–80% in the rhabdovirus VSV, with which much of this work has been done), resulting in progressive attenuation at the start of each new gene, so that genes farthest from the promoter produce the lowest levels of mRNAs. The polymerase L protein is furthest away, while the (structural) nucleocapsid N protein is closest and therefore made (as it is needed) in the largest amounts—the virion contains approximately 50 times more N protein molecules than L protein molecules.

When the capsid protein is present in sufficiently large amounts, it assembles around the forming RNA, suppressing the attenuation and switching synthesis to full-length copies of the genome.

In the case of influenza virus (*Orthomyxoviridae*), the virus actually requires cellular RNA synthesis, since it cannot make the 5′ cap structure. The influenza virus has a special function which removes caps from nascent cellular mRNAs and incorporates them into viral mRNAs. A similar process is used by the *Bunyaviridae*, which have segmented ambisense genomes.

Influenza virus requires the viral mRNAs to be synthesized in the nucleus to allow capping, followed by transport to the cytoplasm for translation. This rather complex process appears to be unique to influenza virus, and may represent a way of down-regulating cellular protein synthesis. The influenza viral polymerase is also required to make full-length copies of

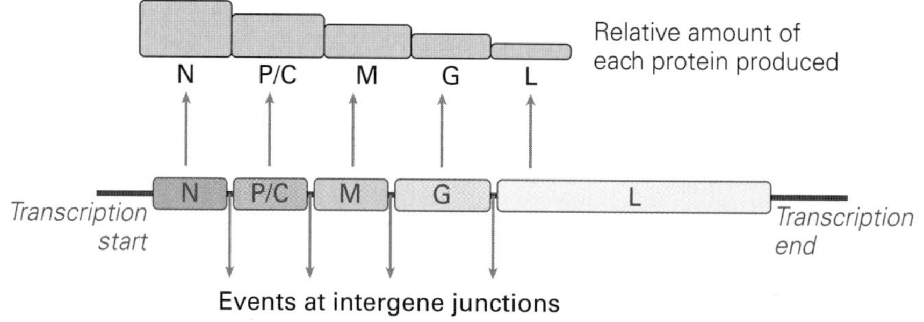

Figure 3.14
Regulation of transcription of mRNA for negative-sense single-stranded RNA viruses (*Mononegavirales*). Binding of the N (and P) proteins to the RNA results in readthrough of intergene junctions, producing full-length monocistronic mRNAs. P and C proteins are produced from alternate reading frames within the same mRNA. Regulation by "stop-start" at intergene junctions for vesicular stomatitis virus (*Rhabdoviridae*) is shown here. Similar mechanisms are used in other members of the *Mononegavirales*. At the intergene junctions, polyadenylation occurs by "stuttering" at UUUUUUU sequences. Only 70–80% of transcriptase continues to the next gene; the remaining 20–30% falls off, terminating transcription. Capping of downstream mRNA followed by transcription to the next intergene junction produces monocistronic mRNA.

the genome which are, in turn, used to make progeny genomes. Influenza shows another feature common to viruses with small genomes in that some of the gene segments code for multiple proteins using overlapping reading frames and splicing of mRNAs (see Figure 1.11).

Unusually among the ssRNA viruses of vertebrates, influenza also has to deal with particular issues arising from its segmented genome. While segmentation can assist with regulation of mRNA levels (and is important in genetic recombination), the virus has to ensure that all eight segments are included in the virion.

There is only limited evidence for the sorting of genome segments to ensure that a full set is incorporated into the forming virus particle. An alternative proposal is that, rather than sorting the genome segments, the virus may rely on encapsidating RNAs at random, with only those containing the 'full set' being fully infectious.

If only eight segments were encapsidated randomly, only 0.24% of virions would contain a full set. However, encapsidation of 12 segments would give approximately 10% of virions with the full set, rising to approximately 15% with 14 segments encapsidated. However, many target cells may be infected by more than one virion, increasing the chances for successful infection. The *Bunyaviridae* seem to use this method, encapsidating a mixture of genome segments, though in their case the situation is less demanding, since they only have three segments.

Viruses with ambisense genomes, including the segmented *Arenaviridae* and *Bunyaviridae*, appear to replicate in a manner broadly similar to the negative-strand viruses, using a viral polymerase to make mRNAs rather than having a genome that is directly translated.

ssRNA (positive-sense) viruses that use a DNA intermediate

Retroviruses synthesize RNA genomes from RNA genomes via a dsDNA intermediate. This 'reverse flow' of information from RNA to DNA gives the retroviruses their name (*retro* in Latin translates to backward in English). The *Retroviridae* are the only viruses of this type that are known to infect humans, although the *Caulimoviridae* of plants and the predominantly intracellular *Metaviridae* and *Pseudoviridae* infecting fungi and invertebrates have similar replication cycles.

Retroviruses contain an enzyme, reverse transcriptase, that is central to their replicative cycle. The retroviral genome contains a long unique sequence between two long terminal repeats (LTRs) and contains three main genes, *gag* (core proteins), *pol* (polymerase/reverse transcriptase), and *env* (envelope glycoproteins). The human retroviruses are among those referred to as complex retroviruses (a group that also includes the foamy viruses), since they produce a number of additional small proteins (*vif, vpx, vpr, vpu, tat, rev*, and *nef* in HIV-1). Analogs of these proteins may be present, but are not as well characterized in most retroviruses of animals.

On entering the cell, the retroviral genome is reverse-transcribed into a dsDNA intermediate. Rather than existing as a circular DNA molecule free within the cell (an episome), the retrovirus dsDNA integrates into the cellular DNA using a viral integrase. It then remains there, being transcribed and replicated by the cell itself. In this state it is referred to as provirus. The proviral DNA can be expressed very efficiently and give rise to a productive infection, or it can remain 'silent'. This situation prevents clearance of the virus by the immune system since a silent, integrated DNA presents no targets for the immune system. Even inactive integrated provirus can have significant effects on the cell, particularly on adjacent genes. It is by this mechanism, as well as by expression of viral oncogenes (see Section 4.8), that the induction of retrovirus-associated cancers can occur.

However, the retroviral genome is small, and oncogenes in the retroviral genome are present in place of essential genes in almost all retroviruses. The exception to this is Rous sarcoma virus of chickens, which can carry an oncogene as well as its full set of replicative genes. Apart from this isolated case, a retrovirus carrying an oncogene will have lost genes essential for independent replication and will require a co-infecting complete 'helper' virus (see Section 2.4). This requirement casts doubt on the ability of such viruses to be a significant cause of cancers outside the laboratory, and it is likely that integration-based effects are of more importance epidemiologically.

Activation of the proviral DNA to begin producing viral mRNAs, genomic RNAs, and proteins can occur in response to a variety of stimuli, such as activation of a host immune cell. Some retroviruses, notably the foamy viruses (subfamily *Spumavirinae*), appear to proceed directly to productive infection in some types of cell.

Once the retroviral genome has made copies of itself and has made the viral mRNAs which, in turn, have made the viral proteins, these components are assembled into the virus particles.

Very unusually, retroviruses are also diploid, since each virion contains two copies of the viral genome linked at their 5′ ends. Full-length retroviral genomes may be copied from both copies of the genome, with the transcriptase jumping the gap between them. This may increase the variability of replication still further. It is notable that HIV, a member of the *Retroviridae*, does show extremely high variability and that this helps it to evade the activities of both the immune response and antiviral drugs.

3.4 VIRAL REGULATION OF CELLULAR ACTIVITIES

Beyond the immediate effect of changing infected cells into sources of progeny virus and (in some cases) killing them, a number of strategies exist to favor viral replication over the normal activities of the cell.

Some viruses need cellular DNA synthesis in order to obtain replicative enzymes for their own use. The very simplest viruses (*Parvoviridae*) may rely on infection of dividing cells (parvovirus B19) or on provision of such functions by a helper virus (adeno-associated viruses), but this places severe limits on their ability to complete their infectious cycle. Slightly larger viruses such as the *Polyomaviridae* and the *Papillomaviridae* are able to drive the cell into S phase, a pre-division state in which DNA synthesis occurs. This makes the necessary enzymes available, but if the virus replication does not kill the cell this can result in uncontrolled cell division, cellular transformation, and a possible role in the formation of cancers (see Chapter 4).

One way of disrupting cellular metabolism is to alter the activity of transcription factors such as eIF-2 or eIF-4 that are involved in the initiation of protein synthesis by the control of their phosphorylation state. This can have profound effects on the ability of the cell to assemble amino acids into proteins. A wide range of viruses including the *Adenoviridae*, *Herpesviridae*, *Orthomyxoviridae*, *Picornaviridae*, *Poxviridae*, *Retroviridae*, and *Rhabdoviridae* use this method. A more direct effect is seen with the degradation of eIF-4 by the *Picornaviridae* or the *Rhabdoviridae*.

Transfer of caps from cellular to viral mRNAs ("cap snatching") is used by the *Bunyaviridae* and the *Orthomyxoviridae*, while alteration of the cellular machinery to prefer variant viral mRNAs is seen with the apparent production of uncapped mRNAs late in infection with the *Reoviridae*.

Other viruses, such as the more complex *Herpesviridae*, produce a number of factors that influence the activities of the host cell. One example is the herpes simplex virus *vhs* (virion host shut-off) protein, carried in the tegument (extracapsid protein matrix) of the virion. The *vhs* function is an endoribonuclease that destabilizes mRNAs, causing them to be degraded. Unsurprisingly, given their complex nature, the herpesviruses are also thought to use multiple mechanisms of inhibiting cellular synthesis, including for example interference with RNA splicing by the protein ICP27, and enhanced degradation of host proteins mediated by the ICP0 protein.

These effects exist at multiple levels. For example, the effect of *small interfering RNAs* (**Box 3.3**) is still being elucidated in virus infection, and this approach appears to be used by both cells and viruses.

The confrontation is not one sided, and cells have developed a variety of mechanisms to try to control virus infections. In many cases, these involve the death of the infected cell. Such responses include programmed cell death (apoptosis), the interferon response, or interactions with the immune system. These systems, and the viral responses which control them, are covered in more detail in Chapter 4. Both viruses and cells use multiple mechanisms simultaneously in this battle, reflecting the innate complexity of biological systems. This is perhaps instructive for the use of antiviral approaches as discussed in Chapter 6.

Box 3.3 Small interfering RNA (siRNA)

Small interfering RNAs (siRNAs) are small, double-stranded RNA molecules (typically 20–25 nucleotides in length) that appear to have multiple functions, particularly in relation to the control of gene expression. They were first reported in 1999 as being involved in post-transcriptional gene silencing in plants. Additional work identified the mechanism of *RNA interference* (RNAi) as being more widely applicable. RNAi is a sequence-specific, post-transcriptional gene silencing mechanism operating in animals and plants, based on the presence of double-stranded RNA (dsRNA) homologous in sequence to the silenced gene. This is now an important experimental tool and therapeutic applications are also under investigation. RNAi is described in more detail in Chapter 6.

3.5 VIRUS ASSEMBLY AND RELEASE

In general, the more complex the virus, the more complex the structures and thus the more complex the assembly pathways involved in actually putting it together.

Poliovirus provides an example of a "simple" virus. The polio virion contains five proteins, four of which make up the capsid (see Section 1.2). The fifth protein (VPg) is linked to the viral genome and appears to act as a primer for RNA synthesis. If produced in a cell-free translation system, the four proteins, VP1–VP4, can assemble themselves into a capsid. In the infected cell, the proteins assemble themselves into precursor forms (pentamers) which are then used in capsid formation. The genomic RNA is either inserted into preformed capsid or condenses with the proteins during capsid formation, although the available evidence favors the latter. Adenoviruses have similar (icosahedral) capsid morphology to poliovirus, but have much larger genomes and a far more complex structure (see Section 1.2) and pathway of assembly. Additional **scaffolding proteins** (proteins required for assembly but not present in the final form) are required at two stages. A 100 kD protein is required for the formation of the hexamer subunit (itself a trimer), while at least two further proteins are required to support assembly of the procapsid, the precursor to the capsid, into which the DNA is inserted.

Once assembled, the viruses must leave the cell. In the case of many bacteriophages infecting prokaryotic cells, this simply involves the cell bursting as a result of the virus infection. The situation is usually more complex in eukaryotic cells.

Viral protein synthesis can only take place in the cytoplasm, and the majority of viruses are assembled there. With some viruses, such as the *Herpesviridae* or the *Orthomyxoviridae*, viral proteins migrate to the nucleus, and are there assembled into nucleocapsids.

Enveloped viruses then acquire their membrane component from a cellular membrane. A variety of membranes may be used, depending upon the nature of the infecting virus. For the *Herpesviridae* this appears to be the nuclear membrane; enveloped virions are then transported to the cell

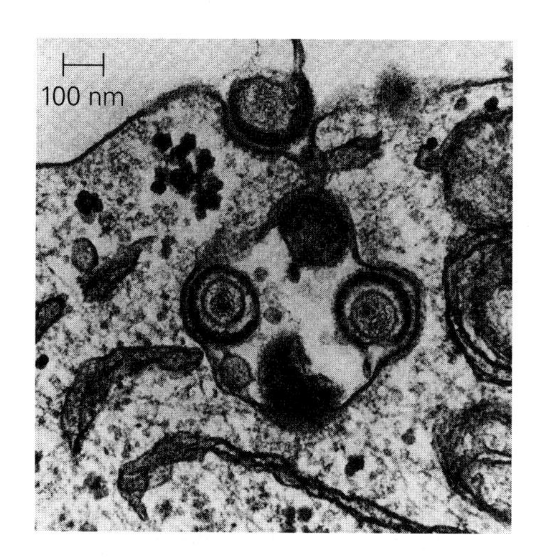

Figure 3.15
Transport of a virus in an intracellular vacuole. Late-stage maturation of varicella-zoster virus, with transport of the virus to the cell surface in a vacuole after acquisition of an envelope derived by budding through the nuclear membrane (human herpesvirus 3, *Herpesviridae*). Courtesy of Charles Grose, Department of Pediatrics, Universtiy of Iowa Hospitals, USA.

surface in vesicles (**Figure 3.15**). In contrast, despite nuclear assembly of the multiple nucleocapsid segments, the *Orthomyxoviridae* acquire their envelope from the plasma membrane at the cell surface. Other viruses can use the membranes of the Golgi apparatus (*Bunyaviridae*) or endoplasmic reticulum (*Flaviviridae*), but the plasma membrane appears to be the most common source.

Viral proteins of *Paramyxoviridae* become associated with specific areas of the plasma membrane which are enriched with cholesterol and sphingolipids and known as **membrane rafts**. The paramyxovirus glycoproteins accumulate in modified sections of the cellular membrane, and appear to prefer membrane rafts. The tails of the viral glycoproteins protrude into the cytoplasm, where they interact with the viral M protein, which accumulates on the inside of the membrane at this point and mediates interaction with the completed nucleocapsid. The whole complex then pouches out of the plasma membrane and buds off into the extracellular environment.

Retroviruses are another group of viruses that exit from cells by budding. The nucleocapsid is assembled either in the cytoplasm or at the plasma membrane (depending on the specific retrovirus), and the core then buds out of the cell surrounded by a section of modified plasma membrane (**Figure 3.16**).

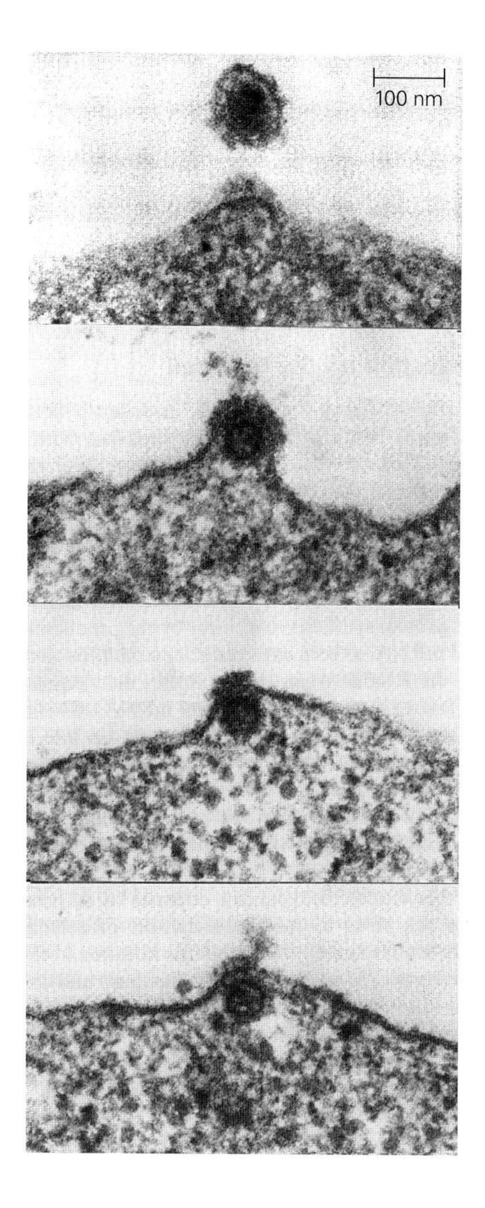

Figure 3.16
Budding from the cell surface. The sequential stages in the budding of a member of the *Retroviridae* from the infected cell; association of the viral nucleocapsid with a modified section of the cell membrane, followed by pouching out of that section of the membrane, then release of the enveloped virus particle. Courtesy of Ian Chrystie, St Thomas' Hospital, London.

Non-enveloped viruses are usually released by cell lysis. However, some can produce infectious virus without lysing the cell. This suggests that a transport mechanism, probably involving vacuoles derived from cellular membranes, must exist. 'Bursting' (i.e. non-active lysis) of a dying cell does occur and can also produce infectious virus particles.

Key Concepts

- Initial interaction of a virus with its host cell is mediated by surface proteins (often glycosylated) on the virus and by proteins and other structures on the host cell. These allow the virus to select the nature of its target cell with great precision.
- Some viruses use just a single receptor, while others can use a range of different receptors. Some viruses require a series of receptors, of increasing specificity.
- Once bound to the outside of the host cell, viruses are internalized by a range of mechanisms, from fusion with the cell plasma membrane to encapsulation in vacuoles. There are various methods of escape from these vacuoles into the cytoplasm, which are often dependent on acidification as part of an intracellular digestion process.
- Viruses may replicate at various locations within the cell. Most DNA viruses replicate in the nucleus (with the exception of the nucleocytoplasmic large DNA viruses) while most RNA viruses replicate in the cytoplasm.
- Different viral genes are expressed at different stages of replication. Those expressed earlier on are often regulatory, or enzymes involved in the replication process. Genes expressed later are generally structural proteins or those involved in viral assembly.
- Seven basic strategies exist based on genome type (dsDNA, ssDNA, dsRNA, +sense RNA, −sense RNA, RNA via DNA, and DNA via RNA) but these are complicated by mixed ss/ds genomes, ambisense (+/−) RNA, segmentation of genomes, and a range of other factors.
- Simple DNA viruses rely on the presence of or induce the production of cellular DNA synthetic machinery, while more complex viruses carry genes for the enzymes they need.
- RNA viruses need to carry or produce their own RNA-dependent RNA polymerase, since this enzyme is not present at useful levels in cells. Positive-sense RNA viruses can do this by using their viral genome as an mRNA when they enter their host cell, while negative-sense RNA viruses (that require transcription to produce mRNAs) have to carry a preformed RNA polymerase within the virus particle.
- Viruses with segmented genomes use a variety of strategies for making sure that all necessary segments are packaged in the virion. Reoviruses include specific packaging signals, while orthomyxoviruses such as influenza may combine limited sorting with the packaging of a large enough number of segments that random chance ensures that enough virions contain the full set.
- Viruses alter the cell to favor the production of progeny viruses, and repress cellular and immunological controls that attempt to prevent this.
- Enveloped viruses acquire their membrane from a cellular source, either the plasma membrane or an internal membrane.
- Non-enveloped viruses are often released by active cell lysis or simply the bursting of a dying cell.

DEPTH OF UNDERSTANDING QUESTIONS

Hints to the answers are given at http://www.garlandscience.com/viruses

Question 3.1: What are the nucleocytoplasmic large DNA viruses (NCLDVs), and do they infect humans?

Question 3.2: What are ambisense genomes, and what effect do they have on virus replication and classification?

Question 3.3: Among viruses infecting humans, segmented genomes are only found in RNA viruses. Explain why.

Question 3.4: What is a nuclear pore, and what is its significance for virus infection?

FURTHER READING

Chandra R (1997) Picobirnavirus, a novel group of undescribed viruses of mammals and birds: a minireview. *Acta Virol.* 41, 59–62.

Fields BN & Howley PM (2007) Virology, 5th ed. Lippincott Williams & Wilkins, Philadelphia.

Gottwein E & Cullen BR (2008) Viral and cellular microRNAs as determinants of viral pathogenesis and immunity. *Cell Host Microbe* 3, 375–387.

Grassmann R & Jeang KT (2008) The roles of microRNAs in mammalian virus infection. *Biochim. Biophys. Acta* 1779, 706–711.

Harborth J, Elbashir SM, Bechert K et al. (2001) Identification of essential genes in cultured mammalian cells using small interfering RNAs. *J. Cell Sci.* 114, 4557–4565.

Harrison SC (2008) Viral membrane fusion. *Nat. Struct. Mol. Biol.* 15, 690–698.

Miller S & Krijnse-Locker J (2008) Modification of intracellular membrane structures for virus replication. *Nat. Rev. Microbiol.* 6, 363–374.

Nicholls JM, Chan RW, Russell RJ et al. (2008) Evolving complexities of influenza virus and its receptors. *Trends Microbiol.* 16, 149–157.

INTERNET RESOURCES

Much information on the internet is of variable quality. For validated information, PubMed (http://www.ncbi.nlm.nih.gov/pubmed/) is extremely useful.

Please note that URL addresses may change.

All the Virology on the WWW. http://www.virology.net (general virology site)

The Big Picture Book of Viruses. http://www.virology.net/Big_Virology/BVHomePage.html

ViralZone. http://www.expasy.ch/viralzone/ (general molecular and epidemiological information, along with virion and genome figures for all viral genus and families)

CHAPTER 4
Immune Response and Evasion

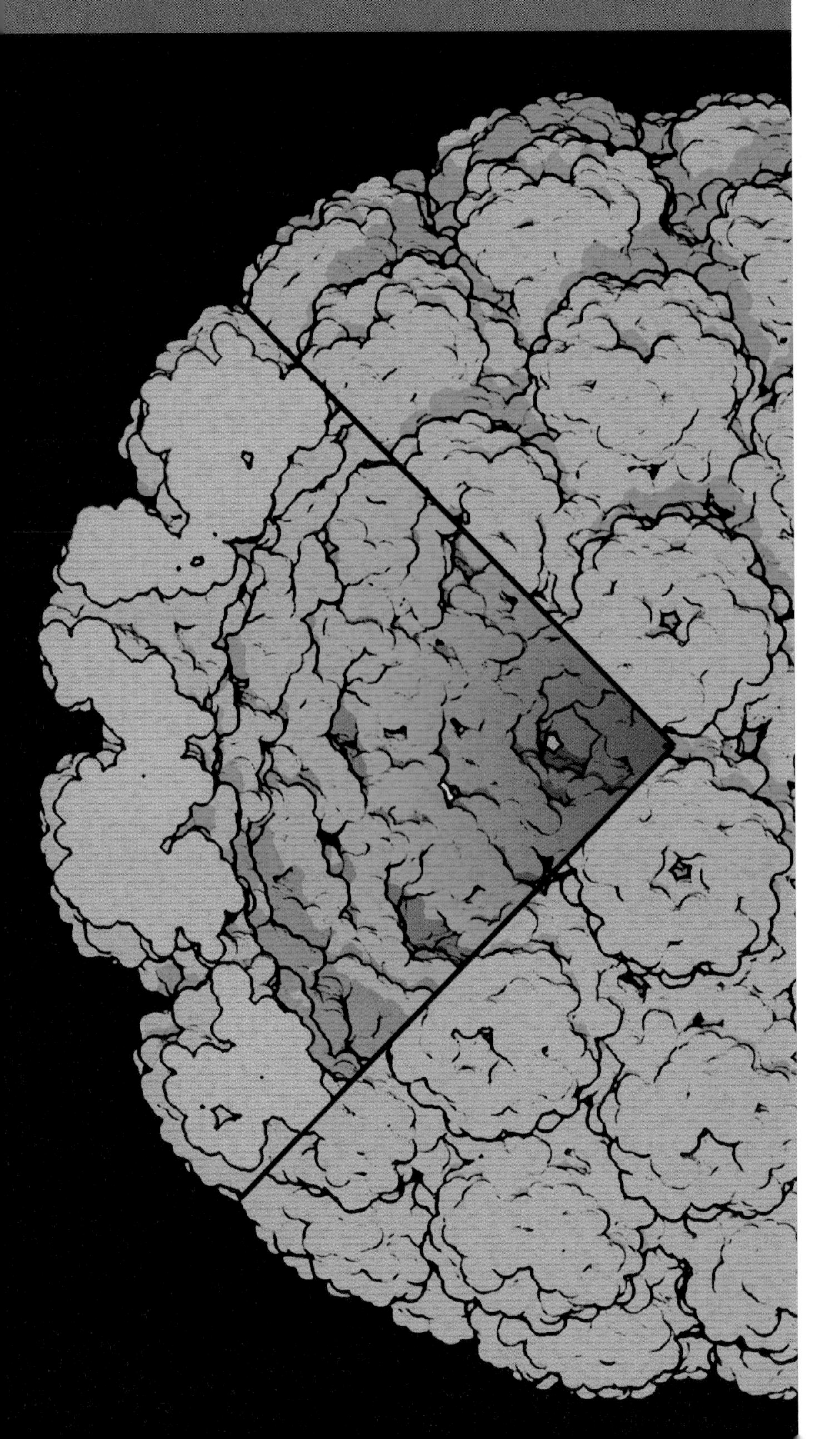

INTRODUCTION

Once a virus has encountered and infected its host, the outcome is hugely influenced by the immunological defense that the host mounts in response. Infection of cells in culture lacks this important element and is thus a poor reflection of the complex situation that occurs in the host organism.

About the chapter opener image
Simian virus 40
(Courtesy of the Research Collaboratory for Structural Bioinformatics Protein Data Bank and David S. Goodsell, The Scripps Research Institute, USA.)

The immune response to virus infection is key to protecting the infected host, but paradoxically can also be the cause of many of what are thought of as the symptoms of virus infection such as inflammation, fever, and even pneumonia. In many cases, these symptoms actually help to limit the infection. Fever can produce temperatures that are not optimal for virus replication, while inflammation summons the immune response to the infected site. Even when its effects are perceived as part of the disease, the inhibition of virus replication by the immune system limits the spread of infection and alters the course of disease, helping to limit and control it.

Without an effective host immune response, many virus infections would eventually be fatal. This is seen in the killing of cultured cells by viruses, but is also seen in highly immunosuppressed individuals, where infections that are normally nonpathogenic can be serious or even lethal.

There is a wide range of possible outcomes of a viral infection. Complete resolution and elimination of the infecting virus is the ideal resolution for the host and is a common outcome. In contrast, an uncontrolled virus infection can kill the host, and inappropriate immune responses may be involved in such an outcome, as with hantavirus pulmonary syndrome (see Section 4.7). In many cases, the virus may be able to establish a long-term infection, enhancing its potential for transmission to a new host. In some cases, the infected cells can develop altered growth characteristics (see Chapter 1) and can then, if they can evade the immune system and maintain their growth, go on to cause cancers (*oncogenesis*).

The human immune response to virus infection consists of two main parts. The *innate immune system* mounts a rapid, generic response by recognizing **non-self** molecules—those which are different to those of the host—on the surface of invading viruses. This rapid response is then followed by the *adaptive immune system*, which is slower to develop but has the ability to recognize specific pathogens and target them for destruction. There are two strands to the adaptive response: the *serological* (or *humoral*) *system*, which is mediated by antibodies produced by B cells, and the *T-cell-mediated system*. The main properties of the innate and adaptive immune response are summarized in **Table 4.1**.

Table 4.1 The innate and adaptive immune responses

Immune response	Type	Speed	Receptors	Effectors	Effect
Innate	Non-cellular	Fast	Preexisting (encoded in the genome)	Cytokines (interferons), complement	Localized response to infection, stimulation of general immune response
	Cellular	Fast	Preexisting (encoded in the genome)	Macrophages and other phagocytic cells, NK cells	Engulfment of pathogens, release of cytokines
Adaptive	Non-cellular	Slow (on first exposure)	On B cells, produced by genetic rearrangement and amplified by clonal selection. Bind to antigens via B-cell receptor and present via MHC-II pathway	Antibodies (IgG, IgM, IgA)	Targeting of immune responses, direct neutralization, immune memory
	Cellular	Slow (on first exposure)	On T cells, produced by genetic rearrangement and amplified by clonal selection. Bind to antigens presented by MHC-I and MHC-II pathways	T cells ($CD4^+$, $CD8^+$)	Cytotoxic, immunostimulatory, and immune memory

Box 4.1 On the naming of cells

T cells move from the bone marrow to the thymus, where they mature. Thus, they are named T (for thymus) cells. However, B cells mature in the blood and the spleen, and are actually named from early studies in chickens, where they were shown to mature in an organ known as the Bursa of Fabricius, which has no parallel in mammals.

All of the cells that drive and control the immune response, both the phagocytic cells of the innate system and the lymphoid cells that develop into B and T cells (see **Box 4.1**), derive from pluripotent hematopoietic stem cells in the bone marrow (**Figure 4.1**).

4.1 THE INNATE IMMUNE RESPONSE

The **innate immune response** involves both cellular and non-cellular elements. It produces a rapid response, based on recognition of generic, non-self molecular structures on the surface of invading pathogens. While it cannot adapt to recognize new targets, it is able to trigger signaling cascades that produce a powerful, often localized response to infection, and is also important in shaping any subsequent adaptive response.

Recognition: Toll-like receptors

When the innate immune system was first characterized, it became clear that it responded to non-self **antigens**, and that certain molecular structures associated with pathogens (pathogen-associated molecular patterns, PAMPs) could trigger responses. More recently, specific activation pathways have been identified. The best characterized of these is that of the **Toll-like receptors**. These are a family of at least ten transmembrane proteins, which is named for the Toll receptor first identified in the fruit fly. Toll-like receptors (TLRs) recognize specific PAMPs. TLR4, the first to be identified, recognizes the lipopolysaccharide of Gram-negative bacteria, while TLR2, -3, -7, -8, and -9 are known to recognize viral structures. TLR3, for example, recognizes double-stranded (ds) RNA and appears to be involved in the activation of the interferon response.

Toll-like receptors are connected to the NF-κB transcription factor pathway that appears to be common to most multicellular organisms. This triggers expression of a range of genes involved in adaptive immunity and the inflammatory response.

Signaling: cytokines and interferons

Cytokines are small regulatory proteins that are involved in many aspects of virus infection and act as cell signaling molecules over short distances. They are produced by a range of cell types and control many aspects of both the innate and the adaptive immune responses. More than 50 cytokines have been identified to date, divided into four structurally and functionally related subgroups; hematopoietins, tumor necrosis factors (TNFs), interferons, and chemokines (**Figure 4.2**).

At the site of infection, a specific subset of cytokines is typically produced by macrophages (see below). One effect of these cytokines is dilating, permeabilizing, and increasing adhesion in local blood vessels. This allows white blood cells to migrate from the blood into the tissues. Chemokines, a closely related group of cytokines that are released early in infection, assist in this

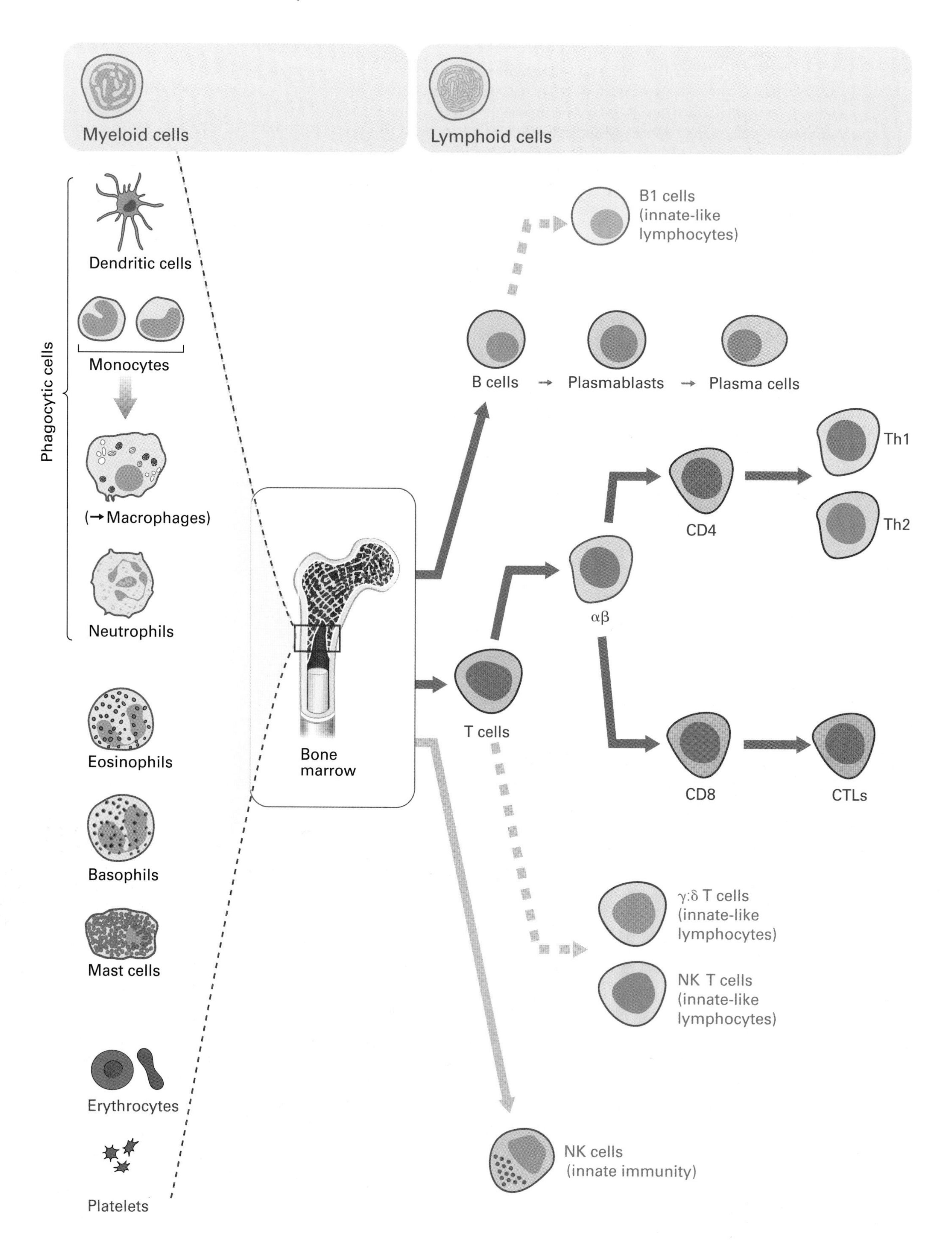
Myeloid cells
Lymphoid cells
Phagocytic cells
Dendritic cells
Monocytes
(→Macrophages)
Neutrophils
Eosinophils
Basophils
Mast cells
Erythrocytes
Platelets
Bone marrow
B1 cells (innate-like lymphocytes)
B cells → Plasmablasts → Plasma cells
Th1
Th2
CD4
αβ
T cells
CD8
CTLs
γ:δ T cells (innate-like lymphocytes)
NK T cells (innate-like lymphocytes)
NK cells (innate immunity)

Figure 4.1
Main lineages of cells by pluripotent hematopoietic stem cells in the bone marrow. Myeloid cells include the three types of phagocytic cell; dendritic cells, monocytes (which mature into macrophages), and neutrophils. Dendritic cells and macrophages are responsible for the presentation of viral antigens to the lymphoid cells that control the adaptive immune response. Eosinophils are important in defense against parasitic infections, while basophils and mast cells appear to be concerned with allergic responses. Both platelets and erythrocytes lack nuclei and are not involved in the immune response. Erythrocytes (also known as red blood cells) are responsible for oxygen carriage in the blood, while platelets are concerned with blood clotting. In the lymphoid lineage, NK cells and innate-like lymphocytes are part of the innate (non-adaptive) immune response early in infection (shown in blue). Cells involved in the adaptive immune response are shown in red. T cells are responsible for controlling the immune response and for cell killing, while B cells mature into antibody-producing plasma cells.

process. Their primary role is to attract leukocytes (white blood cells) to the site of infection or injury.

The cytokine-mediated leakage of components of the blood into the tissue produces the classical symptoms of inflammation; redness, raised local temperature, pain, and swelling (*rubor*, *calor*, *dolor*, and *tumor*) that were first noted by Celsus almost two thousand years ago. Though often seen as a negative event by the sufferer, these contribute very significantly to the clearance of an infection by focusing the effectors of the immune system at the site of the challenge, and are an important element of the early immune response.

One of the most significant groups of cytokines in the context of virus infection is the **interferons**. Interferon was discovered in 1957 by Isaacs and Lindenmann, although work by Nagano and Kojima three years previously on what they termed "virus inhibitory factor" is now recognized as an independent identification of the same effect. For many years, interferon was

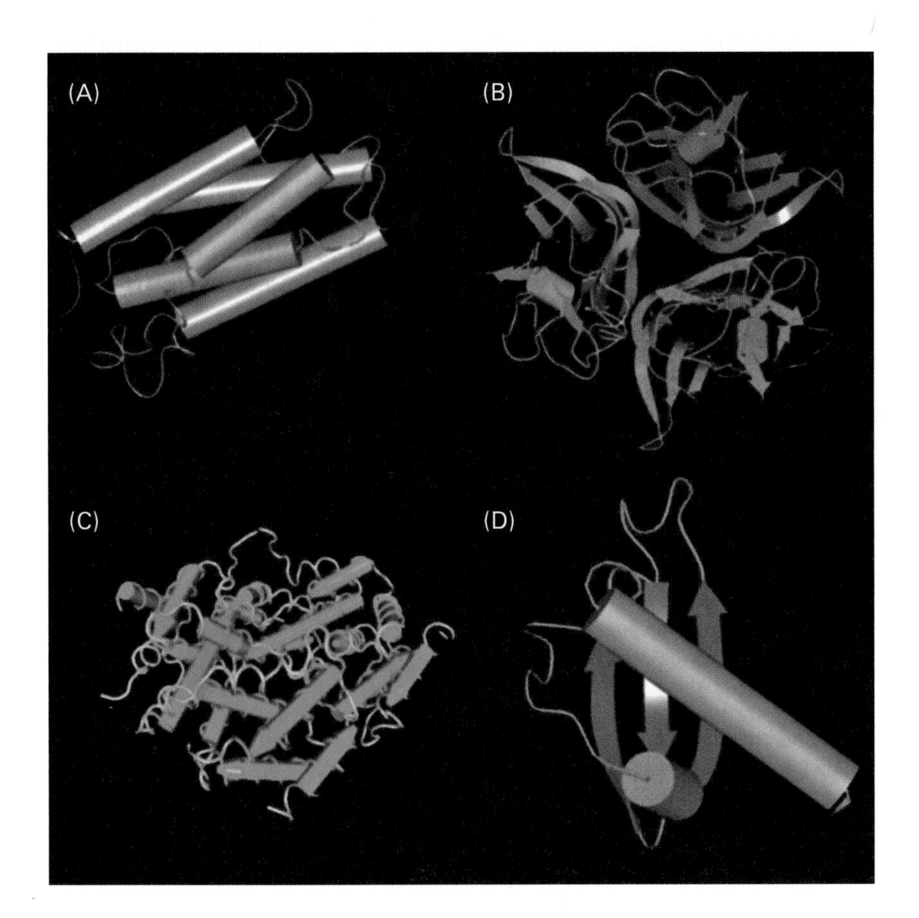

Figure 4.2
The four families of cytokines. Cytokines are small regulatory proteins produced by a wide range of cells, and are involved in signaling between cells, generally over short distances. No uniform naming system has as yet been developed, but the more than 50 cytokines discovered to date appear to be divided into four structurally and functionally related subgroups: hematopoietins, tumor necrosis factors, interferons, and chemokines. This figure shows the structures of representative members of these four families: (A) interleukin 4 (hematopoietin); (B) tumor necrosis factor (trimer); (C) interferon γ; (D) CXCL8 (chemokine). Parts A, B, and D from Janeway CA, Travers P, Walport M & Shlomchik MJ (2005) Immunobiology, 6th ed. Garland Science. Part C courtesy of Nevit Dilmen under GNU Free Documentation License, Version 1.2.

Table 4.2 Basic properties of the interferons

Interferon	Producer cells	Amino acids	Inducers
Alpha (α) (type I)*	Dendritic cells	143 per chain (homodimer)	dsRNA Virus infection
Beta (β) (type I)*	Fibroblasts	166 (monomer)	dsRNA Virus infection Bacterial components Cytokines
Gamma (γ) (type II, immune)	T cells NK cells	166 (monomer)	Antigens Mitogens Cytokines

* Interferon alpha is produced from a family of at least 15 related genes in humans, and interferon beta from a single gene, although other species may vary.

thought likely to be the first effective drug against virus infections, much as antibiotics had been 25 years earlier for bacteria. However, it took some time for the role of interferon in the complex feedback systems of the immune response to be understood and it is only relatively recently that its use has become an accepted part of antiviral chemotherapy.

There are three basic types of interferon (α, β, and γ) as well as some minor species (interferon ω, interferon γ subtypes). Interferons α and β (type I interferons) are specifically antiviral in nature and are produced by a wide variety of cells, while interferon γ (type II interferon) is produced later and in response to different stimuli by the cells of the immune system (**Table 4.2**).

Interferons induce an **antiviral state** (**Table 4.3**) in the infected cell and are also released, alerting nearby cells to the presence of the infecting virus, thus preparing them to induce similar effects in response to subsequent viral infection. The interferon response is very rapid, appearing within hours of viral infection, and is an important element of pre-specific immunity. They make cells resistant to virus infection by a wide range of effects, resulting in the shutting down of protein synthesis. Interferon can also induce the expression of factors that induce the programmed death of the cell (apoptosis, see Section 4.5).

Table 4.3 The antiviral state: major responses triggered by interferons

Response type	Effector	Mechanism	Effect
Intracellular (autocrine)	2′,5′ oligo-A synthetase	Ribonuclease L activation	Degradation of RNA (viral and cellular)
	Protein kinase R	Phosphorylation of eukaryotic initiation factor 2 (eIF-2) and other proteins	Shutdown of protein synthesis
	M_x protein	Inactivation of viral functions	Inhibition of RNA virus replication
	Multiple, including TRAIL and protein kinase R	Apoptosis	Cell death
Extracellular (paracrine)	Increased synthesis of MHC-I proteins	Increased presentation of antigens to adaptive cellular immune response	Increased killing of virus-infected cells
	Activation of NK cells		Increased killing of virus-infected cells

Phagocytosis

Macrophages are specialized phagocytic cells, which form the first line of defense against many pathogens (**Figure 4.3**). When they encounter pathogens, including viruses, they engulf and destroy them, then present the digested viral proteins to activate the immune system via the MHC-II pathway (see Section 4.3). Macrophages also produce a characteristic pattern of cytokines that initiate a localized immune response, including the migration into the tissues of neutrophils, another class of phagocytic cells, from the blood. They also produce a range of protective chemicals, including nitric oxide, which aid in defense against bacteria and viruses. Other cells involved in **phagocytosis** as an immediate response to infection include multiple subsets of **dendritic cells**, which also help to activate the adaptive immune response. It should be noted that a number of pathogens have adapted not only to reduce the effects of phagocytosis but even to use it as part of the infectious process.

NK cells

NK (natural killer) **cells** are a major cellular element of the innate immune response. These are cytotoxic lymphocytes which lack a T-cell receptor (TCR) (see Section 4.3). They are targeted to the site of virus infection by cytokines. They perform their cell-killing function by the release of perforin, which disrupts the integrity of the target cell membranes, as well as the release of granzymes (serine proteinases) which induce programmed cell death (apoptosis, see Section 4.5) of the target cell. Usually, NK cells

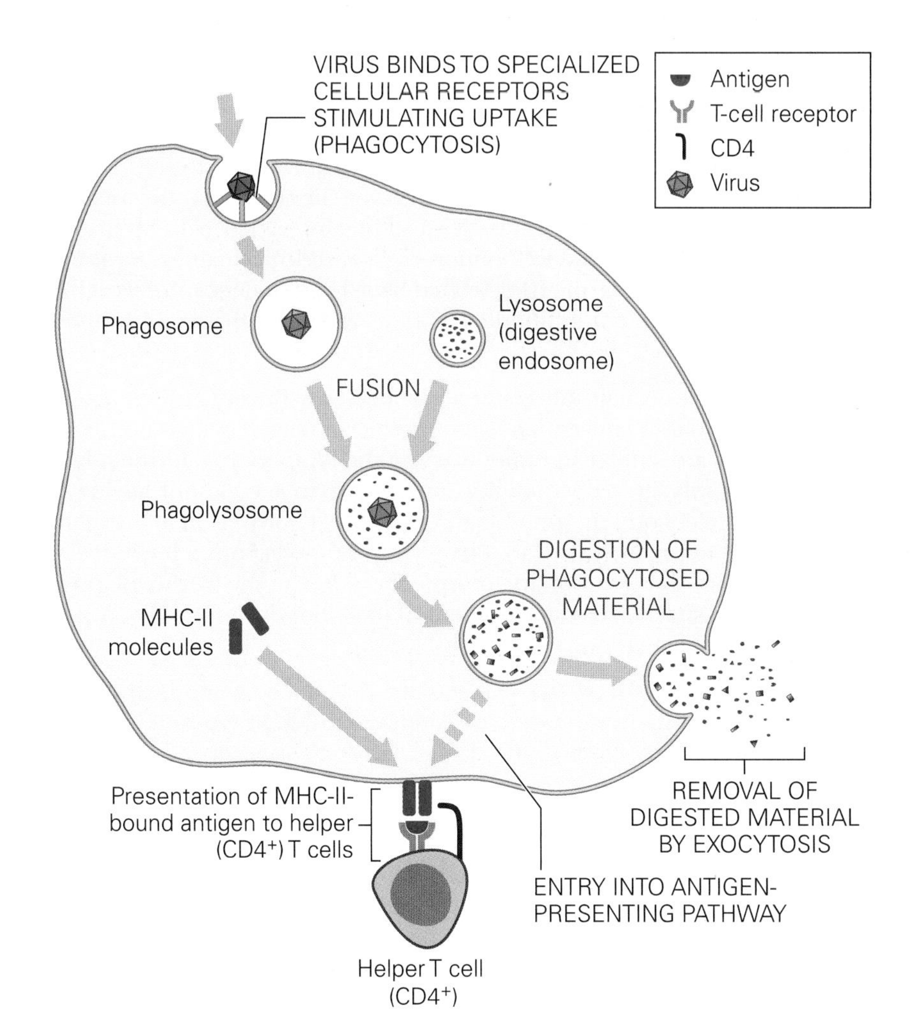

Figure 4.3
The processes involved in phagocytosis.

are only weakly cytotoxic but in the presence of α- or β-interferon or the cytokine interleukin 12, NK cell activity is enhanced greatly. They may also be targeted by bound antibodies, killing the cells that are marked out by antibody binding in a process known as antibody-dependent cellular cytotoxicity (ADCC). NK cells are controlled by a complex balance of activating and inhibitory receptors. An important inhibitory signal is the presence on potential target cells of autologous (**self**) major histocompatibility complex class I (MHC-I) molecules, while activating receptors include the NK receptor complex (composed of C-type lectin homologs) and the KIR (killer cell immunoglobulin-like receptor) proteins.

Viral proteins may also influence NK cell function, either directly or by inhibiting the expression of the MHC-I protein that is involved in cell killing. A range of viruses have such an inhibitory effect, apparently as a way of limiting the cell-mediated immune response, but the lack of MHC-I molecules at the cell surface can actually enhance killing by NK cells. However, this effect of NK cell activity is moderated by a second consequence of virus-induced interferon production: up-regulation of class I MHC gene transcription, which may serve to balance the stimulation of NK cell activity by interferons. NK cell activity declines within a few days of virus infection as the specific immune responses appear in the host, and it has been suggested that they represent an evolutionary "hangover" from before the development of the adaptive T-cell response.

Innate-like lymphocytes

Other pre-adaptive lymphocytes exist, notably the innate-like lymphocytes (ILLs), which include the epithelial γ:δ T cells. These are named for their variant receptor structure—they bear a T-cell receptor (see Section 4.3) containing γ and δ chains unlike the α:β chain heterodimer of the normal T-cell receptor. They appear to recognize specific antigens without the involvement of the MHC proteins. Another variant form of T cells is the NK T cells. These are a specialized subset of T cells with very limited variability in their receptor structure that also express NK cell markers. NK T cells appear to recognize lipid antigens and to function as a source of cytokines. A variant form of B cells also exists, the B1 cells that have restricted receptor variability and can produce IgM antibodies without requiring the usual support from T cells.

In general, ILLs are small subsets of lymphocytes with very limited receptor variability, that in consequence are present in relatively high numbers. Although they are similar in many ways to the lymphocytes of the adaptive response, this limited variability allows them to act without having to undergo the selection and amplification process known as clonal expansion, which is required for the adaptive lymphocyte response. It is unclear whether they are an evolutionary intermediate step on the way to adaptive immunity, or a specialized development of that adaptive immunity targeting specific pathogen types.

As with antibody-mediated NK cell activity, ILLs once again illustrate the difficulties in setting firm borders between the different types of immune response.

The complement system

Complement is the name given to a complex set of proteins that are present in normal blood. Although first identified as enhancers of the antibody activity by producing cytotoxic effects where antibody is bound (see Section 4.2), the complement system is also a significant part of the innate immune response.

Complement is activated by three pathways: by antigen-bound antibodies (the classical pathway), by the specific binding of a lectin to mannose sugar residues on bacteria or viruses (the MB lectin pathway), or by specific structures on the surfaces of some pathogens which stabilize the initial enzymes of the activation process (the alternative pathway).

Activation by any of these pathways initiates a similar series of enzymic reactions, each of which amplifies the previous stages (referred to as a cascade). This means that a small initial response can become very large by the time the complement cascade is completed. Activated complement can act in one of three ways:

- Binding to the surface of pathogens and making them better targets for phagocytosis by cells bearing complement receptors (opsonization)
- Providing a chemokine-like (pro-inflammatory) attractant activity for phagocytes
- Creating pores in membranes via the membrane attack complex (often targeted by bound antibody)

Balancing the inherent amplification of the complement system, activated complement tends to be restricted to the surfaces on which the cascade was initiated. This is an important control of such a powerful and destructive process.

4.2 THE SEROLOGICAL IMMUNE RESPONSE

The **serological immune response** is mediated by antibodies (more properly, **immunoglobulins**), which are glycoproteins adapted to circulation in the blood, which have the ability to bind to specific molecular structures on their target antigen.

Antibodies have a Y-shaped structure, with conformationally variable regions at the tips, which bind to antigens (**Box 4.2**); the other, invariant end of the molecule signals to bring about immune effector functions. Antibodies are produced by activated B cells, especially by the fully differentiated form, plasma cells. These are large, highly specialized cells which are adapted for very high levels of protein synthesis.

Antibody structure

The basic structure of an antibody (**immunoglobulin**) is shown in **Figure 4.4**. There are four polypeptide chains, two light (L) and two heavy (H). Antibodies have a mostly generic (constant) structure, tipped by variable regions that enable the antibody to bind to a specific antigen. Within the variable regions, binding activity is concentrated in three hypervariable regions formed of paired regions in the light and heavy chains. These **complementarity-determining regions** (CDR1, CDR2, and CDR3) combine to form the unique antigen-binding site, with CDR3 being of particular importance. Although most antigens are proteins, antibodies can interact with a wide range of molecules, including lipids and sugars. Binding is noncovalent.

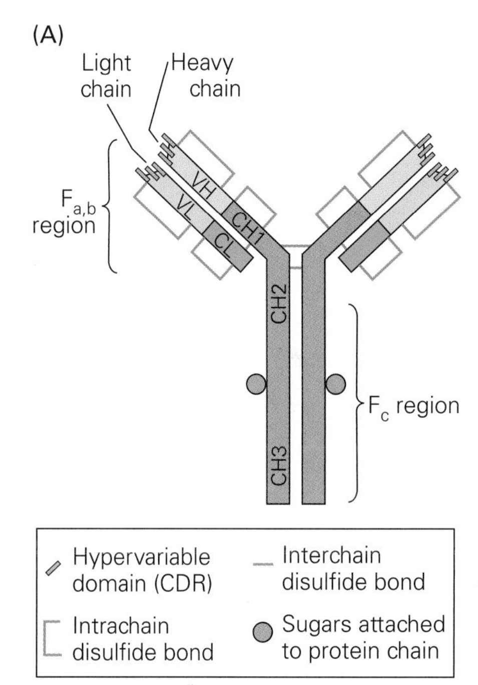

Figure 4.4
The structure of immunoglobulin G. (A) The Y-shaped molecule comprises two heavy (H, orange) and two light (L, green) chains. The 'stalk' has a constant structure (known as the F_c region), while the tips of the arms are variable ($F_{a,b}$ region). Within the variable regions are three hypervariable complementarity-determining regions (CDR1, -2, -3). (B) Ribbon diagram of IgG2 antibody structure, with heavy chains shown in blue and red, light chains in green and yellow.

Box 4.2 Antigens and immunogens

An antigen is defined as a molecule bound by an antibody. More broadly the term is also used for any molecular structure that is recognized by the immune system. While antigens are usually proteins, they may contain other structures such as sugars. An antigen that stimulates an adaptive immune response is referred to as an immunogen.

The five types of immunoglobulin

There are five basic types of immunoglobulin produced in humans (**Table 4.4**). Immunoglobulin G (IgG) dominates the antibody response in serum and undergoes adaptive evolution through the production of variant forms to produce antibodies capable of more effective (higher affinity) antigen binding. Immunoglobulin M (IgM) does not undergo evolution, but exists in the blood as a pentamer, with five joined IgM molecules and a small, joining (J) chain. Immunoglobulin A (IgA) is present in the blood, but is also secreted by vesicular transport through epithelial cells onto mucosal surfaces. This form of IgA is dimeric (again involving a joining chain) and forms an important element of immunity at mucosal surfaces (see Section 4.4).

Table 4.4 Functions and properties of the immunoglobulins

	IgG	IgM	IgA	IgD	IgE
Major functions	Humoral and extravascular immunity, anamnestic responses	Agglutination, defense against bacteremia, primary immune response, receptor on B-cell surface	Secretory form mediates immunity on mucosal surfaces	Receptor on B-cell surface	Anaphylaxis, allergy
Molecular weight (kD)	146–165	970 (pentamer)	160 (dimer 390)	184	188
Subclasses	4	1	2, plus secretory dimer	1	1
Carbohydrate content (number of carbohydrate molecules)	3	8	12	13	12
Chain types					
Heavy	γ_{1-4}	μ	α_{1-2}	δ	ϵ
Light	All immunoglobulin types have both κ and λ light chains				
Domains in heavy chain	4	5	4	4	5
Disulfide bonds between heavy chains	2 to 11	2	1	1	2
Mean serum concentration (mg ml^{-1})	13.5	1.5	3.5	0.03	0.00005
Half-life in serum (days)	7 to 21	10	6	3	2
Complement activation					
Classical	– to +++	+++	–	–	–
Alternative	–	–	– to +	–	–
Neutralization	++	+	++	–	–
Opsonization	± to +++	+	+	–	–
Placental transfer to fetus	+++	–	–	–	–
Antibody-dependent cellular cytotoxicity (ADCC)	– to ++	–	–	–	–

Note that all antibody types can exist in membrane-bound or secreted forms. The former contain a short transmembrane domain that is missing from the secreted form.

Immunoglobulin D (IgD) exists at low levels in serum, but its main function appears to be as a component of the receptor that activates the proliferation of B cells in response to antigenic stimulus. The final type, immunoglobulin E (IgE), exists at extremely low levels in serum, and appears to be mainly involved with allergic and anaphylactic reactions.

The B-cell receptor and the proliferative response

Naive B cells have a cell surface receptor (the B-cell receptor, BCR) that is formed of monomeric, membrane-bound IgD or IgM. When and if a B cell encounters its target antigen, binding to the BCR causes clustering of the bound receptor complexes, inducing phosphorylation and activation of cellular signaling cascades. Co-stimulation is often required to generate a strong proliferative signal, provided by helper T cells (see Section 4.3). As well as providing the initial proliferative stimulus, antigen bound to the BCR is also internalized and cleaved into peptides which are transported back to the cell surface in complexes with class II MHC molecules (**Figures 4.5 and 4.6**). These are recognized by the T-cell receptor (TCR) on helper T cells (see

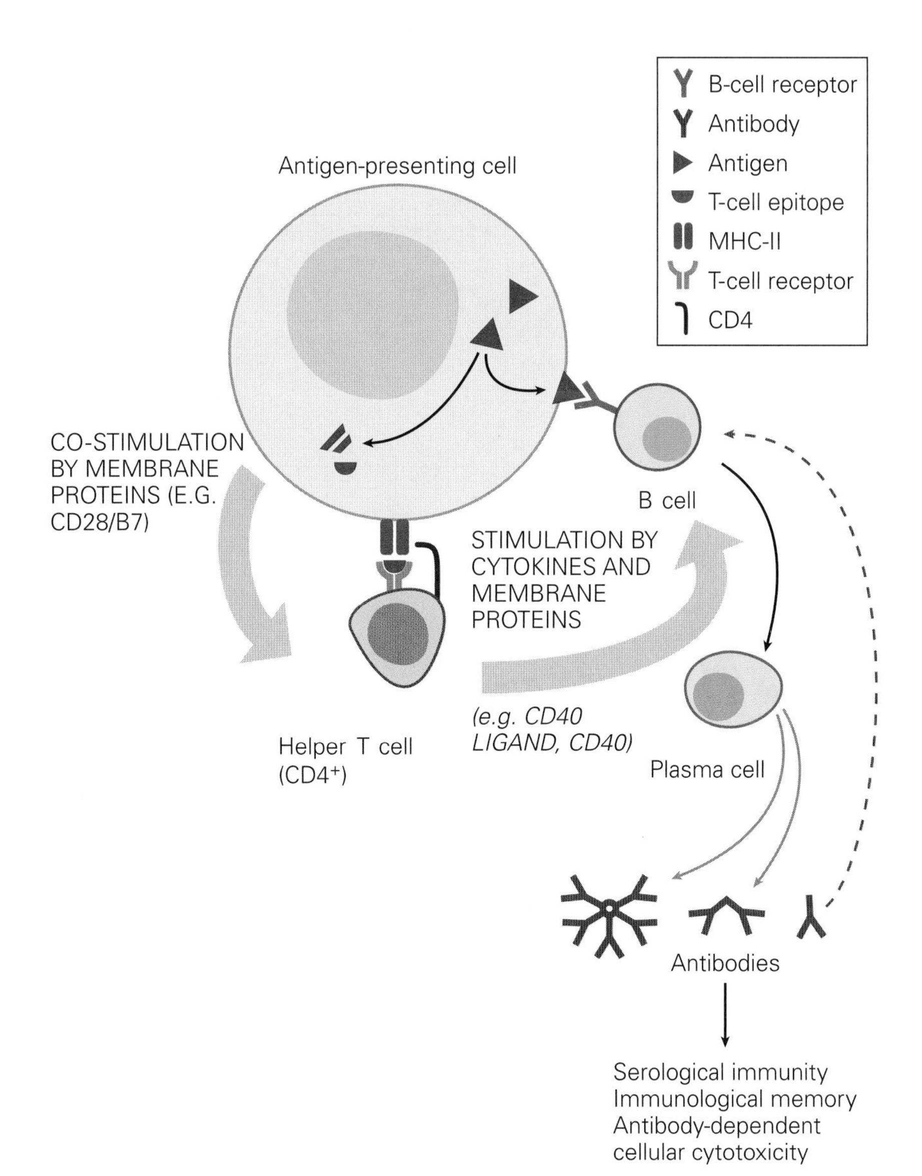

Figure 4.5
The basic features of the antibody response.

Section 4.3 for details of MHC-II–TCR signaling). Cytokines (notably interleukin 4, IL-4) produced by the T cell also provide a proliferative stimulus.

Despite this typical requirement for a complex system of co-stimulation to induce proliferation of the B cell, some antigens do exist which can produce such effects without T-cell assistance. These generally have highly repetitive structures, such as those found on bacterial cell walls.

Generation of antigenic diversity

The diversity of antigen binding arises from the process by which antibodies are formed. In the innate immune system a limited range of pre-programmed structures are recognized. However, with antibodies the range of possible targets is far, far greater. Rather than each being coded for by a specific gene (which would require rather too much DNA to be practical and would not allow the flexibility to cope with novel antigens), the variable regions are encoded by a wide range of small genetic elements. During B-cell maturation, multiple variable elements are spliced together. The types of genetic elements involved are shown in **Table 4.5**.

The net effect of this is a huge range of possible structures for the variable regions, which are produced essentially at random, so there is potentially an antibody for almost any antigen. Cells that produce antibodies

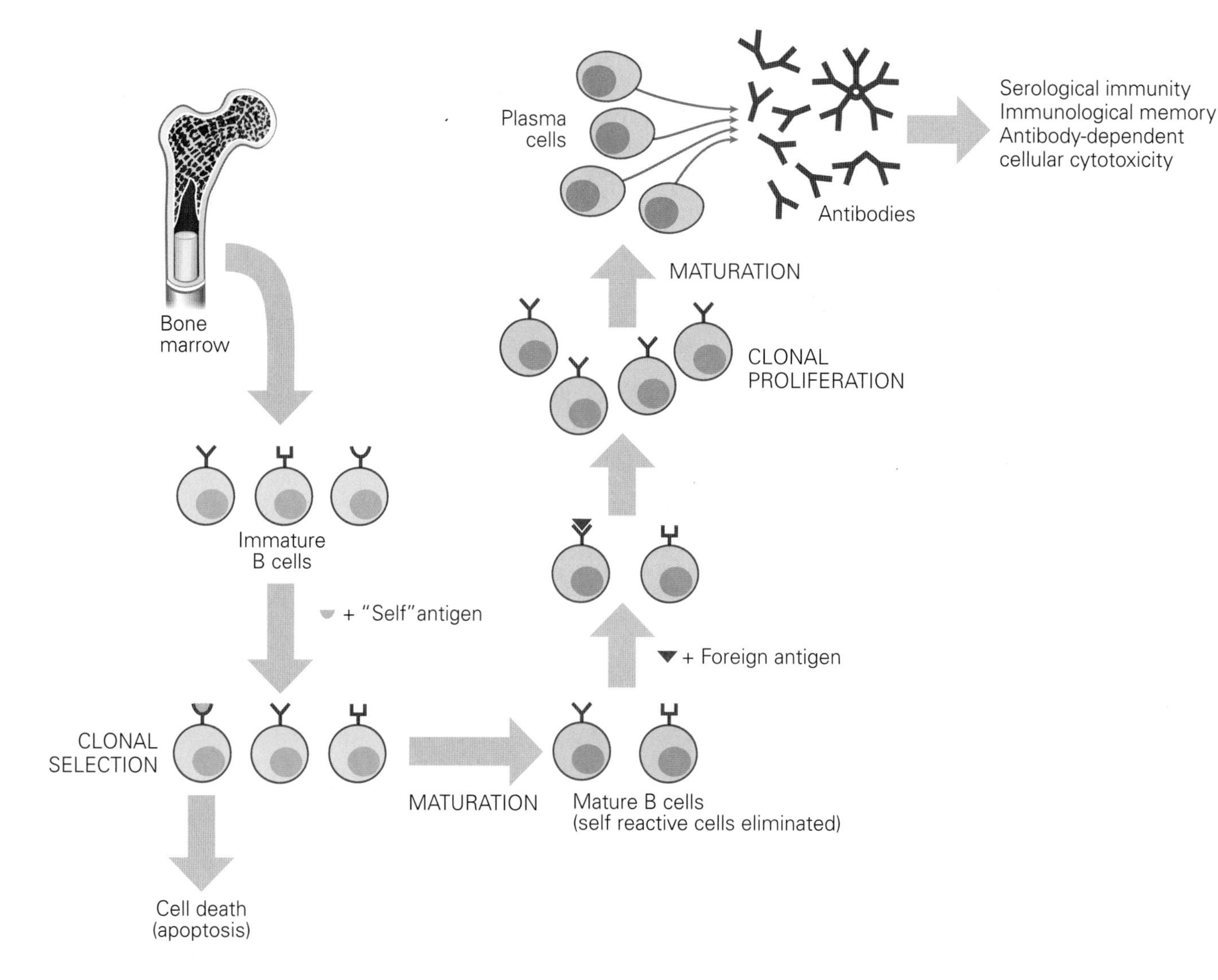

Figure 4.6

B-cell maturation. B cells are produced in the bone marrow and then migrate to the blood and the spleen, where they mature. Each B cell encodes a specific receptor for a particular antigen. Cells recognizing 'self' antigens undergo programmed cell death (apoptosis). When a receptor is engaged by its specific antigen, the B cell divides and forms a clone of daughter cells with identical receptors. Only B cells which bind antigen are stimulated to proliferate.

Table 4.5 Genes coding for immunoglobulin chains*

Chain type	Chain	Chromosomal location	Variable segments	Joining segments	Diversity segments	Constant segments
Light	κ	2	40 (may be duplicated)	5	—	1
Light	λ	22	30	4	—	4
Heavy	All**	14	40	6	25	2α 1δ 1ϵ 4γ 1μ

* The size of individual genetic segments varies widely, from 6 bases for some diversity segments to over 2,000,000 bases for some constant segments. ** Heavy chain constant regions (which determine the immunoglobulin type) are produced from complex sets of genetic segments located downstream from the variable regions. The type of constant region attached is varied by gene splicing during isotype switching.

recognizing **self** molecules of the host organism are destroyed during maturation, a process known as **clonal selection**, and the remainder are released to await possible stimulation (see Figure 4.6).

Even after a B cell encounters its triggering antigen and proliferates, further mutation occurs, resulting in the production of still more variation, and resulting in antibodies of increased affinity. Multiple changes are induced in the variable regions of the immunoglobulin genes by enzymically induced point mutations, a process known as somatic hypermutation. Most of these result in antibody that binds less effectively to antigen (of lowered **affinity**) and these cells are eliminated in favor of those few that bind more effectively (of increased affinity). This results in production of more effective antibodies in a process known as **affinity maturation**. The affinity of antibodies in the blood can be measured, and the presence of low-affinity antibodies shows that the immune response is in its early stages.

Alongside the maturation of affinity, the classes of immunoglobulin that are produced alter, by splicing of the genes coding for the variable regions to different constant regions lying nearby on the genes of the producing cell. IgM is the first type of antibody to be produced, but this then changes to IgG and IgA. Precisely which isotypes (from the four IgGs and the two IgAs) are produced is dependent upon the precise pattern of helper T cells and cytokines that are controlling the response. It should be noted that the binding specificity of the antibody to antigen is unaltered, but the consequences of that binding are changed by the different types of immunoglobulins produced, which has a profound effect on the nature and efficacy of the immune response. The presence of specific antibody is often used to determine whether a patient has been infected by a particular virus. Demonstration of specific IgG can show immunity resulting from a previous infection, while IgM shows that a recent infection has occurred.

The immunoglobulin response is extremely complex because of the enormous diversity within the system. Within each individual there may be as many as a billion different antibody molecules, each generated by a series of highly variable events, and each specific for a different antigen. The possible diversity generated by the processes of antibody formation is estimated to be a hundred times higher still. However, almost all of these antibodies will never encounter their cognate antigen, and the B cells carrying receptors specific for such antigens will never have the opportunity to proliferate.

B-cell memory

Following affinity and isotype maturation, the fully mature antibody-producing plasma cells move to the bone marrow, while other B cells become **memory cells**, able quickly to produce large amounts of high-affinity antibodies in response to another challenge with a familiar antigen. This production of a strong, rapid, and effective immune response by memory cells to the reappearance of specific antigens underlies the adaptive immune system.

However, when a novel antigen is encountered that is similar (but not identical) to one which has previously engendered an immune response, the response of the memory cells to the familiar antigen may swamp that to the new one with antibodies that are less than optimally effective. This process, referred to as "original antigenic sin," may limit the effectiveness of the response to new antigens. A similar effect appears to occur with T cells.

B-cell epitopes

B cells recognize antigens in their native form. For viruses, this is frequently the outer proteins of the viral particle, such as the glycoproteins of enveloped viruses (which are often also present on the surface of infected cells). Proteins that generate an immune response are referred to as 'immunogenic'. In general, an immunogenic protein contains several distinct sites that can be bound by the B-cell receptor (BCR). Each of these antigenic sites is called a determinant or an **epitope**.

B-cell epitopes may consist of a short stretch of amino acids (a *linear epitope*) or, more often, a cluster of amino acids from different regions of the antigenic protein, which are brought close together only when the molecule acquires its correct three-dimensional structure (a *discontinuous or disperse epitope*, also referred to as a *conformational epitope*, **Figure 4.7**). B-cell epitopes may also contain non-protein elements such as lipids or sugars.

An important concept in serological immunity to virus infection is the hierarchy of the response. Although a virus has multiple antigens, the immune response is not equally distributed among these. Many potential epitopes may be located in relatively restricted areas of specific proteins, typically those that are exposed on the surface of a cell or of a virus. These highly immunogenic regions may result in the production of very high levels of

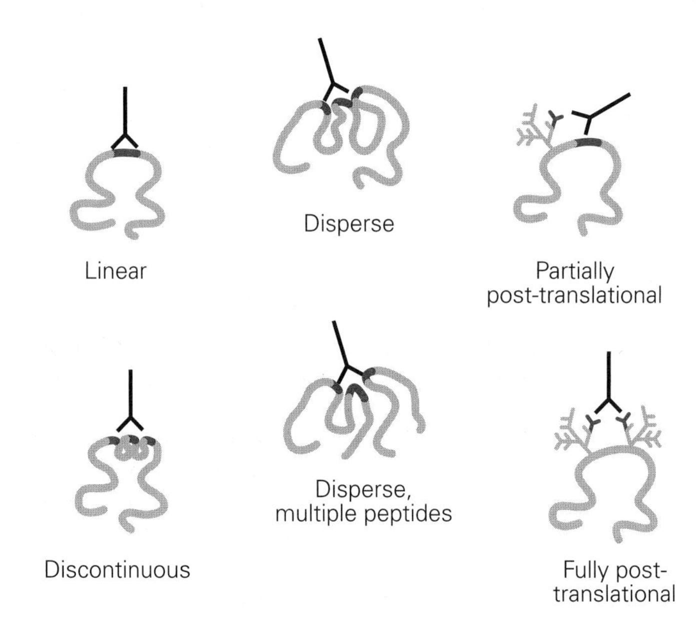

Figure 4.7
Types of B-cell epitope. Linear (epitope is a continuous region of amino acids in the protein); discontinuous (epitope is made of multiple closely located but separate regions of amino acids); disperse (epitope is made of multiple distantly located separate regions of amino acids, may be on different peptides); post-translational (part or all of epitope is formed of post-translational (non-peptide) elements, usually sugars attached to the peptide chain).

differing antibodies, and are referred to as the immunodominant regions of the protein.

At present, there are only limited ways to predict which will be the preferred antigens for any virus. Instead, the antibody responses to individual antigens are assessed and compared in sera taken from patients with typical viral infections. Over time, following viral infection, antibodies to some viral proteins will disappear while those to other antigens will remain detectable for many years. These issues must be considered when designing assays of serological immunity to a particular virus, or producing vaccines or therapeutic antibody preparations (see Chapters 5 and 10).

Denaturation of a protein with detergents or other reagents or in denaturing assays like Western blotting (protein electrophoresis in the presence of the detergent sodium dodecyl sulfate (SDS), followed by transfer to a membrane and reaction with antibodies) may destroy conformational epitopes, but linear epitopes are resistant to denaturation. It is often the case that biologically important viral epitopes are conformational, and this includes many of those involved in neutralization of enveloped viruses. This limits the use of denaturing assay systems in examining the serological immune response. Relatively non-denaturing (but often more complex) systems such as radioimmune precipitation can be used to examine conformational epitopes.

The effects of antibody binding

Specific recognition of a viral protein by the antibody, whether on the virus particle or the surface of an infected cell, can result in direct inhibition of viral infectivity (**neutralization**) if an essential viral function is prevented or physically blocked by the bound antibody. Protein–protein interactions may also mean that antibody binding has effects inside the virus.

However, specific neutralization by antibody is essentially a laboratory situation, since in the body the F_C (constant) region of a bound antibody acts as a flag for specific elements of the immune system. The effects of such binding include:

- Enhanced uptake by phagocytic cells which engulf and digest antibody-coated virus (**opsonization**)
- Antibody-dependent cellular cytotoxicity (ADCC), where NK cells are targeted by F_C receptors expressed on their membranes to kill cells to which antibody is attached
- Activation of the complement system (by the antibody-dependent 'classical pathway'), which has multiple functions, including forming the membrane attack complex which is able to create lethal pores in membranes where appropriate antibody is attached
- Mediating inflammation and phagocyte recruitment

The significance of the serological response

The serological immune response is very important in the control of some viral infections, notably those viruses that are present at high levels in a cell-free form (such as poliovirus). Cell-mediated immunity, which can target infected cells and the viruses inside them, is thus more significant for viruses that are usually associated with cells (such as members of the *Herpesviridae*). Poliovirus (*Picornaviridae*) stimulates a serological response consisting of IgG and IgM in the blood as well as a local (secretory IgA) response in the gut. The IgG response (which can neutralize the virus directly) then persists for the lifetime of the individual and can prevent further infection.

In addition to active serological immunity, where antibody is synthesized by the infected individual, it is also possible to give 'passive' serological immunity by the injection of antibody preparations (see Section 6.6).

4.3 THE CELL-MEDIATED ADAPTIVE IMMUNE RESPONSE

The innate immune system and antibody-mediated immunity rely on recognition of surface molecules on invading microorganisms. By contrast, T cells respond to peptides produced by cleavage of proteins within cells that are then presented on the surface of cells in association with self **major histocompatibility complex** (**MHC**) molecules.

The nature of the presenting molecule determines the effect of antigenic presentation to T cells. The two main pathways of antigen presentation, the MHC-I and the MHC-II pathways, involve different effector mechanisms and produce different responses. The MHC-I pathway predominantly activates *cytotoxic T lymphocytes* (CTLs), while the MHC-II pathway mainly activates *helper T cells* that facilitate the immune response.

Antigens produced in or introduced into the cytoplasm are processed there and presented in association with MHC-I molecules (**Figure 4.8**), which call in cell killing to target both the presenting cell and others presenting the same antigen. Cells using this pathway thus identify themselves as infected and in the process help to develop a specialized cytotoxic response

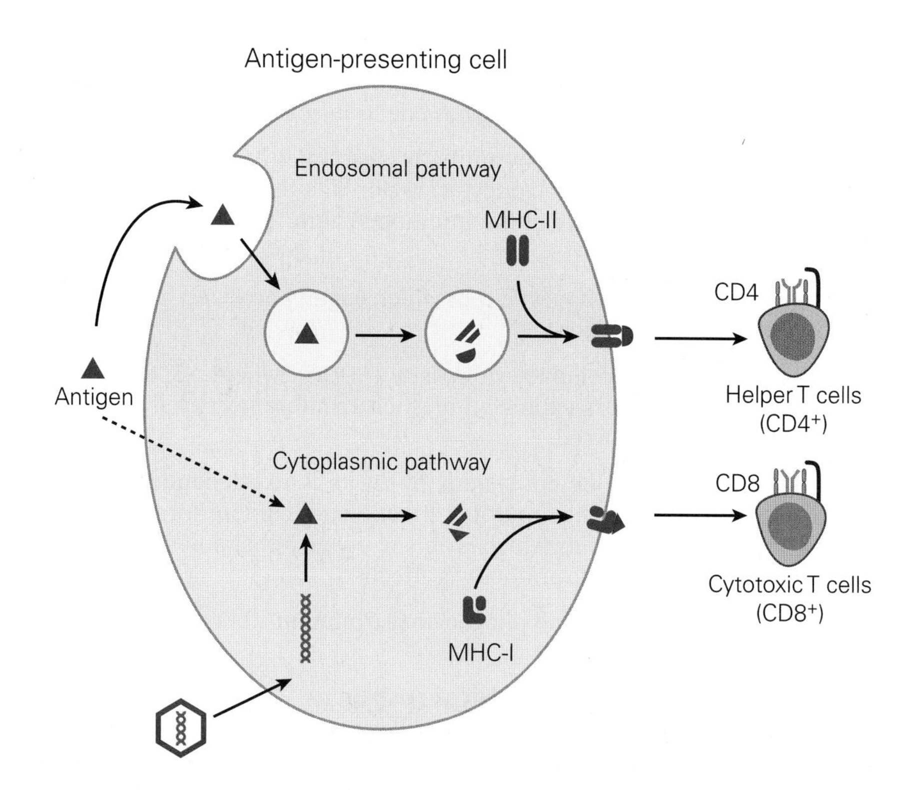

Figure 4.8
Processing and presentation of T-cell epitopes. Antigens arising in the cytoplasm are bound and presented by the MHC-I complex and activate cytotoxic T cells. Extracellular antigens processed via the endosomal pathway are bound and presented by the MHC-II complex and activate helper T cells.

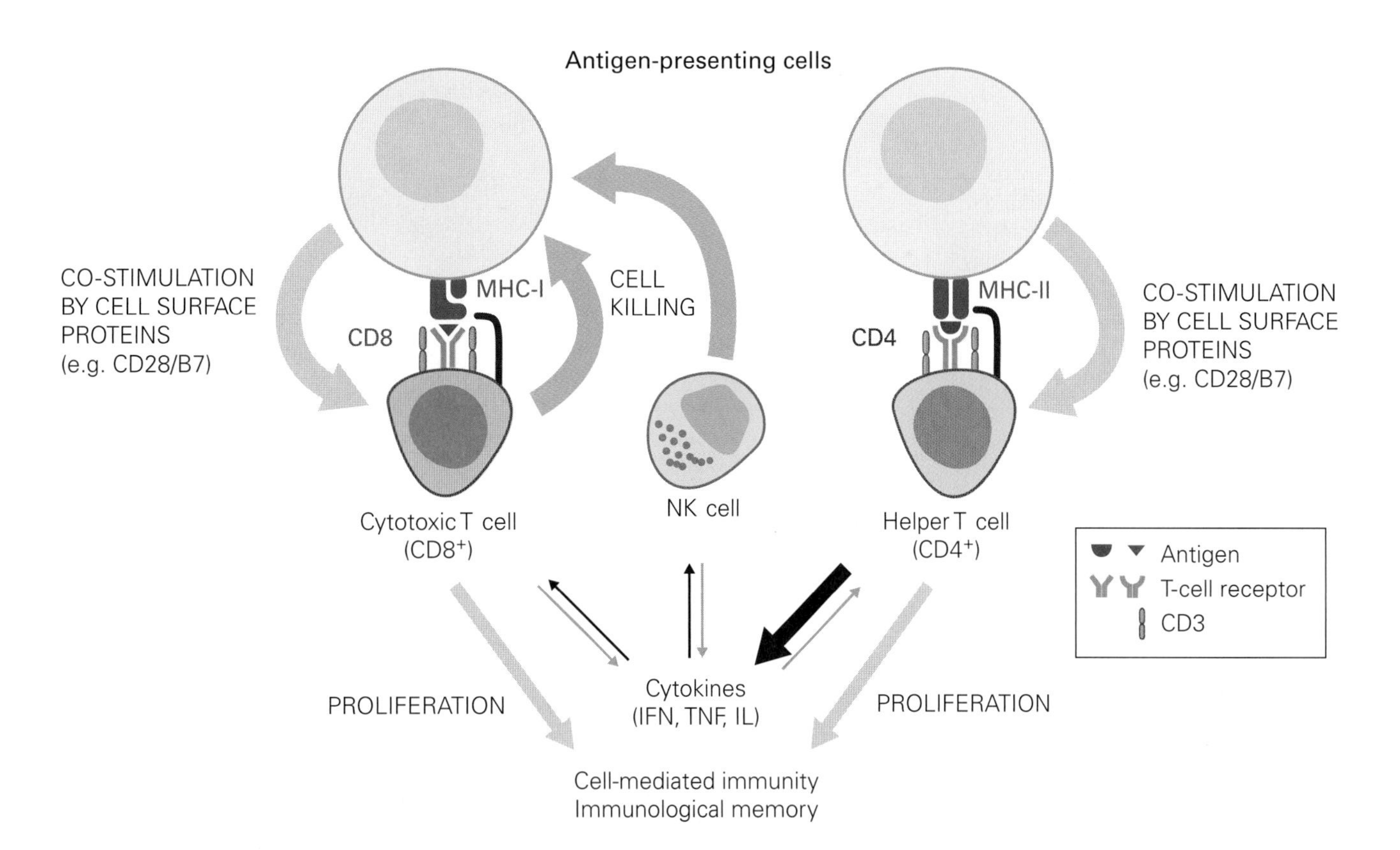

Figure 4.9
Basic features of the cell-mediated immune response.

targeting the presented antigen (**Figure 4.9**). This enables the immune system to target viruses and other pathogens even when they are inside the cells of the host.

In contrast, antigens taken up from the extracellular space by cells of the immune system are processed in degradative endosomes and deliver an immunostimulatory signal in association with MHC-II molecules. Cells using this pathway thus serve to identify potential pathogens and bring these to the attention of the immune system without targeting themselves for destruction.

The range of epitopes that can be presented for recognition by T cells is huge, and the necessary variability occurs in both the MHC proteins (see **Box 4.3**).

Unlike antibodies, where binding is specific to one molecular structure, mature MHC proteins can bind a range of different peptides that match the basic characteristics of the peptide-binding pocket of that particular protein. The binding sites for antibodies are usually on the surface of a

Box 4.3 The MHC proteins

The MHC proteins (Figures 4.10 and 4.11) are produced from a family of highly variable genes in the human leukocyte antigen (HLA) region on chromosome 6. This region contains more than 200 genes, among them a wide variety of cell surface proteins. The specific HLA proteins are unique to individuals, although many are shared among populations with similar genetic heritage. They act as a form of cellular fingerprint, and are central to the identification of the cells of the body (self) against those from other (non-self) sources.

In each individual, MHC-I and MHC-II molecules are each produced from a basic family of three genes. For MHC-I, the HLA-A, HLA-B, and HLA-C produce the single variable α chain, while the invariant β_2-microglobulin component of the MHC-I complex is produced from a gene on chromosome 15. For MHC-II, the HLA-DP, HLA-DQ, and HLA-DR genes produce the variable α and β chains. Additionally, the HLA-DR gene often contains an extra β chain, producing four types of MHC-II molecule. Variation is further increased by the presence of different genes on each parental chromosome in almost all cases.

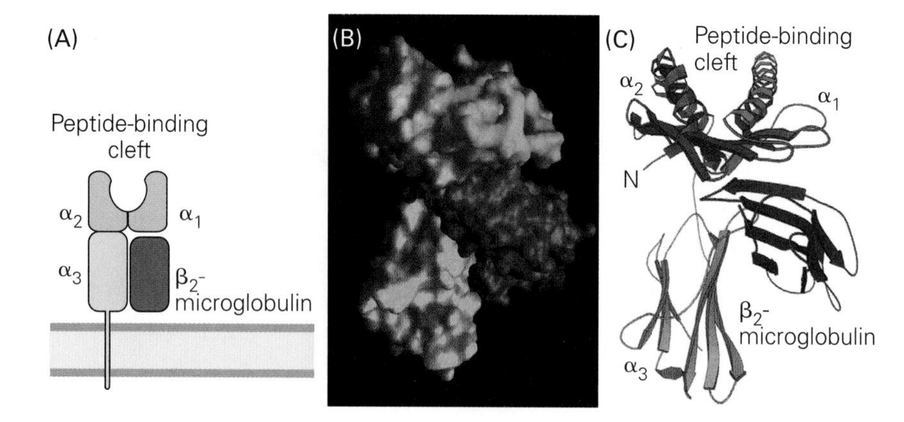

Figure 4.10
Structure of the MHC-I molecule.
(A) Schematic diagram (colors of domains correspond to B and C). (B) Three-dimensional structure in space-filling form (papain-cleaved form lacking membrane anchor). (C) Ribbon diagram corresponding to B. Adapted from, Murphy K, Travers P & Walport M (2008) Janeway's Immunobiology, 7th ed. Garland Science.

protein, virus, or cell, but T-cell reactive peptides may come from almost anywhere within a protein. Multiple T-cell epitopes may be present on one protein, and are distinct from the B-cell epitopes which induce the antibody response.

The structure of the MHC-I molecule is shown in **Figure 4.10**. The deep binding groove can accommodate a peptide of 8–10 amino acids in length that fits certain structural requirements. These are produced by digestion of antigenic proteins in the proteasome, a proteinase complex in the cytoplasm of the cell, and then pass through the endoplasmic reticulum (ER) where they encounter MHC-I complexes. Until recently, it was thought that antigens had to be synthesized within the cell in order to be entered into this pathway, but it is now clear that exogenous proteins which are entered into the cytoplasm by osmotic shock, fusogenic liposomes, or some vaccine adjuvants (see Chapter 5) can also be presented via the MHC-I pathway. The antigen–MHC-I complexes are presented on the cell surface.

The size of peptide presented by MHC-II molecules is both larger and less rigid than with MHC-I molecules since (unlike MHC-I) the MHC-II-binding groove is open-ended (**Figure 4.11**). The peptide is usually at least 13 amino acids in length, but can be far longer. However, it will usually be trimmed to a maximum of 17 amino acids after binding to MHC-II. Unlike the MHC-I pathway, processed antigens presented by the MHC-II pathway do not need to be synthesized within the presenting cell or entered into the cytoplasm by specialized systems. Rather, these are usually proteins taken up by specialized cells and processed within degradative endosomes. MHC-II molecules are targeted to these endosomes by a specialized MHC-II-like protein, the invariant chain (Ii), which also blocks the peptide-binding groove until this is loaded in the endosome.

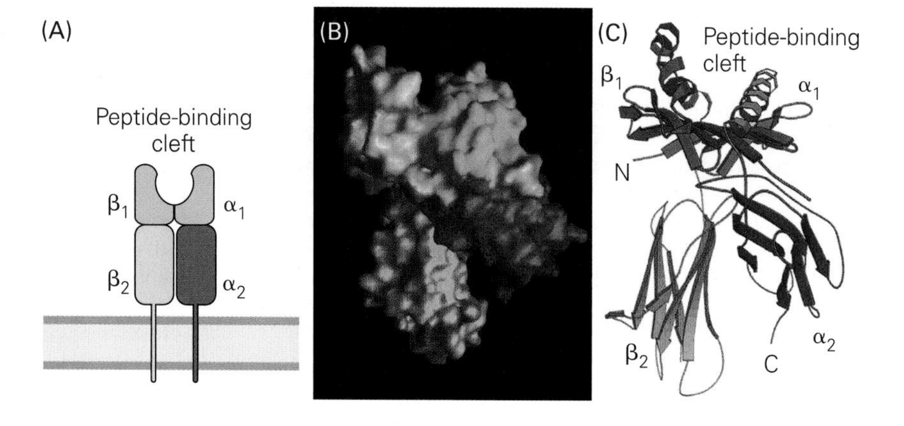

Figure 4.11
Structure of the MHC-II molecule.
(A) Schematic diagram (colors of domains correspond to B and C). (B) Three-dimensional structure in space-filling form. (C) Ribbon diagram corresponding to B. Adapted from, Murphy K, Travers P & Walport M (2008) Janeway's Immunobiology, 7th ed. Garland Science.

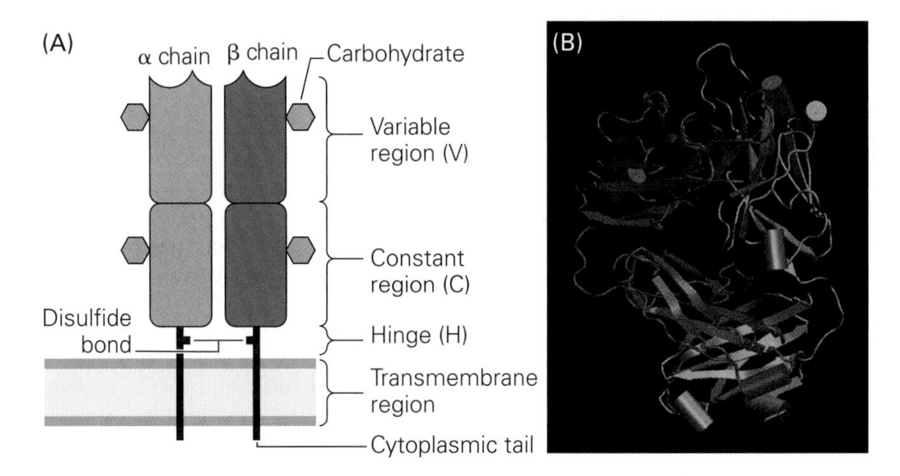

Figure 4.12
Structure of the T-cell receptor. (A) Schematic diagram (colors of domains correspond to B). (B) Three-dimensional structure in ribbon view. Adapted from, Murphy K, Travers P & Walport M (2008) Janeway's Immunobiology, 7th ed. Garland Science.

T-cell activation

T-cell recognition of the peptide–MHC complex on antigen-presenting cells is mediated by the T-cell receptor (TCR; **Figure 4.12**). TCRs are similar in structure to the antibody $F_{a,b}$ region (see Section 4.2) and, like antibodies, have binding regions that are highly variable in nature. This variability results from the combined effects of multiple genetic rearrangements and transcriptional mechanisms during production of the TCR molecule. As with antibodies, there are three complementarity-determining regions (CDRs), although in the T-cell receptor only one of these (CDR3) plays a major role in antigen binding.

TCRs bind to MHC–peptide complexes, causing clustering of the bound TCRs and activating cellular signaling systems. However, this binding alone provides only a weak stimulus to the T cell. The activation signal is greatly amplified by binding of the TCR–MHC complex to other specific receptors on the T-cell surface. For MHC-I this is the **CD8** molecule, while for MHC-II the **CD4** molecule serves this purpose. One other important co-stimulatory element is the binding of CD28 on the naive (unstimulated) T cells to the B7 proteins, which are present on the surface of antigen-presenting cells. This is required for the T cell to begin proliferating. In an elegant demonstration of the feedback controls of the immune system, a close relative of CD28, CTLA-4, is induced by this process and also interacts with B7, but much more strongly than CD28. CTLA-4/B7 binding then shuts down the activation signal and actually limits further proliferation, preventing uncontrolled T-cell proliferation.

The TCR–MHC complex does not transfer signals into the T cell directly, but instead associates with the **CD3 complex**. Formed of invariant transmembrane proteins, the CD3 complex activates cellular kinase enzymes which phosphorylate (and thus activate) a complex cascade of cellular signaling molecules, transmitting an activation signal to the T cell.

Certain proteins have the ability to stimulate the TCR directly, without the need for presentation by MHC molecules. By interacting with existing MHC-II–TCR complexes, superantigens can induce very high levels of T-cell response and consequently high levels of cytokines. The resultant immune response can be highly damaging. Superantigens are typically bacterial toxins, and while they are thought to exist in some viruses (such as the *Rhabdovirus* rabies and the *Herpesvirus* EBV), their role and nature are far less clear than for bacterial superantigens.

The MHC-I pathway: cytotoxic T cells

MHC-I proteins are present on almost all cells and antigen presentation by the MHC-I pathway is, in most cases, limited to proteins actually synthesized within the presenting cell, and so the MHC-I pathway is a method for cells to call in a T-cell response if they become infected.

Presentation of antigen with MHC-I molecules will only activate those T cells with TCRs matching the expressed MHC-I molecule (*MHC restriction*). Successful interaction activates T cells with the CD8 surface marker protein ($CD8^+$ T cells), which are mostly **cytotoxic T lymphocytes (CTLs)**. These then kill the presenting cell by mechanisms that appear to be similar to those used by NK cells, involving cytokine synthesis, perforin release, and the induction of apoptosis by granzyme proteinases (see Section 4.1). CTLs, by killing virus-infected cells, prevent the spread of virus and protect the human host from further harm—most of the time.

However, there are situations where T cells can cause direct harm, such as in intracerebral infection of mice with lymphocytic choriomeningitis virus (LCMV; *Arenaviridae*). In the absence of a competent T-cell response, a persistent, non-lethal LCMV infection is established. In the presence of competent T cells, intracerebral LCMV infection is highly destructive, and kills the mouse. In human disease, T cell-mediated cytotoxicity is believed to be significant in viral hepatitis. The cytotoxic T-cell response is very powerful, and plays an important role in controlling virus infection. Despite this, it can be very damaging and of necessity is tightly controlled, for example by the need for co-stimulation, the absence of which can lead to tolerance of the expressed antigen, and by the presence of feedback systems which moderate the activity of the response.

The MHC-II pathway: helper T cells

Peptide–MHC-II complexes are recognized by T cells with the CD4 surface marker protein ($CD4^+$ T cells) via their TCR–CD3 complex. The MHC-II proteins are generally present only on a limited range of antigen-presenting cells associated with the immune system, although some other cell types are able to express them in specific circumstances, including keratinocytes in the skin. Cells of the immune system presenting antigens by the MHC-II pathway pick up foreign proteins for presentation to other elements of the immune system, rather than identifying themselves as infected, so it would be inappropriate for them to be killed. Instead, the main type of T cell activated by this pathway is **helper T cells**. The importance of helper T cells to the immune response is shown by the profound immunosuppression (and consequent vulnerability to infections) that results from the selective destruction of these cells by the human immunodeficiency virus (see Section 4.6). In response to antigenic stimulation, these cells proliferate and secrete cytokines, which activate both the antigen-presenting cell and other cells of the immune system. Helper T cells and the cytokines that they produce are vital to the activation of many components of the immune system, including NK cells, CTLs, and B cells. Interferon γ produced by helper T cells also up-regulates MHC-II expression on cells, including some types of cell that do not generally express these proteins. However, not all cytokines are stimulatory. The cytokine tumor necrosis factor β (TNF-β, also known as lymphotoxin), which has a role in the clearance of bacterially infected cells, is inhibitory for B cells and actually kills activated T cells.

While it is possible to list the major activities of the cytokines produced by these cells, it is important to note that each has multiple functions, and that the interactions of cytokines within the immune system are a hugely complex area of study far beyond the range of this chapter.

Th1 and Th2

Helper T cells exist in two broad classes. Inflammatory T cells (**Th1**) are associated with the cell-killing and inflammatory responses of the immune system, including an important role in the activation of macrophages. They also activate macrophages to destroy the pathogens they have internalized, and stimulate specific subclasses of antibodies that enhance macrophage uptake (referred to as opsonization). **Th2** cells activate B cells and the general serological (antibody) response although, as noted above, Th1 cells play a role in controlling the production of specific types of antibody.

Activation of a Th1 or Th2 often leads to that element of the response becoming dominant, so that either cellular/inflammatory or serological responses predominate. The cytokines produced by Th1 or Th2 responses differ, and are often used to determine the nature of the response. For example, the presence of IL-12, IL-27, TNF-α, TNF-β, and interferon-γ indicates a Th1 response, while IL-4, IL-5, IL-6, IL-9, IL-10, and IL-13 indicate a Th2 response. These cytokines are not simply markers; they both stimulate the relevant arm and inhibit the other arm of the Th response.

Another relevant factor is the nature of the peptide stimulating the response. Weakly binding antigenic peptides can stimulate a Th2 response, while high levels of strongly binding peptides can favor a Th1 response.

Cytokines produced by T cells are also involved in delayed-type hypersensitivity. These cytokines increase the permeability of capillaries at the site of infection, helping other elements of the cellular immune response to get to where they are needed, along with serum proteins such as fibrin. These effects produce a characteristic local **induration** (stiffening) of the tissue at the site of infection, initially observed in the local reaction to tuberculosis (a bacterial disease), which is referred to as delayed-type hypersensitivity. This localized, typically cutaneous inflammatory response is also thought to be important in mediating local reactions to viral infections, such as the formation of vesicular lesions.

T cells: suppression and memory

T cells are produced in the bone marrow but mature in the thymus, located in the thoracic cavity above the heart. During this period, maturing T cells are extensively tested to determine their suitability. Those that do not bind to the body's own, self MHC molecules are eliminated—since MHC binding is essential for T-cell function. Also eliminated are those that bind to inappropriate self antigens. It is estimated that 98% of T cells are eliminated at this stage (**Figure 4.13**).

Levels of specific T cells rise many times during an active infection. However, once the infection is resolved, such high levels are no longer required, and many of these are removed by programmed cell death (apoptosis, see Section 4.5). Activated T cells express a protein known as Fas ligand at the cell surface. This binds to another T-cell protein known (perhaps unsurprisingly) as Fas. Binding of Fas ligand to Fas activates an apoptotic pathway. While the population of T cells is expanding in the early stage of the immune response, the T cells are resistant to the effects of this binding. However, the longer that they remain activated, the more sensitive they become to the effects of Fas ligand binding, and the more effective such binding becomes

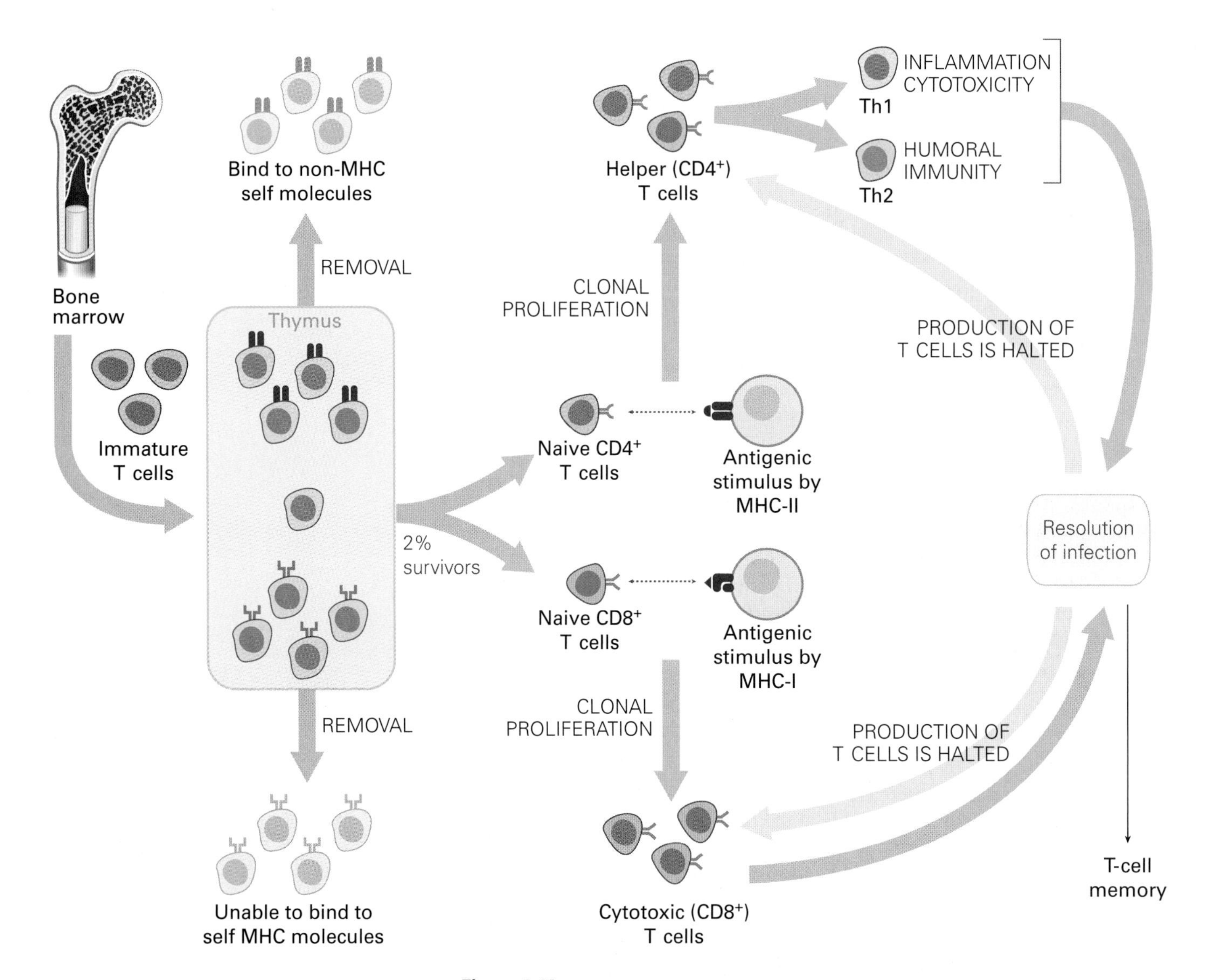

Figure 4.13
Pathways of the T-cell response. T cells are produced in the bone marrow but mature in the thymus. T cells that do not bind to the body's own, "self" MHC molecules are eliminated as are those that bind to inappropriate self antigens. Naive T cells stimulated by binding to antigen-presenting cells proliferate and form helper T cells and cytotoxic T cells. Once an infection is resolved, production of further T cells is stopped, except for a small number of memory T cells that can be reactivated if the same antigen is encountered in future.

in driving them down the apoptotic (cell death) pathway. This is yet another illustration of the feedback systems that regulate the immune response.

Even after the culling of activated T cells, some do persist at levels 10–1000 times higher than the level prior to the initial infection. This subset contains **memory T cells** that express some, but not all, of the markers associated with active T cells, including the Bcl-2 protein that is associated with extended cell survival.

Subsets of memory T cells have different reactivation characteristics. Effector memory cells mature into activated T cells that produce cytokines and effector cells that move into infected tissues, while central memory T cells respond more slowly and are specialized to move to lymphatic tissues.

The presence of memory T cells allows a rapid and directed response to later challenge with the same antigen. However, in at least some cases, continued antigenic re-stimulation by low-level infections or persisting antigen may also play a role.

4.4 COMPARTMENTALIZATION OF THE IMMUNE RESPONSE

The immune response is not identical throughout the body. Distinct compartmentalization exists, with the two main compartments being systemic and mucosal immunity. Systemic immunity encompasses the full range of the immune response, and is directed against infections that have spread through the organs of the body or into the blood. These sites are normally sterile and the full power of the immune system is used against such agents. However, there are sites within the body where specialized responses occur, and to which the cells of the immune system are targeted by tissue-specific signaling mechanisms. These include membranes within the body cavities (the peritoneum surrounding the intestines and the pleura around the lungs), the skin, and the mucosal surfaces. The mucosal surfaces (mucosae) are of particular significance, and the immune response at this location is referred to as mucosal immunity.

Mucosal immunity

Mucosal immunity is the variant and more limited immune response that is present on many of the surfaces of the body, including the respiratory, alimentary, and genital tracts. As such, it is often in the first line of defense against infection, but it must be extremely selective since it is exposed to a very high level of foreign material compared to that seen deeper within the body. The mechanisms of immune **tolerance**, whereby the response to an antigen is damped down rather than amplified, play a key role in this. Tolerance is due to a number of factors including **anergy**. This is when T cells show a limited response to subsequent antigenic stimulation due to the lack of necessary co-stimulatory signals at their first encounter with an antigen. Tolerance is important in such settings in order to avoid inappropriate immune responses to harmless proteins such as those in food.

In some locations, mucus is secreted and this can in itself be a defense by entrapping small particles before they get into the body. Mucus containing such material is then cleared by methods such as ciliary action, coughing, or sneezing. The effectiveness of this barrier is illustrated by the need for influenza virus (*Orthomyxoviridae*) to have present on the virus a neuraminidase (NA) enzyme that can cut it free of mucus that it has adhered to.

Mucosal immunity also involves elements of the adaptive immune response. Since the mucosal surfaces are often the first part of the body to encounter new pathogens they are extremely important in developing the immune response to such agents. Specialized regions of lymphatic tissue such as the Peyer's patches underlying the small intestine, the appendix, the tonsils, and the adenoids work with other lymphoid tissues such as the mesenteric lymph nodes to generate a strong and flexible cellular response involving such elements as B cells, T cells (including high levels of innate-type γ:δ T cells), and neutrophils. Cells are directed to the correct locations by tissue-specific signaling systems that include both cellular adhesion molecules and chemokines. Alongside this cellular response, antibodies are also present. The dimeric form of IgA is secreted directly onto mucosal surfaces and is a major component of mucosal immunity, although other types (in particular pentameric IgM) are also present.

Stimulating strong mucosal immunity has long been a target in vaccinology, since the possibility exists that infection could be entirely prevented by such means. This is important for viruses such as HIV (*Retroviridae*) and herpes simplex virus (HSV; *Herpesviridae*) since, once they have entered the body, they are able to establish latent infections within specific tissues from which clearance is effectively impossible with current technologies. However, generating mucosal immunity that is able to block such infections completely (**sterilizing immunity**) remains extremely challenging.

4.5 APOPTOSIS

Apoptosis is a programmed pathway of cell death in which the cell dies from within and is removed, typically by phagocytes. This provides a safe way of removing unwanted cells without the toxic effects of necrosis (the uncontrolled death and decay of cells; **Box 4.4**). Apoptosis is used to destroy cells once they are no longer needed or when they are considered dangerous. Almost all cells are programmed to undergo apoptosis following the reception of appropriate signals and, in many cases, external stimuli (including cytokines and a complex range of signaling mechanisms) are required to prevent apoptosis from occurring. Apoptosis can be triggered by a number of events, including many associated with virus infection, such as disruption of the cell cycle. It is also possible for apoptosis to be induced from outside the cell, and this is a major method for controlling pre-malignant cells within the body. Apoptosis provides an important mechanism by which cytotoxic T cells and NK cells can kill their targets (see Sections 4.1 and 4.3), and by which redundant activated T cells are removed once they are no longer needed.

Box 4.4 Apoptosis versus necrosis

Apoptosis involves a cell degrading itself from within, digesting its DNA and signaling phagocytes to clear up the debris. Necrosis does not. Necrosis results when a cell dies unexpectedly, for example from oxygen starvation or the effects of toxins. While the intracellular events during necrosis superficially resemble those of apoptosis, there is not the same pre-programmed clearing-up procedure. As a result, inflammation and enhanced toxicity can result, with the potential for destroying large areas of tissue.

All known apoptotic pathways appear to use a common effector system, working via a cascade of **caspases** (aspartic acid-specific cysteine proteinases). One well-characterized activation system is that of the Fas protein on T cells, which is activated by binding to specific Fas ligand described in Section 4.3. Binding to Fas ligand induces Fas to form trimers. Clustering of the rather dramatically named "death domains" on Fas and other associated proteins recruits and activates the caspase system which, in turn, triggers DNA digestion.

The effects of apoptosis involve shrinking of the cell, clumping and breaking of the nuclear DNA (often into a characteristic 'ladder' of fragments of differing sizes), break up of the nucleus, disruption of the cell membrane and, finally, breakdown of the cell into "apoptotic bodies," which are cleared by phagocytes responding to cellular signals expressed in the later stages of apoptosis. Several cellular oncogenes (see Section 4.8) are involved in apoptosis, notably the p53 tumor suppressor gene product, which suppresses cell proliferation and can induce apoptosis. There are also proteins that have an anti-apoptotic effect. For example, the proteins of the bcl-2 family appear to be involved in preventing apoptosis. This effect may be important in establishing long-term immunological memory, but is also seen in B-cell lymphomas. External stimuli such as co-stimulatory proteins may also be involved in preventing apoptosis.

Apoptosis is an important mechanism in the control of cell proliferation and the prevention of uncontrolled cell growth. However, it has become clear that it is also the major means by which 'redundant' cells are removed from the body, including many cells of the immune system. Apoptosis is also central to normal cellular development and to the fine tuning of cellular differentiation.

4.6 EVASION OF IMMUNE SURVEILLANCE BY VIRUSES

The ability to evade the immune response can enhance the ability of that virus to replicate in the host and, as a result, select for viruses with this ability. This is necessarily a highly complex field, and space does not allow more than a brief review of some of the better-studied systems, which are summarized in **Table 4.6**.

Active immune evasion

Many viruses interfere actively with the immune system, preventing the proper functioning of some aspect of the immune response. Some mechanisms are relatively common, presumably reflecting the importance of the mechanisms affected in controlling virus infection. Other mechanisms have only been reported for one type of virus.

One key mechanism for any but a directly cytopathic virus is likely to be inhibition of cellular apoptosis (see Section 4.5), since this form of induced cellular suicide is a very common consequence of virus infection. Such mechanisms have been identified for members of the *Adenoviridae*, *Herpesviridae*, and *Poxviridae*, and are likely to be considerably more widespread. Another mechanism is interference with the interferon-induced protein kinase that inhibits protein synthesis in infected cells, which is seen with viruses from at least seven families.

Table 4.6 Examples of immune evasion mechanisms by viruses

Active	Infection of immune cells	*Herpesviridae, Paramyxoviridae, Picornaviridae, Retroviridae*
	Interference with complement function	*Herpesviridae, Poxviridae*
	Interference with MHC-I presentation	*Adenoviridae, Herpesviridae, Paramyxoviridae, Poxviridae*
	Interference with MHC-II presentation	*Herpesviridae*
	Inhibition of NK cells (via MHC-I homolog)	*Herpesviridae*
	Interference with Toll-like receptor function	*Poxviridae*
	Interference with cytokine production or function	*Adenoviridae, Herpesviridae, Hepadnaviridae, Papillomaviridae, Poxviridae*
	Interference with interferon function	*Adenoviridae, Filoviridae, Hepadnaviridae, Herpesviridae, Orthomyxoviridae, Papillomaviridae, Picornaviridae, Polyomaviridae, Poxviridae, Reoviridae, Retroviridae*
	Interference with apoptosis	*Adenoviridae, Herpesviridae, Poxviridae*
	Interference with inflammatory response	*Poxviridae*
Passive	Antigenic drift	RNA viruses, particularly: *Orthomyxoviridae, Retroviridae*
	Antigenic shift (segmented genomes only)	*Orthomyxoviridae*
	Molecular mimicry	*Herpesviridae, Paramyxoviridae*
	Masking of virus	*Herpesviridae, Retroviridae*
	Latency	*Herpesviridae, Parvoviridae, Retroviridae*

The above list is incomplete, and represents only those virus types studied. In particular, additional mechanisms are known in viruses infecting animals.

One common mechanism is interference with MHC-I presentation, limiting the cytotoxic T cell response, by a variety of methods. However, a lack of MHC-I will in itself stimulate NK cell cytotoxicity, but at least one virus (cytomegalovirus herpesvirus; CMV) produces an MHC-I analog to prevent this. Interference with cytokines or complement also appears common.

Other mechanisms are less common. The production of an enzyme that reduces inflammation by inducing the production of immunosuppressive steroid hormones appears to be restricted to the *Poxviridae*, while EBV herpesvirus unusually interferes with MHC-II presentation, apparently reflecting its infection of the cells of the immune system.

Specific examples of direct interference with the immune system are too numerous to describe in detail, but the herpesviruses deserve mention, and indeed have been described by Banks and Rouse as "immune escape artists." Strategies used by different members of the family include down-regulation of MHC-I (cytomegalovirus), interference with peptide transport for MHC-I presentation by herpes simplex virus type 1 (HSV-1), production of a cytokine synthesis inhibitor analogous to IL-10 together with interference with MHC-II presentation by Epstein-Barr virus (EBV/HHV-4), and interference with apoptosis by Kaposi's sarcoma virus (KSV/HHV-8).

Cytomegalovirus appears to have taken this process to extremes. It inhibits MHC-I expression, function, and transport; interferes with the preparation of peptides for loading onto MHC-I molecules; produces inhibitory cytokine analogs; mimics cellular proteins; and, perhaps most remarkably of all, wraps itself in a cellular protein that not only helps to hide it from the immune system, but also inhibits MHC-I presentation since the protein it sequesters for this purpose is β_2-microglobulin—an essential part of the MHC-I molecule.

HIV and the immune system

Direct interference with the immune system is taken to extremes by HIV (*Retroviridae*). After infecting the host, this virus targets itself to the tissues where lymphocytes are actually produced and then uses the CD4 molecule as its initial receptor. This means that as part of its replication HIV infects and kills $CD4^+$ T cells, thus disabling the helper T cells which play a central role in supporting the activities of the immune system. This removal of the helper T cell function along with other effects progressively destroys the functionality of the immune system, eventually producing the massive immunosuppression that underlies the acquired immune deficiency syndrome (AIDS). The destruction of $CD4^+$ T cells involves both virus killing of cells and other effects, including apoptosis. It appears that the HIV proteinase may be directly involved in cleaving the Bcl-2 (anti-apoptosis) protein, in direct contrast to the anti-apoptosis mechanisms observed for other viruses. HIV also has other powerful effects on the immune response. For example, the two viral proteins (nef and tat) decrease MHC-I expression, weakening the ability of cells to present antigens to cytotoxic ($CD8^+$) T cells.

Passive immune evasion

Although many viruses have evolved 'active' methods of interfering directly in the immune response, 'passive' alternatives exist.

In the case of influenza virus, the surface glycoproteins undergo rapid mutation and the resulting changes in their epitopes mean that the targets which are recognized by the immune system as influenza virus proteins may not be present in a mutated virus. This is *antigenic drift*, and is common to many RNA viruses due to the high mutation rate of RNA genomes (see Chapter 1). Viruses with segmented genomes, including influenza, can also acquire whole new genes when multiple viruses infect the same cell, producing the rapid and dramatic changes known as *antigenic shift*. Antigenic shift permits the immune evasion which allows the characteristic worldwide pandemics of influenza. Antigenic drift and shift are discussed in more detail in Chapter 1.

Another interesting approach is molecular mimicry, where viral antigens mimic epitopes present on host proteins. Examples include proteins of measles virus (*Paramyxoviridae*) and of cytomegalovirus (CMV; *Herpesviridae*). Clearly, the production of antibodies reacting with self proteins can cause serious problems for the host organism (including a range of **autoimmune diseases**), and many systems exist to avoid this (**Box 4.5**). Thus, the virus recruits the aid of host systems to prevent the production of specific immunity.

Box 4.5 Autoimmunity

When the immune response is directed against the cells of the body rather than invading pathogens, this is referred to as autoimmunity. Such a response can be hugely damaging, and prevention of autoimmunity is one reason for the complex system of feedback controls inherent in the immune response. While many viruses (particularly herpesviruses) have been proposed as causes of autoimmunity, often due to mimicry of cellular antigens by the virus, precise mechanisms remain to be proven.

Viral latency represents another way for viruses to 'hide' from the immune system (see Chapter 1). While the virus itself cannot replicate during latency, the ability to become latent allows the virus to remain present until the host's immune responses are depressed and then to reactivate, allowing a productive infection. This approach is used by herpesviruses, and the resulting reactivation disease can be both severe and highly infectious, allowing the virus to spread to new hosts.

With picornaviruses, the structure of virus proteins hides essential cell-binding regions of the viral surface protein in a "canyon" (actually, a small cleft) on the virus surface that is small enough to restrict the ability of antibodies to reach the essential amino acids that it contains.

A combined effect

Some forms of immune evasion contain elements of both active and passive approaches. The production of F_C receptors by herpesviruses both conceals the virus by coating it in immunoglobulin, and may interfere with immunoglobulin function as a result. In another example, HIV adsorbs regulatory proteins from the complement system that then interfere with complement function.

Many of the above systems are very limited in the advantage that they confer on the virus. However, it should be remembered that several systems may combine to produce a greater effect, and that in the infected host, even a slight advantage can be enough to allow a productive virus infection rather than one controlled by the immune response.

4.7 HOST GENETIC FACTORS

It is clear that the immune response mounted by any individual is based on complex variables, and is thus unique. It is also known that variation in the genetic elements of the immune system can result in different response to infection. It is impossible in a book of this length to cover all aspects of this, but one example may help to explain the situation.

Hantavirus pulmonary syndrome (HPS) is a disease found in various forms in North and South America. It is caused by a variant of hantaviruses (*Bunyaviridae*) that spread to humans from New World rodents. HPS produces a damaging and often lethal infection that produces severe pneumonia that seems to occur after virus replication, and to be mediated by an inappropriately high level of immune response to the infection. However, there seem to be major differences in how it affects different populations. In one Paraguayan outbreak, of 17 cases all but one were among Mennonite settlers of European descent, even though they were heavily outnumbered by the local Guarani-speaking indigenous native (Indian) population in the area of the outbreak. The same phenomenon was true in a hantavirus outbreak in Chile, where almost all cases were among Spanish speakers rather than the Mapuche-speaking natives of the area.

For the Paraguayan outbreak, blood samples confirmed an extremely high level of exposure among the Guarani-speaking native population. Antibody testing confirmed that 50% had been exposed to the virus, as against 2–6% of the local Europeans. But the latter accounted for 94% of cases.

It was study of the HLA types (a gene cluster coding for highly variable cell surface and immune marker proteins; see Box 4.3) present in the two populations that provided the explanation. There is a far greater risk of death from hantavirus in people with one particular subtype of HLA B35; Europeans with HLA B35 are usually the B35-01 subtype, but while HLA B35 is very common among the native population, they are only very rarely the B35-01 subtype. It appears that this represents an evolutionary adaptation to life with a deadly endemic disease. Native populations that have lived in an area for thousands of years have had time for their HLA types to adapt to their environment. In North America, a similar effect is seen among the Navajo, who live in the same areas as another tribe, the Hopi. However, the Navajo were part of the Athabascan migrations from 1350 onward, and have been in the area for at most seven hundred years, against thousands of years for the Hopi and other Puebloan tribes. Again, risk seems higher among the incomers who have not yet had the time to adapt, and still have the "wrong" HLA type for this particular disease.

4.8 VIRAL ONCOGENESIS

As detailed in Chapter 1, transformation of infected cells is an unwanted side effect of some abortive virus infections, and a common by-product of some other viral infections. The effects of cell transformation are a loss of control over replication, usually resulting in more aggressive growth of the transformed cells. Transformation results from virus-induced alterations to cellular genes and the expression of viral **oncogenes**—genes associated with cellular transformation and **oncogenesis** (the formation of cancers).

Transformed cells show a range of properties associated with cancers, including the loss of normal growth limitations and controls, the expression of unusual surface proteins, and genetic abnormalities including the insertion and expression of viral genes or gene fragments.

Although transformation may be a first step toward oncogenesis, and some transformed cells can induce cancers in animals (though others cannot do so), many other factors are involved and transformation appears to be an early step in a multi-stage process. Additional factors include the immunological monitoring systems that normally ensure that transformed cells are destroyed, notably the cytotoxic T-cell response (see Section 4.3). Other functions are also required, including the necessary housekeeping functions needed to support growth. These include the ability to induce angiogenesis—the formation of new blood vessels to support the growth of the cancerous tissue.

In many cases, specific viral functions associated with transformation and oncogenesis have been identified, and these are summarized in Table 4.7. However, as might be expected, not all transforming viruses are oncogenic.

In the case of virus-infected transformed cells there may be specific virus-encoded functions that aid in the evasion of the immune response (see Section 4.6). While the primary aim of such functions is to aid virus replication, they can also have the effect of protecting transformed cells from the immune response that would normally control them.

The role of cellular tumor suppressor genes

Also very important at this stage are the activities of key **tumor suppressor genes** of the host cell such as the p53 or retinoblastoma (Rb) genes. For example, p53, which has been described as the "guardian of the genome," is critically important in insuring normal cell function through roles in DNA stability and repair, control of the cell cycle, and initiation of apoptosis.

Loss of p53 function thus releases many of the controls that prevent aberrant cell function. The p53 protein, a proline-rich 393 amino acid protein, is activated in response to cellular stress or damage. Excess p53 protein is removed by binding to the MDM2 protein, which targets it for proteolytic degradation via the ubiquitin pathway (**Box 4.6**).

The Rb protein is important in regulating the cell cycle, and controls entry into cell division. As with p53, mutation of the Rb gene is strongly linked to the development of cancer.

Viral oncogenes

With viruses of the genus *Oncornavirus* (family *Retroviridae*), transformation can result from transduction; the transfer from cell to cell within the viral genome of an **oncogene** (cancer-associated gene). Oncogenes are genes of cellular origin that are not involved in virus replication. At least 30 genes in eight families are known to be carried by transducing retroviruses. Only a very few retroviruses (including various forms of the Rous sarcoma virus of chickens) can carry an oncogene within a fully functional virus. In all others, essential regions of the viral genome are replaced, requiring co-infection by a complete virus to permit replication. For this reason, it has been suggested that many of the transducing retroviruses identified would not be able to replicate outside the laboratory, and thus do not cause significant infections.

Alternatively, integration of the virus DNA into the host DNA can result in loss of control over viral genes, or alter the regulation of nearby cellular genes, producing similar effects. It is important to realize that such transformation can occur even when the virus DNA is inactive, and that the initial difference between stable integration and oncogenic integration may be very small.

Other viruses may also be transforming and/or oncogenic by a range of effects (**Tables 4.7** and **4.8**). Viruses of the families *Herpesviridae, Polyomaviridae, Papillomaviridae*, and *Hepadnaviridae* are all associated with transformation of cells and with oncogenesis, while the *Adenoviridae* are transforming but not known to be oncogenic in humans. With many transforming viruses such as the *Adenoviridae, Polyomaviridae*, and *Papillomaviridae*, viral regulatory (early) proteins may affect cell growth directly. These and other effects are summarized in Table 4.7.

It is now clear that viruses are very important in a range of human and animal cancers, but that, in many cases, other factors (known and unknown) are required for oncogenesis. However, prevention of viral infection by limiting the spread of such viruses, often by vaccination, provides a real hope for preventing some cancers, notably those associated with the human papillomaviruses or the hepatitis B virus, for which vaccines are now available.

Box 4.6 The ubiquitin pathway

The ubiquitin pathway is found in almost all eukaryotic cells. By attaching four or more copies of the small (76 amino acid) ubiquitin protein to another protein molecule, the cell targets that protein to proteasomes, where the ubiquitinated protein is proteolytically degraded. Attachment of fewer than four ubiquitins seems to control other processing pathways.

Viruses typically redirect cellular pathways, and ubiquitination is no exception. For example, the E6 protein of cancer-associated papillomaviruses can direct ubiquitination and subsequent degradation of the cellular p53 protein, removing this important tumor suppressor protein from infected cells.

Table 4.7 Mechanisms of transformation/oncogenesis in viruses infecting humans

Virus family	Virus	Mechanism
Adenoviridae (not oncogenic in humans)	Adenovirus	E1A protein targets Rb protein, inhibits MHC-I presentation E1B protein targets p53 tumor suppressor protein Fragments of viral genome integrate into host chromosomes
Herpesviridae	Epstein-Barr virus (HHV-4)	Constitutive up-regulation of viral promoters (including c-myc) by EBNAs 2, LP, 3A, and 3C and LMP-1 activates multiple cellular signals via TNF receptor-associated factors (including NF-κB) Co-factors include dietary intake of nitrosamines (nasopharyngeal carcinoma) or areas with endemic *Plasmodium falciparum* malaria (Burkitt's lymphoma)
Herpesviridae	Kaposi's sarcoma virus (HHV-8)	LANA (ORF 73 protein) interaction with Rb (and Rb regulators) as well as possible interaction with p53, cellular kinases (notably GSK-3β), and effects on cellular transcription v-Cyclin (ORF 72 cyclin homolog protein) promoting cellular proliferation ORF 71 (v-FLIP) interference with apoptosis Production of kaposin proteins promoting transformation
Hepadnaviridae	Hepatitis B virus	Possible *trans*-activation by pX protein Insertional effects on cellular oncogenes (including c-myc in animal models) Possible *trans*-activation by truncated pre-S proteins Co-factors include dietary intake of aflatoxins and nitrosamines, hepatitis C virus infection Chronic virus-induced inflammation may play a role
Flaviviridae	Hepatitis C virus	Core protein modulation of transcription, interaction with p53, and effects on cell cycle Possible effects of nonstructural proteins on cell proliferation Chronic virus-induced inflammation may play a role
Papillomaviridae	Human papillomavirus types 16 & 18	E6 protein interaction with p53 protein and at least 12 other cellular factors E7 interaction with Rb protein and at least 15 other cellular factors Effects of other early proteins on cell cycle Insertional effects on cellular oncogenes (non-oncogenic types maintained as episomes, oncogenic types integrate genome fragments) Co-factors include smoking and possible role for HSV-2 infection
Polyomaviridae	Merkel cell polyomavirus (JC virus, BK virus)	Small and large T antigens (early proteins) interact with multiple cellular factors including p53 and Rb and are associated with cellular transformation Possible insertional effects
Retroviridae	Human T-lymphotropic viruses types 1—4	*trans*-Activation by HTLV Tax protein Insertional effects on cellular oncogenes Integration of viral genome as essential part of viral life cycle For transducing (usually defective) retroviruses at least eight different families of host-derived oncogenes are carried by individual virus types

EBNA, Epstein-Barr virus nuclear antigen; LMP-1, latent membrane protein 1; NF-κB, nuclear factor κ B; LANA, latency-associated nuclear antigen; HTLV, human T-cell leukemia virus.

Table 4.8 Viruses implicated in cancer

Virus family	Human tumors	Animal tumors	Tumor types	Agents associated with human cancers
Adenoviridae	No	Yes	Solid tumors	None known
Flaviviridae	Yes?	No	Hepatocellular carcinoma	Hepatitis C virus (HCV)
Hepadnaviridae	Yes	Yes	Hepatocellular carcinoma	Hepatitis B virus (HBV)
Herpesviridae	Yes	Yes	Lymphomas, carcinomas, and sarcomas	Epstein-Barr virus (EBV, HHV-4) Kaposi's sarcoma virus (KSV, HHV-8)
Papillomaviridae	Yes	Yes	Papillomas and carcinomas	Human papillomavirus (HPV)
Polyomaviridae	Yes	Yes	Solid tumors	Merkel cell polyomavirus (JC, BK viruses?)
Poxviridae	No	Yes	Myxomas and fibromas	None known
Retroviridae	Yes	Yes	Hematopoietic cancers, sarcomas, carcinomas, leukemias	Human T-cell leukemia viruses (HTLV)

Key Concepts

- The immune system is responsible for controlling infections and for removing viruses from the body. Many of the symptoms associated with virus infections such as fever and inflammation are in fact due to the activity of the immune system, and aid in the resolution of infection.
- The immune response can be broadly divided into two types: innate, and adaptive. The innate system provides a rapid response to a limited number of signals, including pathogen-associated molecular patterns (PAMPs), while the adaptive system can respond to billions of possible antigens, but takes more time to respond to an initial infection.
- The adaptive immune response falls into two broad classes: the cellular system comprises helper and cytotoxic T cells and is important in controlling viruses that are predominantly intracellular; the serological system involves antibodies, produced by B cells, which are important in the control of viruses that are present in the blood.
- Antibodies bind specifically to one antigen. Antibodies are produced by differentiated B cells, and fall into five broad types: IgA, IgD, IgE, IgG, and IgM. IgG is the dominant antibody in serum.
- Antigens are processed into peptides and presented in conjunction with MHC-II molecules to stimulate a helper T cell response. Antigens produced within the cells of the body will generally be processed into peptides and presented in conjunction with MHC-I molecules to stimulate a cytotoxic T cell response.
- The immune system relies on a very complex series of checks and balances. In particular, most signals to activate the immune system require confirmatory (co-stimulatory) signals to produce a strong response. In some cases, an unconfirmed stimulus may even damp down the response, inducing tolerance to a particular antigen.
- Activation and control of the immune system and its responses is mediated by a range of highly complex signaling pathways. Central to these are the cytokines, a wide range of extracellular messenger molecules that fall into four broad families, including the interferons which are of particular relevance to virus infection.
- Once the immune response has resolved an infection, activated effector cells are removed, often by programmed cell death (apoptosis), while specialized memory cells persist to provide for a rapid and powerful response to re-exposure to the same infection.
- In order to replicate, viruses must evade or counter the effects of the immune system. Many viruses have evolved complex and powerful mechanisms to achieve this. However, the immune system also has mechanisms to deal with such threats. The balance between the two is continually changing.
- Transformation of cells by viruses causes loss of cellular growth controls and may be the first step toward oncogenesis, the formation of cancers. Additional steps are needed; for such cells to form cancerous tissues they must both evade the immune system and support such growth.

DEPTH OF UNDERSTANDING QUESTIONS

Hints to the answers are given at http://www.garlandscience.com/viruses

Question 4.1: Why do we have both innate and adaptive immunity?

Question 4.2: Why do innate-like lymphocytes challenge the conventional division into innate and adaptive immunity?

Question 4.3: Why do antibodies have F_c (constant) regions when the variable regions are what binds to their target?

Question 4.4: What is the difference between MHC-I and MHC-II presentation?

Question 4.5: If cellular transformation removes controls over cell growth, why does it not always lead to cancer?

FURTHER READING

Arrand JA & Harper DR (1998) Viruses and Human Cancer. BIOS Scientific Publishers, Oxford.

Banks TA & Rouse BT (1992) Herpesviruses—immune escape artists. *Clin. Infect. Dis.* 14, 933–941.

Biron CA & Sen GC (2007) Innate responses to viral infections. In Fields Virology, 5th ed. (DM Knipe, PM Howley eds). Lippincott Williams & Wilkins, Philadelphia.

Boccardo E & Villa LL (2007) Viral origins of human cancer. *Curr. Med. Chem.* 14, 2526–2539.

Braciale TJ, Hahn YS & Burton DR (2007) The adaptive immune response to viruses. In Fields Virology, 5th ed. (DM Knipe, PM Howley eds). Lippincott Williams & Wilkins, Philadelphia.

Campbell K (2010) Infectious causes of cancer: a guide for nurses and healthcare professionals, Wiley, Chichester.

Dayaram T & Marriott SJ (2008) Effect of transforming viruses on molecular mechanisms associated with cancer. *J. Cell. Physiol.* 216, 309–314.

Murphy K (forthcoming 2012) Janeway's Immunobiology, 8th ed. Garland Science, New York.

Vossen MT, Westerhout EM, Söderberg-Nauclér C & Wiertz EJ (2002) Viral immune evasion: a masterpiece of evolution. *Immunogenetics* 54, 527–542.

INTERNET RESOURCES

Much information on the internet is of variable quality. For validated information, PubMed (http://www.ncbi.nlm.nih.gov/pubmed/) is extremely useful.

Please note that URL addresses may change.

Immunology Link. http://www.immunologylink.com/ (a multifunctional immunology resource)

Inside Cancer. http://www.insidecancer.org/ (multimedia guide including basic biology)

Microbiology and Immunology On-line at the University of South Carolina. http://pathmicro.med.sc.edu/book/immunol-sta.htm (online textbook, open access)

CHAPTER 5

Vaccines and Vaccination

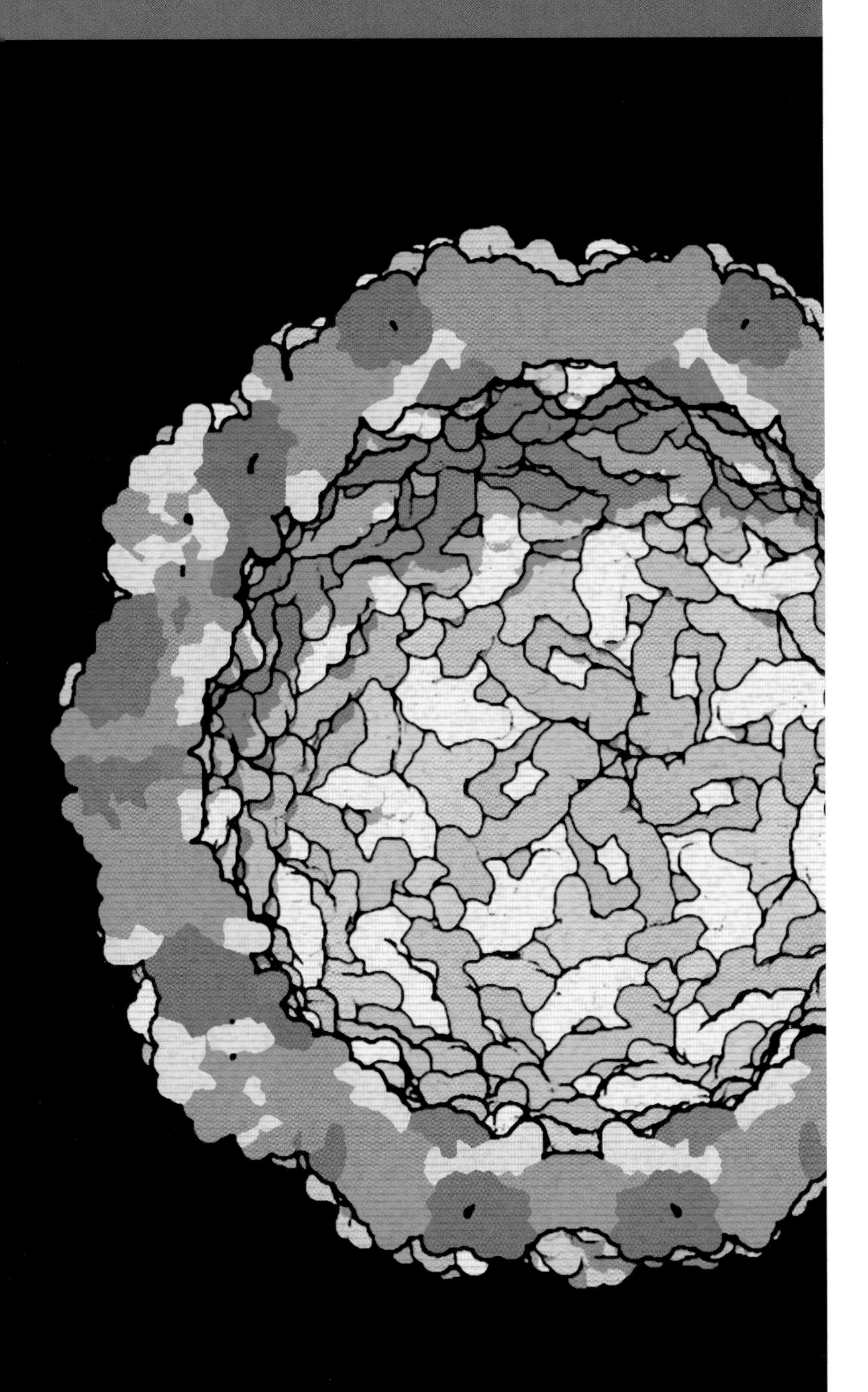

INTRODUCTION

Vaccination uses a weakened form of a disease agent to protect against natural infection. This may be an altered form of the natural infection, a less pathogenic close relative, or even just part of the natural agent. It may be combined with a potentiating chemical, an adjuvant, that enhances the resultant immune response.

About the chapter opener image
Picornavirus
(Courtesy of the Research Collaboratory for Structural Bioinformatics Protein Data Bank and David S. Goodsell, The Scripps Research Institute, USA.)

Vaccination is one of the great success stories of human health. Smallpox, once described as "the most dreadful scourge of the human species" has been eliminated, and polio looks likely to follow soon. Despite this, controversy surrounds both vaccines and vaccine ingredients, and science continues to be unequal to the task of persuading all of society of the benefit of vaccination. Since vaccines are given to healthy people, the benefits are often not clear to the recipient, while the possibility of side effects is far more significant. Thus vaccine development requires the utmost care. But when it works, the benefits to both vaccine recipients and to populations as a whole can be very real indeed. Vaccination remains the only approach that has been able to eliminate entirely a virus that was a major cause of human disease. However, many other viruses, including HIV, present much more challenging targets and success with these may be a long way off.

5.1 THE ORIGINS OF VACCINATION

The first use of a virus to protect against infection was believed to have been in ancient China more than 2000 years ago, where dried material from smallpox scabs was blown into the nose. This generally induced a mild form of smallpox, protecting against the natural infection, which could be far more severe.

By 1700, a variation of this method was in use in the Ottoman Empire, where the scab material (crusts) taken from mild cases of smallpox was inoculated through the skin, restricting the resultant disease still further. As described by Voltaire, this involved "inoculating in the Body of a Child, a Pustle taken from the most regular, and at the same Time the most favourable Sort of Small-Pox that could be procur'd."

Variolation

This process, known as **variolation** (or inoculation), was by no means harmless—it relied on giving smallpox to the recipient, but without any defined way of ensuring that this remained limited. Reports indicate that 1–2% of recipients actually died of smallpox following variolation, a rate which would be completely unacceptable for any vaccine today. Against this, smallpox was still a very serious problem at the time, infecting half or more of the population and causing mortality of 20–30% or even higher. Even in those who survived the infection, permanent scarring was common—so much so that a face free of smallpox scars was considered a real sign of beauty. In such a setting, the practice of variolation was the lesser of two evils.

In fact, two forms of smallpox existed in parallel, caused by differing strains of the *variola* virus; the milder form, *variola minor*, and the more severe *variola major*. Mortality for the two forms was very different—20% or more for *variola major*, but only 1–2% for *variola minor*. As Voltaire noted, crusts for use in variolation were generally taken from those with the milder form. Thus, the mortality from variolation was not hugely different to that from the natural disease.

Variolation was observed by Lady Mary Wortley Montagu, the wife of the British Ambassador to the Ottoman Empire, in Constantinople (now Istanbul). In 1718 she arranged the variolation of her son. Her support was a major factor in introducing the practice to England. Following an initial experiment with six condemned prisoners, variolation became popular in England and in other European countries.

Introduction to the Americas occurred at around the same time, although a group of Boston clergymen recorded some of the earliest objections to

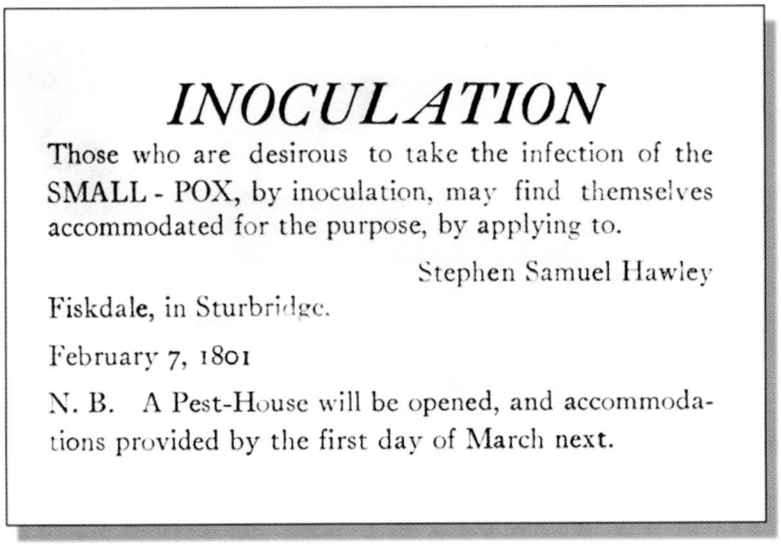

INOCULATION

Those who are desirous to take the infection of the SMALL - POX, by inoculation, may find themselves accommodated for the purpose, by applying to.

Stephen Samuel Hawley

Fiskdale, in Sturbridge.

February 7, 1801

N. B. A Pest-House will be opened, and accommodations provided by the first day of March next.

Figure 5.1
A sign from 1801 offering variolation. From, CDC Public Health Image Library (http://phil.cdc.gov/).

vaccination, reportedly saying that this was "bidding defiance to Heaven itself, even to the will of God." Many early objections were similarly based on theological objections to altering the natural course of such events. Lady Mary's own physician is recorded as having objected to her variolation of her son on the grounds that "this was an unchristian Operation, and therefore that it cou'd succeed with none but Infidels."

These attitudes persisted even after vaccination (see below) was accepted. While Pasteur was proving the value of rabies vaccine in the late nineteenth century, there are reports of other patients still being treated with holy relics.

Variolation remained popular through the eighteenth and nineteenth centuries (**Figure 5.1**). In places, it persisted even longer. The last reported instance of variolation occurred in 1965 in rural China—when supplies of smallpox vaccine were disrupted, older methods were used in their place. Doses of material for variolation could be found in India even into the 1970s (**Figure 5.2**).

Figure 5.2
Variolation (live smallpox) inoculation material, India, 1970s. From, CDC Public Health Image Library (http://phil.cdc.gov/).

Vaccination

Although variolation was in common use, its risks were recognized and safer alternatives were sought. Multiple observations indicated that exposure to cowpox (notably in milkmaids during milking of cows) was linked to protection from smallpox. In 1774, the Dorset farmer Benjamin Jesty used exposure to cowpox to treat his wife and children during a smallpox epidemic.

Despite these limited and local uses, it was not until Edward Jenner, a Gloucestershire doctor, carried out a series of experiments that the practice became widespread. In 1796, Jenner treated an eight-year-old boy with cowpox material taken from a local milkmaid. The boy was later exposed to smallpox by variolation. No disease resulted from this, and Jenner went on to prepare a publication on 23 cases of such treatment which was published in 1798 under the title "An Inquiry into the Causes and Effects of the Variolae Vaccinae; a Disease Discovered in some of the Western Counties of England, Particularly Gloucestershire, and Known by the Name of The Cow Pox." The high profile of this work is reflected in the fact that we not only know the name of the boy (James Phipps) and of the milkmaid from which the cowpox came (Sarah Nelmes), but even the name of the cow that she milked (Blossom).

Jenner took on a high profile in promoting the new vaccination, named for the Latin term for cowpox, *variolae vaccinae*. Subsequently, the term came to mean the use of a different agent to protect against disease—either a related organism, as with cowpox, or a weakened form of the original (wild-type) agent.

For almost 50 years, vaccination was practiced in England alongside traditional variolation. The success of vaccination was shown by the high offices and awards given to Jenner, including those as diverse as the title of Physician Extraordinary to King George IV and the minting of a special medal by Napoleon in 1804. More significant was the compulsory use of vaccination under a series of Vaccination Acts passed in the United Kingdom during the nineteenth century. Napoleon, by now Emperor of the French and never one to give up a military advantage, made vaccination compulsory in the French army in 1805.

Jenner is quoted as saying of vaccination that "the annihilation of the Small Pox, the most dreadful scourge of the human species, must be the final result of this practice." He was right, but it was to take almost two hundred years to achieve this (see Section 5.10).

5.2 CURRENT VACCINES

While smallpox vaccine was the first to be developed by many years, the nineteenth and in particular the twentieth century saw a wide range of vaccines come into use. Vaccination is now a common strategy against viral diseases and has led to the elimination of one major human disease, smallpox, which will be described in Section 5.10. Recent work suggests that rinderpest, an economically important disease of cattle, may also have been eliminated through vaccination.

However, despite the undoubted successes of vaccine development, many important human viral diseases remain for which no vaccine is available. These include the human immunodeficiency virus (HIV, *Retroviridae*) and hepatitis C (*Flaviviridae*), for which there is not even a candidate vaccine that appears likely to be effective in preventing disease. Thus a great deal of development work is continuing.

Types of vaccine

There are four basic formulations used in currently available viral vaccines, summarized in **Figure 5.3**. These are:

- Attenuated live viruses
- Whole inactivated (killed) viruses
- Purified subunits (proteins, frequently glycoproteins) of the virus
- Purified proteins produced from cloned viral genes

Vaccines based on all of these approaches are currently available for human use, and are listed in **Table 5.1**. This is not an exhaustive list, but shows the major products now in use. The advantages and disadvantages of these approaches are summarized in **Table 5.2**.

The newer strategies of expressing viral genes in live vectors or in DNA expression vectors have not yet produced a vaccine available for general use in humans, but are in veterinary use and are the subject of a great deal of work and are likely to be of importance in the near future. These are described in detail in Section 5.4.

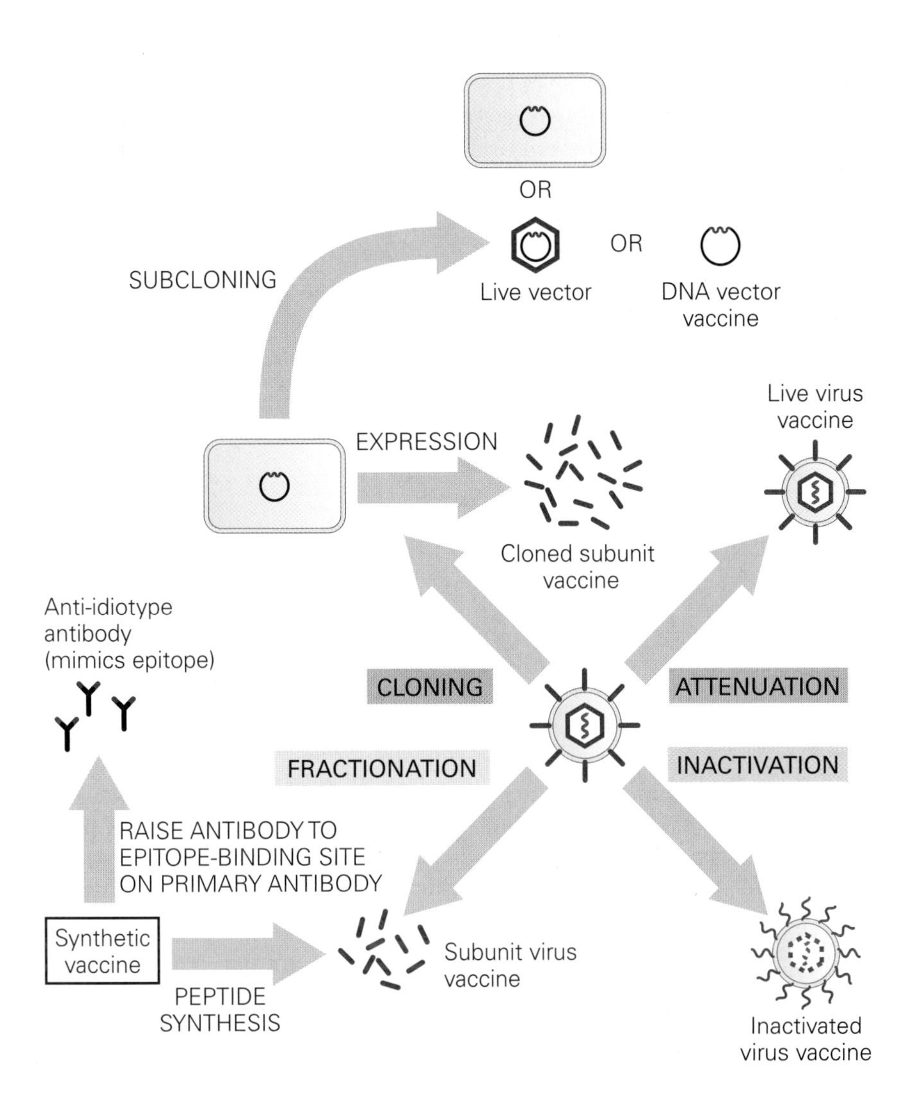

Figure 5.3
Different approaches to vaccine production.

Live vaccines

Traditional vaccines are based on live viruses in which the ability to cause disease is weakened or attenuated by induced mutation. The mutations are induced by passaging the virus in semi-permissive or atypical animal cells (or animals) and/or under altered conditions (commonly, at lower temperatures), selecting variants of the virus that were altered from the optimum for replication in the original host. Nevertheless, they are still able to replicate (if at a reduced level), and are still immunogenic. The changes that are induced are essentially random and historically attenuated viruses have been selected on the basis of their phenotype, with little understanding of the genotypic events underlying the observed changes. While the sequencing of viral genomes is now illuminating the genetic changes in these vaccine viruses, the changes themselves came via a distinctly less molecular approach.

Live vaccines can be highly immunogenic and are able to activate both the serological (via MHC-II) and cell-mediated (via MHC-I) arms of the immune system (see Chapter 4). They may also generate *mucosal immunity* (see Section 5.6), which is important in protection against many viral infections. However, they are not free of problems. While an attenuated virus can be selected so as not to cause significant disease in the normal population, not all potential vaccinees will have normal immune function. As well as natural variation in the population, immunosuppression can result from infections (e.g. HIV), from immunosuppressive drugs (for example, in transplant recipients or cancer patients), or even from the aging process.

Table 5.1 Current vaccines

Live attenuated	Inactivated virus	Purified subunit	Cloned subunit
Adenovirus[1]	Hepatitis A	Influenza	Hepatitis B
Influenza	Japanese encephalitis		Human papillomavirus
Measles	Polio		
Mumps	Rabies		
Polio	Tick-borne encephalitis		
Rotavirus			
Rubella			
Smallpox			
Varicella			
Yellow fever			

[1] Not currently manufactured.

Thus, it is very important to ensure that recipients have an acceptable level of immune function before administering a live virus vaccine. For example, the live varicella vaccine (Oka) has very strict rules governing its use in cancer patients, while the new live influenza vaccine (FluMist®) may not be used in anyone over 49 years of age.

Another concern is that a live vaccine must remain live. This generally requires the maintenance of a refrigerated environment during the whole of the distribution and delivery of the vaccine—the "cold chain." While not an issue in developed countries, there are many areas of the world where this is a real challenge. Encapsulation of the vaccine virus may avoid this need, but this has yet to be proven in the clinic.

An example: oral polio vaccine

The oral polio vaccine (OPV) provides an excellent illustration of the development and use of a traditional live attenuated vaccine. It was developed in the 1950s and is a mixture of three serotypes of human poliovirus (types 1, 2, and 3), attenuated by repeated passage in cell culture. It is well tolerated, gives long-lasting immunity, and is highly protective. It has been used worldwide since 1963 with major benefits for public health. However, all live virus vaccines are subject to a range of concerns (see Table 5.2), and these are well illustrated by OPV.

After administration, the viruses present in OPV can revert to virulent form. Reverted virus can then be transmitted by the vaccinee, causing further disease. In rare instances (approximately one per million doses) OPV can cause vaccine-associated paralytic polio (VAPP).

Sequencing of the genomes has shown that the type 3 vaccine strain, the strain most commonly associated with reversion to virulence, has only 10 of 7431 bases altered, of which only three lead to amino acid changes. By contrast, the type 1 vaccine strain has 55 base changes, with 21 leading to amino acid changes. It is thus possible to see why the type 3 virus is more prone to reversion. The molecular basis of type 3 reversion is not yet fully defined, although a change at nucleotide position 472 appears to be particularly associated with reversion to virulence.

Table 5.2 Advantages and disadvantages of vaccine types in current use

Vaccine type	Advantages	Disadvantages
Live	Development and production straightforward Low cost Cell-mediated (and mucosal) immunity stimulated Can be highly immunogenic, giving long-lasting protective immunity (mimics natural disease) Fewer inoculations than other systems required	May cause (generally mild) disease Reversion to virulence may occur Virus shedding—can infect others Unstable, requires continuous refrigeration (cold chain) Contamination with other viruses possible Contains viral genome, which may be pathogenic or oncogenic in some systems
Inactivated	Development and production straightforward Low cost No reversion or spread of virus Stable, may not require cold chain Low risk of live virus contamination	Culture system required to produce virus in large quantities Cultured virus may not have antigens in the same conformation as the natural virus Immunogenic enhancers (adjuvants) may be required for immunogenicity Frequent revaccination often required Contains viral genome, which may be pathogenic or oncogenic in some systems
Purified subunit	Immune response can be targeted to part of the virus No reversion or spread of virus Stable, may not require cold chain Very low risk of live virus contamination	Production complicated High costs Culture systems required to be able to produce virus in large quantities Adjuvants required for immunogenicity Frequent revaccination often required
Cloned	Can be used where virus cannot be grown in culture Low production costs Production relatively simple to scale up Antigens can be optimized or combined No reversion or spread of virus Stable, may not require cold chain No risk of live virus contamination	Development is expensive and complex Adjuvants may be required for immunogenicity Frequent revaccination often required

Biotechnology offers a number of routes to improving even a preexisting vaccine, and this is well illustrated by OPV. Knowledge of the genome sequences of the viruses present in OPV could allow mutations associated with non-reversion of the types 1 and 2 strains to be introduced into the type 3 genome, adding to the random attenuation of the original vaccine, where random changes were induced by altered culture conditions. An alternative approach is to clone type 3 antigen genes on to a core of a less problematical poliovirus strain, representing a limited form of the live vector approach, discussed in more detail in Section 5.4. In live vaccines in general, the mutations causing attenuation are usually single base changes (point mutations), which can revert easily. Mutations that involve deletion or addition of groups of bases, while rarer, are less likely to revert. It is now possible to use genetic manipulation to introduce specific deletions of such groups of bases, making reversion far less likely.

The poliovirus genome is RNA and, as a result, it has a relatively high mutation rate (see Chapter 1). The possibility of mutations in the seed stocks used to prepare the virus has always been a concern but has now been greatly reduced by cloning the whole genome of the seed strains and maintaining them as DNA copies (**complementary DNA** or **cDNA**). If these copies are introduced into cells, the virus can be produced. This procedure is referred to as genetic stabilization.

Defined attenuation

By understanding the molecular factors underlying virulence, it should be possible to make defined alterations in specific genes that will **attenuate** almost any virus. For example, DISC (disabled infectious single cycle) vaccines are now under development for multiple viruses. One or more essential genes are deleted and the virus is grown in cells expressing the missing gene(s). The resulting, fully infectious virus is used as a vaccine. It can infect cells within the body and replicate, thus calling in all elements of the immune response; however, the virus produced is not infectious since the essential gene is not present and cannot spread any further. DISC vaccines for genital herpes simplex infection (HSV-2) lacking the glycoprotein H (gH), ICP27, or the ICP8 genes have been under evaluation for some time but have yet to prove effective—Whitley and Roizman referred to them in 2002 as showing a "hint of efficacy."

The first vaccines with defined attenuations to reach the market have been for viruses with multipartite genomes. The segmented nature of the genomes of these viruses allows the production of reassortant viruses containing mixtures of genes from different sources without the use of genetic manipulation technologies, which can add significant extra regulatory issues to the process of gaining approval for use.

Live vaccines with reassortment-based defined attenuations are now available for both influenza (FluMist®, MedImmune) and rotavirus diarrhea (Rotateq®, Merck). These use a mixture of genes created by co-infecting cells and selecting natural recombinants. In the case of FluMist®, six genes are from an attenuated influenza B, while two are the surface glycoproteins of the influenza strain being targeted. These are selected using a combination of growth characteristics and immunological techniques. In the case of RotaTeq®, five reassortant viruses are present with the majority of genes coming from a bovine rotavirus, along with single immunogenic genes from human rotaviruses.

Since many workers believe that live attenuated vaccines can provide a broader and more protective immunity than do subunit vaccines, intensive studies are under way to identify and modify virulence determinants in additional viruses, including HIV. Defined attenuations are likely to form the basis of many future live vaccines.

Inactivated vaccines

As with live vaccines, whole inactivated (killed) vaccines have traditionally owed little to molecular virology. Produced by culturing and then chemically inactivating (killing) the virus, they have been very successful in controlling many viral diseases. Since they do not replicate in the host they are less immunogenic than live vaccines, requiring higher dosing and often also needing additional, booster doses to generate useful levels of immunity. The latter raises the issue of patients who do not return for all of the required doses, and may thus only be partially protected. They are also unable to stimulate all types of immunity, most notably the MHC-I-mediated cell-killing arm of the immune system (see Chapter 4). This limits

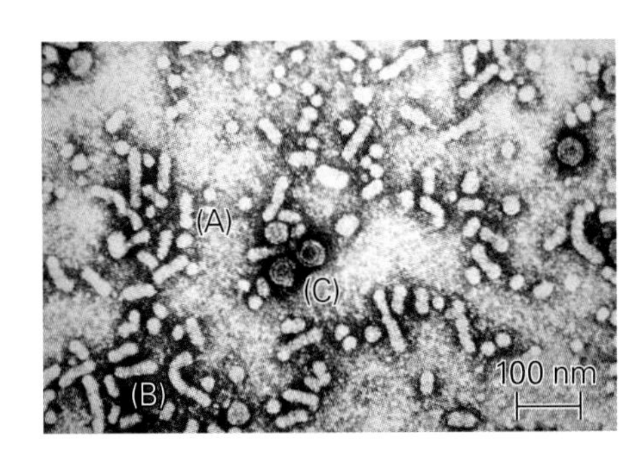

Figure 5.4
Particles present in the blood with hepatitis B. (A) 17–25 nm spherical HBsAg particle; (B) extended tubular HBsAg particle; (C) 42 nm virion (Dane particle). From, CDC Public Health Image Library (http://phil.cdc.gov/).

their efficacy to those viruses where a serum antibody response is effective. While various methods for stimulating cell-mediated immunity using non-live vaccines are under evaluation, the availability of such approaches in the clinic is severely limited (see Section 5.3). A positive feature of such vaccines is that they are not able to cause infections—making them preferred in situations where this could be problematic.

A careful balance is needed in the inactivation process. It is essential to ensure that the process is not too extreme, because this could reduce the immunogenicity of the vaccine to unacceptable levels, compromising its effectiveness. But it is, of course, extremely important to ensure that all of the virus present is inactivated. Failing to ensure complete inactivation led to serious problems in the early days of vaccine production (see Section 5.10).

The concentration and nature of the inactivating agent (such as formaldehyde or β-propiolactone) are critical, but so is insuring that the virus preparation is consistent, without aggregations of virus that could prevent inactivation of those at the center. This problem can be offset by the use of an attenuated viral strain to prepare the vaccine, but current requirements specify a high level of certainty about such processes.

Subunit vaccines

While the use of whole inactivated virus is usual, not all viral proteins are immunogenic. Antibodies tend to react to proteins on the exterior surface of the virus or of virus-infected cells. The presence of other viral proteins may dampen the immune response to the protective antigens and, in some cases, those internal proteins may lead to undesirable effects. As a result, some vaccines have been produced by purification of immunogenic viral components, including those for influenza (HA and NA proteins) and hepatitis B (HBsAg protein).

The hepatitis B virus grows poorly in cultured cells, so the first vaccine used hepatitis B surface antigen (HBsAg) which was purified from the blood of carriers of the virus, where high levels were present. There are two forms of HBsAg particle; small (17–25 nm) spheres, and extended tubular forms (**Figure 5.4**). These can be present at up to 10^{13} particles per milliliter of blood. Extensive purification and treatment removed the larger, infectious (42 nm) viral particles prior to use as a vaccine.

Cloned subunit vaccines

The original hepatitis B vaccine has now been superseded by vaccines using cloned HBsAg produced in yeast cells (Engerix-B®, made by GlaxoSmithKline, and Recombivax HB®, made by Merck), avoiding any problems of residual infectivity. These, together with Gardasil® (Merck) and Cervarix™ (GlaxoSmithKline) for the prevention of human papillomavirus infection, are the only cloned subunit vaccines currently available for use in humans.

For both hepatitis B and papillomavirus, poor growth of the virus in cell culture prevents the use of classical purified subunit approaches. Both vaccine types use *yeast expression systems* (see below and Section 9.1) along with ability of the assembled proteins to form virus-like particles that will be recognized by the immune system. For hepatitis B these are similar to those seen in the blood, while for human papillomavirus these are formed of the viral L1 protein. All are adsorbed onto alum to potentiate the immune response, although in the case of Cervarix™ this also contains the immunostimulatory component monophosphoryl lipid A. Additional materials that potentiate the immune response in this way are referred to as adjuvants (see Section 5.3).

The use of cloned viral genes expressed in heterologous cells simplifies the production of immunogens on an industrial scale, and also allows manipulation and optimization of protein immunogenicity. For those viruses that do not grow or grow very poorly in tissue culture, such as papillomaviruses and the herpesvirus EBV, cloned viral genes may offer the only viable approach.

When cloned viral genes are to be expressed, it is important to note that not all of the resultant proteins are equally effective. Proteins expressed in prokaryotic or eukaryotic expression systems may be used as antigens, but prokaryotic systems, while relatively straightforward to use, produce proteins which are likely to differ from the natural protein, particularly in terms of solubility and post-translational modification (see Chapter 9). Eukaryotic systems are more complex but are capable of producing proteins more like those of the virus itself, which are often more immunogenic. At present, all such systems use yeast, which is a simple eukaryote. Techniques to scale up more complex eukaryotic systems for industrial vaccine production have been under active development for many years, as discussed in more detail in Chapter 9.

Routine post-translational modifications in eukaryotic cells include a wide variety of processes, including cleavage, excision, *O*-linked and *N*-linked glycosylation, sulfation, phosphorylation, acetylation, and fatty acylation. These occur alongside other events such as protein folding (which is a complex process that may require specific guiding *chaperone* proteins), and association with other proteins, membranes, or other structures. Thus, the key to determining whether an expressed protein is usefully immunogenic is simply to use it to stimulate immunity, initially in animal systems, and then to see what immunity is induced. Although there is some understanding of the basic factors, predicting immunogenicity simply from the genetic sequence is not yet possible given the range of other factors involved.

HIV: the center of attention

The human immunodeficiency virus (HIV; *Retroviridae*) continues to provide an as yet unanswerable challenge to vaccine development. In order for a vaccine to be effective, the immune response that it stimulates must provide protection against infection, or at least against the damaging consequences of such infection. Unfortunately, the number of experimental HIV vaccines that have failed to deliver protective immunity continues to rise. This reflects the inability of naturally induced immunity to protect against the progression of infection, except in a relatively small (and intensively studied) group of "long-term survivors."

As well as attacking the immune system directly, HIV has the ability to integrate into the host genome and thus to maintain itself within the host while presenting no targets for an immune response. Furthermore, HIV is a rapidly evolving virus, and many different geographical variants (clades) exist. In many cases it is difficult even to regard these as being the same target; certainly many trials use antigens from a range of these clades to broaden immunity.

The HIV Vaccine Trials Network (HVTN) records 13 completed and 15 current trials. The former include the high-profile failure of Merck's 3000-patient STEP trial in 2007. Given the level of interest in developing an HIV vaccine, it is not surprising that almost all of the approaches listed in this book are being evaluated in attempts to obtain a protective immune response against HIV, and that this represented the main area of activity in antiviral vaccine development in 2009.

After so many disappointments, a vaccine appeared still to be a long way off. One news commentator, writing about the failure of the STEP trial for HIV late in 2007, noted 1984 claims that a vaccine was "about two years" away, while summarizing the 2007 situation as "At this point, it looks like an AIDS vaccine remains decades away, if one is ever to be found." But there are no certainties, as recent events have shown.

In September 2009, the RV144 trial of two anti-HIV vaccine candidates that had not produced a protective response individually showed promise with a combined approach. The trial, sponsored by the US and Thai governments, involved over 16,000 people. Vaccine recipients were treated with an ALVAC live recombinant vaccinia virus expressing three HIV proteins followed by a multivalent formulation containing the clade B and clade E forms of the viral gp120 glycoprotein with an alum adjuvant (see Section 5.3). The combined treatment involved six vaccinations over a six-month period and was intended to induce both cell-mediated and serological immunity. In the treated group, 51 of 8197 people became infected with HIV during the following three years, against 74 in the group that received an inert placebo instead of the vaccine. This represented a 31.2% decrease in infections—the first time a candidate HIV vaccine has produced a significant benefit.

Of course, a treatment that leaves the recipients with the fully two-thirds of the usual chance of becoming infected (and does not appear to moderate disease in those that do become infected) is far from ideal. But it could be the first step on a long pathway to effective vaccination against HIV.

5.3 ADJUVANTS

Vaccines consisting of purified protein (rather than live or whole inactivated viruses) frequently require the presence of an adjuvant to induce a protective immune response, since proteins in solution may be poorly immunogenic. An **adjuvant** has several distinct functions:

- Presentation of antigen in a particulate form, enhancing uptake by antigen-presenting cells
- Direct stimulation of the immune response, often mediated by cytokines or antigen-presenting cells
- Localization of antigen to the site of inoculation, also known as the depot effect, providing an intense local immune stimulation rather than a diffuse generalized stimulus
- Targeting of antigens to particular pathways

The relative importance of each of these effects varies with different adjuvants. Of the available adjuvants, only a range of aluminum salts generally referred to as alum (including microparticulate aluminum hydroxide gel, aluminum phosphate, and aluminum potassium sulfate) are widely licensed for use in humans. Antigenic proteins are adsorbed on to the surface of the alum particles, and the depot effect is thought to be important for alum-mediated immunogenesis. However, alum does not appear to modulate the immune response directly and is generally regarded as a weak adjuvant, with a bias toward the stimulation of only an antibody response.

(A)

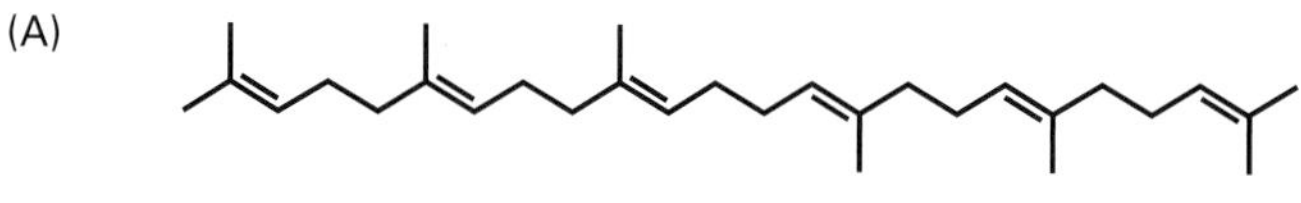

(B)

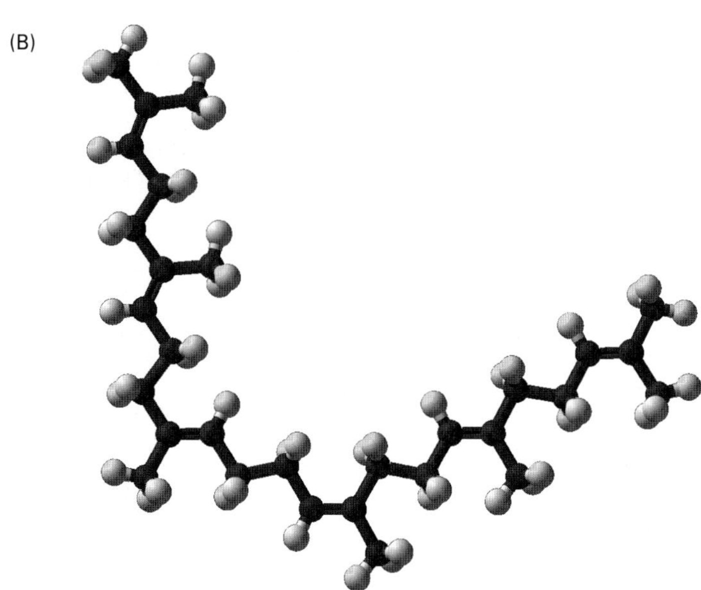

Figure 5.5
The structure of squalene. (A) Chemical structure. (B) Ball-and-stick representation of the squalene molecule as determined by X-ray crystallography.

Squalene is an oil, originally extracted from shark livers, that has attracted a great deal of interest as a component of adjuvants (**Figure 5.5**). A mixture of squalene oil emulsion and the surfactants sorbitan trioleate and Tween 80, known as MF59, is licensed as part of an influenza vaccine in some European countries, and reports in the medical literature indicate enhanced immunogenicity along with a good safety profile after use in millions of patients. However, MF59 is not licensed for use in the United States, and the reported use of MF59 in experimental vaccines has led to considerable controversy. Claims of apparent autoimmune effects (including claims of involvement in toxicity among Gulf War veterans who received such vaccines) have been refuted by further work, but it is hard to remove such associations in the public mind once they have been made.

These concerns did not stop the use of MF59 in a vaccine developed for use against influenza H1N1(2009), popularly known as swine flu. With a new pandemic under way, the need for a vaccine was seen as urgent, and one large pharmaceutical company decided to use it. Other vaccine manufacturers did not.

When the results of the first clinical trials came in, it was clear that the MF59 adjuvant, together with 7.5 μg of influenza surface protein (produced in cell culture) induced protective levels of antibodies within 21 days in 80% of recipients. However, another vaccine, without adjuvant, using twice the level of influenza protein from the traditional source of infected hens' eggs, was reported to have achieved protective antibody levels in 98% of recipients within the same 21-day period. The MF59 adjuvant vaccine was also associated with a higher level of soreness at the site of injection, and the use of MF59 has already resulted in a number of adverse comments. In this case, the use of a high-tech vaccine with cultured antigens combined with a controversial adjuvant could prove problematical for the manufacturer, if a conventional approach can produce equivalent or better results.

It is clear that any adjuvant, by its very nature, has risks attached to its use. Stimulating the immune system is not entirely beneficial, since there is always the risk of damaging effects arising from such activation (for example, see details of the TGN1412 trial in Section 6.3). This has limited the progression of many scientifically interesting adjuvants into the clinic.

The components of an adjuvant

Immunization of animals for the purpose of raising antibodies has traditionally used Freund's adjuvant, which contains killed mycobacteria in a water-in-oil emulsion. The water droplets of the emulsion serve to give a particulate character, and the mycobacterial component is a potent immunomodulator—rather too potent, since it is the cause of significant toxic effects, making it unacceptable for human use. Freund's adjuvant was developed when understanding of molecular events was very limited, and the nature of the immunomodulatory components of this and other crude adjuvants was poorly understood. More recent studies identified muramyl dipeptide and lipid A as bacterial components mediating an immune response in crude adjuvants. These were still too toxic for use in humans, but less toxic derivatives such as muramyl tripeptide-phosphatidylethanolamine (MTP-PE) have been produced which show reduced toxicity while retaining immunomodulatory effects. However, even these have yet to be approved for use in humans.

Many experimental adjuvants have been evaluated in animal systems and in experimental work in humans. Such adjuvants tend to consist of three components: particulate elements (including emulsions such as that produced with squalene), stabilizers, and immunomodulators (such as MTP-PE). However, other than alum and the limited availability of MF59 and monophosphoryl lipid A, none are yet available for general use. As with vaccines themselves, many promising developments based on increasing understanding of events at the molecular level have not yet carried through to practical uses.

Studies using individual cytokines as adjuvant components have often now been replaced by work on the Toll-like receptors that mediate innate immunity (see Chapter 4). However, understanding at this level is very limited and is highly subject to the complexities of the immune system.

Liposomes and cochleates

Liposomes are lipid bilayer vesicles with an aqueous core that can be used to deliver a wide variety of substances, for example in the transfection of DNA into cells. They can also allow the membrane insertion of viral antigens, to form composite particles—virosomes. Liposomes, virosomes, and the complex lipid rolls known as cochleates are under intensive investigation as delivery systems for vaccine antigens.

T cells and adjuvants

One clear advantage of some second-generation adjuvants is their ability to stimulate a cytotoxic ($CD8^+$) T-cell response (see Chapter 4). These cells are stimulated by antigen presented in combination with proteins of the MHC-I system expressed by all cells and, until relatively recently, it was accepted that proteins presented via the MHC-I/$CD8^+$ pathway had to be synthesized inside the presenting cells. This was perceived as an important problem for any non-live (subunit or inactivated) vaccine directed against a disease where cell-mediated immunity was important, which includes many virus diseases, since it was thought that these antigens would only be presented to $CD4^+$ T cells via the MHC-II pathway and would not induce a significant cytotoxic T-cell response. There is now convincing evidence that some adjuvants (including the squalene emulsion used in MF59, but not including alum) can allow proteins to enter the MHC-I-mediated pathway, apparently by carrying antigenic proteins directly into the cytoplasm. This has generated a great deal of interest in such systems, for example in immune-stimulating complexes (ISCOMs) containing the tree bark derivate QS21. Such systems are likely to aid the development of effective subunit vaccines, but only once such adjuvants are licensed for use. Despite a great deal of work, very few have made it through to market. Even when this happens, as with MF59, regulations in this area mean that they are not licensed for general use, but only as part of a particular antigen-adjuvant mix. This limits the commercial drivers for adjuvant development.

The lack of progression of advanced adjuvants from laboratory to clinic has been widely perceived as disappointing. This reflects both the complexities and limits of clinical development, and the effectiveness of existing vaccine formulations, many of which contain innate immunostimulatory activities.

5.4 APPROACHES TO VACCINE DEVELOPMENT

When considering the current efforts to develop antiviral vaccines, it is important to make a distinction between research aimed at developing a vaccine to protect against infection or against disease, and the research aimed at increasing understanding of the effects and mechanisms of protective immunity. Much of the work that is taking place, even within industrial laboratories, falls into the second category and, while such work may provide the basis for future generations of vaccines, direct clinical applications are still remote.

The potential properties of some of the numerous vaccine types in development are shown in **Table 5.3**. The use of purified antigenic proteins produced from cloned viral genes is the approach used in commercially available vaccines for hepatitis B and human papillomaviruses. In both of these, as well as in other experimental approaches, the viral proteins are expressed in yeast-based systems so as to form virus-like particles (VLPs, generally spherical particles within the normal size range for viruses) to enhance their immunogenicity. While best established for non-enveloped viruses, similar approaches are under development for enveloped viruses such as

Table 5.3 Properties of vaccine types in development

Vaccine type	Adjuvant requirement	Cell-mediated immunity	Multivalency[a]
Live (defined attenuation[b])	–	+++	–
Antigenic proteins[c]	++	±[d]	+
Peptide	+++	±[d]	+++
Fusion vector	+[e]	±[d]	++
Anti-idiotype[f]	±	±[d]	–
Live vector[g]	–	+++	++
Gene vector[h]	–	++	++
DNA vector[i]	–[j]	++	±

This table provides a general guide only. Individual cases may vary. [a] In a single system. Mixing of vaccine preparations can provide multivalent vaccines for any approach. [b] Including defective viruses that can only undergo a single round of replication (DISC). [c] May be purified or cloned, full length or fragments. [d] Depends on adjuvant system. [e] Peptide or protein is attached to a carrier protein, which may have innate adjuvant activity. [f] Antibody molecules that mimic viral antigens. [g] Virus genes expressed in another virus or bacterium. [h] Virus genes expressed in another virus or bacterium that is incapable of complete replication. [i] Purified plasmid DNA expressing viral gene under eukaryotic promoter control. [j] Methods for potentiating DNA uptake may be used.

HIV, where the term **pseudovirion** is used to refer to an assembled structure resembling the viral capsid. However, the envelope glycoproteins are typically the main antigens of enveloped viruses, and the value of the pseudovirion approach remains to be proven.

Another approach is to use peptides, either synthesized from cloned genes or produced chemically, which correspond to known epitopes of viral antigens (see Section 5.8). This approach is of particular interest for the induction of cell-mediated immunity, since the natural presentation of proteins to T cells is in the form of peptide fragments. Peptides are less suitable for the induction of antibodies: despite the ability to select peptides corresponding to highly immunogenic linear epitopes, they have a very limited ability to present more complex epitopes (see Chapter 4), and lack any post-translational processing. There are also issues of potential resistance with any immune response produced to such a small region of a viral protein, and many peptide approaches recognize this, and deal with it (at least partially) by presenting multiple peptides in a single vaccine.

While there is some evidence that peptides may be directly immunogenic, most approaches using peptides also rely on some additional method to increase immunogenicity. This may be an advanced adjuvant, or fusion to an immunogenic protein—a fusion vector approach. In a fusion vector system, the gene coding for the peptide is linked directly to a gene for a carrier protein. The protein synthesized by this hybrid gene contains the peptide as part of a modified carrier molecule. A less commonly used alternative approach is to fuse a synthetic peptide directly to the carrier protein. Both approaches have been used in clinical trials. Where a fusion vector is used, the carrier is often selected for its adjuvant effect (see Section 5.3).

Examples have included a yeast system, where fusion with the yeast Ty p1 protein permits assembly into 70 nm VLPs, or the direct fusion of smaller synthetic peptides to keyhole limpet hemocyanin. With the latter, it is difficult to understand just how a protein from such an obscure source came to be used, but it has in fact been very valuable in such studies. As well as inducing antibodies, both of these proteins may be able to enter peptides into the MHC-I/CD8^{+} (cytotoxic T cell) pathway, presumably by entering protein into the cytoplasm, as discussed in Chapter 4.

A variation on peptide vaccines is the use of antibody molecules raised against antigen-binding sites of antiviral antibodies (**anti-idiotypic antibodies**), which mimic the conformation of the viral antigen. This method does not appear to hold much immediate promise for viral vaccination and may be more applicable where an antigen cannot be produced by cloning, for example with the polysaccharide epitopes which distinguish some cancer cells.

All of the above approaches rely on the presentation of nonreplicating proteins to the immune system. An alternative approach is to express the gene coding for the antigen in a virus or bacterium that can make the antigen inside the vaccinee. Bacterial vectors under consideration have included *Mycobacteria*, *Listeria*, *Salmonella*, and *Shigella*, while a broad range of vector viruses have been evaluated, as will be described in the next section.

Viral vectors

A **vector virus** is attenuated, either in the classical, random way or by defined attenuation involving the deletion or site-directed mutation of viral genes. The inserted gene is expressed in the vaccinee's cells as part of the vector virus. The protein produced in this way acts as the immunogen, stimulating both cell-mediated and serological immunity.

Some examples of viruses that have been evaluated for use as vectors are shown in **Figure 5.6.** There is a wide variation in the size and complexity of both the viruses and the viral genomes used.

Viruses with large genomes (e.g. poxviruses, adenoviruses, herpesviruses) allow for the insertion of multiple genes, potentially generating immunity to multiple viruses. However, larger viruses also present more targets to the immune system, generating responses to both the viral vector and to the cloned antigen. This can result in undesirable immunity to the vector, and can also weaken the reaction to the cloned antigen. Smaller viruses present fewer targets but allow for the insertion of less genetic material. While it is possible to insert fragments of genes coding for a series of immunogenic peptides, which has been described as the string-of-beads approach, this restricts the immunity that can be developed.

Another problem is that some of the live vectors under study are viruses to which a significant section of the population has already been exposed. An elevated immune response to the vector could thus occur even when a novel vaccine is used, boosted by preexisting immunity to the vector or to a related virus, and this could result in damping of the immune response to the novel (vaccinating) antigens. Even where immunity does not already exist (as would be the case with many fusion vectors), the first such vaccine to be used might reduce the response to the novel components of any future vaccination using the same vector system. While some workers discount this possibility or even suggest a potentiating effect on immunogenicity, the matter has not as yet been resolved.

Of the viruses shown in Figure 5.6, replication-deficient poxviruses and adenoviruses are the most used, particularly for experimental HIV vaccines. Both types of virus are large, allowing the inclusion of multiple genes, and

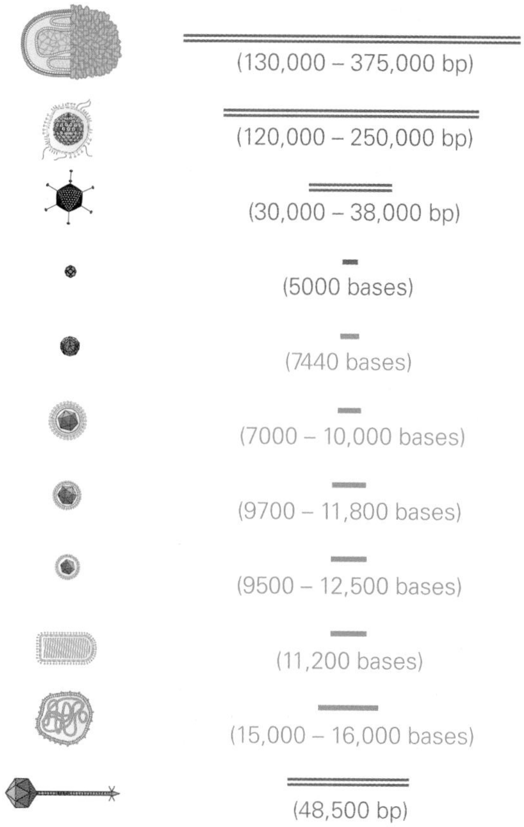

Virus Type	Virus Morphology	Genome Size	Genome Type
Poxviridae		(130,000 – 375,000 bp)	dsDNA
Herpesviridae		(120,000 – 250,000 bp)	dsDNA
Adenoviridae		(30,000 – 38,000 bp)	dsDNA
Parvoviridae		(5000 bases)	ssDNA
Picornaviridae		(7440 bases)	ssRNA
Retroviridae		(7000 – 10,000 bases)	ssRNA
Togaviridae		(9700 – 11,800 bases)	ssRNA
Flaviviridae		(9500 – 12,500 bases)	ssRNA
Rhabdoviridae		(11,200 bases)	ssRNA
Paramyxoviridae		(15,000 – 16,000 bases)	ssRNA
Siphoviridae		(48,500 bp)	dsDNA

Figure 5.6
Viruses used as live vaccine or gene vectors.

Table 5.4 Disabled poxvirus vaccine vectors

Name	Virus	Basis of attenuation	Approved uses	Clinical trials as recombinant vector (target examples)
MVA	Vaccinia	Vaccine strain with loss of 10% of genome after repeated passage in chicken cells	Smallpox vaccine	HIV, papillomavirus, cancers
NYVAC	Vaccinia	Loss of 18 open reading frames from 6 regions of the genome	None	HIV, Japanese encephalitis, malaria
ALVAC	Canarypox	Avian virus cannot productively infect mammalian cells	None	HIV, Japanese encephalitis, cytomegalovirus, rabies, cancers
TROVAC and others	Fowlpox	Avian virus cannot productively infect mammalian cells	None	HIV

All listed viruses undergo very limited replication in humans.

problems with vector antigens can be reduced by using other antigenic presentation systems for booster vaccinations—the so-called "prime-boost strategy." This is intended to stimulate different elements of the immune system by the use of the different vaccination strategies, but moderation of vector effects is another advantage of such an approach.

Much work with poxvirus vectors uses vaccinia, the virus used as the vaccine for smallpox. However, the strain used as a vaccine vector in early work was frequently a virulent strain (such as vaccinia WR) rather than an accepted vaccine strain. During the 1990s, the focus of such work moved on to a variety of replication-incompetent poxvirus vectors (**Table 5.4**) and this work has contributed much to understanding of the potential for live vector approaches. However, all are themselves highly immunogenic, and even the canarypox ALVAC has been shown to induce a strong immune response to the carrier virus. This is a real problem for such approaches, as noted above.

Adenoviruses have relatively large genomes and mild pathogenicity even in their unaltered form. They are also able to be given orally or intranasally and are an attractive system for the development of live vectors, particularly since adenovirus vaccines have been used extensively by the US military. Replication-incompetent adenovirus vectors are currently in use in HIV vaccine trials. However, while adenovirus vaccines seem to be safe, there are once again some concerns over the induction of immunity to the vector virus—particularly after a reported potential linkage of this with the potential for raised levels of HIV infection in the recent STEP trial. One approach that is being evaluated is to make hybrid adenoviruses using some of the 51 human serotypes (strains) that are less likely to encounter existing immunity.

Another vector that has been investigated is the attenuated (live) Oka vaccine strain of varicella-zoster virus (VZV, human herpesvirus 3). Multiple genes have been expressed in this system, but its usefulness is limited by several factors. It is a classically attenuated virus, and while the basis of its attenuation is now better understood it lacks the extensive attenuation of viruses such as the vaccinia MVA and NYVAC. In addition, it is known to establish latent infection following vaccination. Finally, although now licensed for general use in the USA and some other countries, even the unmodified virus is not yet universally approved.

As with herpesviruses, retrovirus vectors present considerable problems since they are capable of establishing latency, integrating genes into the host DNA. While desirable for gene therapy, this is less desirable for vaccines. With retroviruses there are also concerns over the possible oncogenic effects of the vector, especially following reports of leukemias following the use of retroviruses in gene therapy work (see Chapter 7).

Picornavirus (polio) vectors could allow oral administration, but the small size of the poliovirus genome could be problematic and would almost certainly require the expression of peptide rather than protein antigens—200 amino acids has been reported as the maximum size for an inserted protein.

The even smaller adeno-associated virus (AAV, *Parvoviridae*) vector system would also have this problem. In this case, no viral genes are present in the vector—the whole of the coding genome is replaced by the insert. Concerns over the routine integration of AAV into the cellular genome have been reduced by the finding that this results in maintenance of the vector as an extrachromosomal genetic element when used in this way.

Replication-deficient viruses: gene vectors

As noted above, a common approach to the use of live virus vectors is the use of replication-defective variants that cannot replicate fully in the vaccine recipient. The key here is to allow enough replication to stimulate a strong immune response but not enough to cause disease.

Viruses such as the MVA and NYVAC forms of vaccinia or the DISC herpesviruses are grown in cell lines expressing viral genes or otherwise able to allow viral growth despite the missing functions of the virus. The viruses produced in such systems may infect other cells but are not then able to produce infectious progeny virus. Such systems are sometimes referred to as gene vectors rather than live vectors due to their very limited replication in the vaccine recipient. In fact, some live vectors now under study fall into this category since they are also incapable of full replication when used as vaccine vectors. For example, ALVAC is actually a canarypox virus and cannot replicate fully in mammals.

While these viruses cannot replicate fully and can only undergo a single cycle of infection, viral (and insert) proteins are produced within the cells of the vaccinee. Thus, both live and gene vectors can stimulate the full range of cell-mediated immunity (including MHC-I/CD8$^+$ cell-mediated immunity), as well as serological and innate immune responses.

As well as potential uses as vectors, replication-defective viruses may be used to stimulate immunity to the parental virus. This is essentially a defined attenuation as outlined in Section 5.2.

Nucleic acid vaccination

The ultimate reduction of a vector system is to use a naked (protein-free) DNA molecule that expresses the desired genes. However, the world is full of enzymes that degrade nucleic acids—this is why evolution has provided a protein shell around viral genomes. Thus, it came as a major surprise when it was reported in 1993 that injection of mice with plasmids expressing the influenza nucleoprotein (NP) gene elicited an immune response that was sufficient to protect mice from infection.

Work in this area expanded rapidly due to the simplicity and potentially wide applicability of the technique, which involves the construction of a plasmid coding for an antigen under the control of a strong promoter (e.g. the CMV immediate-early promoter). This is then introduced, usually into muscle tissue, and appears able to induce the full range of immune responses while avoiding any vector-associated responses, since there are no proteins to which an immune response can be mounted. The finding that DNA can produce immunogenic proteins directly also emphasizes the need to avoid DNA contamination of protein-based vaccines.

Nucleic acids are usually degraded quickly within the body, and it was initially estimated that only one in a million plasmids enter cells and express the viral gene. The use of DNA vaccines was referred to as a revolution in vaccinology, and seemed to make possible a far more rapid progress from genetic information to candidate vaccines than was previously the case. Multiple delivery systems were evaluated, including conventional injection, small gold beads fired from a helium-powered Gene Gun (**Figure 5.7**), and direct application of the DNA to mucosal surfaces.

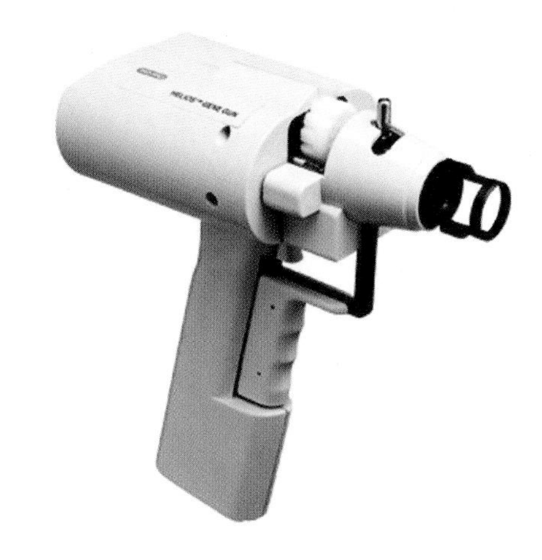

Figure 5.7
The Bio-Rad Helios "Gene Gun." Courtesy of Bio-Rad Laboratories, Inc.

However, as with many areas of vaccinology based on advanced technologies, results to date have been less impressive than was originally hoped. Despite very promising results in mice, levels of immunogenicity in humans have been disappointing. Two DNA vaccines have been approved—one for protecting horses from West Nile virus (*Flaviviridae*), and the other for protecting farmed salmon from infectious hematopoietic necrosis virus (*Rhabdoviridae*). DNA vaccines for HIV continue to be developed and promising initial results have been reported for influenza, but as yet the revolution has not been carried through and no DNA vaccine for use in humans has yet reached the market.

One interesting approach now being evaluated is to insert the DNA coding for the desired antigens (along with the required eukaryotic promoters) into a virus that normally infects bacteria—a bacteriophage. The bacteriophage is then introduced into the patient. Being a virus-sized particle, it is then taken up by the cells of the immune system, where the insert DNA can be expressed. The bacteriophage DNA itself cannot be expressed in eukaryotic cells, making this effectively an extreme form of the gene vector approach. Initial studies using bacteriophage lambda (*Siphoviridae*, **Figure 5.8**) are currently under way.

5.5 TAILORING OF THE IMMUNE RESPONSE TO VACCINATION

It is clear that different infections are controlled by different balances of the individual elements of the immune system (see Chapter 4). For some, an antibody response is effective, but for many others additional elements are needed. As always with biological systems it is very unlikely that a single infection has a single definitive response. The immune response to a particular infection is both complex and highly variable.

For example, it is clear that reactivation of herpesviruses can cause disease even in the presence of high levels of neutralizing antibody in the blood. In this particular case, the tightly cell-associated nature of the virus (including direct spread from cell to cell by means of fusion into giant cells, or syncytia) is thought to require the cell-mediated cytotoxic elements of the immune system to control the infection.

Conversely, viruses that exist free in the blood, such as hepatitis B, may be susceptible to direct antibody-mediated control. Thus, it can be seen that the specific nature of the immune response induced by a vaccine must be appropriate to the pathogen.

A great deal of work is currently under way aimed at identifying the correlates of immunity for viral and other diseases, with the intention of specifically stimulating either type of responses or by precise combination activating multiple arms of the immune system in the way which is optimal to prevent disease.

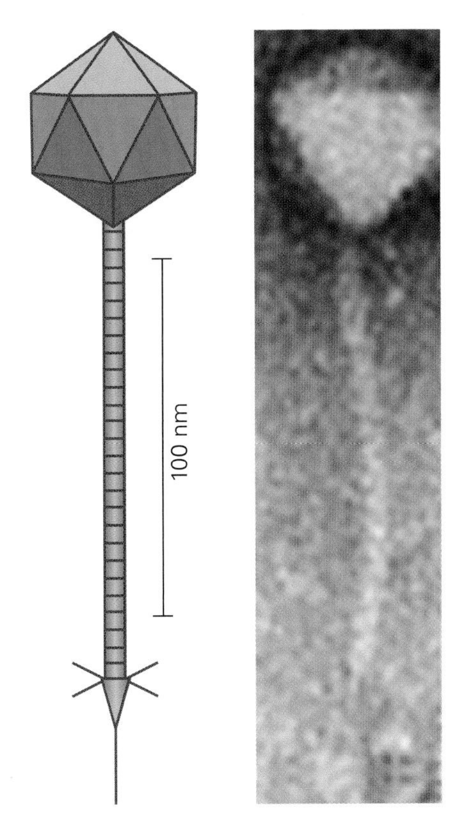

Figure 5.8
The structure of bacteriophage lambda. Courtesy of BigDNA Ltd, UK.

Such vaccine strategies are best illustrated by current work with HIV. HIV continues to be a demanding vaccine target, although there are tantalizing clues with natural immunity in some patient groups that a protective immunity may be possible. Thus, much current work uses combinations of delivery systems (including live vectors, DNA vectors, gene vectors, and subunit vaccines) for the initial vaccination ("prime") and subsequent vaccinations ("boost") (see Section 5.4). Of 28 past and current HIV vaccine trials recorded by the HIV Vaccine Trials Network, 12 used this prime-and-boost approach. However, it should be noted that the immune system is highly complex, and attempts to provide an effective combination, especially for a rapidly moving target like HIV, are likely to prove challenging.

Against the optimistic projections for tailoring of the immune response must be set the informed opinion of one very senior vaccinologist. Dr Philip Minor of the United Kingdom's National Institute for Biological Standards and Control writes in the *Encyclopedia of Life Sciences*, "It is a commonly shared illusion that the basis for protection by vaccination is understood either in terms of immunology or pathology. In the author's opinion this is universally erroneous, the effects of vaccines being only understood empirically as they are used." In the opinion of the author, Dr Minor is making a valid point.

5.6 ALTERNATIVE DELIVERY SYSTEMS

Most vaccines have traditionally been administered by inoculation. However, oral delivery of live polio vaccine, mimicking the natural route of infection for poliovirus, has shown the increased ease of administration by this route.

Oral administration of vaccines

Oral delivery is effective at stimulating mucosal immunity in the gut, which is thought to be very important in protecting against many infections.

For a subunit vaccine, specialized systems must be used, since the gut is optimized to digest proteins rather than mount an immune response to them. However, cloned proteins from hepatitis E virus (which normally spreads via oral intake of the virus) assembled into virus-like particles (see Section 5.2) have been reported to generate antibody responses when given orally. This unexpected finding that a subunit vaccine can be effective when given in this way may have significant implications.

A range of approaches to getting protective vaccination by oral ingestion are under development, including (not unreasonably) the use of vectors based on live poliovirus. Both vaccinia (*Poxviridae*) and adenovirus (*Adenoviridae*) may be administered orally, although neither causes pathogenic infections by this route. In the former case, an encapsulated recombinant rabies/vaccinia vaccine is contained in bait capsules laid for wild animals, including foxes and raccoons, and it is thought that the virus may enter the blood via cuts in the mouth. In the latter, in a vaccination program restricted to the US military, enterically coated tablets containing freeze-dried virus were used (the program has now ceased). These tablets released virus into the intestine, giving an asymptomatic disease that induced a protective immune response.

A recent development is an influenza vaccine (FluMist®, MedImmune) that is given by nasal spray. This reflects an increasing trend to use vaccines that can be given by a route mimicking the natural process of infection. The idea behind this is that such vaccines will then induce immunity at the sites where it will best counter the infecting virus.

Mucosal immunity

Mucosal immunity is an area under intensive investigation, since it has the potential to stop a virus from even entering the body (*sterile immunity*), the ultimate aim of vaccination.

Immunity at the mucosal surface involves secretory IgA (see Chapter 4) as well as low levels of infiltrating antibodies and cells from the blood. Live vaccines appear able to stimulate mucosal immunity, while inactivated vaccines are less able to do so.

It is possible to stimulate mucosal immunity directly by inoculation of vaccine onto almost any mucosal surface (including the oral, respiratory, and genital areas). There is good evidence that mucosal immunity may be required for optimal protection against viruses that infect via such routes. However, there is some evidence that the resultant immunity may be restricted to the specific site of inoculation, possibly requiring the use of separate vaccination for each site that is at risk of infection. Some of these are likely to prove highly unpopular if used in a clinical setting. Despite this, studies on optimizing vaccine delivery via almost every mucosal surface are currently under way, often in combination with systemic vaccination.

Interestingly, a Venezuelan equine encephalitis virus (VEE, *Togaviridae*) vector system has been reported to stimulate high levels of mucosal immunity even when given by injection. While the mechanisms involved are not understood, this approach could be promising for the generation of mucosal immunity if the other issues related to the use of this vector are resolved.

Slow release

Slow release systems are also under investigation for vaccine delivery. These are extensions of the existing depot effect provided by adjuvants. Such systems would prevent the need for multiple vaccinations by delivering antigen over a period of weeks from within a biodegradable implant at the site of the initial vaccination. Many problems remain in this area, such as how to ensure the stability of the antigen in a warm environment perfused by body fluids, and the possibility of severe localized immune reactions.

It is clear that many of the vaccines now under development will be tailored to produce a spectrum of immunity providing maximum protection against a particular route of infection by an individual pathogen. This will involve careful selection and evaluation of multiple antigens, vector systems, adjuvants, and routes of delivery. All of this will need to be based on a more complete understanding of events at all levels. It is debatable whether such understanding exists at a useful level at the present time, as discussed in Section 5.5.

5.7 THERAPEUTIC VACCINATION

While vaccination traditionally is considered as a method of preventing disease (**prophylaxis**), it is also possible to use vaccination to moderate the effects of a pathogen that is already present (therapy). Some vaccines, such as that for rabies, may be used after exposure, but this relies on establishing an immune response before infection becomes established rather than being a true therapeutic vaccination.

There are a number of diseases where the organism responsible remains present at low levels or in an inactive form and causes disease at a later date; examples include herpesvirus infections and papillomavirus-induced cancers of the cervix. In addition, there are many ‘slow’ infections where symptoms develop a long time after infection, including the development of AIDS after HIV infection.

In all of these cases, the use of a vaccine to boost immunity has the potential to help reduce progression of the infection, severity of disease, or frequency of the recurrence of symptoms. Such vaccination may also reduce virus shedding and the potential for infection of new individuals, thus benefiting the population as a whole. In an extreme form, there may be no symptomatic benefit for the vaccinated individual, the benefit of vaccination being seen exclusively in reduced levels of infection within the community (altruistic vaccination). This of course does bring ethical challenges, especially concerning any side effects of the vaccine for the recipient.

Even though they are difficult targets owing to the nature of the infection, some studies have suggested that the use of vaccines for herpes simplex virus (HSV), HIV, and a range of papillomaviruses may have the ability to reduce recurrent disease. The DISC vaccine approach under investigation for vaccination against HSV (see Section 5.2) is a possible candidate for this approach.

It is also possible to use classical vaccines in this role, and extensive studies have been undertaken on the use of the live varicella vaccine to boost immunity in the elderly with a view to preventing zoster (which is caused by reactivation of VZV). This is now licensed by the FDA as Zostavax®, for use in patients aged 50 or older.

It is likely that therapeutic vaccinations will play a significant role in disease control in the future.

5.8 EPITOPE STRUCTURE IN VACCINE DEVELOPMENT

It has been known for a long time that not all regions of a protein are equally immunogenic. Those regions that are recognized by the immune system are known as epitopes (see Chapter 4), and it is these that are the targets of vaccine development.

B-cell epitopes activate the production of antibodies, which appear to be raised against proteins in their native configuration. It should be noted that many systems for detecting antibodies use denatured antigen and therefore favor linear epitopes. Even binding an antigen to a solid phase will result in the loss of some discontinuous epitopes.

For viruses, B-cell epitopes may be neutralizing or non-neutralizing. Neutralizing epitopes interfere directly with the processes of infection (receptor binding, entry, uncoating), and the binding of antibody to such epitopes is enough to render the virus non-infectious. Non-neutralizing epitopes do not prevent viral infectivity directly, but rather rely on other effector arms of the immune system (notably antibody-dependent killing by NK cells, enhanced phagocytosis, and the complement system) to target the F_c region (see Chapter 4) of particular subclasses of antibody that have been bound to viral proteins.

T-cell epitopes are rather different, since the processing of antigens for presentation to T cells involves their digestion within the cell into short oligopeptides that are then presented to the immune system (see Chapter 4). Thus, they can come from the internal proteins of the virus and from locations within the viral proteins themselves, rather than being located on the virus surface. This means that mapping using short synthetic peptides, often in large array-based systems, is particularly suited to the identification of T-cell epitopes.

Identification of epitopes

While traditional vaccines based on live or inactivated whole viruses contained multiple epitopes by default, this is not true of some biotechnology-based vaccines under development, which can contain only fragments of

Table 5.5 Epitope mapping

	Value for epitope types			
	Linear[a]	Discontinuous[b]	Disperse[c]	Post-translational[d]
Prediction from amino acid sequence[e]	+	±	±	±
Prediction from protein structure calculations	++	+	±	±
Immunoassay of proteolytic fragments	+++	+	–	++
Immunoassay of partial clones	+++	+	–	– to ++[f]
Pepscan mapping using synthetic peptides	+++	–	–	–
Crystallography of antigen–antibody complexes	+++	+++	+++	+++

[a] Epitope formed by amino acid residues in adjacent positions on peptide chain (includes T-cell epitopes). [b] Epitope formed by localized protein folding; also referred to as a conformational epitope. [c] Epitope formed from different regions of protein or from multiple proteins by complex folding; also referred to as a conformational epitope. [d] Epitope dependent upon post-translational processing events such as glycosylation or proteolytic cleavage. [e] Characteristics such as hydrophilicity may indicate surface availability for immunoreactivity. [f] Dependent upon degree of post-translational modification in expression system used.

viral antigens. For such vaccines it is essential to identify the immunogenic regions or epitopes within an antigen. Similarly, antigens prepared for use in diagnostic immunoassays are increasingly produced using cloning or synthetic methods. Locating epitopes can involve many different approaches, summarized in **Table 5.5**.

Of these approaches, only X-ray crystallography is optimized for the detection of discontinuous or disperse epitopes. However, this technique is laborious and technically demanding, and other approaches are frequently used. Increasingly these are based on *in silico* modeling—prediction from computer-based simulations.

In some cases, concentrations of epitopes are apparent on particular areas of a protein. This is well illustrated by the V3 loop region of the HIV gp120 glycoprotein (**Figure 5.9**). This is a short (34 to 36 amino acid) highly variable region that contains a principal neutralizing determinant of gp120. In addition to a powerful neutralizing B-cell (antibody) epitope, it contains epitopes for helper T cells and CTLs, as well as other less well-characterized determinants.

Due to this concentration of immunogenic sites, the V3 loop came under intensive study for use in vaccines during the 1990s. Indeed, some early approaches to the development of an HIV vaccine used the presentation of multiple V3 loops on a single vector (the so-called "V3 octopus"), but given its highly variable nature, the changeable pattern of its glycosylation (referred to as the "glycan shield"), and the folding of the mature gp120 protein (which hides the critical cell-binding region until an initial receptor binding has occurred), it is an extremely challenging target.

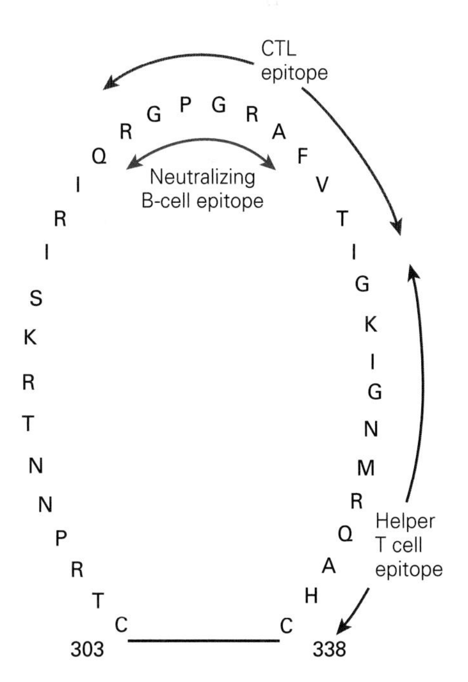

Figure 5.9
Schematic representation of HIV gp120 V3 loop epitopes.

5.9 SOCIAL OPPOSITION TO VACCINATION

Vaccination is an invasive medical procedure and is normally undertaken on healthy individuals in order to reduce the risk of a future disease rather than to treat a current one. In consequence, attitudes toward vaccination are often mixed.

Even in its early years, vaccination attracted considerable controversy. Anti-vaccination campaigns and organizations were quite common in both Europe and America in the nineteenth century. The situation was exacerbated by compulsory vaccination for smallpox in the United Kingdom and

Figure 5.10
James Gillray cartoon. This cartoon from 1802 reflects attitudes to Jenner's use of a cowpox-based vaccine against smallpox.

elsewhere. As well as James Gillray's memorable cartoon from 1802 (**Figure 5.10**), the dramatist George Bernard Shaw has been quoted as referring to vaccination as "a horrible reversion to the most degraded and abominable forms of tribal ritual." Looking back from after the elimination of the "global scourge" of smallpox, such attitudes may seem odd. However, they are still with us today.

The Anti-Vaccination Leagues of the nineteenth century have their equivalents in the Web sites of campaign groups and individuals opposed to vaccination, whether because of a belief that an individual vaccine is linked to a particular damaging effect, or because of objections to some component of multiple vaccines. Examples are not hard to find.

Something in the mixture

The campaign against the adjuvant MF59 in the USA (see Section 5.3) is one example of concerns about a component of a vaccine, even though extensive use in influenza vaccine in Europe along with testing in the United States has shown no safety issues, and this same adjuvant was used in the Novartis swine flu vaccine in 2009.

Similarly, the use of the organomercury compound thimerosal as a preservative and antibacterial agent in vaccines has attracted a great deal of controversy, despite support for its safety from the Food and Drug Administration (FDA), Centers for Disease Control and Prevention (CDC), and the World Health Organization (WHO). Overdoses of such compounds can of course cause toxicity, but it is also easy to forget that vaccines contain preservative for a reason (see **Box 5.1**).

Developed countries have now gone over to (more expensive) single-dose vaccine containers that can be used safely even when free of preservatives, but use of preservatives continues in the developing world, for compelling reasons.

The bottom line

Cost also remains an issue in the developed world, with worrying indications during 2009 that insurers in the USA were not paying the full costs of vaccination, leading some health care providers to cease offering routine childhood vaccinations. One community physician reported to be cutting back in this way was quoted as saying "It doesn't do me any good. I am losing money," summarizing his position as "no margin, no mission." Vaccination in such cases ends up being left to the public clinics, but vaccinations may just be missed instead. A spokesman from the CDC noted that when a similar situation arose between 1989 and 1991, many parents did not have their children vaccinated. This was reported to have led to 55,000 cases of measles, 11,000 hospitalizations, and 123 child deaths. The consequences for both individual and public health of an apparently simple financial decision can be severe. And then, of course, it is always possible to generate other reasons.

Polio vaccine conspiracy theories

Polio eradication has been a goal of the World Health Organization since 1988. By 2003 there were only six countries left where polio was actively circulating. The highest number of cases was in Nigeria. Then religious leaders in the north of the country declared that polio vaccine contained contraceptive agents as part of a "conspiracy to de-populate the developing world." Even though tests in Nigeria indicated otherwise, the state government of Kano withdrew from the vaccination program. The ban lasted eleven months, until vaccine supplies from Indonesia were declared acceptable. But even this relatively short interruption was to have consequences. Polio cases rose: in 2004 there were 789 cases in Nigeria; in 2005, 830; in 2006, 1129. Cases began spreading within Nigeria and into other countries. Eradication was set back once again, and paralytic polio claimed yet more victims.

In the early days of inoculation against smallpox, religious leaders tried to stop the treatment. Nearly 300 years later, it was still happening with polio.

One small study

Such events are not restricted to the developing world. In 1998 a paper based on a study of 12 patients was published in the prestigious British medical journal *The Lancet*, proposing a link between the measles–mumps–rubella triple vaccine (MMR) and childhood autism. This was based on what were reported as sudden and dramatic changes in this group following vaccination. An editorial in the same issue questioned the validity of this small study. The controversy was fed by concerns about adverse effects from whooping cough vaccine in Britain during the 1970s. Demand for single vaccinations for measles, mumps, and rubella soared, though the multiple visits required to complete such courses could leave children open to infection, even if this more complex regimen was completed successfully.

One often-cited suggestion was that the MMR vaccine "overloads" the immune system by using three live viral vaccines in one dose. This ignores the fact that three viruses together contain far fewer antigens than does a single bacterium. It should also be noted that the age at which the MMR vaccine is given is also the age at which autism is commonly diagnosed. Two independent events occurring around the same time can often be seen as linked, especially when parents are quite understandably looking for causes for something as serious as a diagnosis of autism.

Many scientific papers, including large meta-analyses combining the results of multiple studies, followed on from the initial *Lancet* report, looking at far larger numbers of patients. They found no such association. While one section of the British press, backed up by pressure groups, published more claims about adverse effects from MMR, other sections produced detailed criticisms of the work itself, of the way in which it was carried out, and of

Box 5.1 Additives can have a useful purpose

Sir Graham S. Wilson wrote in his book *The Hazards of Immunization* of a case where preservatives were not used:

"In January 1928, in the early stages of an immunization campaign against diphtheria, Dr. Ewing George Thomson, Medical Officer of Health of Bundaberg, began the injection of children with toxin-antitoxin mixture. The material was taken from an India-rubber-capped bottle containing 10 mL of TAM. On the 17th, 20th, 21st, and 24th January, Dr. Thomson injected subcutaneously a total of 21 children without ill effect. On the 27th a further 21 children were injected. Of these children, eleven died on the 28th and one on the 29th."

A Royal Commission investigation found that "The consideration of all possible evidence concerning the deaths at Bundaberg points to the injection of living *Staphylococci* as the cause of the fatalities."

They further concluded that "biological products in which the growth of a pathogenic organism is possible should not be issued in containers for repeated use unless there is a sufficient concentration of antiseptic (preservative) to inhibit bacterial growth."

Box 5.2 Herd immunity

Unsurprisingly, individuals tend to focus on whether a vaccine will protect them. Epidemiologists, on the other hand, know that even a vaccine that protects only partially can have important effects. By raising the general level of immunity in the population (herd immunity), it is possible to ensure that when a virus is shed from an infected individual the balance of probability is that it will not find a new susceptible host. This interruption of transmission means that, over time, the virus will be eliminated from the population. It is this high level of herd immunity that has, for example, resulted in the elimination of polio from the Western hemisphere.

This also means that a decision by a substantial section of the community not to vaccinate, whether for religious, social, or other reasons, does not just affect that section of the community. If the general level of immunity drops below a specific, critical level then the virus will begin circulating again. This is why vaccination levels need to be kept high—not vaccinating affects everyone.

the motivations behind it. In 2004, 10 of the 12 original authors published a formal retraction of the findings.

On 25th May 2010, the London *Times* newspaper reported that Dr Andrew Wakefield, the first author of the original *Lancet* paper, had been struck off the medical register. The *Times* stated that "after nearly three years of formal investigation by the General Medical Council (GMC), Dr Wakefield has been found guilty of serious professional misconduct over 'unethical' research that sparked unfounded fears that the vaccine was linked to bowel disease and autism."

Despite all of this, public opinion now associated the MMR vaccine with adverse effects. The fact that measles is a killer disease and that the period since vaccination was introduced had seen a more than 80% drop in cases did not seem to register with the public.

As with the rise in whooping cough seen after vaccination concerns in the 1970s, rates of measles in Britain began to rise. A BBC headline reported that the country was "in grip of measles outbreak," and also reported the first death in Britain due to measles for 14 years. Underlying this is the loss of herd immunity (see **Box 5.2**), which is particularly important for measles. However, despite rejection of the original narrow study on the basis of widespread evidence, public opinion could not be calmed easily. Ten years after the original scare, vaccination numbers were again rising—but too late to stop an avoidable death and many more unnecessary cases of this dangerous but preventable disease.

It is often said that scientists are the clerics of the modern age. It seems some of the old habits of such a role come with the title, and that even developed countries can react illogically when a scare story hits home.

5.10 PRACTICAL ISSUES AND OUTCOMES OF VACCINATION

While the use of vaccines can be controversial, this is rarely the case when they are seen as effective controls for a deadly disease. It is only when fear of the disease is almost eradicated that other concerns come to the fore.

The eradication of a disease

The situation in the early eighteenth century was very different to that today. Voltaire, writing in 1733, noted that "Upon a general Calculation, threescore Persons in every hundred have the Small-Pox. Of these threescore, twenty die of it in the most favourable Season of Life, and as many more wear the disagreeable Remains of it in their Faces so long as they live." One-fifth of the population dying, with another fifth seriously scarred.

At that time the best preventive was variolation, with a mortality rate of 1–2%, against 20% or more for smallpox. Once a safer approach (vaccination) became available, this was no longer the case. However, vaccinia can cause a pox-like illness—mild compared to that from variolation, but still clinically significant.

Historically, between 14 and 52 people per million receiving vaccinia smallpox vaccine would experience serious side effects, with 1–2 of them dying as a result. To put this in context, it is estimated that between 1900 and 1978 half a billion people were killed by smallpox. In that setting, vaccination using vaccinia virus made good sense.

In 1966 the World Health Organization (WHO) decided to try to eradicate smallpox using vaccination as the main tool (**Figure 5.11** and **Box 5.3**). This was a very large undertaking, unprecedented at the time. Challenges such as getting the vaccine to remote regions and into war zones were met, despite a very limited budget. Success came in 1975 with the last case of *variola major* in a Bangladeshi girl, followed in 1977 by the last case of *variola minor* in a Somali cook.

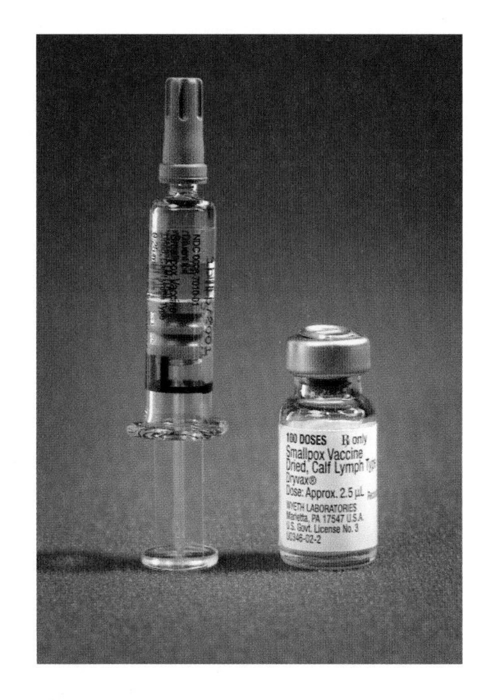

Figure 5.11
Smallpox vaccine (Dryvax®, Wyeth). From, CDC Public Health Image Library (http://phil.cdc.gov/).

It took another three years to confirm that these were indeed the last natural cases, a situation complicated by the escape of smallpox virus from a laboratory in Birmingham, UK in 1978 which caused the last smallpox death—a medical photographer at the University of Birmingham.

By some estimates, half a billion people were killed by smallpox in the twentieth century alone. The elimination of such a lethal disease provides the strongest possible evidence of the benefits of vaccination and remains one of the greatest successes of medicine. Jenner's often-cited prediction that "the annihilation of the Small Pox, the most dreadful scourge of the human species, must be the final result of this practice" was finally proven true.

Following on from success with smallpox, in 1988 the WHO next targeted polio, aiming to eradicate it by the year 2000. The target was missed, but the number of countries where polio is actively circulating had been reduced

Box 5.3 The unknown virus

The vaccine that eliminated smallpox was not cowpox. At some time after Jenner's work and before the tools for molecular characterization became available in the twentieth century, the cowpox virus was replaced by something else. The vaccinia virus that has been used now for many years is not cowpox. In fact, no-one knows what it is. It is clearly a pox virus, and is closely enough related to smallpox that it serves as an effective vaccine (see **Figure 1**), stimulating immune responses that protect against smallpox. There is some evidence that vaccinia may have come from buffalo, but there are also suggestions that it is a horse pox. Were it a new vaccine subject to modern approvals, there is little doubt that vaccinia would not be authorized for use due to this lack of knowledge of its origins. However, it is this poorly understood virus that eliminated smallpox.

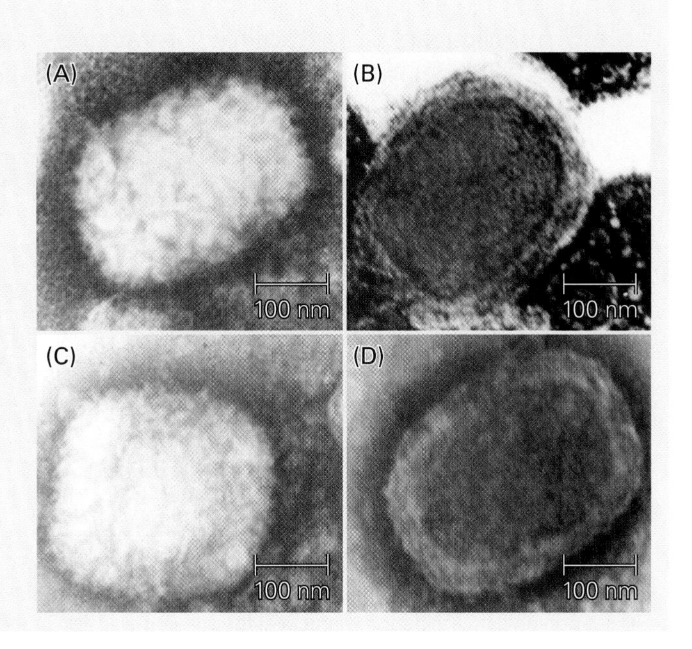

Figure 1
Smallpox and vaccinia viruses. (A) Mulberry (M) form of smallpox (variola) with visible surface proteins. (B) Capsular (C, membrane-covered) form of smallpox (variola). (C) Vaccinia, M (mulberry) form. (D) Vaccinia, C (capsular) form. From, CDC Public Health Image Library (http://phil.cdc.gov/).

from 125 to 4 (Nigeria, India, Pakistan, and Afghanistan), while the number of cases worldwide has been cut from 350,000 in 1988 to 1997 cases in 2006. One of the three major strains of polio has already been eradicated—the last case of type 2 poliovirus infection occurred in 1999. Eradication is likely, provided that social factors do not intervene (see Section 5.9).

In order to eradicate a viral disease by vaccination, it must have certain characteristics. As can be seen from **Table 5.6**, many virus diseases do not fit the profile for elimination using current technologies. While smallpox has been eliminated, eradication of polio is proving more difficult than expected. The high level of subclinical infections with poliovirus necessitates high levels of vaccination to break the infectious cycle, and this is proving challenging in some areas. In such cases it is necessary to raise the general level of immunity in the population (herd immunity, see Box 5.2) to the point where the virus can no longer circulate effectively.

Measles is also listed for eradication, but there is no realistic prospect of eliminating many other viral diseases including chickenpox, influenza, or HIV within the foreseeable future.

Risk and benefit

Given what is now known about the benefits of vaccination, there can be no real doubt that there are very significant health benefits for this approach, both for the individual and for the wider community. However, this is not to say that vaccination is always entirely safe. When analyzing the potential benefit, it is critically important to compare the risk of being vaccinated with the risk of not being vaccinated.

Today, with smallpox eliminated, even the relatively slight risk arising from the use of vaccinia vaccine is seen as unacceptable, and mass vaccination has ceased. Despite this, there are still groups who are vaccinated, for example in the US military to protect against threats of bioterrorism. In these groups vaccinia can cause side effects, and can also spread from vaccine recipients to those around them. As a result, safer alternatives are sought, now that vaccinia has eliminated this "most dreadful scourge of the human species."

Table 5.6 Indicators of suitability for vaccine-based eradication of virus infection

	Smallpox (eradicated 1977)	Polio (eradication under way)	Measles (targeted for eradication)	Chickenpox (controlled by vaccination, where used)	Influenza (moderated by vaccine)	HIV (no vaccine available)
No carrier state[1]	Yes	?	Yes	No	Yes	No
No subclinical infection[2]	Yes	No	?	?	Yes	No
Stable viral antigens[3]	Yes	Yes	Yes	Yes	No	No
No animal reservoir[4]	Yes	?	Yes	Yes	No	Yes
Effective vaccine[5]	Yes	Yes	Yes	Yes	?	No
Effectively controlled by natural immunity	Yes	Yes	Yes	Yes	?	No

[1] Virus causes acute disease and is then eliminated from the body. [2] Inapparent infections can cause cases to be missed, helping the spread of infection. [3] Antigenic variation weakens vaccine efficacy. [4] Re-infection of humans from animal reservoirs can prevent elimination unless vaccination is practical for the animal hosts. [5] Producing long-lasting immunity.

A similar situation exists with polio vaccine. While an inactivated "Salk" form of the vaccine (IPV) was the first to be developed, the predominant form to be used since the 1960s has been the trivalent "Sabin" live oral vaccine (OPV). OPV can cause paralytic polio approximately once for every million doses given. Against this, polio paralyzed 350,000 people worldwide in 1988, and nearly 58,000 in the United States alone in the epidemic year of 1952. Many of the survivors lost the ability to breathe unaided, and could spend the rest of their lives requiring breathing support from an external pressure respirator—an "iron lung" (**Figure 5.12**). Vaccination against polio made sense at the time. Now the use of OPV has eliminated the disease in the Western hemisphere and seems poised to do so worldwide. Where polio is an established and circulating (endemic) disease, a one in a million risk makes sense. But the last case of "wild" polio in the United States was in 1979, while vaccine-associated paralytic polio (VAPP) resulting from the use of OPV has caused 5–10 cases in most years since that time (**Figure 5.13**). Unsurprisingly, since 2000 the United States has switched to exclusive use of the inactivated (IPV) form of the vaccine. It is generally accepted that this is less immunogenic, and it requires injection rather than being given orally. However, it is an unavoidable fact that in the United States in the last quarter of the twentieth century the main risk was the vaccine itself.

Figure 5.12
An iron lung. Emerson respirators, more commonly referred to as "iron lungs," were used by polio victims whose ability to breathe was compromised by the disease. This particular machine was used for almost 50 years by a polio victim. From, CDC Public Health Image Library (http://phil.cdc.gov/).

Vaccination in infancy

While some (usually live) vaccines should not be given to the elderly or to the immunosuppressed, because of the possibility of disease due to the

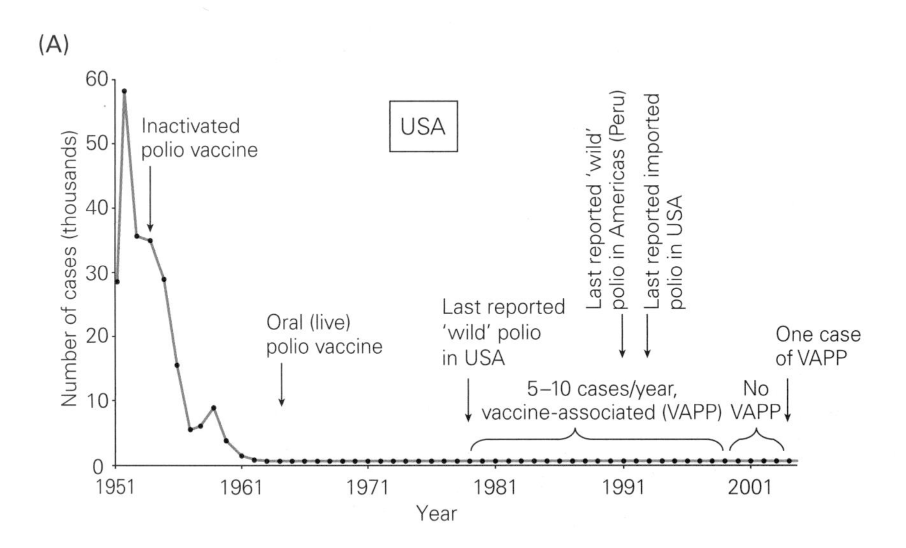

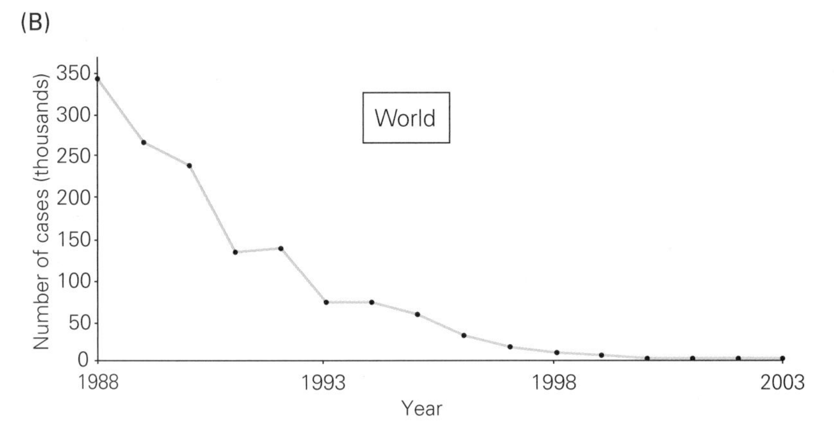

Figure 5.13
Numbers of reported polio (poliomyelitis) cases. (A) Cases in USA. (B) Global cases.

generally weakened immune system in the vaccine recipient, vaccination of infants is a more complex issue. It is clearly desirable to vaccinate children as early as possible in order to prevent disease. However, during the first few years of life the multiple elements of the immune system are still maturing. As well as a varying ability of the immune system to respond to a vaccine in the very young, there is also the complicating factor of maternal antibodies, transferred to the fetus *in utero*, which can limit the ability of the infant to mount an immune response following vaccination. These maternal antibodies persist for the first year of life, limiting the efficacy of vaccination during this period. As a result, vaccination of infants is strictly controlled, with defined schedules based around best evidence for efficacy.

Box 5.4 The Cutter Incident

When the inactivated polio vaccine was licensed in 1955 under pressure from high levels of polio in the community, a mass vaccination program began immediately. Within two weeks, five million doses of polio vaccine were sent out from five laboratories. Then, cases of polio in vaccinated children began to appear. Almost all were traced back to the vaccine made by Cutter Laboratories. Despite following the correct guidelines for manufacture, live poliovirus had survived the inactivation process. In a way that would be unacceptable today, the starting material was a fully virulent strain. If not inactivated, it could (and did) cause high levels of paralytic disease. The consequences were that over 100,000 children were infected who, in turn, infected even more; 70,000 developed muscle weakness, 164 developed severe paralysis, and 10 died. While often cited as a disaster, the Cutter Incident needs to be seen in context. It was the result of early production processes very different to those in use today, being pushed through in the middle of a polio epidemic. It should be remembered that three years earlier, at the peak of the epidemic, there were 58,000 cases of paralytic polio in the United States alone.

Production issues

Apart from unavoidable risks arising from the use of vaccines to counter far more damaging diseases, there are also risks arising from failures in the production process. Early batches of the inactivated polio vaccine were found to contain live, virulent virus due to flaws in the production process. Unfortunately, they were only found after they infected 100,000 children—an event known as the Cutter Incident (**Box 5.4**).

The Cutter Incident was to have profound effects on the whole pharmaceutical industry. It was this terrible series of mistakes (and the 60 lawsuits arising from it) that created the idea of no-fault liability. This in turn made such activities too risky for small companies. In 1957, 26 companies were making 5 vaccines in the United States. By 2004, only 4 companies were making 12 vaccines. The Cutter Incident was a major element in creating the current system, where a few big pharmaceutical companies dominate the market. Later, yet another problem with the polio vaccine was identified.

Endogenous growth

Since live vaccines are grown in animal or human cells, contaminating or endogenous (literally, produced from within) viruses can be a problem. This was observed with early batches of polio vaccine, in which simian virus 40 (SV40, *Polyomaviridae*, **Figure 5.14**), a highly transforming virus in some systems, was also produced by the monkey cells used to grow the virus. Many early batches of the inactivated polio vaccine contained live SV40, since it was not inactivated by the treatments used to inactivate the poliovirus. It is also believed to have been present in some live (OPV) preparations from that time.

It is estimated that 10–30 million of the 98 million Americans vaccinated may have received SV40-contaminated vaccine in the period 1955–1963. Given the properties of SV40, there was a great deal of concern about this. However, long-term monitoring has shown no SV40-associated disease in recipients of contaminated vaccine. Though SV40 DNA has been found in some cancers, in particular those of the brain, no causal role has been established. On this occasion, the bullet was (probably) dodged.

Viral contamination of such a widely used vaccine indicated the need for great care in the use of animal cells for vaccine preparation. Advances in testing, particularly at the nucleic acid level, have contributed to a much greater ability to prevent such incidents, and cell types used are now very closely monitored by the relevant authorities. It is telling that many issues since this time have been identified early in the development of vaccines. Unfortunately, examples still occur, such as the recent identification of (apparently harmless) porcine circovirus (*Circoviridae*) material in multiple rotavirus (*Reoviridae*) vaccines.

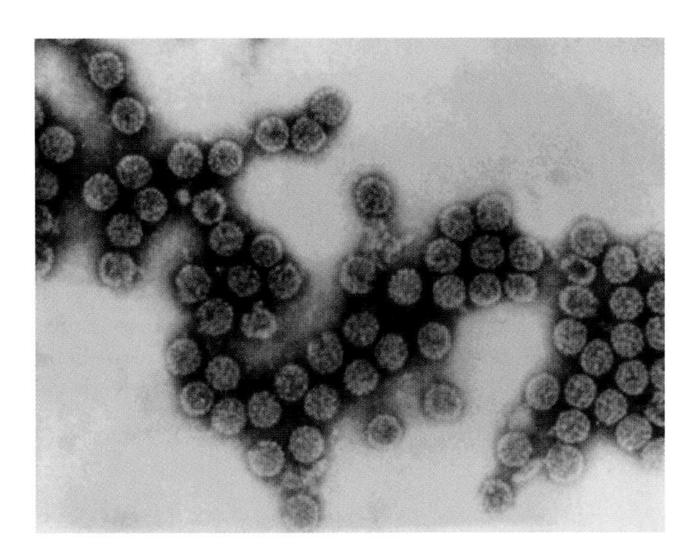

Figure 5.14
SV40 (*Polyomaviridae*), a contaminant of early polio vaccines. From, CDC Public Health Image Library (http://phil.cdc.gov/).

The wrong type of immunity

In 1969 it was reported that an experimental vaccine containing inactivated respiratory syncytial virus (RSV, *Paramyxoviridae*) actually increased the severity of subsequent infection. With RSV, formalin treatment during vaccine preparation altered the immunogenicity of the viral glycoproteins, and the resulting immunity was directed toward a non-neutralizing, inflammatory response that potentiated RSV disease.

Potentiation of disease by the wrong type of immunity is also seen with many other viruses, including measles (*Paramyxoviridae*), Epstein-Barr virus (EBV, *Herpesviridae*), SARS (Severe Acute Respiratory Syndrome), Coronavirus (*Coronaviridae*), and Dengue virus (*Flaviviridae*). There seems to be a particular need for concern where a virus infects cells of the immune system that bear F_C (antibody) receptors, since bound antibody that does not neutralize the virus can then help it to get inside its target cell.

In the recent STEP large-scale HIV vaccine trial, more vaccinees became infected (49/914) than placebo-treated controls (33/922). Not much of a difference, perhaps, but worrying. Particularly so since there seemed to be some association with the strength of reaction to the adenovirus vector used.

Direct effects

There is also the possibility of direct pathogenic effects: one major concern with subunit vaccines based on the HIV gp120 glycoprotein is that this protein has been implicated in the induction of apoptosis (programmed cell death) in cells of the immune system. This is clearly not a desirable property for a vaccine to exhibit, and it is necessary to consider the potential effects of any antigen to be used in this way.

In a case with more clinical relevance, Rotashield®, the first approved vaccine against rotavirus (*Reoviridae*), was withdrawn 14 months after its approval by the FDA. The vaccine was a tetravalent live vaccine with human genes expressed in a simian rotavirus. Following licensing, 15 cases of intussusception (folding of the bowel, causing obstruction) were reported in the first 1,500,000 vaccinees. Further analysis identified the risk as 1 in 5000, which was considered unacceptably high. No mechanism by which the vaccine might cause intussusception has been proven, although several are suggested. Although further analysis has suggested that the actual rate may be lower than that originally suggested, the vaccine is no longer in use in the USA.

Monitoring safety

As can be seen from the problems outlined above, it is essential that vaccines are evaluated thoroughly for safety. In clinical trials, the issue of safety is addressed before efficacy against the target disease is even considered. In addition to antigen effects, toxicity associated with experimental adjuvants and any other vaccine component also needs to be considered, emphasizing again the need for careful testing of novel vaccine formulations.

Monitoring for safety continues even after a vaccine is approved for use. In the case of Rotashield®, reports to the Vaccine Adverse Event Reporting System (VAERS) operated jointly by the FDA and CDC identified the issues. VAERS exists specifically to monitor problems arising from vaccine use, since adverse effects may occur at levels too low to be detected in even the largest clinical trial.

Future potential

In considering the future of human vaccination, it is useful to look at veterinary vaccines. Here, some of the more advanced technologies are already in use. Two DNA vaccines and nine recombinant viral vectors are approved for use in North America, alongside the traditional live and inactivated vaccines. While use in animals does not necessarily lead on to use in man, these must be seen as promising signs. However, when considering the many scientifically intriguing areas of vaccine development, it must be borne in mind that the relatively crude attenuated live vaccines from the last century saved many lives, eliminated one global killer, and are closing on another. In the veterinary field it was a traditional vaccine that now seems to have eliminated rinderpest.

However vaccines are produced, they remain one of the great success stories of modern health care.

Key Concepts

- Vaccination, the use of a weakened form of a disease agent to protect against natural infection, has a 2000-year history. The modern era of vaccination began in 1796 with the work of Edward Jenner.
- There are four basic types of vaccine in current use: live attenuated (a weakened form of the original virus), inactivated ("killed" virus), subunit (proteins purified from the original virus), and cloned subunit (viral proteins expressed in another system, often yeast cells).
- Many advanced biotechnological approaches are under development, but as yet the traditional vaccine formulations remain the most widely used.
- HIV is now (and is likely to remain as) the greatest challenge for viral vaccine development, not least since the type of immunity needed to protect against infection remains undefined.
- Vaccine efficacy may be potentiated by adjuvants, co-administered chemicals that enhance the immune response. Despite much promise, only a very few are licensed.
- New routes and methods for the administration of vaccines have the potential to shape the immune response, though understanding of the basis of effective immunity is needed. The sheer complexity of the immune system makes this a challenging area for research.
- DNA vaccines, naked DNA molecules expressing antigens of choice, promise a rapid and simple vaccine development pathway, but have yet to deliver for humans.
- The use of therapeutic vaccination (the use of vaccines to control existing disease rather than to prevent new disease) is blurring the lines between vaccines and antiviral drugs, as is the use of immunomodulators and antibody preparations as antiviral agents.
- Vaccination is the only human intervention to eradicate a major disease, smallpox. Two more virus-induced diseases, polio and measles, are targeted for eradication.
- Vaccine use always has been controversial and remains so today. In large part this is because vaccines are generally given to healthy people to prevent rather than to cure a disease, so that the benefit to the recipient is not always apparent.

DEPTH OF UNDERSTANDING QUESTIONS

Hints to the answers are given at http://www.garlandscience.com/viruses

Question 5.1: Why is vaccination controversial in the face of its real successes in controlling human disease? How can this be changed?

Question 5.2: All vaccines have side effects, which must be outweighed by the risks of the real disease. What level of side effects is acceptable?

Question 5.3: What is an adjuvant? Why has the development of adjuvants been slower than anticipated?

FURTHER READING

Dyer C (2010) Wakefield was dishonest and irresponsible over MMR research, says GMC. *BMJ* 340, 593.

Gerber JS & Offit PA (2009) Vaccines and autism: a tale of shifting hypotheses. *Clin. Infect. Dis.* 48, 456–461.

Graham BS & Crowe JE Jr (2007) Immunization against viral diseases. In Fields Virology, 5th ed. (DM Knipe, PM Howley eds). Lippincott Williams & Wilkins, Philadelphia.

Minor PD (2007) Viruses. In Encyclopedia of Life Sciences. John Wiley & Sons, Chichester. http://www.els.net/

Whitley RJ & Roizman B (2002) Herpes simplex viruses: is a vaccine tenable? *J. Clin. Invest.* 110, 145–151.

INTERNET RESOURCES

Much information on the internet is of variable quality. For validated information, PubMed (http://www.ncbi.nlm.nih.gov/pubmed/) is extremely useful.

Please note that URL addresses may change.

CDC Vaccines & Immunizations. http://www.cdc.gov/vaccines/

DNA Vaccine. http://dnavaccine.com/

Encyclopedia of Life Sciences. http://www.mrw.interscience.wiley.com/emrw/9780470015902/els/article/a0003386/current/abstract (article on disease eradication)

HIV Vaccine Trials Network. http://www.hvtn.org/ (details of activities in the development and testing of HIV vaccines)

National Institute of Allergy and Infectious Diseases Vaccines Research Center. http://www.niaid.nih.gov/about/organization/vrc/Pages/default.aspx

United Kingdom Health Protection Agency Vaccination Immunisation. http://www.hpa.org.uk/infections/topics_az/vaccination/vacc_menu.htm

VaccinePlace. http://vaccines.com/

World Health Organization Vaccines. http://www.who.int/topics/vaccines/en/

CHAPTER 6
Antiviral Drugs

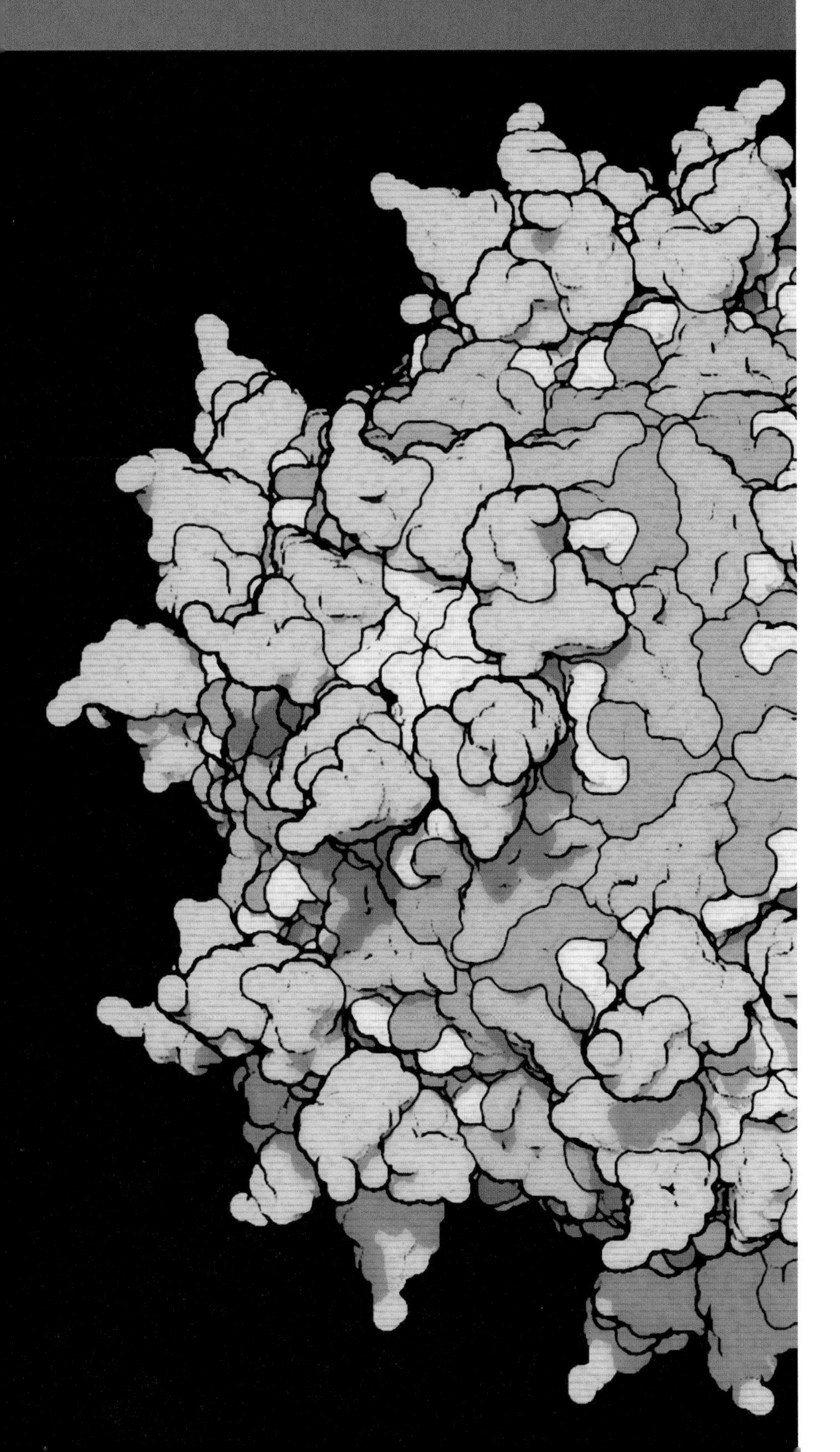

INTRODUCTION

Antiviral drugs have come a long way in the 50 years since their first discovery. The technology has moved from early drugs with marginal effects and significant toxicity to precisely designed inhibitors of specific viral functions. Similarly, the underlying science has moved from simple screening of potential compounds to exquisitely detailed understanding of viral targets. Currently there are over 40 licensed antiviral drugs from 11 broadly defined classes, targeting viruses from eight families. However, the classical challenges of antiviral drug development remain: most drugs affect only one or a few related viruses, toxic side effects remain for many current drugs, and the chance of failure for any drug candidate remains high. The development of resistance to a newly developed drug can be rapid. Understanding how these issues can be tackled underlies the field of antiviral drug development.

About the chapter opener image
Rhinovirus
(Courtesy of the Research Collaboratory for Structural Bioinformatics Protein Data Bank and David S. Goodsell, The Scripps Research Institute, USA.)

6.1 EARLY ANTIVIRAL DRUG DEVELOPMENT

The first antiviral drugs were derived from the sulfonamide antibiotics in the early 1950s, when thiosemicarbazone compounds were found to be active against poxviruses. This raised the possibility of antiviral chemotherapy against smallpox (*Poxviridae*), which was still a major cause of human disease at that time. Further efforts identified methisazone, a synthetic thiosemicarbazone, which appeared to be effective in preventing smallpox and in controlling complications of vaccination with the live vaccinia virus vaccine. With the thiosemicarbazones, the ratio between the toxic dose and the effective therapeutic dose (known as the **therapeutic index**) was small and, in time, vaccination was to prove far more effective at bringing smallpox under control (see Chapter 5).

In the following years, other antiviral compounds were identified, but toxicity meant that many of them were restricted in their use. For example, the anti-herpes drugs trifluorothymidine and idoxuridine were used against keratitis caused by herpes simplex virus (HSV, *Herpesviridae*), but they were too toxic to be given systemically (**Table 6.1**). Instead, they were administered directly to the surface of the eye, which allowed high localized concentrations of drug at the area of need. Thus, these drugs could only be used where the site of infection was directly accessible. It was not until the development of antiviral drugs that could be given systemically that infections within the body could be addressed with acceptable levels of safety. Herpes simplex can cause a life-threatening encephalitis, and the first antiviral drug that could be used to treat it was vidarabine (adenosine arabinoside) which is given intravenously but is metabolized (and inactivated) rapidly once within the body. Vidarabine, in turn, was replaced by aciclovir (Zovirax®), which had superior activity and could be given orally. However, standard dosing was high and had to be taken at very regular intervals. As a result, aciclovir was then replaced for many uses by its valine ester, valaciclovir (Valtrex®), which has improved oral bioavailability. This illustrates both the ongoing nature of drug development and the limitations of early drugs.

6.2 TOXICITY

It should be noted that concerns over toxicity are not unique to antiviral drugs. Many cancer therapies still raise such concerns, while the treatment of bacterial infections before the development of antibiotics also used relatively toxic compounds. One 1911 textbook notes that for treatment of syphilis the patient "has a deposit of mercury settle all over his body" (**Figure 6.1**), which was done in order to avoid the toxic effects resulting from the

Table 6.1 Evolution of antiviral drugs in clinical use against herpes simplex infection

Date	Drug(s)	Route	Toxicity	Efficacy	Specificity
1960s	Idoxuridine, trifluorothymidine	Topical only	High	Moderate	Low
1970s	Vidarabine	Injection Topical	Low	Moderate	Moderate
1980s	Aciclovir, penciclovir	Topical Injection Oral (low efficiency)	Low	High	High
1990s	Valaciclovir, famciclovir	Optimized for oral use (high efficiency)	Low	High	High

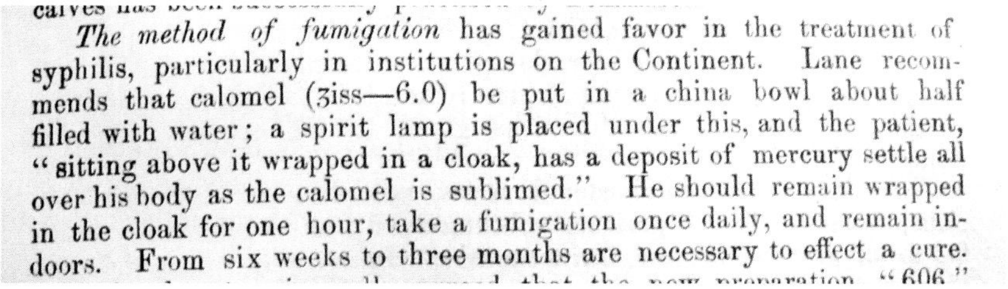

The method of fumigation has gained favor in the treatment of syphilis, particularly in institutions on the Continent. Lane recommends that calomel (ʒiss—6.0) be put in a china bowl about half filled with water; a spirit lamp is placed under this, and the patient, "sitting above it wrapped in a cloak, has a deposit of mercury settle all over his body as the calomel is sublimed." He should remain wrapped in the cloak for one hour, take a fumigation once daily, and remain indoors. From six weeks to three months are necessary to effect a cure.

Figure 6.1
The method of fumigation for the treatment of syphilis with mercury.
From, Anders JM (1911) A Text-book of the Practice of Medicine, 10th ed. WB Saunders, Philadelphia.

concentrated administration of the drug to any one site. Given our current knowledge of the toxic effects of mercury this may seem odd, but it was the best approach with the treatments available at the time. It should also be borne in mind that our own therapies are likely to seem equally limited a hundred years from now.

Toxicity with antiviral drugs results in large part from the overlap between viral and cellular metabolism (**Figure 6.2**); many processes in viral replication use cellular pathways and enzymes. A great deal of effort has gone into identifying specific functions which are essential to the virus but which are either not present in or not essential to the cells of the host organism. Even where specific viral targets exist, similarities between viral and cellular metabolism may mean that inhibition of cellular activities can cause toxic effects. A good example of this are the nucleoside analogs used against human immunodeficiency virus (HIV, *Retroviridae*). These antiviral drugs resemble nucleosides but generally lack the 3′-hydroxyl residue in the sugar ring to which the next nucleotide will be attached. Thus, they terminate DNA synthesis when incorporated in a forming DNA chain. The viral RNA-dependent DNA polymerase (reverse transcriptase) incorporates them far more efficiently than the cellular DNA polymerases, which is the basis of their selective antiviral effect. However, while this is true of the nuclear polymerases, the mitochondria of the cell contain their own DNA polymerase (pol γ), which is less selective. Incorporation of these drugs by the mitochondrial polymerase is a major element in observed toxicities. This reflects the complexity of the cellular environment, where a multitude of factors have to be taken into consideration that may not be apparent from biochemical studies *in vitro*.

6.3 DEVELOPMENT OF ANTIVIRAL DRUGS

There are many ways to develop a drug, but there are two basic approaches to the identification of candidate molecules for initial testing: high-throughput screening of extensive libraries of candidate compounds, and rational design where a molecule is optimized to interact with a known target structure. It should be noted that these approaches can be complementary

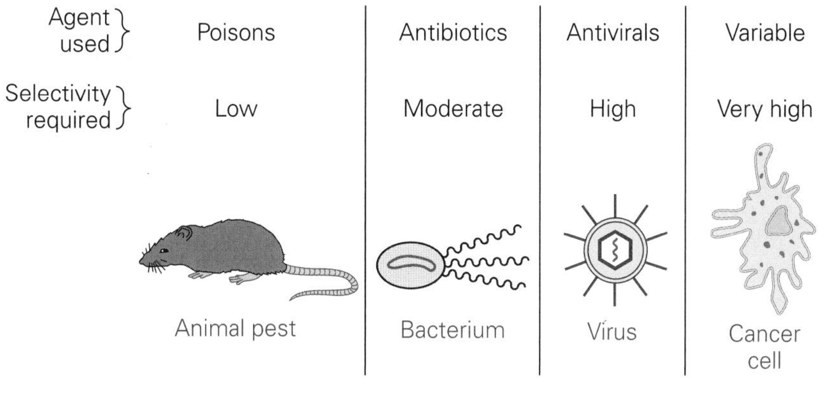

Agent used	Poisons	Antibiotics	Antivirals	Variable
Selectivity required	Low	Moderate	High	Very high
	Animal pest	Bacterium	Virus	Cancer cell

Figure 6.2
Comparison of selectivities required for antiviral drugs and other agents.

rather than mutually exclusive, and are frequently combined in the production of effective drugs.

Once identified, all drug candidates must go through a rigorous program of testing, moving from the laboratory into *in vivo* systems and then into clinical trials in humans. While time-consuming and expensive, this highly regulated process is intended to provide solid evidence for both safety and efficacy before a drug is licensed for general use.

The development pathway: preclinical, clinical, and beyond

All compounds being developed as novel drugs are required to follow a similar development pathway (**Figure 6.3**).

Initial target identification and computer modeling generates large numbers of possible candidates; for example, there are estimated to be over three million potential targets arising from the human genome project alone.

Initial testing, often using high-throughput technology, will highlight a smaller number of candidates. Typically a particular enzyme or pathway will be selected for development, and inhibitors identified. This stage can involve a very large amount of work to develop a **lead compound** for further testing.

Following *in vitro* testing and identification of the lead compound, *in vivo* studies are carried out. Despite significant and ongoing progress in using systems such as cell culture, this stage of the work will usually involve animal systems, since this kind of data is required both by regulatory authorities and to assess effects in whole body systems (for example, where the immune system is involved). Where no usable animal system exists (for example, with VZV, where the only animal to show disease similar to that in humans is the gorilla) it may be possible to substitute *in vitro* data, but this is rare.

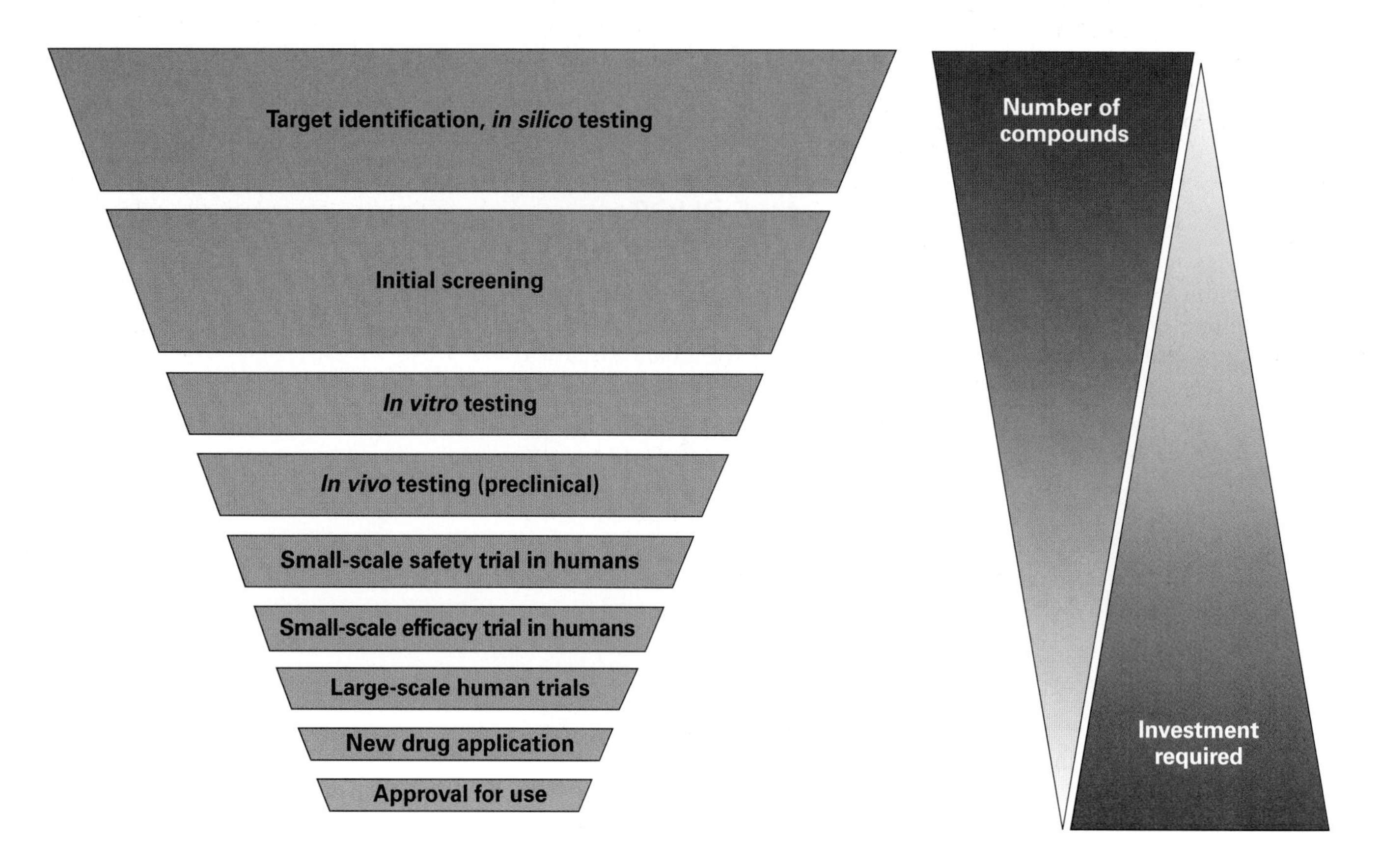

Figure 6.3
The development pathway for novel drugs.

While studies in nonhuman animals can provide useful data, their predictive value for human uses is necessarily limited. So, following preclinical studies, **clinical trials** are carried out in order to evaluate the safety and efficacy of a drug candidate in humans (**Box 6.1**). In the United States, very small-scale "first in human" studies with low dosing levels of drug are sometimes referred to as *phase 0 clinical trials*. The usual first step, however, is *phase 1 clinical trials*, which are small-scale trials in healthy volunteers. Typically they will involve 20 or more patients, though some may involve much lower numbers. They address safety and aspects of drug metabolism (referred to as ADME: absorption, distribution, metabolism, and excretion), but do not provide data on how well the drug works (efficacy data) since the target condition or infection is not present. Patient numbers are typically low at this stage. Occasionally a larger-scale safety trial may be required, referred to as a *phase 1b clinical trial*. If acceptable safety and drug metabolism are indicated by the trial results the drug can then progress to the next stage.

Phase 2 clinical trials are small-scale (typically 20–100 participants) trials in patients with the target condition, and are intended to test the efficacy of the drug. Where there are sound reasons why data from healthy volunteers would not be meaningful they may sometimes be combined with phase 1 trials in a so-called *phase 1/2 clinical trial*, but again this is not usual. As with phase 1 trials, a larger *phase 2b clinical trial* may be undertaken to provide further evidence of efficacy.

Following a successful phase 2 trial, a drug can then enter a large-scale (typically 500–1500 patients, but sometimes much larger) *phase 3 clinical trial*. This is an efficacy trial, conducted in patients with the target condition, but also generates data on less common side effects that might not be seen in earlier, smaller trials.

After successful phase 3 trials (usually two are required), the company developing the drug may apply for approval of the drug for sale (in the USA, a New Drug Application or NDA; in the European Union, a Marketing Application Approval or MAA). This is a lengthy process, addressing many aspects of the production, presentation, and marketing of the drug. It may also be followed by other reviews such as those carried out by the National Institute for Health and Clinical Excellence (NICE) in the United Kingdom before adoption by the state-funded National Health Service (NHS).

Even after marketing of a drug, it is possible for additional clinical trials (known as *phase 4 clinical trials*) to be requested if results of using the drug raise issues of concern or the possibility of new applications.

Production of a drug can also be an issue. At some point (often after completion of phase 2 trials) production will be transferred from the laboratory to the industrial level. The scale-up of production will require different skills, and there can be many pitfalls. For example, oseltamivir is produced by a long synthetic process from a base compound (shikimic acid) isolated from Chinese star anise. Given fears over avian influenza, many countries were stockpiling oseltamivir and there was a significant shortage during 2005, since the manufacturers were already using 90% of the available crop. These stockpiles were then available with the new pandemic of H1N1 ("Mexican") influenza starting in 2009.

Box 6.1 The clinical trials process

Phase 0
A relatively new term used in the US for studies of reduced dosing levels in small numbers of patients. Not yet in general use

Phase 1
Initial "first in human" safety and metabolism study in healthy volunteers, small number of participants (typically 4+ patients)

Phase 1b
Expanded phase 1 study (preceding study may be referred to as phase 1a)

Phase 1/2 study
Combined "first in human" study of safety and efficacy, small number of participants

Phase 2a
Study of efficacy in patients with target condition or infection, small number of participants (typically 20+ patients)

Phase 2b
Expanded study of efficacy in patients with target condition or infection, larger number of participants (typically 100+ patients)

Phase 3
Large-scale study of safety and efficacy in patients with target condition or infection (typically 500+ patients, maybe more than 10,000)

NDA/MAA
New Drug Application (US) or Marketing Application Approval (EU), application for permission to sell and market a new drug

Phase 4
Post-licensing trial to address issues of concern or potential new applications

High-throughput screening

The classical approach to drug development involved testing large numbers of molecules for activity in the desired system. Such screening is now often carried out by specialist units or companies holding extensive libraries of compounds, using automated equipment handling nanoliter (or even smaller) volumes. This allows testing of libraries of hundreds of thousands or even millions of compounds within a reasonable time scale, although the assays are of necessity rather limited. For example, assays may be restricted to one type of solvent for the compounds under assay by the liquid-handling technology involved, even though varying the solvent can have significant effects on activity. High-throughput screening allows identification of compounds without any underlying reason to suspect that they might be active, and can include entirely novel compounds isolated from natural sources such as marine life or land invertebrates.

Rational design

An alternative approach is to model the intended target and to use computer software to identify molecules likely to have the desired activity. Testing of theoretical structures using computer modeling is referred to as *in silico* testing, and may also be used at later stages of development to investigate potential activities or problems.

The first step to such design is usually to resolve the structure of the target molecule using high-resolution techniques such as **X-ray crystallography**, where the diffraction of beams of X-rays is used to determine the arrangement of atoms within a crystalline test material. The crystalline form of a biological molecule may have a variant structure (as was observed in early work with DNA), so the results of such studies must be interpreted with care. However, they are able to provide precise details of conformation, for example during target binding. This allows potential interactions to be mapped out and molecules designed to fit with the target. Potential inhibitors may also be based on molecules that mimic the substrate halfway through its enzyme-catalyzed reaction, the so-called **transition state mimetics**, such as the HIV protease inhibitor ritonavir.

However, even a molecule which binds to its target perfectly will need considerable further development to make it suitable for therapeutic use. This involves multiple elements such as:

- Increasing specificity (increasing activity, decreasing toxicity)
- Stabilizing the molecule (minimizing degradation)
- Facilitating entry into cells
- Optimizing uptake by the desired route of administration
- Refining partition to target sites within the body

All of these, together with many other elements specific to the drug in question, will have to be addressed during the long development process from an inhibitor to a drug.

Examples of molecules that have been designed using such systems include the protease inhibitors (for HIV as shown in Table 6.2, and under development for other viruses), neuraminidase inhibitors (oseltamivir and zanamivir for influenza), and binding inhibitors (pleconaril, in clinical trials for rhinoviruses).

An excellent example of rational design was the design of inhibitors of the HIV protease. In molecular studies of HIV replication, many aspects of virus metabolism were characterized. Among these was a protease function essential for the formation of mature (infectious) virus particles. The protease was identified as an unusual enzyme, both on the basis of its structure and that it appeared to rely on the presence of an aspartic acid residue in its active site. The apparent scarcity of such proteases in normal cells meant that this presented a promising target for the development of an antiviral drug, as inhibition of its function was thought less likely to interfere with cellular metabolism.

Inhibitors based on the target peptide cleaved by the HIV protease were developed, and these were then refined by chemical modifications giving known alterations in the properties of the inhibitors until a highly active and specific inhibitor was produced. The relative structures of the natural peptide substrate of the HIV-1 protease and saquinavir, the first of the protease inhibitor drugs to be licensed for the treatment of HIV, are shown in **Figure 6.4**. The modifications were selected to make this inhibitor interact very efficiently with the active site of the viral protease and to inhibit its function as a result of this high avidity, as well as to enhance its suitability for use as a drug. The value of this approach is shown by the fact that there are now ten such drugs in clinical use.

Not all virus targets are of equal value. This is shown by the failure of efforts to develop equivalent drugs targeting the herpesvirus maturational protease, the structure of which is more similar to cellular proteases, with a serine residue in the active site.

Similar approaches underlie the development of the neuraminidase inhibitors for the treatment of influenza infections. However, it should not be imagined that this approach is universally successful. Pleconaril, hailed by some media sources as the cure for the common cold, was in clinical trials from 1997 to 2002. It was based on a rational design of an inhibitor to block the receptor-binding pocket known as a canyon on the surface of the virus. This canyon conceals the receptor-binding site from the immune system, and was identified as a promising target for small-molecule drugs. Pleconaril was the first such drug to be submitted to the US Food and Drug Administration (FDA) for approval. It was rejected because, first, it had a very limited effect on the symptoms of the common cold and second, it had an undesirable interaction with oral contraceptives—something that could not be predicted from molecular studies. As of 2010, pleconaril has been reformulated and is once again in clinical trials, but with no guarantee of success.

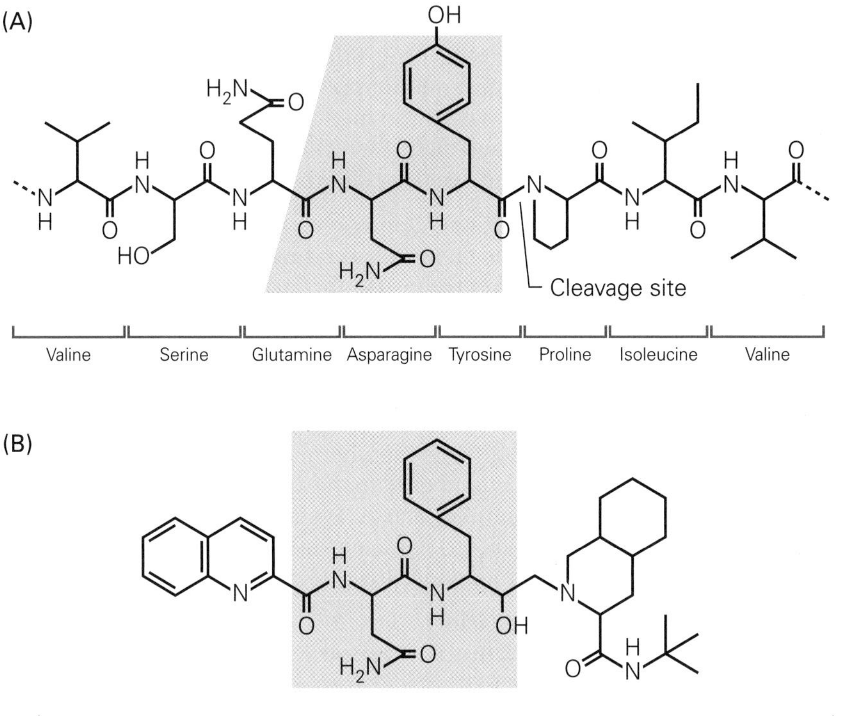

(A)

Cleavage site

Valine Serine Glutamine Asparagine Tyrosine Proline Isoleucine Valine

(B)

Figure 6.4
Structures of the natural substrate of the HIV-1 proteinase (A) and the proteinase inhibitor saquinavir/Invirase (B). Note similarity of regions in blue box.

(A)

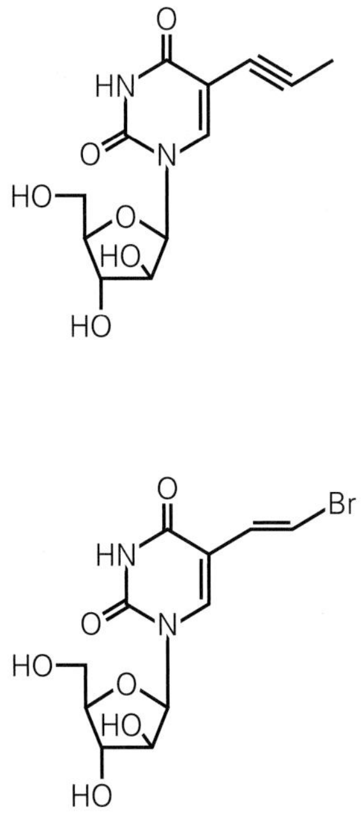

(B)

Figure 6.5
Examples of nucleoside analogs with enhanced activity against varicella-zoster virus that have failed during development. (A) Netivudine: 1-[β-D-arabinofuranosyl]-5 propynyluracil; (B) Sorivudine; 1-β-D-arabinofuranosyl-E-5-[2-bromovinyl]uracil.

Development from existing drugs

Development of many drugs, including antivirals, involves the extension of approaches known to work in one area to another, related area. An example of this is the attempted development of nucleoside analogs with improved activity against those herpesviruses (CMV, EBV, and VZV) that are less inhibited by available drugs.

Although this can be seen as a lower-risk approach, success is by no means guaranteed. In the case of VZV, two candidate drugs that initially showed promise (**Figure 6.5**) ran into severe problems, and the contrast between the two illustrates some of the ways in which drugs can fail.

Netivudine was 10 times more active against VZV than aciclovir, but failed relatively quietly. Long-term preclinical toxicity assays in rodents identified concerns and the drug was withdrawn from testing before any human trials were undertaken.

Sorivudine was 10,000 times more active against VZV than aciclovir. However, a metabolite of the drug inhibits metabolism of fluoropyrimidines by the enzyme dihydropyrimidine dehydrogenase. Fluoropyrimidines are synthetic compounds, so this is not normally a concern. However, some of the patients treated with sorivudine in a Japanese trial were also being treated with fluoropyrimidine drugs, including 5-fluorouracil, which is an anticancer agent also used **off label** (use in an individual patient based on specific need rather than general licensing for that application) for papillomavirus infections. The toxicity which built up resulted in the deaths of three patients during trials. The drug was authorized for use in Japan in part because the basis of the interaction was understood. Monitoring identified a further 23 cases of severe toxicity after approval of sorivudine, with 16 deaths. Adding to the situation, serious concerns were identified concerning interactions with an antifungal fluoropyrimidine drug, flucytosine. Sorivudine was withdrawn from use in Japan. A subsequent application for FDA approval was turned down due to concern over these interactions. The development of sorivudine was then stopped and all further trials cancelled. This illustrates a high-profile, costly, and late-stage drug failure.

Rather more successful evolutionary approaches underlie the development of other drugs, for example the chain-terminating nucleoside analogs and the proteinase inhibitors currently in use for HIV. These have produced seven and ten licensed drugs, respectively. In both cases, the first such drugs to be licensed (zidovudine and saquinavir), although still in use, have been supplemented by newer drugs with more desirable characteristics.

Another approach to enhancing existing drugs are attempts to make these drugs more effective by modifying the chemical structure to get larger amounts of the orally administered drug to the site of infection (to increase its oral bioavailability); examples active against herpesviruses include valaciclovir (derived from aciclovir), famciclovir (derived from penciclovir), and valganciclovir (derived from ganciclovir) (**Figure 6.6**)

With valaciclovir and valganciclovir, an attached valine residue greatly enhances oral bioavailability, but is removed within the body so that the delivered antiviral drug is actually aciclovir or ganciclovir. Thus, these are

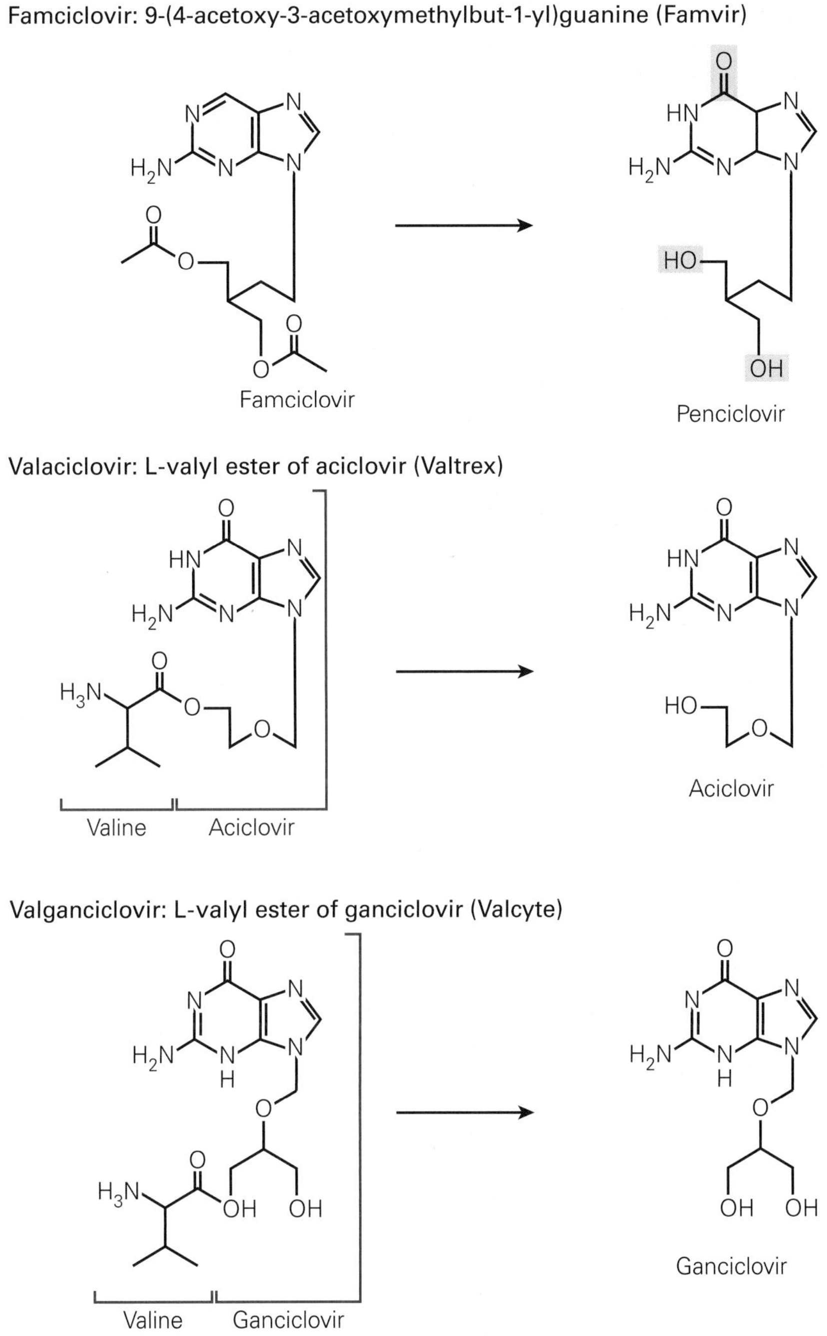

Famciclovir: 9-(4-acetoxy-3-acetoxymethylbut-1-yl)guanine (Famvir)

Famciclovir → Penciclovir

Valaciclovir: L-valyl ester of aciclovir (Valtrex)

Valine | Aciclovir → Aciclovir

Valganciclovir: L-valyl ester of ganciclovir (Valcyte)

Valine | Ganciclovir → Ganciclovir

Figure 6.6
Examples of improved nucleoside analog-based prodrugs with improved oral bioavailability (from comparison of serum levels with oral dose levels) for use against herpesviruses. Elements of the prodrugs shown on the left are removed to create the active form, shown on the right. Famciclovir has 50-times higher oral bioavailability than penciclovir (structural changes from famciclovir are shown by blue boxes). Valaciclovir has 4-times higher oral bioavailability than aciclovir, as the valine residue is removed within the body. Valganciclovir has 11-times higher oral bioavailability than ganciclovir, as the valine residue is removed within the body.

prodrugs (precursor forms, converted to the active drug). The parent drugs aciclovir and ganciclovir are also prodrugs of the active triphosphate form which is then incorporated into forming DNA. The prodrug famciclovir is the diacetyl ester of 6-deoxypenciclovir and is converted to penciclovir (and then to the triphosphate form). Again, while penciclovir is poorly absorbed by the oral route, famciclovir has good oral bioavailability.

The cost of drug development

Progress to a marketed drug is both lengthy and expensive. Results from one 2001 study showed a less than 2% chance of a compound in late-stage *in vitro* testing actually reaching the market, with an average cost of over

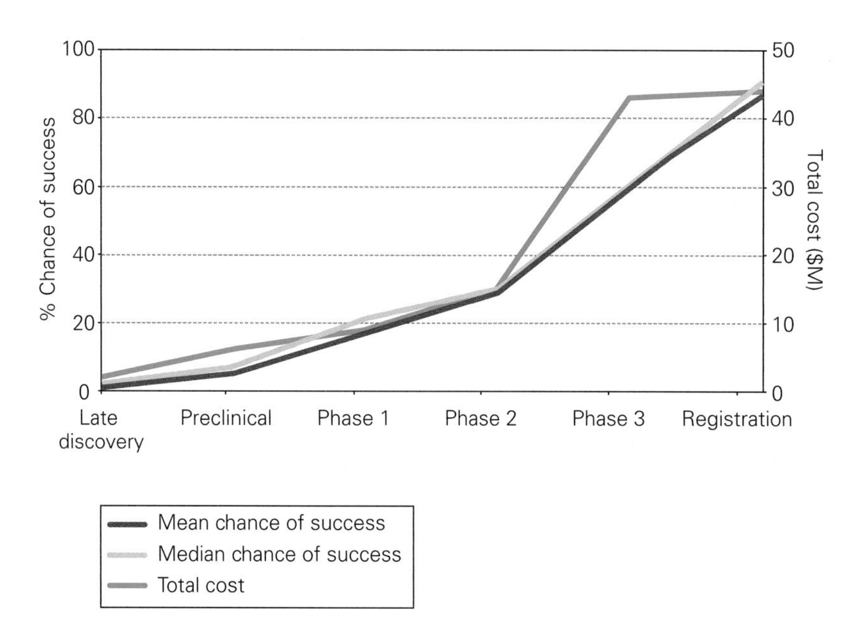

Figure 6.7
Example of the success probabilities and costs for the development of novel drugs. Data from the Report of the Pharmaco-Economics Working Group of the Rockefeller Foundation Microbicide Initiative (2001).

$50 million (**Figure 6.7**) for progression to approval for sale, most of which is spent on large-scale trials in the later stages of development.

Opponents of high drug prices see the substantial income generated by successful drugs, but drug companies are quick to point out the high costs of development (and the even higher cost of failure) when challenged on the apparently high price of their products.

Failures can happen at any and all stages of the development process. For example, it is important to note that animal models need not give results that can be extrapolated fully into human systems. A recent and very high-profile example of this was the immune activator TGN1412 (see **Box 6.2**).

Box 6.2 TGN1412: A bad news story

TGN1412 was a humanized monoclonal antibody intended for use against B-cell leukemia and arthritis, which entered initial safety trials in humans in March 2006, following animal testing in rabbits and monkeys. It targeted human immune activators, cross-linking T-cell receptors to produce strong immunostimulation. However, promising results in animal immune systems did not correspond to the response produced in humans. Even the low doses used in initial trials produced massive immune activation in the six human test subjects, apparently involving an overwhelming cytokine response and producing profound and damaging effects. The effects of the reaction to the drug were severe, and all six participants required lengthy and intensive hospitalization. Long-term damage to their health appears to have resulted from the trial.

This outcome illustrates that simple extrapolation from animal data is not reliable. A review of procedures then took place, and additional precautions are now in place for candidate drugs of this type. TeGenero, the German company that developed TGN1412, closed down as a result of the trial outcome.

6.4 CURRENT ANTIVIRAL DRUGS

Figure 6.8 shows the stages of viral infections amenable to antiviral intervention with examples of drugs that operate at each stage. Currently available antiviral drugs are detailed in **Tables 6.2** and **6.3**. While many of these are *nucleoside analogs*, which are mistaken for DNA components by the viral enzymes and thus inhibit viral DNA synthesis, there are now many different classes.

The effects of the **non-nucleoside reverse transcriptase inhibitors (NNRTIs)** are to inhibit the DNA polymerase by binding away from the active site in a region known as the NNRTI pocket. This is similar to the mode of action of the anti-herpes drug foscarnet, which binds to the pyrophosphate-binding site rather than to the active site of the DNA polymerase.

Protease inhibitors are analogs of peptides rather than nucleic acids, and block maturational protein cleavages which results in the release of non-infectious virus. In a similar development to that of NNRTIs, non-peptide analog proteinase inhibitors are also under development, most notably bevirimat (PA457), which similarly inhibits maturational protease activity, but is actually a derivative of betulinic acid.

A more recent development are the **fusion inhibitors**, which inhibit virus entry to cells. Both docosanol (for herpesviruses) and enfuvirtide (for HIV) work in this way.

Antiviral agents targeting influenza virus do not fit in to any of the above categories. Amantadine and rimantadine function by blocking an ion channel

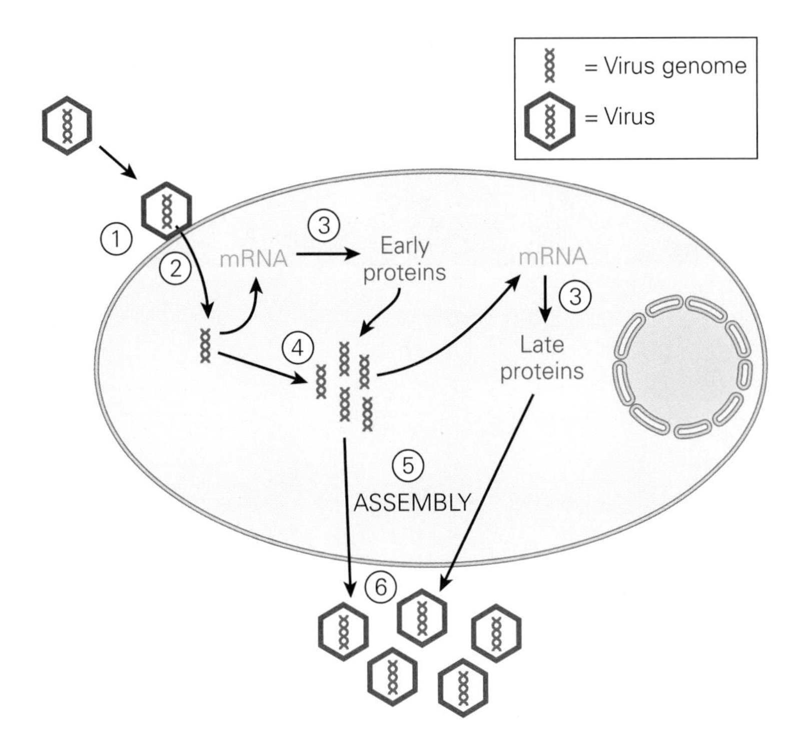

	Stage	Examples
1	Binding to viral receptor	Pleconaril, palivizumab
2	Penetration of cell	Amantadine, enfuvirtide, n-docosanol
3	mRNA function	Interferon
4	DNA synthesis	Nucleoside analogs, nucleotide analogs, NNRTIs
	or	
	RNA synthesis (genomic)	Ribavirin
5	Viral assembly	Protease inhibitors
6	Transport and release of virus	Neuraminidase inhibitors (influenza), amantadine (some influenza strains)

Figure 6.8
Stages of viral infection available to antiviral intervention.

that is required for uncoating of the virus (and, in some cases, for maturation). The **neuraminidase inhibitors** function by interfering with release of influenza viruses from the cell surface, leaving the virus attached to sialic (neuraminic) acid residues on cellular glycoproteins. As well as the current neuraminidase inhibitors oseltamivir (Tamiflu) and zanamivir (Relenza), two novel neuraminidase inhibitors (laninamivir and peramivir) have been licensed in Japan, though not yet in the United States. Laninamivir is an aerosolized drug with a long duration of activity within the body, allowing a single dose to replace a multi-day course of the other drugs in this class, while peramivir is administered intravenously, facilitating its use in seriously ill patients.

The mode of action of some early antiviral drugs is less well defined, since they were identified experimentally, rather than designed to act on a specific molecule. The anti-herpes agents idoxuridine and trifluridine are incorporated into the forming nucleic acid but do not terminate chain formation; they appear to function by disrupting chain formation after their

Table 6.2 Antiviral drugs in current use: targets and types by virus family

Drug type	*Arena-viridae* (Lassa fever)	*Flavi-viridae* (Hepatitis C)	*Hepadna-viridae* (Hepatitis B)	*Herpes-viridae* (HSV, VZV, CMV, HHV-8)	*Orthomyxo-viridae* (Influenza)	*Papilloma-viridae* (HPV)	*Paramyxo-viridae* (RSV)	*Retro-viridae* (HIV)
Nucleoside analog		1	2	9			1	7
Nucleotide analog			1	1				1
Ribonucleoside analog	1	1		1			1	
NNRTI								4
Protease inhibitor								10
Fusion inhibitor				1				1
Receptor antagonist								1
Integrase inhibitor								1
Ion channel blocker					2			
Neuraminidase inhibitor					4			
Immuno-modulator		1	1	2		3		
Oligo-nucleotide				1				
Monoclonal antibody							1	
Glycosylation inhibitor				2				
TOTAL	1	3	4	17	6	3	3	25

Note that among those listed above, some individual drugs are active against multiple classes of virus (see Table 6.3) and thus are included more than once.

incorporation. Others such as vidarabine and ribavirin act by as-yet undefined mechanisms. Although it has been suggested that ribavirin (which works against a wide range of RNA viruses) may induce enough errors in the replication of the viral RNA genome to cause an **error catastrophe**, where copying of the genes from generation to generation (which is already error-prone in RNA viruses, see Chapter 1) drops below sustainable levels, this is now not thought to be the case, and it appears inhibition of cellular nucleic acid metabolism may be the key event.

One class of antivirals with necessarily complex functions are those that affect the immune system. These include imiquimod, inosine pranobex, and, of course, interferon which is itself an element of the *innate immune system*. Palivizumab is a *humanized monoclonal antibody* (see Section 6.6)

Table 6.3 Antiviral drugs in current use by type (see Appendix for additional information)

Virus target	Drug type and mode of action	Drugs
Human immunodeficiency virus (*Retroviridae*)	Nucleoside analog, reverse transcriptase inhibitor	Abacavir, didanosine, emtricitabine, lamivudine, stavudine, zalcitabine, zidovudine
	Nucleotide analog, reverse transcriptase inhibitor	Tenofovir
	Non-nucleoside reverse transcriptase inhibitor (NNRTI)	Delavirdine, efavirenz, etravirene, nevirapine
	Protease inhibitor, disrupts maturational proteolysis	Atazanavir, darunavir, fosamprenavir[1], indinavir, lopinavir, nelfinavir, ritonavir, saquinavir, tipranavir
	Fusion inhibitor, blocks virus entry	Enfuvirtide
	Chemokine receptor antagonist, blocks virus entry	Maraviroc
	Integrase inhibitor, blocks viral genome integration	Raltegravir
Herpes simplex virus (*Herpesviridae*)	Nucleoside analog, DNA polymerase inhibitor	Aciclovir, famciclovir, penciclovir, valaciclovir
	Nucleoside analog, disrupts DNA synthesis	Idoxuridine, trifluorothymidine
	Ribonucleoside analog, disrupts DNA synthesis	Vidarabine
	Fusion inhibitor, blocks virus entry	Docosanol
	Immunomodulator, stimulates immune system	Inosine pranobex
	Non-nucleotide DNA polymerase inhibitor	Foscarnet
	Glycosylation inhibitor; disrupts glycosylation, entry, and uncoating	Tromantadine
Varicella-zoster virus (*Herpesviridae*)	Nucleoside analog, DNA polymerase inhibitor	Brivudine, famciclovir, valaciclovir
Cytomegalovirus (*Herpesviridae*)	Nucleoside analog, DNA polymerase inhibitor	Ganciclovir, valganciclovir
	Nucleotide analog, DNA polymerase inhibitor	Cidofovir
	Oligonucleotide, inhibits expression of viral genes	Fomivirsen
Human herpesvirus 8 (*Herpesviridae*)	Immunomodulator, cellular and systemic effects	Interferon alfa
Papillomaviruses (*Papovaviridae*)	Immunomodulator, stimulates immune system	Imiquimod, inosine pranobex
	Immunomodulator, cellular and systemic effects	Interferon alfa
Hepatitis B virus (*Hepadnaviridae*)	Nucleoside analog, reverse transcriptase inhibitor	Entecavir, lamivudine
	Nucleotide analog, reverse transcriptase inhibitor	Adefovir dipivoxil
	Immunomodulator, cellular and systemic effects	Interferon alfa
Influenza A virus (*Orthomyxoviridae*)	Ion channel blocker, interferes with virus entry and uncoating	Amantadine, rimantadine
Influenza A and B viruses (*Orthomyxoviridae*)	Neuraminidase inhibitor, inhibits extracellular mobility	Oseltamivir, zanamivir, laninamivir, peramivir
Respiratory syncytial virus (*Paramyxoviridae*)	Monoclonal antibody; neutralizing antibody, uptake inhibitor	Palivizumab
Hepatitis C virus (*Flaviviridae*)	Immunomodulator, cellular and systemic effects	Interferon alfa
Multiple RNA viruses (*Paramyxoviridae, Arenaviridae, Flaviviridae*)	Ribonucleoside analog, mode of action unconfirmed	Ribavirin

[1]Prodrug of and replacement for amprenavir.

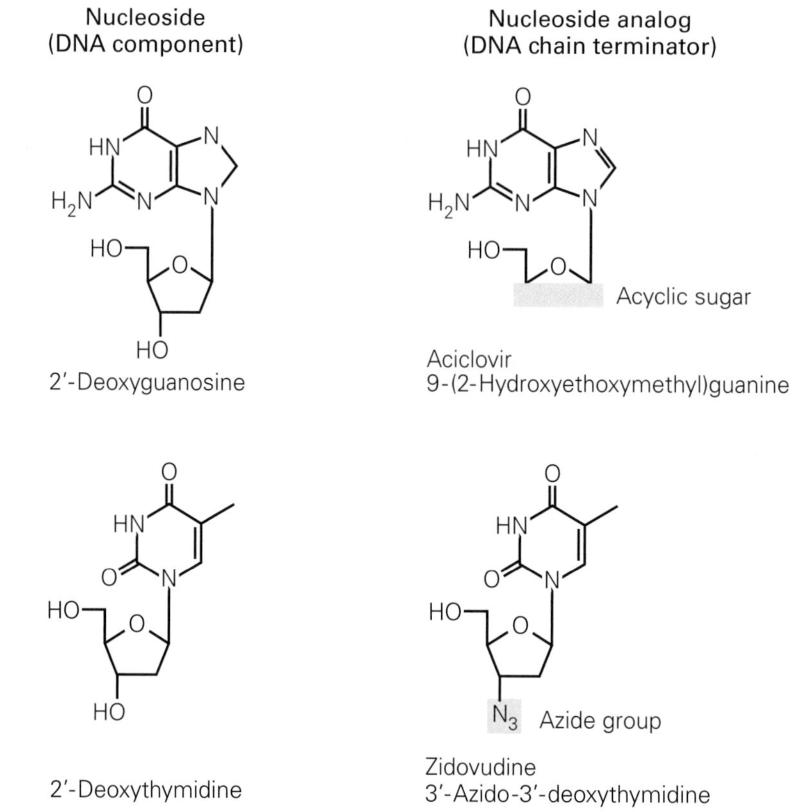

Figure 6.9
Nucleosides and nucleoside analogs.

Nucleoside (DNA component)

Nucleoside analog (DNA chain terminator)

2′-Deoxyguanosine

Aciclovir
9-(2-Hydroxyethoxymethyl)guanine

Acyclic sugar

2′-Deoxythymidine

Zidovudine
3′-Azido-3′-deoxythymidine

N_3 Azide group

that binds to the F (fusion) surface glycoprotein of respiratory syncytial virus and thus inhibits entry to the cell, while also providing targeting of the immune system to the virus itself. However, it needs to be given before infection (prophylaxis) rather than as a therapy. While palivizumab is the first monoclonal antibody targeting a virus to be licensed for use, other monoclonal antibody therapeutics are in use, for example against psoriasis where a range of such agents is available. Monoclonal antibodies and interferons are covered in more detail in Section 6.6.

Finally, fomivirsen is a 21-base antisense oligonucleotide with a phosphorothioate modification (substitution of a sulfur for an oxygen atom) that inhibits nuclease degradation. It was designed to inhibit expression of the immediate early genes of CMV. Direct injection into the eye gets round the problem of the bioavailability of this large molecule. However, while an excellent rationale exists for how this agent works, recent results have suggested that other pathways may be involved.

Many other functions, both virus specific and (more rarely) cellular, which are essential to the virus are being studied for possible antiviral uses.

Nucleoside analogs: understanding the mechanism

Aciclovir (9-[2-hydroxyethoxymethyl]guanine, also formerly known as acyclovir) is a **nucleoside analog**, similar to the DNA component guanosine but with an acyclic sugar group (see **Figure 6.9**). A more detailed analysis of just how such drugs work provides a useful background to understanding the mechanisms of action of antiviral drugs in general.

Aciclovir is effective against HSV, which causes genital and oral herpes, and to a lesser extent against VZV, which causes chickenpox and shingles. As mentioned above, aciclovir itself is a *prodrug*; it is converted within the body into the active form—in this case, aciclovir triphosphate. The selectivity of aciclovir is based on the presence in herpesvirus-infected cells of a viral enzyme, thymidine kinase (TK), which initiates the phosphorylation of aciclovir, converting it to the monophosphate form. Cellular kinase enzymes then continue this process, producing aciclovir triphosphate (**Figure 6.10**). The cellular thymidine kinase is more selective (by a factor of approximately 3000 times) and will not phosphorylate aciclovir at significant levels, leaving it in the inactive prodrug form. In addition, the cellular enzyme is only produced by cells at specific stages of their life cycle, but the viral thymidine kinase is routinely produced by HSV since it allows them to replicate in cells without waiting for the cellular enzyme to appear. Aciclovir triphosphate is then used by the viral DNA polymerase as a component of the growing DNA chain. Cellular polymerases are approximately 100 times

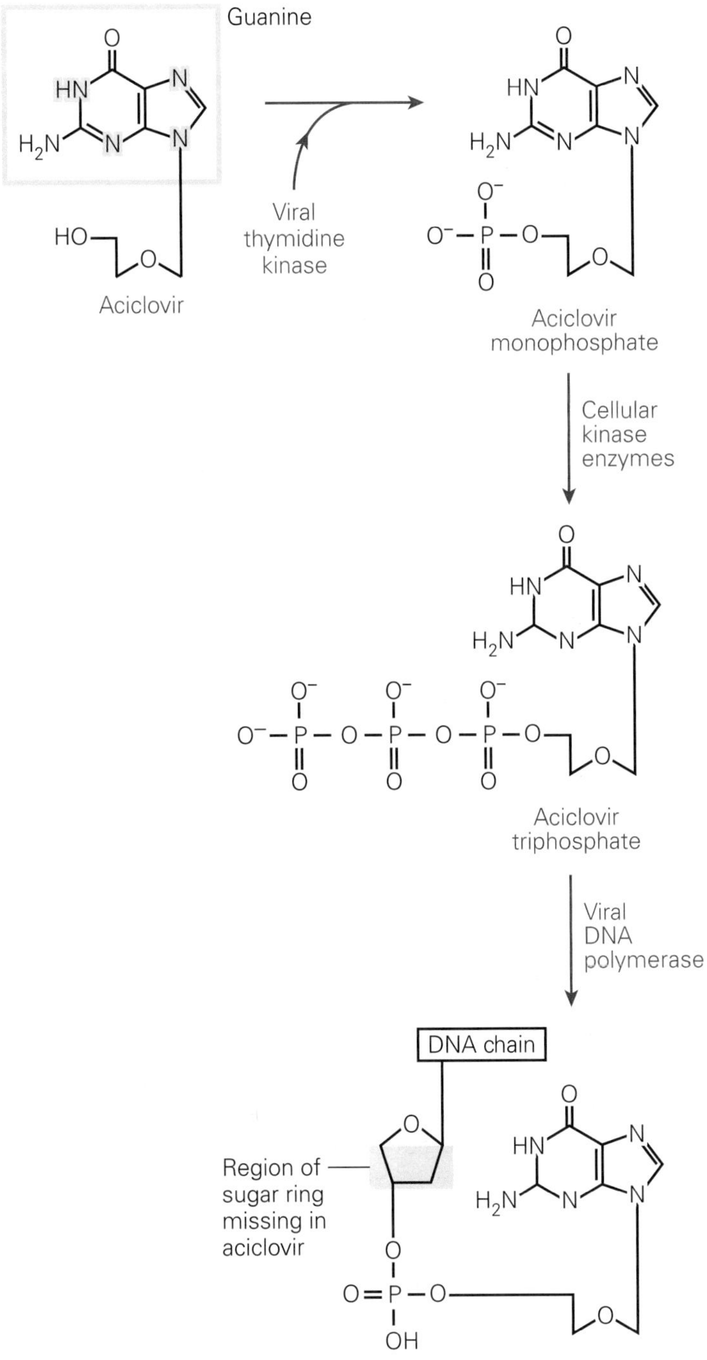

Figure 6.10
The activation of aciclovir by sequential phosphorylation prior to insertion into the forming DNA chain.

less likely to incorporate aciclovir triphosphate into DNA. This two-stage selectivity (producing 300,000-fold selectivity) results in a low level of toxicity for aciclovir. When aciclovir is added to the DNA chain, the lack of a large part of the deoxyribose sugar (and, specifically, of the 3′-hydroxyl group on the sugar ring to which the next nucleotide is added) means that no further bases can be added, and DNA synthesis is terminated (Figure 6.10). This effect means that aciclovir, along with many other nucleoside analogs lacking the 3′-hydroxyl group, is an obligate chain terminator. In the case of aciclovir there is an additional effect, since when trying to add the next nucleotide the viral DNA polymerase becomes trapped in an inactive complex. As a result of the requirement for viral enzymes to activate and incorporate aciclovir, the drug is effective against the virus while showing very low toxicity to uninfected cells.

Another nucleoside analog drug that illustrates this principle is the anti-HIV drug zidovudine (3′-azido-3′-deoxythymidine, azidothymidine, AZT; see Figure 6.9). Unlike aciclovir, zidovudine is phosphorylated to the triphosphate form by cellular kinase enzymes. As noted in Section 6.2, most cellular DNA polymerases do not incorporate zidovudine triphosphate into the growing DNA chain, but the mitochondrial DNA polymerase γ does appear to do so. Studies have shown that zidovudine is incorporated at much lower levels than some other nucleoside analogs, notably didanosine and zalcitabine, but mitochondrial toxicity remains a concern. The HIV reverse transcriptase enzyme is less selective, and will incorporate zidovudine triphosphate at higher levels of efficiency. The azide (N_3) group on the deoxyribose sugar ring is present (Figure 6.9) instead of the hydroxyl group that the DNA would normally be extended from, and blocks the addition of further bases to the DNA chain, terminating DNA synthesis. However, the HIV reverse transcriptase (RNA-dependent DNA polymerase) also contains a proofreading activity and is able to remove incorporated zidovudine, which is important in drug resistance. This activity is inhibited by the anti-HIV (NNRTI) drug nevirapine, and this appears to be one reason why a combination of these drugs is effective.

6.5 NUCLEIC ACID-BASED APPROACHES

Nucleic acid antivirals can be both highly specific and relatively simple to produce. If a gene sequence is known, it could in theory be possible to produce a specific inhibitor within a few days. Another reason for the high level of interest is the possibility that nucleic acid drugs need not rely on disruption of active viral metabolism. As a result, they could allow the targeting even of inactive viruses, and this approach has been the subject of investigation for many years. However, this approach has yet to fulfill that early promise. Although industry estimates suggest that the whole sector was worth $1.2 billion in 2010 across all indications, this is very limited by current biopharmaceutical standards. Despite this, the potential of such drugs continues to generate significant interest in this area.

Oligonucleotides

Therapeutic oligonucleotides may be synthetic and introduced from outside the cell, or they may be produced from expression vectors—DNA constructs that often contain viral sequences and which drive the expression of inserted genes introduced into the cell (see Chapter 9 for more detail). However, the latter approach would involve both genetic manipulation and a form of gene therapy, and is thus highly experimental.

If synthetic oligonucleotides are used these are often modified to enhance their activity. One common modification is the inclusion of sulfur to replace

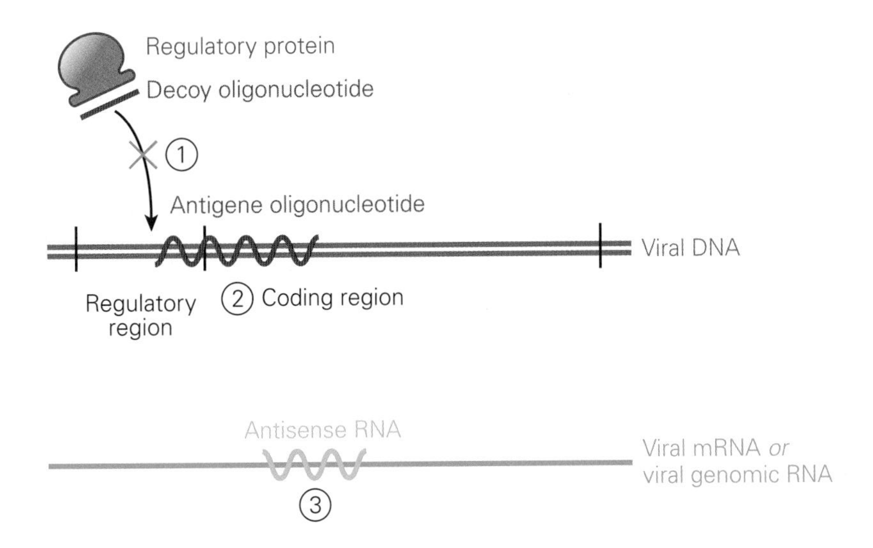

Figure 6.11
Nucleic acid-based approaches to antiviral drug development. (1) Decoy oligonucleotides; (2) antigene inhibition; (3) antisense inhibition. Small interfering RNAs (siRNAs) and micro-RNAs (miRNAs) function at multiple levels, notably the targeting of expressed mRNAs for destruction via bonding of siRNAs.

one of the oxygens in the phosphate group, referred to as **phosphorothioate modification**. This stabilizes the oligonucleotide, which is important since these molecules are readily degraded by nucleases. There are three main approaches, summarized in **Figure 6.11**:

- **Antisense inhibition**, where the target is a single-stranded nucleic acid (mRNA or viral RNA)
- **Antigene inhibition**, where the target is a double-stranded DNA
- **Decoy oligonucleotides**, where the target is a regulatory protein

With antisense strategies, an RNA complementary to a viral (genomic, replicative, or most often messenger) RNA is introduced into an infected cell, where it can form a dsRNA duplex by base pairing with the viral genomic or messenger RNA. Simply by locking up the viral RNA in a duplex, its function can be inhibited. Alternatively, the RNA duplex may be digested by dsRNA-specific cellular nucleases, which can result in the production of siRNAs (see below). dsRNA is also a potent inducer of a localized interferon response, which can induce antiviral effects in the cell.

Antigene oligonucleotides can be produced by similar routes, but bind to the DNA itself, forming a short region where the classical double helix of DNA is replaced by a triple helix, which can be stabilized by the presence of abnormal bases in the oligonucleotide or the chemical linkage of intercalating agents (which slide between the chains of a double helix) to the molecule. Such a structure appears to inhibit gene transcription as well as interactions with regulatory proteins in the region of the triple helix. While effects have been apparent in some systems with this strategy, they remain highly experimental.

Decoy RNAs, which correspond to regulatory regions of the genome or transcripts from it, bind the limited supply of regulatory proteins and prevent their function. Studies have targeted regions including the TAR (trans-activation response element) and RRE (Rev-response element) of HIV. The viral trans-activating protein Tat binds to the TAR region, preventing termination of transcription, while binding of the viral Rev protein to RRE allows mRNA transport. Both of these functions are essential for viral replication, and disruption can exert a strong antiviral effect in experimental studies.

Ribozymes

Ribozymes present an alternative approach to the therapeutic use of nucleic acids. They act as RNA-only enzymes, capable of cutting RNA molecules at specific sequences without the need for proteins.

Their name derives from combining the terms ribonucleic acid and enzyme. They were first identified from work with the RNAs of the protozoan *Tetrahymena*. Further work was performed using the RNA of viroids, sub-viral pathogens of plants which consist only of intricately folded RNA which can cleave itself—which is helpful since viroids do not produce any proteins. Ribozyme cleavage can involve one RNA molecule cutting itself, or the cutting of one RNA within a structure containing two RNA chains. Cleavage by ribozymes occurs at precise sequences (due to the requirement for matching sequences of bases), and it is possible to design synthetic ribozymes that will cut at a sequence specific to the genes of a target virus, giving a very high selectivity. While ribozyme substrates appear to be limited to RNA (owing to the lack of the reactive 2′-hydroxyl group in DNA, see Figure 3.10) they are able to interfere with any RNA stage of viral replication. Many viruses have RNA genomes, and ribozymes can inactivate the viral genome itself. Even DNA viruses must use mRNA intermediates, which can be targeted by ribozymes.

Ribozymes have been shown to suppress transformed cell characteristics with EBV and papillomaviruses, and have been investigated for activity against a range of viruses, including some very early clinical work with HIV in the 1990s. However, ribozymes are large molecules, posing significant problems of delivery, and it is now believed that the majority of potential ribozyme binding sites may be blocked by folding of the target RNA.

RNA interference: siRNAs and miRNAs

RNA interference (**RNAi**) is a radical development in the field of nucleic acid inhibition first identified in plants in the early 1990s, and later in the nematode roundworm *Caenorhabditis elegans*. While it was known that dsRNA could be degraded by cellular enzymes, it became clear that dsRNA forms the target of a specific enzyme, known as *Dicer*, that cuts the RNA into fragments of 20–25 nucleotides. These are then converted to single-stranded form and incorporated into a protein—nucleic acid complex known as the **RNA-induced silencing complex** (RISC). These **small interfering RNA** (**siRNA**) molecules recognize and base pair with identical sequences in larger RNA molecules, which are then degraded by the RISC. Thus, gene expression of specific mRNAs can be shut down. RNAi underlies many different reported mechanisms of gene expression inhibition, including *post-transcriptional gene silencing*, *transgene silencing*, and *quelling*. It is also the basis for the apparent activity of many antisense molecules, and siRNAs themselves appear to form part of a larger class of **micro-RNAs** (**miRNAs**). Early studies have had to be adapted to take account of this apparently universal phenomenon, challenging many of the accepted concepts in nucleic acid drug development in general, and antisense technology in particular.

RNAi is now under investigation for use against a range of viruses, including hepatitis A, hepatitis B, HIV, influenza, measles, and respiratory syncytial virus. There is significant potential for future therapeutic applications, but whether it will be possible to deliver effective drugs based on this approach remains to be proven.

Approved drugs

Despite many years of research and development activity, only one nucleic acid drug has been licensed for use: fomivirsen. This antisense oligonucleotide stabilized with phosphorothioate modifications is specific for the immediate early genes of CMV (a herpesvirus). Fomivirsen also illustrates some of the limitations of nucleic acid antivirals.

First, because of the large size and polar nature of the molecule, it cannot be taken orally or even systemically. Since the target infection is retinitis caused by CMV, the drug is inoculated directly into the eye. While CMV retinitis was a relatively common and extremely serious condition in the early days of the AIDS epidemic, it is now far less common due to the success of combination therapy with multiple anti-HIV drugs (see Section 6.8) and this route of administration is by its very nature rather demanding.

Second, while fomivirsen is very much a rational design, intended to target specific CMV genes, there is considerable dispute about whether it actually does act in this way, or whether some other effect is responsible.

6.6 IMMUNOTHERAPIES

The prophylactic and therapeutic use of human immunoglobulins (**passive immunity**) has a long history. They provided some of the earliest protective antiviral treatments. They are still in use for post-exposure prophylaxis with some viruses, including such diseases as Ebola where no other treatment is available. Even where disease is not prevented, the symptoms may be moderated. However, obtaining sufficient supplies of immune serum has always been a problem (as well as extremely expensive), and there is now a trend away from using human-derived material, due both to the problems of obtaining sufficient material and to the possibility of undetected (and possibly even unknown) pathogenic viruses being present. Despite this, such preparations remain available for use for a wide variety of viral diseases, including measles, hepatitis A and B, chickenpox, polio, and rabies.

There are several approaches to the production of specific antisera, and such approaches are increasingly based on advanced biotechnology.

Production of specific antibodies

The original approach to generating antibodies to specific targets was the production of *monospecific antisera*. These owe little to molecular technology, relying on the harvesting of antibody produced by an inoculated animal in response to immunization with the chosen antigen combined with a suitable adjuvant. The antigen used must be a single protein in order to produce a truly monospecific response.

Following immunization, the animal (usually a rabbit or guinea pig, although larger animals may be used) is bled and the antibody response assayed. When high titers of antibody are detected, monospecific antisera are harvested. In an alternative approach, if chickens are immunized, "IgY" antibody may be extracted from egg yolk.

Monospecific antisera are directed against a single protein but they will, of course, contain a wide range of different types of antibody directed against multiple epitopes of the antigen (and carrier). This can result in a high degree of noise when such antisera are used in detection systems. It was only with the advent of monoclonal antibody technology that it became possible to produce antibodies directed against a single epitope.

Monoclonal antibodies

The classical technique of **monoclonal antibody** production as initially developed by Köhler and Milstein in 1975 is shown in **Figure 6.12**. In this approach, immunization is substantially as for monospecific antiserum production, except that mice (or rats) are the usual animals of choice.

Antibody production is monitored but, rather than taking serum from the animal, the spleen is removed and the antibody-producing plasma cells are extracted. On their own they would only grow for a short while in culture. However, plasma cells that have lost the controls that restrict their growth are available, from the plasma cell cancers known as myelomas.

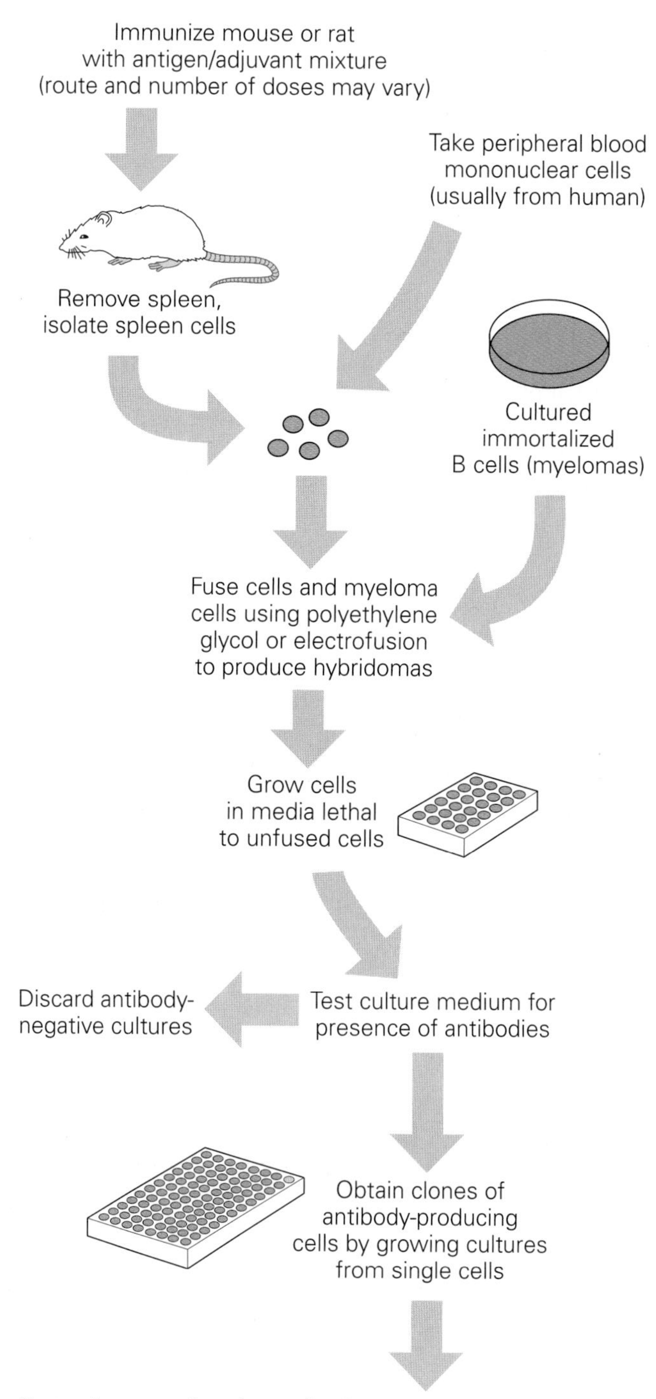

Figure 6.12
Monoclonal antibody production (classical method).

In the technique developed by Köhler and Milstein, the plasma cells from the immunized animal are immortalized by fusion with myeloma cells from mice or rats to produce long-lasting antibody-producing cells known as hybridomas. Antibody-producing hybridomas are separated by dilution and culture so that each population grows from a single cell. As a result, it is also quite possible to use multiple antigens and derive monoclonal antibodies to these multiple antigens from the same immunizations. Once hybridomas producing a desired monoclonal antibody are available, useful quantities of antibody can be prepared by any of a number of approaches, summarized in **Table 6.4**.

Monoclonal antibodies are now an indispensable part of science and medicine. The technology has had a revolutionary impact on many areas of science, not least on virology. This importance was recognized in the award of the 1984 Nobel Prize in Physiology or Medicine to Jerne, Köhler, and Milstein "for theories concerning the specificity in development and control of the immune system and the discovery of the principle for production of monoclonal antibodies." The number of applications for this technology continues to increase.

Humanization

Monoclonal antibodies can be produced in effectively unlimited quantities and with almost any desired specificity. However, while monoclonal antibodies are now invaluable for preparative, diagnostic, and imaging work, there are limits on their use in therapy. Since they are mouse (or rat) proteins, the human immune system will mount a strong response to the foreign antigen, removing it from circulation and limiting its future use.

The problem of an immune reaction to mouse antibodies can be countered by the use of **humanized monoclonal antibodies**, where the binding regions of the mouse antibody are inserted into the structure of a human

Table 6.4 Production of monoclonal antibodies

Method	Advantages	Disadvantages	Comments
Ascitic fluid: harvest fluid from hybridoma-induced tumor in mice or rats	Low cost, high activity	Low volumes produced Ethical issues exist around use of live animals Some hybridomas may not form ascites May contain impurities Variable yield	Original method, now rarely used
Culture fluid: harvest medium from hybridoma cultures	Low cost, simple	Low concentrations of antibody may require purification	Research use only
Bioreactor: uses cultured hybridoma cells in commercially available bioreactors	Continuous production, may produce high yield and volumes	Expensive and technically demanding High set-up costs Hybridomas may require adaptation	Commonly used for large-scale production
Recombinant: uses recombinant expression of antibody genes, usually in a bioreactor system	Highly flexible, continuous production, may produce high yield and volumes	Complex, expensive, and technically demanding High set-up costs Hybrid antibodies may have limited activity	Genetically modified product

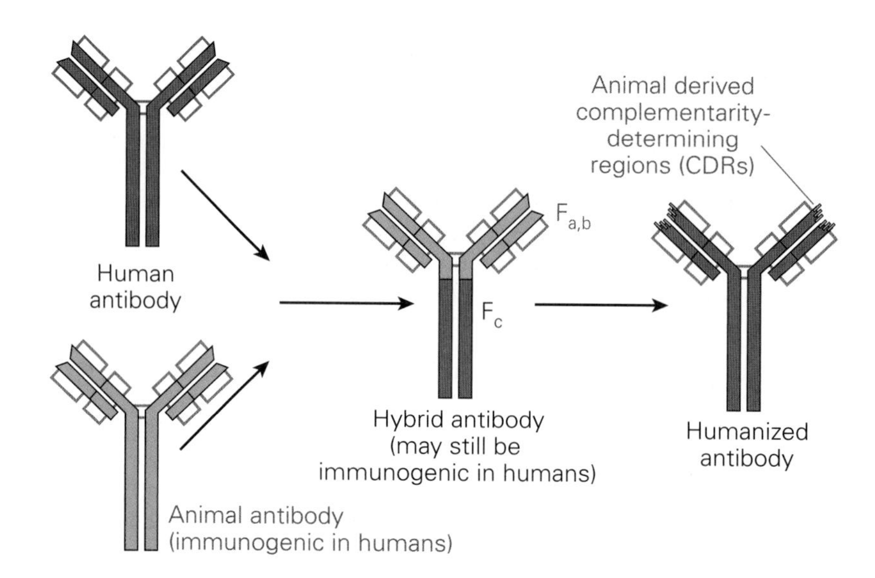

Figure 6.13
Humanization of an antibody.

antibody. As shown in **Figure 6.13**, there is also the halfway stage of a hybrid mouse—human antibody.

With hybrid antibodies, the F_c region of the mouse antibody is replaced with that of a human antibody. The hybrid molecule is less immunogenic than is the fully murine antibody, but still retains the potentially immunogenic murine $F_{a,b}$ region.

In a fully humanized antibody, only the much smaller antigen-binding sites (known as complementarity-determining regions, or CDRs) are retained, reducing immunogenicity still further. The process is, of course, extremely complex, and there are many constraints on just which antibodies may be used to provide the basic structure for the humanized antibody.

Even after full humanization, the human part of the antibody may not be an exact match to the recipient's immune system, so some immunogenicity is possible, but the very limited amount of murine material present reduces this to a workable minimum.

Both hybrid and humanized antibodies are now on the market, but great care needs to be taken in their use, since animal studies will not necessarily produce the same effects as use in humans. This is in part because the human F_c region has specific effects on the immune response (such as targeting other immune cells to the bound antibody) that may not work in the same way in another species. This difference underlay the tragic events of the TGN1412 trial (see Box 6.2).

One of the first humanized monoclonal antibodies to reach the market is palivizumab, marketed by MedImmune as Synagis®. Palivizumab is a humanized monoclonal antibody that binds to the F (fusion) surface glycoprotein of respiratory syncytial virus (RSV, *Paramyxoviridae*) and thus inhibits viral entry, while also providing targeting of the immune system to the virus itself. Other monoclonal antibody therapeutics are in use, for example against cancers, psoriasis, and immune dysfunction. These include the high-profile drug Herceptin®, that binds HER2, a receptor for epidermal growth factor (EGF) that is found on some breast cancers. The total market for therapeutic monoclonal antibodies was estimated at $16 billion in 2006, and is growing rapidly.

Human monoclonal antibodies

Subsequent work made possible the direct production of monoclonal antibodies from circulating B cells, fused to produce hybridomas (see Figure

6.12). This also made possible the direct production of monoclonal antibodies from human sources, which initially looked to be a very promising approach. However, human myeloma cells do not have the same immortalizing properties as those from mice. Fusion with mouse myelomas will produce immortalized cells, but these hybrids tend to lose the human chromosomes. Treating human cells with EBV (*Herpesviridae*) can immortalize them, but the process lacks specificity. Work to develop effective immortalization techniques for human cells continues, but at present the market is dominated by humanized rather than human monoclonals.

Recombinant antibodies

An alternative approach is to use immunoglobulin mRNA purified from mouse spleen cells or hybridomas. This can be expressed using recombinant vectors. A useful example is those based on filamentous bacteriophages, particularly M13 or fd. This latter system is available commercially as the recombinant phage antibody system (RPAS, Pharmacia Biotech), and avoids the use of live animals or eukaryotic cell culture. Using recombinant methods, it is possible to produce very large libraries of antibodies or, in the case of the RPAS system, antibodies expressed on the bacteriophage surface. These may be derived from immunologically naive sources (reflecting the general range of antibodies present), or may be taken after immunization or disease. In the latter case, the resultant "biased" library will be enriched for antibodies binding to the agent to which an immune response was present. Although there may be issues with the strength of binding of the resultant antibodies, it is also possible to mutate or recombine the antibody genes and express them to allow selection of antibodies with increased affinity, mimicking the affinity maturation that occurs during the natural immune response.

Micro-antibodies

A final approach is the chemical synthesis of antibody fragments, usually based around known binding-site structures. Technical limitations prevent the synthesis of full-sized antibodies, or even of $F_{a,b}$ fragments, but peptides corresponding to the CDR sites have been shown to have antiviral activity. Of course, these cannot target the immune system in the same way as a fully functional antibody, but peptides as small as 17 amino acids may be able to neutralize viruses if targeted to the correct epitope.

Targeting using monoclonal antibodies

While monoclonal antibodies alone target elements of the immune system, it is possible to enhance the effects of passive immunity by conjugating biologically active substances to a specific antibody. Examples of such molecules include toxins, radioactive compounds, cytokines, or antiviral drugs. Relatively toxic agents can be used because effective concentrations are achieved only at the site of antibody binding. For example, bacterial toxins or ricin (a lectin from the castor bean) have been considered for such applications: the natural cell-binding ability of these toxins is blocked or deleted, and they are conjugated to antibody to provide binding only to the target cells. The toxin—antibody conjugates are then taken up by the cell, killing it. It is also possible to link enzymes that activate drug precursors (prodrugs) to antibodies. The enzyme is then only present where the target antigen is bound by the antibody—enzyme complex, and the prodrug is only converted to the active form at those locations. This is known as antibody-directed enzyme prodrug therapy (ADEPT).

Interferons

The interferons are proteins that activate multiple elements of the cellular immune system, and also induce an *antiviral state* within cells that makes them resistant to virus infection (see Chapter 4). The three basic types (α, β, and γ) have unique properties and thus different therapeutic applications. Interferons are part of the *innate immune system* and play an important and potent role in protection against virus infections. It should also be noted that other elements of the innate immune system and other cellular protective states such as the heat shock response are also the targets of investigation.

The direct antiviral effect of interferon was reported before effective antiviral drugs were available, and this raised the hope that interferons would prove to be useful as antiviral agents. Studies were carried out using purified interferons, but interferon is naturally produced in very small amounts and it was extremely difficult and expensive to obtain sufficient quantities. A great deal of effort went into the cloning and expression of interferon genes in order to permit detailed studies and to support the anticipated uses of interferon.

Unfortunately, it soon became clear that interferons were not the magic bullet that had been hoped for. Studies with respiratory viruses showed that interferon was effective if administered before infection. In practice, this would require regular dosing when there was a risk of viral infection. In a useful illustration of the problems that can occur with prophylaxis (preventive use), it was found that interferon itself irritates and damages the nasal epithelium, producing symptoms similar to those of viral infection. Thus, a complex and expensive drug actually brought on the very symptoms that it was being used to avoid.

With the knowledge that interferon is actually a potent immunoregulatory agent rather than a direct inhibitor of virus replication, this does not come as a surprise. Interferon produces a wide range of effects which underlie such a reaction.

Following these disappointing results, much of the interest in interferon as an antiviral agent was lost, although interferon α has been used successfully elsewhere, in particular in the treatment of a range of cancers. It is reportedly also used in Eastern Europe to prevent respiratory infections, although this uses doses well below those believed to be active and the mechanism and benefits are unclear.

Interest in the antiviral uses of interferon α was revived following studies with hepatitis B and C viruses that showed positive effects. Despite the similar names (which date from an era when viruses were named for symptoms rather than for what they actually are), hepatitis B is a member of the *Hepadnaviridae*, DNA viruses that use reverse transcriptase and an RNA intermediate, while hepatitis C is a member of the *Flaviviridae*, true RNA viruses that do not have a DNA stage in their replication. It is important to remember that interferon acts by inducing an antiviral state in cells and by activating the immune system. Thus the effect of variations in viral metabolism is less than it is with a direct inhibitor of viral metabolism such as a nucleoside analog or a protease inhibitor.

Treatment of hepatitis B with interferon typically uses both interferon and nucleoside analog drugs, while treatment of hepatitis C uses interferon and ribavirin. The use of ribavirin raises the responder rate from 15–20% of patients that was seen in early clinical uses of interferon α to over 50%.

Another development that has made interferon therapy more useful is the development of pegylated interferons, in which interferon α is complexed with polyethylene glycol to prolong the time it remains in circulation. This reduces the need for injecting the drug from three times weekly (or even daily) to once per week. There are six marketed interferons, of which three are combined with polyethylene glycol. Non-pegylated commercially available interferons are IntronA (replacing Viraferon) and Roferon-A, while the pegylated forms are Pegasys, PegIntron, and ViraferonPeg.

It should be noted that other immunostimulatory approaches may involve the interferon response, as with the use of imiquimod against papillomaviruses.

While interferon therapy has not lived up to early hopes for a magic bullet against viral infections, it does now have an established and valuable role in this sector, alongside a wide range of applications in cancer therapy and in other areas.

6.7 RESISTANCE TO ANTIVIRAL DRUGS

When antibiotics were first developed in the 1930s and 1940s, they were seen as the answer to bacterial infections. They held this position for 50 years, but by the 1990s it was becoming clear that bacteria had changed faster than the antibiotics that were used to control them. Antibiotic-resistant superbugs are now a fact of life in all hospitals, and a range of new technologies is having to be developed to try to control these damaging infections. Antiviral drugs have been available for a much shorter time, but already resistance to many of these drugs is a matter for concern (**Figure 6.14**).

Resistance to anti-HIV drugs

The problem has been most apparent with HIV, which appears to be a highly adaptable virus. Resistance to anti-HIV drugs has been a significant problem in clinical practice almost from the first use of drugs to control the infection.

Resistance to the nucleoside analog zidovudine appears to result from a very wide range of mutations within the viral reverse transcriptase gene, typically single base changes, such as a change of methionine to leucine at position 41 within the polypeptide chain (written using the single-letter code for amino acids as M41L) which produces a fourfold rise in resistance to zidovudine. Complete loss of reverse transcriptase function is lethal, so mutant enzymes must remain functional, although they may have reduced levels of activity. The different mutations can also complement one another; if T215Y (threonine to tyrosine at residue 215) is present along with M41L, resistance is increased 60-fold.

Contrasting with this, resistance to different nucleoside analogs can induce mutants that block resistance. Mutations conferring resistance to

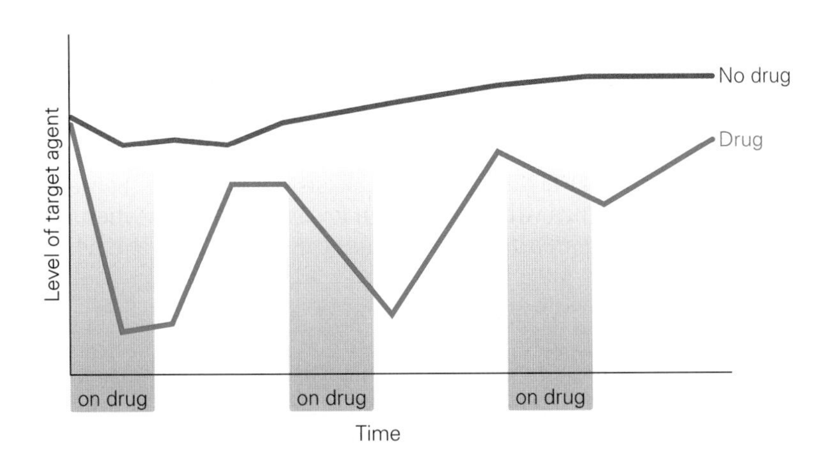

Figure 6.14
The development of resistance showing reduced effect after multiple courses of an initially effective drug.

didanosine (L74V) or lamivudine (M184V), for example, can block resistance to zidovudine.

Resistance can involve additional selectivity, preventing phosphorylation or the incorporation of nucleoside analog drugs into the forming DNA chain, or enhanced proofreading activity, allowing incorporated drug to be removed.

Unlike nucleoside analogs, NNRTIs bind away from the active site in a region known as the NNRTI pocket. Since mutations in this site are less critical than those within the active site (where major changes can lead to loss of polymerase activity), resistance to NNRTIs arises much more rapidly, effectively preventing their use except in combination with other drugs. This use of multiple drugs has resulted in widespread prevention of the almost inevitable progression to late-stage AIDS that was seen in the early days of the AIDS epidemic.

Resistance to anti-herpesvirus drugs

Among the anti-herpesvirus drugs, clinical drug resistance is far less of a problem, although it does occur at significant levels. Resistance to aciclovir most commonly involves mutation in conserved regions of the thymidine kinase (TK) gene, resulting in complete loss of function. Unlike a viral DNA polymerase, TK is useful but not essential for the virus to replicate. However, it is essential to initiate the phosphorylation of aciclovir (see Figure 6.10). HSV mutants lacking thymidine kinase activity (TK^-) are less pathogenic and do not appear to cause such extensive disease, at least in immunocompetent patients, although they can cause problems in severely immunocompromised patients. TK^- mutant HSV are thought to exist as a small fraction (0.01–0.1%) of the virus population in an infected individual, since they are at a replicative disadvantage compared to TK^+ strains. By inhibiting the replication of viruses with a functional TK, aciclovir will select for such mutants rather than requiring *de novo* mutation. Other, less common aciclovir-resistant mutants may have increased TK substrate specificity (preventing aciclovir phosphorylation) or mutations in the DNA polymerase which prevent the incorporation of aciclovir triphosphate into DNA (although these cannot involve complete loss of polymerase function since this is lethal to the virus). DNA polymerase mutants can also be resistant to the non-nucleoside polymerase inhibitor foscarnet, which can be a problem since this is a main drug of choice for use against aciclovir-resistant strains. There is also some evidence for changes in other viral proteins being involved in aciclovir resistance. Mutations giving a TK^+ aciclovir-resistant phenotype do not appear to occur frequently *in vivo*, possibly because they require very specific mutations to avoid loss of TK function. However, such mutations result in more pathogenic viruses, since TK function is retained, which may be a problem in the future. Similar patterns of resistance have been observed with VZV and with cytomegalovirus, where resistance once again involves either a failure to phosphorylate the inhibitor used or alterations in the DNA polymerase.

Among influenza strains, resistance is seen to all drugs used, and this was a concern in the 2009 H1N1 pandemic. Resistance is also seen among strains of avian influenza which show high levels of mortality when infecting humans and are seen by some as the forerunners of a potential human pandemic.

Minimizing resistance

It is clear that, as with bacterial resistance to antibiotics, viral drug resistance is inevitable. However, many of the mistakes of antibiotic use can be avoided. By using multiple drugs in combination where resistance is likely to occur, by restricting use to situations where there is real clinical need, and by preventing widespread veterinary use of drugs required for human therapy, it is possible to slow the development of resistance in the viral population. Measures are being put in place; for example, in the United States, veterinary use of any licensed anti-influenza drugs was banned by the FDA in 2006. However, reports persist of such uses in other countries. There are far fewer antivirals available than there are antibiotics, and the development of resistant viruses must be contained or, ideally, avoided entirely where this is possible. The hard-learned lessons of the antibiotic crisis must be remembered.

6.8 COMBINATION THERAPY

The giving of multiple drugs simultaneously is nothing new, and antibiotics have been used this way for many years. In what could be regarded as the bacteriological equivalent of HIV, *Mycobacterium tuberculosis* has generated *XDR* (extremely drug resistant) forms and is routinely treated with at least three antibiotics.

With herpesviruses resistant to aciclovir, the majority of drug-resistant strains are of reduced virulence, so that the clinical need for approaches to counter resistance has been relatively limited, and could be addressed with a substitution of another drug such as foscarnet. The situation with HIV was quite different.

The nucleoside analog zidovudine became the first available anti-HIV drug (and the seventh available antiviral) in 1987. That same year, a paper reported on resistant virus appearing in HIV-infected cell cultures treated with zidovudine. By 1989, papers were being published which reported in detail on the issue of zidovudine-resistant HIV in the clinic. Reports indicated that clinical benefits from the use of zidovudine alone began to diminish after one year due to the development of resistant virus. With the non-nucleoside drugs such as nevirapine, the development of resistance was even more rapid.

The concept of combination therapy was thus an apparent solution to the development of resistance, but until more drugs became available in the 1990s this was a largely theoretical possibility. Another problem was that, until 1996, all approved drugs were nucleoside analogs with similar modes of action. While these were used in combination and there was evidence that the mutations required for resistance to one drug could reverse resistance to others, it was not until drugs with entirely different modes of action became available that combination therapy came of age. First with the non-nucleoside reverse transcriptase inhibitors, and then with the protease inhibitors, it became possible to target multiple stages of the viral life cycle.

Highly active anti-retroviral therapy

The use of combinations of drugs against HIV eventually gave rise to *highly active anti-retroviral therapy* (HAART), combinations of drugs with the potential to control HIV infection and prevent progression to AIDS in the majority of infected individuals, at least for those with access to antiviral drugs. Typically, three (or more) drugs from at least two drug classes are used. These are often used at lower doses than in monotherapy (single drug) approaches, and drug toxicity can be reduced as a result.

Early forms of HAART were extremely expensive; costs of over $40,000 per patient, per year were often cited, placing such therapy beyond reach of most people in all but the richest countries. They were also very difficult to comply with. One 1997 report stated that "To gain a durable suppression

Table 6.5 Combined formulations of antiviral drugs targeting HIV

Name*	Content	Agents
Atripla	Efavirenz	Non-nucleoside reverse transcriptase inhibitor
	Emtricitabine	Nucleoside analog
	Tenofovir	Nucleotide analog
Combivir	Lamivudine,	Nucleoside analog
	Zidovudine	Nucleoside analog
Kaletra	Lopinavir	Protease inhibitor
	Ritonavir	Protease inhibitor
Kivexa/Epzicom	Abacavir	Nucleoside analog
	Lamivudine	Nucleoside analog
Trizivir	Abacavir	Nucleoside analog
	Lamivudine	Nucleoside analog
	Zidovudine	Nucleoside analog
Truvada	Emtricitabine	Nucleoside analog
	Tenofovir	Nucleotide analog

* Trade names used here are trademarks (TM) or registered trademarks (®).

of viral replication, near-perfect adherence to dosage, timing, frequency, and food requirements is necessary." This could involve taking exactly the right subset of more than 20 pills every day at precise times, some before food, some with food, some after food, and some in the middle of the night. Unsurprisingly, compliance (and concern over the effects of failure to comply on the progression of disease) became a major issue. However, with the development of compatible drug regimes, HAART has become the main approach to control of infection in HIV-positive individuals.

Combination therapy is further simplified since some anti-HIV drugs are now available in combined forms (see **Table 6.5**). Indeed, one protease inhibitor, lopinavir, is only available in such a form.

Antagonism, addition, and synergy

Combinations of drugs may interact negatively, minimizing or cancelling the effects of one or both drugs, a phenomenon known as interference, or an **antagonism** (**Figure 6.15**). Most drugs come with warnings to avoid specific medicines or foods. In the case of the HIV proteinase inhibitor invirase, there is a list of over 20 substances, including herbal medicines and even garlic. These are to be avoided because of anticipated negative effects arising from the combination. Sometimes these effects can be very serious indeed, as with the potentially lethal toxicity produced by fluoropyrimidine drugs in combination with the VZV inhibitor sorivudine mentioned above.

However, it is possible to identify drugs that work together to counter infection. A combination of drugs may provide significant benefits for two major reasons. Since any mutant would have to develop resistance to two (or

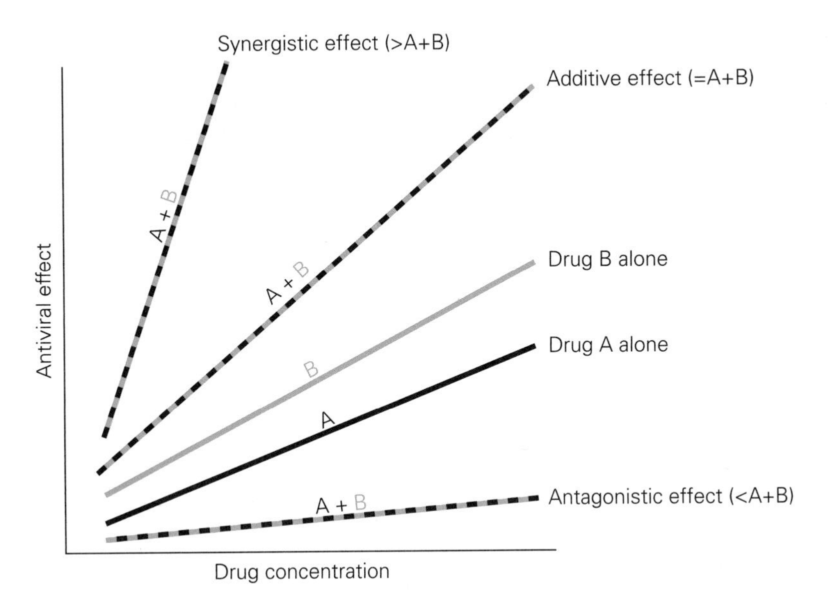

Figure 6.15
Combination therapy. Two drugs (A and B), when given together, can have additive effects (where the combined effect = effect of A + effect of B), a synergistic effect (combined effect > effect of A + effect of B), or an antagonistic effect (combined effect < effect of A + effect of B).

more) drugs simultaneously, this is less likely, and in addition some combinations of drugs may both inhibit the virus, thus combining their beneficial effects (an **additive effect**).

It is also possible for the effects of some drugs to combine to enhance their antiviral effect still further. This is referred to as a **synergistic effect**, with the combined effect being greater than would be expected by simply adding the individual effects of the drugs together (Figure 6.15). Such effects are also seen with combined antibiotic therapy against bacterial diseases. Originally, synergistic combinations were derived without the necessity for a detailed understanding of the underlying molecular events, using extensive testing programs where multiple drugs were tested singly and in combinations at various levels against virus grown in cell cultures. However, there is now increased understanding of the molecular basis of such synergies. For example, the NNRTI nevirapine inhibits the removal of the nucleoside analog zidovudine by the proofreading function of the viral reverse transcriptase enzyme, after it has incorporated zidovudine into forming DNA chains. In the case of ritonavir, a protease inhibitor, there are clear benefits from administering this drug with other drugs of the same class since it inhibits the cytochrome P450 3A4 isozyme that breaks them down, thus prolonging their activity within the body. This is similar to the use of tazobactam to inhibit the breakdown of the antibiotic piperacillin in the antibiotic formulation Tazocin, further underlining the similarities between antibiotics and antivirals.

While the majority of drug combinations are used against HIV, others are available. Combinations of nucleoside analogs with interferon are used for the treatment of hepatitis B, while the use of interferon and ribavirin in combination to treat hepatitis C has tripled the number of patients showing positive outcomes compared to interferon alone, even though ribavirin on its own has a very limited effect. This provides a clear example of synergy, with drugs of different types used together to produce significant clinical benefits.

6.9 LIMITATIONS OF ANTIVIRAL DRUGS

While a lot of effort has gone into developing antiviral drugs and there are life-saving benefits from their use, notably against HIV, they also have some major limitations.

First, unlike most antibiotics, almost all antiviral drugs target one, or at most a very few closely related viruses (see Table 6.2). Exceptions include the nucleoside analog lamivudine, which is effective against both HIV and hepatitis B (which, while unrelated, both use a reverse transcriptase to replicate), and the ribonucleoside analog ribavirin, which is effective against a wide range of RNA viruses. It has been suggested that this wide activity reflects the effect of ribavirin on cellular processes rather than the viruses themselves. Agents which activate the immune system or cellular defenses, such as interferon, do have the potential to be active against multiple viruses, but this has not yet been established in the clinic and interferon is used only against a limited range of viruses at present.

Second, many of the drugs in use cannot cure the infection that they target. Herpesviruses, HIV, and papillomaviruses all have the ability to establish long-term infections within cells where they are not metabolically active enough to be a target for current drugs, which target replicating virus. Thus, these infections may be suppressed, but are unlikely to be cured.

Third, by the time drugs are used, the actual replication of the target virus is often of very limited relevance to the symptoms that the patient wants to see resolved. A very high proportion of such symptoms are actually due to the efforts of the immune system to clear the infection. This includes fevers, excess nasal mucus, many rashes, and even pneumonias. In the case of the extremely dangerous hantavirus pulmonary syndrome, by the time the patient develops the intense, flooding (and often lethal) pneumonia that is characteristic of the final stages of the disease, the virus has already been eliminated.

With cold sores (caused by HSV type 1), the sore itself is an immune response to infection. To inhibit the virus, treatment is best applied when virus replicating at the skin causes a characteristic itching, before the sore even forms. With influenza, treatment at the very early stages of the illness is required for drugs to produce any effect. By the time the average patient has noticed the symptoms and scheduled a visit to their doctor, there is little point in treating the infection.

In fact, many drugs work best when used for prophylaxis (prevention of infection) rather than for therapy (curing an established infection). In the case of palivizumab, the first antiviral monoclonal antibody therapeutic, it is actually intended for use only in prophylaxis, to be injected at monthly intervals throughout the season when infections with its target virus are likely. Of course, long-term prophylaxis often brings its own concerns over cost, possible side effects of long-term treatment, and patient compliance (since patients are more likely to take medication properly when they can see clear benefits).

Treating the symptoms

In the case of cold sores, there is increasing awareness of the need to treat the symptoms as well as the virus. Aciclovir or penciclovir are active against the virus rather than the inflammation, although the cream base may produce some benefits. A treatment has recently been marketed in Europe which targets only shielding and resolution of the sore, dismissing virus inhibition entirely. However, attention has been given to this issue, and a mixture of 5% aciclovir (to inhibit the virus) and 1% hydrocortisone (an anti-inflammatory corticosteroid) is now marketed as Xerese™. Used alone, hydrocortisone has been reported to produce benefits in healing the sore, but also to increase the amount of virus present, showing the value of the inflammation in limiting virus replication. As a result, it would not be approved for this use since it could increase spread of the infection. By combining the two agents in one treatment both aspects of the infection are addressed.

As with any combined treatment, there are potential issues with drug compatibility regarding partition (where they go) and longevity (how long they stay there) of the components. In a possible answer to this issue, some early stage compounds exist with both antiviral and anti-inflammatory properties, but these are a long way from the market.

CONCLUSIONS

Antiviral drug development is a complex and expensive process. Despite this, the availability of drugs has increased sharply, from 6 approved agents in 1987 to over 40 today. Even HIV, which was not identified until 1983, is today treatable with over 20 antivirals from 5 different classes.

It is to be hoped that the lessons of the antibiotic crisis have been learned, and that progress against viral diseases will continue.

Key Concepts

- Antiviral drugs originated in the 1950s, and came of age with the introduction of aciclovir in the late 1970s. Early drugs were relatively toxic and were found by testing large numbers of compounds.
- Similarity and overlap between viral and cellular metabolism means that toxicity is a common problem for antiviral drugs. Newer drugs are less toxic and more convenient to use.
- Computer-aided design of drugs to fit with known molecular structures of targets is now the preferred approach in most situations.
- New drugs face a complex and challenging route to market, through several stages of clinical trials. There are many more failures than successes in drug development.
- The major classes of current antiviral drugs are: nucleoside analogs, nucleotide analogs, non-nucleoside reverse transcriptase inhibitors, protease inhibitors, fusion inhibitors, ion channel blockers, neuraminidase inhibitors, immunomodulators, oligonucleotides, and monoclonal antibodies.
- Passive immunotherapy, using antibody preparations to control viral disease, remains a valuable approach. Its value has been greatly increased by the licensing of the first monoclonal antibody antiviral, palivizumab.
- The use of passive immunotherapy, and of drugs that stimulate immunity, is blurring the line between vaccines and antiviral drugs.
- Despite early promise, nucleic acid-based approaches have been slow to deliver useful drugs, with only one such drug in use. RNA interference may provide a route to such drugs.
- Resistance to antiviral drugs is an important issue for all viruses, but is most significant for HIV due in large part to its rapid mutation.
- Most antiviral drugs have a narrow spectrum of activity, and are active against one or a very few (usually related) viruses.
- The use of multiple drugs simultaneously against the same virus is known as combination therapy and relies on the availability of multiple drugs with different modes of action. Highly active anti-retroviral therapy (HAART) against HIV is a success story of combination therapy.

DEPTH OF UNDERSTANDING QUESTIONS

Hints to the answers are given at http://www.garlandscience.com/viruses

Question 6.1: Drug companies develop drugs for high-value markets, and the cost of manufacture accounts for only a small part of prices, the rest being stated as necessary for the recovery of development costs. If drug company profits were reduced, as some wish to happen, what other method would bring new drugs through the legally required and costly testing process?

Question 6.2: There are many stages of virus infection that are not targeted by current drugs. Why, and which of these functions have the most potential for the future development of therapeutic approaches?

Question 6.3: Why are there few broad-spectrum antiviral agents, unlike the case with antibiotics?

Question 6.4: Why go through the extremely complex processes of antibody humanization?

Question 6.5: Why did interferons not live up to their initial promise as antiviral agents?

FURTHER READING

Bennasser Y, Yeung ML & Jeang KT (2007) RNAi therapy for HIV infection: principles and practicalities. *BioDrugs* 21, 17–22.

Blair E, Darby G, Gough G et al. (1998) Antiviral Therapy. Bios Scientific Publishers, Oxford. (Dated, but still useful in context and as a historical review.)

Coen DM & Richman DD (2007) Antiviral agents. In Fields Virology, 5th ed. (DM Knipe, PM Howley eds). Lippincott Williams & Wilkins, Philadelphia. (Excellent overall review.)

De Clercq E (2004) Antiviral drugs in current clinical use. *J. Clin. Virol.* 30, 115–133.

Diasio RB (1998) Sorivudine and 5-fluorouracil; a clinically significant drug-drug interaction due to inhibition of dihydropyrimidine dehydrogenase. *Br. J. Clin. Pharmacol.* 46, 1–4.

Rottinghaus ST & Whitley RJ (2007) Current non-AIDS antiviral chemotherapy. *Expert Rev. Anti Infect. Ther.* 5, 217–230.

Schubert S & Kurreck J (2006) Oligonucleotide-based antiviral strategies. *Handb. Exp. Pharmacol.* 173, 261–287.

INTERNET RESOURCES

Much information on the internet is of variable quality. For validated information, PubMed (http://www.ncbi.nlm.nih.gov/pubmed/) is extremely useful.

Please note that URL addresses may change.

DrugBank. http://www.drugbank.ca

Encyclopedia of Life Sciences (subscription required for full article). http://www.mrw.interscience.wiley.com/emrw/9780470015902/els/article/a0000410/current/abstract

Medline Plus drug information. http://www.nlm.nih.gov/medlineplus/druginfo/drug_Aa.html

Wikipedia. http://en.wikipedia.org/wiki/List_of_antiviral_drugs (although Wikipedia is open to editing by any user, the antivirals pages are generally of a very high standard)

CHAPTER 7
Beneficial Use of Viruses

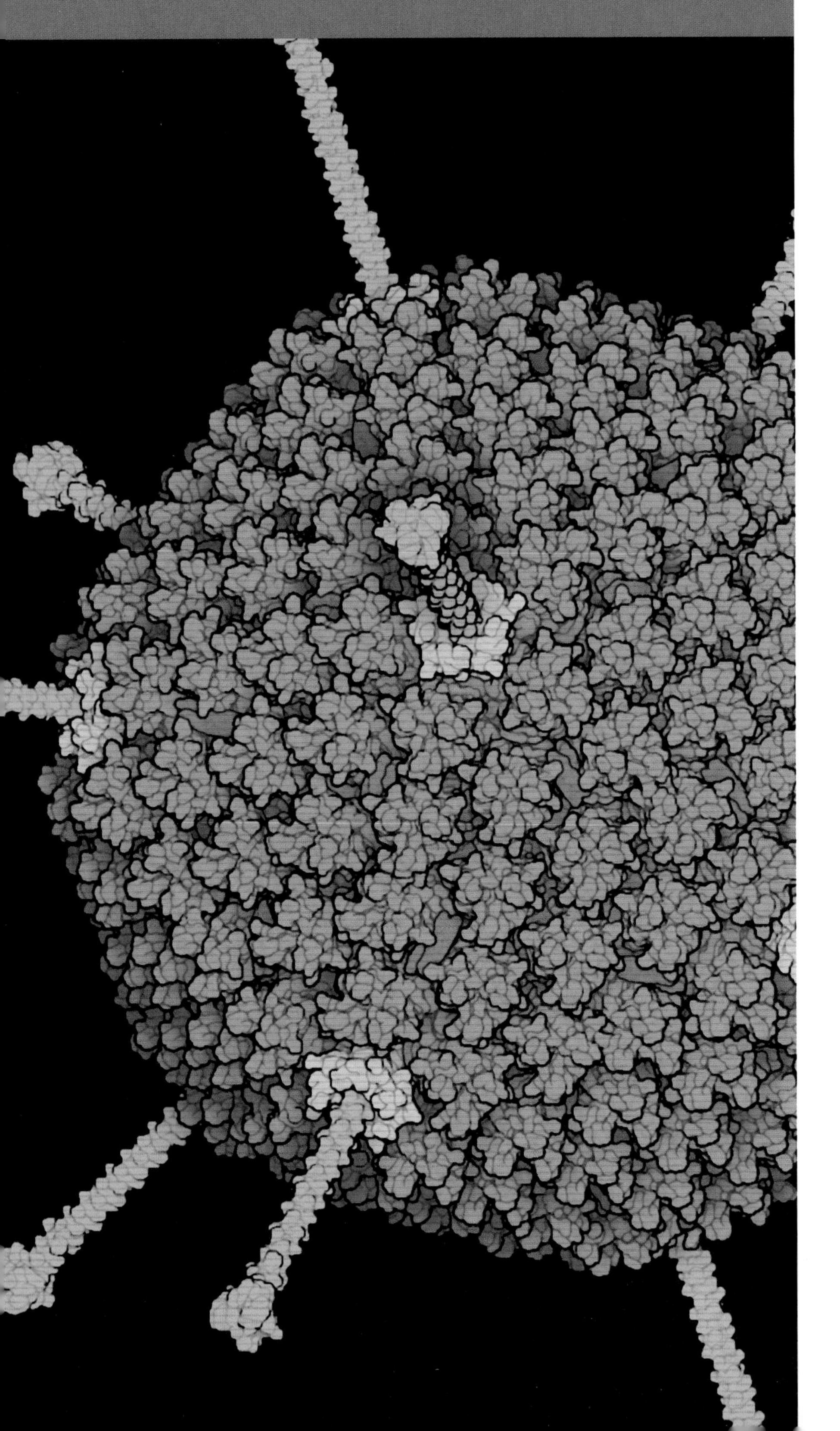

INTRODUCTION

While viruses are primarily known as pathogens, they can also be used for the benefit of humans. The most obvious is in their role in increasing our understanding of biology. The first genome sequences to be revealed were those of viruses (RNA bacteriophage MS2 and DNA bacteriophage ΦX174). The basic technology of genetic manipulation was developed from studies with bacterial viruses, and viral elements are still widely used in such work (described in more detail in Chapter 9). And the reverse flow of genetic information (RNA to DNA) was discovered from work with the *Retroviridae*.

About the chapter opener image
Adenovirus
(Courtesy of the Research Collaboratory for Structural Bioinformatics Protein Data Bank and David S. Goodsell, The Scripps Research Institute, USA.)

There are also a number of ways in which viruses may produce direct benefits for human health. The most obvious of these is as vaccines and vaccine vectors, which are covered in detail in Chapter 5. Vaccines do not simply protect against infection with the same virus. Relatively harmless viruses are often used to provide protection from their more dangerous relatives (for example, the use of vaccinia virus to protect against smallpox, or Shope fibroma virus to protect against myxomatosis) and viral vectors may be used to develop candidate vaccines against a range of diseases both viral and nonviral in nature.

This chapter will describe other ways in which viruses can be beneficial:

- Gene therapy
- Cancer prevention and control
- Control of harmful or damaging organisms, in both agriculture and medicine

7.1 GENE THERAPY

Viruses are routinely used in the genetic modification of model organisms for research purposes. The production of transgenic plants and animals in agriculture has also been established, but germ-line modification of humans has not been attempted for technical and ethical reasons. However, genetic manipulation of somatic cells of individuals has been under investigation for many years. This is known as **gene therapy**.

The key element of gene therapy is the introduction of functioning genes into the cells of a human patient, to express desired functions or to correct defective or non-operational genes within those cells. The original concept behind gene therapy was the treatment of individuals with an inherited genetic disorder, but applications in this area have been limited. The most common target has been cancers, accounting for almost two-thirds of all clinical trials to date (see also Section 7.2). It is also possible to target infectious diseases by introducing specific inhibitory genes, including those producing antisense or small interfering (si) RNAs (see Chapter 6).

The type of condition or disease amenable to gene therapy is one where a single gene (or, potentially, a very few genes) is able to correct the disorder, since it is not possible to introduce enough functioning genes to modify a complex condition resulting from the interaction of many different genetic elements. Another essential requirement is to have an identified and available therapeutic gene (often a functional version of a defective cellular gene), along with some way to deliver it to the cells where it is required.

Delivery systems must be able to introduce DNA into appropriate target cells (**Figure 7.1**). *Ex vivo* gene therapy involves removal of the cells from the body, followed by their treatment and reintroduction. *In vivo* gene therapy is more challenging, and involves targeting of target cells within the body. Examples of current approaches include the lung cells of cystic fibrosis sufferers where a defect in an ion transport protein produces excess mucus and damages the lung surface, and the hematopoietic cells of sufferers from various forms of severe combined immunodeficiency (SCID), which causes profound immunosuppression by preventing the production of lymphocytes. Hematopoietic cells are removed from the patient for treatment before being reintroduced (*ex vivo* gene therapy), while the lung cells of cystic fibrosis sufferers are readily accessible and can be treated *in vivo*. For both of these cases, chemical/mechanical systems such as fusogenic liposomes containing the therapeutic DNA may be adequate. With such systems, once the DNA is within the cell it may integrate into the cellular genome, although this occurs at very low efficiency.

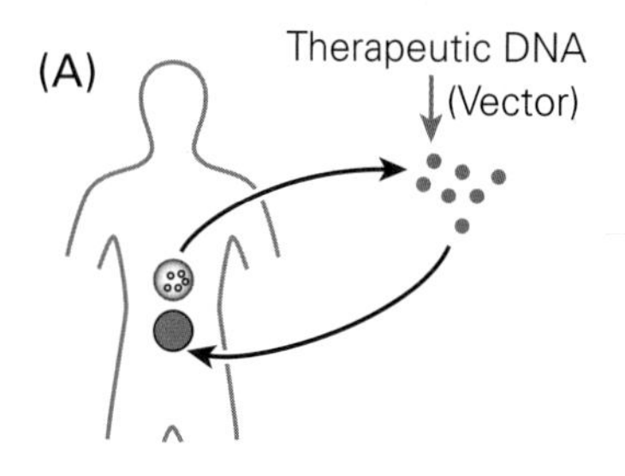

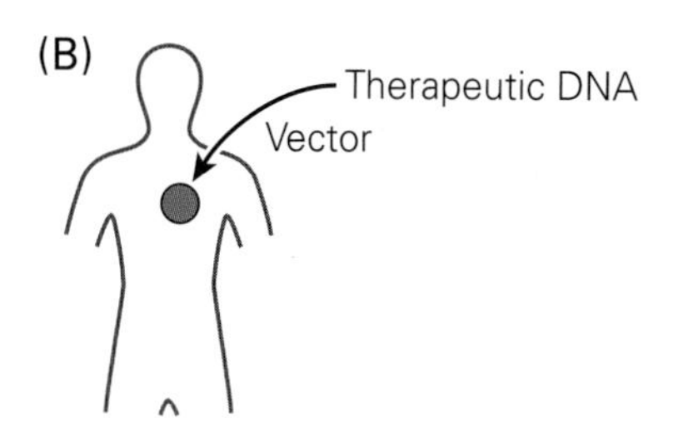

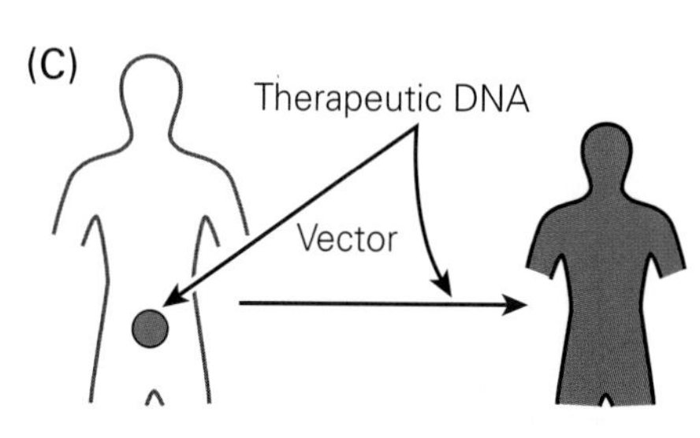

Figure 7.1
Approaches to gene therapy. (A) *Ex vivo* somatic: gene introduced into cells outside the body before reintroduction into the patient. (B) *In vivo* somatic: gene introduced into cells inside the body. (C) Germ line: gene introduced into cells that will produce the next generation. This is currently not used for human therapy.

Virus vector systems

Virus vectors are used for *ex vivo* gene therapy, but are particularly useful where the target cells are in less accessible areas of the body. Viruses provide highly efficient systems for getting foreign nucleic acid into cells, and they are also highly suited to protecting a nucleic acid while transporting it to the required area of the body. Unsurprisingly, the use of viruses in gene therapy has a long history.

Viruses naturally exhibit *cell tropism*, where the requirements of the virus for specific receptors (see Chapter 3) along with other factors can be used to ensure that specific types of cells are infected. While most viruses used for gene delivery will infect many types of cell, careful selection (and, where appropriate, genetic modification) of the virus vector can favor the delivery of the therapeutic gene to the required location.

The main types of viruses being evaluated for use as gene therapy vectors in clinical trials are summarized in **Tables 7.1** and **7.2**. Adenoviruses are widely used as vectors, and can be engineered both to enhance specificity and to minimize unwanted effects. Enhancement of specificity can involve either altering the surface receptors of the virus, or using cell type-specific promoters to control the expression of inserted genes. Approval of the first gene therapy product in Europe (Colybera® for pancreatitis) may occur in 2011. This uses an adeno-associated virus vector.

Complications

While the principles are apparently straightforward, there are many issues that complicate the successful use of gene therapy.

Table 7.1 Viral vectors in use for clinical trials of gene therapy

Vector	Virus family	Number of trials
Adenovirus	*Adenoviridae*	372
Flavivirus	*Flaviviridae*	8
Herpes simplex virus	*Herpesviridae*	51
Measles virus	*Paramyxoviridae*	3
Newcastle disease virus	*Paramyxoviridae*	1
Sendai virus	*Paramyxoviridae*	2
Adeno-associated virus	*Parvoviridae*	67
Poliovirus	*Picornaviridae*	1
SV40 virus	*Polyomaviridae*	1
Vaccinia virus	*Poxviridae*	95
Poxvirus	*Poxviridae*	64
Lentivirus	*Retroviridae*	21
Retrovirus	*Retroviridae*	326
Vesicular stomatitis virus	*Rhabdoviridae*	2
Semliki Forest virus	*Togaviridae*	1
Venezuelan equine encephalitis virus	*Togaviridae*	2
Multiple viruses		37

The remaining 483 trials used nonviral vectors.

Table 7.2 Characteristics of viral vector systems

Virus	Advantages	Disadvantages
Adenovirus type 5 and others (*Adenoviridae*)	Efficient nuclear entry, high levels of expression possible, specialized vectors available, insert size up to 8 kbp (36 kbp in some systems)	May be cytotoxic, immunity to adenovirus may prevent use, narrow host range with some types, safety concerns from previous clinical trials
Adeno-associated virus (*Parvoviridae*)	Nonpathogenic if helper virus not present, infects a broad range of cell types, easy to manipulate ssDNA genome, low immunogenicity, can produce long-lasting expression (vectors necessarily unable to replicate), efficient integration into host genome at defined site	Very limited insert size (5 kb), may be high levels of preexisting immunity
Herpesviruses (*Herpesviridae*)	Well characterized, large viruses, wide choice of insertion sites, inserts up to 10 kb (larger in amplicon or episomal vectors	May be pathogenic, cytotoxic, concerns over latency, may transform cells, limited availability of vectors
Vaccinia (*Poxviridae*)	Wide choice of insertion sites, large inserts possible (25 kbp), some systems allow high-level expression, wide availability	May be pathogenic for humans, risk of early termination with some inserts, introns can be problematical
Moloney murine leukemia virus, Lentiviruses (*Retroviridae*)	High efficiency of gene transfer, efficient integration into host genome, multiple systems available	Concerns over safety and oncogenicity (including leukemia induction in clinical trials), integration at variable sites, limited insert size (8–10 kbp maximum), requirement for actively dividing cells (except Lentiviruses)
Simian virus 40 (*Polyomaviridae*)	Stable high-level expression, low immunogenicity, infects a broad range of cell types, inserts up to 18 kb possible using pseudovirion (viral particle produced *in vitro* with no viral DNA sequences)	Small genome may restrict insert size, concerns over transformation and possible oncogenicity (if viral sequences present)
RNA viruses (*Coronaviridae, Flaviviridae, Paramyxoviridae, Picornaviridae, Reoviridae, Rhabdoviridae, Togaviridae*)	Capability to target specific cell types, high levels of gene expression	Small genomes restrict insert size, high mutation rate from RNA genome, no defined route of insertion into host genome

Specialization of cells (differentiation) may result in variant gene expression. For example, lung cells (targeted in attempts to treat cystic fibrosis) are highly specialized. It is also important to note that the routine elimination of cells by the body (the rate of which varies enormously between cell types) can require relatively frequent repetition of gene therapy since cells expressing the therapeutic protein are not immune to routine replacement by new, untreated cells. Such replacement is relatively frequent in the case of lung cells, but appears to be low enough to permit some therapeutic benefit. However, the extremely high rate of turnover of lymphocytes makes them more difficult targets. By delivering the therapeutic gene into the hematopoietic stem cells which actually produce all of the cell types in the blood, this problem is avoided while the number of cells produced which contain the therapeutic gene is greatly increased. Unsurprisingly, hematopoietic stem cells represent a major target of gene therapy.

Treatment of germ-line cells would overcome some of the problems encountered with attempts to genetically modify differentiated cells of specific tissues. But even the possibility of establishing foreign DNA in the human germ line, and (at least potentially) in every copy of the human genome descended from the treated individual, is something that requires thorough knowledge along with extremely careful consideration. The expanding field of bioethics is concerned with the acceptability of such procedures, among many others.

The limited number of cases where a single gene defect produces serious illness and is amenable to correction using such methods are likely to provide the first examples of human use. However, the humans in question would be permanently altered by these procedures, as yet with little understanding of the long-term effects of such changes. While the technology has been established and transgenic animals produced, as of 2010 there are no serious proposals for human **germ-line gene therapy**.

Once the therapeutic gene is inside the target cell, it must be expressed at an appropriate level. Once again, viruses can provide a route to achieve this. Many viruses, such as retroviruses (*Retroviridae*) or adeno-associated virus (AAV; *Parvoviridae*), have a highly efficient integration step in their life cycle, and foreign genes introduced into the viral nucleic acid may use viral mechanisms to become integrated into the cellular DNA in order to permit stable expression. However, this integration is not without its problems for the *Retroviridae* (see below), while the very small size of the AAV genome limits its utility.

Other viruses (*Herpesviridae, Polyomaviridae*) may be maintained relatively stably as extrachromosomal genetic elements, while others (*Adenoviridae*) may produce more transient expression.

Problems

The potential for gene therapy was originally seen as immense and immediate, but as with genomics progress to useful medicines has been far slower than originally thought. As one commentator noted, "speculation that gene therapy would quickly revolutionise medicine has clearly been wrong ... there was an overoptimistic view of the pace of progress and an underestimation of the problems remaining to be overcome."

In 1999, a patient being treated with an adenovirus vector for an ornithine transcarbamylase deficiency died, with multiple organ failure and a pathology indicating a reaction to the adenovirus vector with which he had been treated. There were also reports of serious side effects in three patients in a related study.

Then, problems arose in trials against X-linked severe combined immunodeficiency disease (X-SCID) in France and in the United Kingdom, using a vector derived from a murine retrovirus. Of eleven patients in France, nine responded to treatment, and seven apparently had long-term positive benefits. But within a few years four patients developed leukemia. Of ten British patients, one developed leukemia. The main cause seemed to be insertional mutagenesis caused by the retroviral vector, with insertion near the proto-oncogene LMO2 promoter in four of the five cases, leading to uncontrolled clonal proliferation of mature T cells. Four patients responded to therapy, but one death was reported.

X-SCID is a major and life-threatening illness, justifying higher levels of intervention than does a milder illness such as that in the adenovirus trial, but this was enough to stop trials using this approach in the USA. Even more worryingly, studies carried out at the Salk Institute in California seemed to indicate that the therapeutic gene in use (IL2RG, the γ chain of the interleukin 2 receptor) could itself promote leukemia in mice. While these data have been challenged, it remains a worrying issue.

After these results, progress was more cautious. Adenovirus vectors were modified to restrict their ability to replicate, while retrovirus vectors were used in more limited ways. It is undeniable that these findings were a major setback for work on gene therapy.

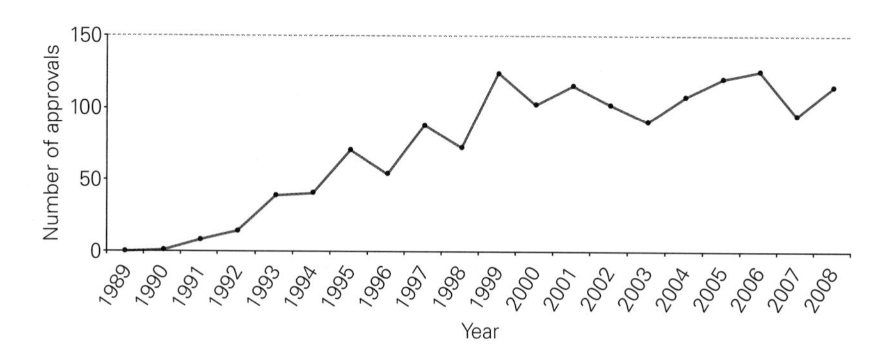

Figure 7.2
Number of gene therapy trials approved by year.

Another potential setback emerged in 2009, when a xenotropic murine leukemia virus–related virus (XMRV) related to the original virus used to create the murine retrovirus vector used in trials was identified, and was suggested to be linked to both chronic fatigue syndrome and, perhaps more worryingly, prostate cancer.

Where are we now?

As will be appreciated, any work with recombinant DNA requires careful assessment of the risks and benefits and of the ethical issues involved. Codes of conduct and supervisory bodies have been established in many countries. Gene therapy, where the intention is to introduce and express recombinant DNA in humans, is one of the most controversial areas. Despite this, by 2009, 1537 protocols had been approved worldwide, rising from one trial in 1989 to 117 in 2006 (**Figure 7.2**); 975 of these trials were in the United States and 348 in the European Union.

Over two-thirds of these trials used viral vectors (Table 7.1), indicating the central role of this approach in gene therapy.

This is an impressive number of trials, but only 65 of these (4.2% of the total) were the large, late-stage phase 3 trials that precede drug approval (see Box 6.1 for an explanation of the phases of clinical trials). Even more worryingly, only one gene therapy product (Gendicine, see below) has been approved for use against cancer—and that only in China. Advexin, a similar product delivering p53 via an adenoviral vector, encountered very serious delays in the United States, eventually resulting not in approval of the product but rather in the failure of the company developing it.

In theory, there is a huge range of biomedical problems which could be addressed by gene therapy. That promise is still there, but has yet to deliver.

7.2 CANCER PREVENTION AND CONTROL

A number of viruses are associated with cancer in humans (see Chapter 4) and these have provided the first instances of the prevention of cancers by vaccination. However, viruses can also have beneficial applications in the control of cancer. Some viruses are innately able to target and destroy cancer cells, while other methods use molecular approaches based on viral vector systems to create specific therapeutics. Approaches which are in use or under investigation are listed in **Table 7.3**.

Vaccines

The most direct approach to using viruses to prevent cancers is simply that of vaccination against viruses that are associated with cancer. Vaccines

Table 7.3 Use of viruses in the prevention or control of cancers

Approach	Mode of action	Examples
Prophylactic vaccine*	Stimulation of immune system to prevent a cancer, typically one associated with a virus infection	Existing subunit vaccines for hepatitis B virus, human papillomaviruses
Therapeutic vaccine*	Stimulation of the immune system to control an existing cancer	Experimental approaches under evaluation, e.g. using adenovirus vectors or papillomavirus DNA
Replication-competent virus	Preferential killing of cancer cells by virus	Adenovirus, Newcastle disease paramyxovirus
Modified replication-competent virus	As above, with enhanced killing of cancer cells	Adenovirus with enhanced receptor binding
Nonreplicating virally derived vector	Transfer into cancer cells of a cytotoxic gene	Rexin-G (retrovirus core with cytocidal cyclin G gene)
Virus-directed enzyme prodrug therapy (VDEPT)	Virus-mediated delivery of enzyme combined with systemic administration of prodrug	Recombinant adenovirus expressing herpes simplex enzyme, plus treatment with ganciclovir

* Vaccine may be any of the current types, including live virus, subunit, vector, or DNA, as discussed in Chapter 5.

for hepatitis B virus (*Hepadnaviridae*; associated with hepatocellular carcinoma) and human papillomavirus (*Papillomaviridae*; associated with cervical cancer) are already available and are in widespread use. Both use selected proteins of the virus (*subunit vaccines*, see Chapter 5). Although many studies have been carried out, these are the only vaccines approved to date by the US Food and Drug Administration (FDA) for the prevention of cancer. Of course, few forms of cancer involve identified viruses.

Therapeutic vaccination is also under evaluation, where vaccines are used to stimulate an immune response in an attempt to control or to eliminate an existing cancer by stimulating specific immunity to the cancerous cells. Cancer cells are derived from and very similar to the cells of the host, and so it is important to avoid stimulating immunity that targets normal host cells. Fortunately, many cancers express characteristic proteins on the surface of malignant cells. These may be normal cell surface proteins that are overexpressed or mutated, markers of the particular cell type present, or proteins associated with embryonic cells (see **Table 7.4**). Where viruses are associated with the formation of the cancer, viral proteins or MHC-associated peptides may be present on the cell surface. It is possible to select any marker of the type of cancer cell that is present and use it to target the immune response whether using a vector virus or by any other route.

Cancers of all types are subject to the effects of the immune system at all stages of growth, and it is only those that can circumvent this (as well as maintain themselves in a viable state) that are able to develop. Thus immunological control of tumors is a challenging prospect. Although a wide range of clinical trials have been carried out, as of 2009 no FDA-approved therapeutic vaccine exists for any cancer.

Virotherapy

It is also possible to use the cell-killing effects of viruses directly, rather than relying on the immune system. A range of viruses have been used in efforts to produce targeted killing of cancerous cells, and the approach as a whole

Table 7.4 Cancer-associated cellular antigens

Surface antigen	Type	Locations
Seen in multiple types of cancer		
Carcinoembryonic antigen (CEA)	Glycoprotein	Fetal tissues and some cancers, e.g. breast, colorectal, pancreatic and stomach cancer, pancreatic cancers as well as some lung cancers
Cancer/testis antigens (e.g. NY-ESO-1)	Multiple proteins	Sperm cells, melanomas, cancers of the brain, breast, colon, lung, ovary, pharynx, and tongue
Gangliosides (e.g. GM3 and GD2)	Glycolipid	Melanomas, neuroblastomas, soft tissue sarcomas, some lung cancers
HER2/neu protein (ERBB2)	Overproduced protein	Breast, ovarian, and other cancers (targeted by Herceptin)
Mucin-1 (MUC-1)	Glycoprotein	Mucus-producing epithelial cells, cancers including breast, colon, pancreatic, prostate, and some lung cancers
p53 protein	Mutated forms of tumor suppressor protein	All types of cell
Restricted to a single type of cancer		
Idiotype (Id) antibodies	Unique antibodies	Produced by B-cell cancers (used as markers for diseases including multiple myeloma and lymphomas)
Mutant epidermal growth factor receptor (EGFRvIII)	Mutated protein	Glioblastoma multiforme
Melanocyte/melanoma differentiation antigens (e.g. gp100, MART1, and tyrosinase)	Multiple proteins	Mature melanocytes and in melanomas
Prostate-specific antigen (PSA)	Glycoprotein, proteinase	Released into semen and (at lower levels) blood, greater production by prostate cancer cells compared to normal prostate cells leads to higher blood levels (used as a marker for prostate cancer)

Adapted from, Cancer Vaccines - National Cancer Institute. http://www.nci.nih.gov/cancertopics/factsheet/Therapy/cancer-vaccines

is known as **virotherapy**. Some of the earliest evidence for virotherapeutic approaches resulted from the observation of beneficial effects in cancer patients associated with naturally occurring virus infections, apparently due to the induction of *innate immune responses* (see Chapter 4). More recent studies have shown that virus infection of cancer cells may enhance both innate and *adaptive immune responses*. In the latter case, responses to both viral and cancer cell antigens may be associated with beneficial effects.

Virotherapeutic approaches have used either natural, unmodified viruses or genetically modified viruses or virus vectors, either of which can produce selective killing of cancer cells (oncolytic activity). Typically, viruses are introduced directly into the cancerous tissue. Systemic administration has also been evaluated, but removal of viruses by the immune system can be a significant problem.

A number of RNA viruses have innate anti-cancer activity, producing higher levels of cytotoxicity in cancerous cells. Several types of RNA virus have been investigated as potential therapeutic agents, including reovirus (*Reoviridae*), vesicular stomatitis virus (VSV, *Rhabdoviridae*), and Newcastle disease virus (NDV, *Paramyxoviridae*). For example, NDV produces severe disease (fowl pest) in chickens but only limited and localized disease in humans. It also grows better in transformed human cells compared to normal (diploid) human cells. This led to its evaluation as a virotherapeutic agent in a number of clinical trials, with apparently promising results. Naturally selected variants with enhanced cytotoxic activity in cancer-derived cells have also been used. However, no RNA virus has yet been approved for use as an anti-cancer therapy.

With DNA viruses, genetically modified viruses have been used. The modification may be to reduce replication, to increase specificity of infection for cancer cells, or to increase targeted cytotoxicity. A range of viruses have been used, including adenoviruses (*Adenoviridae*), herpes simplex virus (*Herpesviridae*), and vaccinia (*Poxviridae*). Some studies have shown limited effects, for example the 0–14% local tumor regression rates seen in trials with Onyx-015 adenovirus (see below). There is evidence that such responses can be increased by the simultaneous use of chemotherapy against the cancer under treatment.

Many different genetic modifications have been evaluated. Two products using this approach are approved and in clinical use in China, although some commentators believe such approval is premature.

Oncorine (H101) (approved in November 2005) is an adenovirus with deletions in the E1B-55 and E3 regions. This makes the virus replicate preferentially in cancer cells.

Interestingly, Oncorine (and the very similar Onyx-015) was originally intended to function by removal of the p53 suppressor activity of the adenoviral E1B-55K protein, preventing its replication in cells with functional p53 (and thus allowing replication in cancer cells, where this is often disrupted). However, more recent studies have shown that p53 inhibition is not linked to the cancer cell-specific activity of these mutants, the molecular basis of which remains unclear.

Gendicine (approved in October 2003) is an E1-deleted nonreplicating adenoviral vector expressing a functional p53 tumor suppressor gene. The virus is not oncolytic (though immunological responses to infection may have this effect), and is actually a form of gene therapy since the introduction of a functional p53 gene to cancers where the existing gene is nonfunctional provides the therapeutic basis for the effect.

Other modifications include the use of selective promoters that favor gene expression in cancerous cells, changes to receptor binding to favor infection of such cells, or the inclusion of effector genes such as interleukin 18 or CD40 ligand to enhance their anti-cancer effect (see Chapter 4). The latter approach is seen as necessary by many commentators, given the limited efficacy of adenoviruses without such enhancements.

Since the use of any live virus in an immunosuppressed patient carries with it an element of risk, other approaches bypass the use of replicating viruses entirely. These use gene vectors which lack functional viral nucleic acid. One example that is currently in clinical trials is Rexin-G, a murine retroviral core that delivers a cytocidal form of the cyclin-G1 protein.

Virus-directed enzyme prodrug therapy (VDEPT)

Viruses may also be used in **virus-directed enzyme prodrug therapy** (VDEPT) to insert into target cells an enzyme that can activate an inactive precursor of a cytotoxic drug (a prodrug) that is administered systemically. Thus, the active, cytotoxic form of the drug is only produced where the relevant enzyme is present and active. For example, an adenovirus expressing the thymidine kinase (TK) enzyme of herpes simplex virus can be combined with systemic administration of ganciclovir, which is converted by the TK to its active form only in cells where this enzyme is present. Alternatively, antibody can be used to provide the cellular targeting function (ADEPT), as described in Section 6.6. A related technique, GDEPT (gene-directed enzyme prodrug therapy) may use virally derived vectors that are incapable of replication.

Table 7.5 Viruses used as pest control agents

Virus type	Number in use	Target
Baculoviruses (various)	13	Caterpillars, sawflies
Oryctes rhinoceros virus	1	Rhinoceros beetle
Myxoma poxvirus	1	Rabbit
Rabbit hemorrhagic disease calicivirus	1	Rabbit

7.3 BIOLOGICAL PEST CONTROL

The use of biological organisms to control damaging pests is broadly known as **biological control**, or biocontrol. Traditionally this has been used in agriculture, but applications exist in the control of agents important to human health as well. There are four basic approaches:

- Predators, which prey on the target species
- Parasites or parasitoids (insects that lay their eggs inside or on the host)
- Pathogens, which cause disease in the target species
- Competing species (antagonists)

Of these, only pathogens are viral in nature. Although they account for a small amount of total pesticide use, viruses are used for the control of multiple species of insects (and have been evaluated for other arthropods such as mites), and also for the control of rabbits (**Table 7.5**).

Biological agents can produce long-lasting effects and in some cases are able to spread among the target population. They have also been recognized as inherently less toxic than conventional pesticides by the US Environmental Protection Agency. However, they account for less than 2% of total agricultural pesticide usage, with total sales in the region of $500 million, compared to more than $25 billion for conventional pesticides (**Figure 7.3**). The limited (though still expanding) use of biological control agents is due to a number of factors, including:

(A)

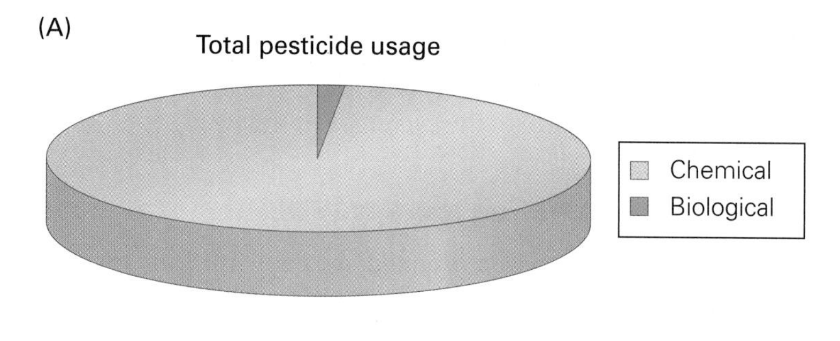

(B)

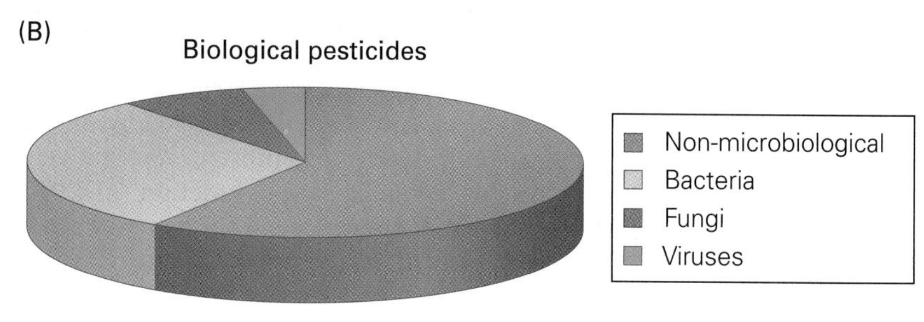

Figure 7.3
Usage of biological pesticides.
(A) Biological pesticides (green) as a proportion of total pesticides used.
(B) Viral pesticides (green) as a proportion of biological pesticides.

- High specificity, limiting the range of insects that can be controlled and often requiring identification of the pest insect before use
- Relatively slow effects, compared to chemical agents, allowing crop damage from an infestation to continue for some days after treatment
- Relatively high cost for the initial treatment, although control may then be long lasting
- Low environmental stability, particularly in sunlight
- Lack of support for their use from large pesticide companies

Microbiological agents account for approximately 40% of expenditure on biological pesticides. The most widespread use for biological pest control is the control of crop-damaging insects. For such applications, the most commonly used microbiological agent is *Bacillus thuringiensis* (BT), which produces a crystalline toxin that is insecticidal across a wide range of species. Some varieties of BT may be also used for control of mosquitoes, and may thus help to prevent viral diseases.

Fungi and viruses are also used, but BT alone accounts for the majority of expenditure on microbiological control agents in agriculture, while viral agents account for less than 10%. Thus it can be seen that the market for virus-based biological controls in agriculture in all applications is limited, and is in total worth only a few tens of millions of dollars each year. This is further divided between more than 20 products.

Within this sector, the main products are baculovirus insecticides (**Table 7.6**).

Table 7.6 Commercial baculovirus insecticides

Strain	Trade names*	Target insects
Anagrapha falcifera NPV	*CLV LC*	Celery looper
Autographa californica NPV	Gusano Biological Pesticide	Alfalfa looper (and other lepidopteran species)
Cabbage army worm NPV	Mamestrin	American bollworm, cabbage moth, diamondback moth, grape berry moth, potato tuber moth
Codling moth GV	*Carpovirusine*, *Cyd-X*, Madex, Granupom, *Virosoft CP4*	Codling moth
Helicoverpa zea NPV	(*Biotrol VHZ*), *Gemstar LC*, *Elcar*, *Heliothis NPV*, *NPH*, *Stellar LC*, *Viron H*	Cotton bollworm, tobacco budworm, tomato fruitworm
Heliothis armigera NPV	Ness-A	Old world bollworm
Lymantria dispar NPV	(*Gypchek*, *Gypsy moth NPV*)	Gypsy moth
Neodiprion lecontei NPV	Lecontvirus	Redheaded pine sawfly
Neodiprion sertifer NPV	(*Neochek-S*, *Preserve*)	European pine sawfly
Orgyia pseudotsugata NPV	*TM Biocontrol 1*, Virtuss	Douglas fir tussock moth
Plodia interpunctella NPV	*Nutguard-V*, *Fruitguard-V*	Indian meal moth
Spodoptera exigua NPV	Ness-E, *Spod-X LC*	Beet army worm
Spodoptera littoralis NPV	Spodopterin	Egyptian army worm

* Trade name in italic where listed by US Environmental Protection Agency. Strain names follow the naming conventions for baculoviruses, where each virus is named for the target insect followed by general type (genus) of the infecting baculovirus; nucleopolyhedrosis virus (NPV) or granulovirus (GV).

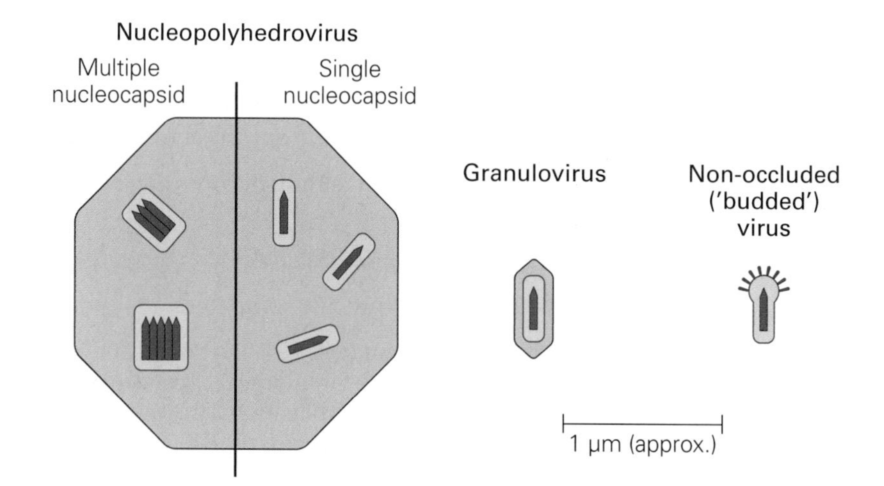

Figure 7.4
Baculovirus structure.
Nucleopolyhedroviruses have a single occlusion body (OB) that contains multiple infectious units while in *Granuloviruses*, each OB contains a single infectious unit. During infection of the host, a non-occluded or 'budded' form of the virus is produced which lacks the occlusion body. From, Harper D (2006) Biological control by microorganisms. In Encyclopedia of Life Sciences. With permission from John Wiley & Sons, Inc.

Viruses to control insect pests

Baculoviruses (*Baculoviridae*) are a large group of viruses that infect insects and other arthropods. All tend to be quite specific in the species that they will infect. Baculoviruses show a high level of environmental stability due to their formation of thick protein shells, know as **occlusion bodies** (OBs), around the nucleocapsid that contains the viral DNA genome (**Figure 7.4**). The family *Baculoviridae* is subdivided into two genera. For the nucleopolyhedroviruses (NPVs), a single occlusion body contains multiple nucleocapsids packed singly or in groups, and is formed of the viral polyhedrin protein. For the granuloviruses (GVs), each OB contains a single nucleocapsid and is formed of the viral granulin protein. A non-occluded form of the virus, referred to as the budded form, is formed during infection of insect hosts, in which it spreads the infection from cell to cell. The OBs also have an outer, carbohydrate-rich coat that helps to protect them from degradation, particularly if eaten by larger animals.

As part of their natural infectious cycle, baculoviruses are eaten by insect larvae. They then infect the cells of the gut and grow there. From these cells, the virus can then spread throughout the body of the insect, destroying it and releasing a new generation of virus from the liquefied remains of the killed larva (**Figure 7.5**).

This pattern of infection is seen in the larvae of butterflies and moths (members of the family *Lepidoptera*), and usually results in death in 4–5 days, although this can take longer under field conditions. In sawfly larvae (suborder *Symphyta*) the infection, while still lethal, is localized to the gut, and virus is shed by defecation and vomiting. Virus is released more rapidly, although less is produced overall.

Oryctes rhinoceros virus (OrV) infects the coconut rhinoceros beetle, a destructive tropical pest. This virus was formerly considered to be a baculovirus, but does not form OBs and is now classified separately. Virus infection is restricted to the midgut of adult beetles, but generates high levels of infectious virus. The beetles have been referred to as flying virus factories, and this helps with the effectiveness of the control process. The virus results in a generalized infection of larvae, which are killed in 9–25 days after infection. Since it does not form an OB, it is less stable than occluding viruses. However, it is capable of long-term persistence in the environment, aided by virus production in the midgut of adult beetles. It has been in use as a control agent since 1967 and appears to produce long-term control. In one field study, OrV reduced the numbers of beetles over a two-year period, then kept numbers low for a further two years.

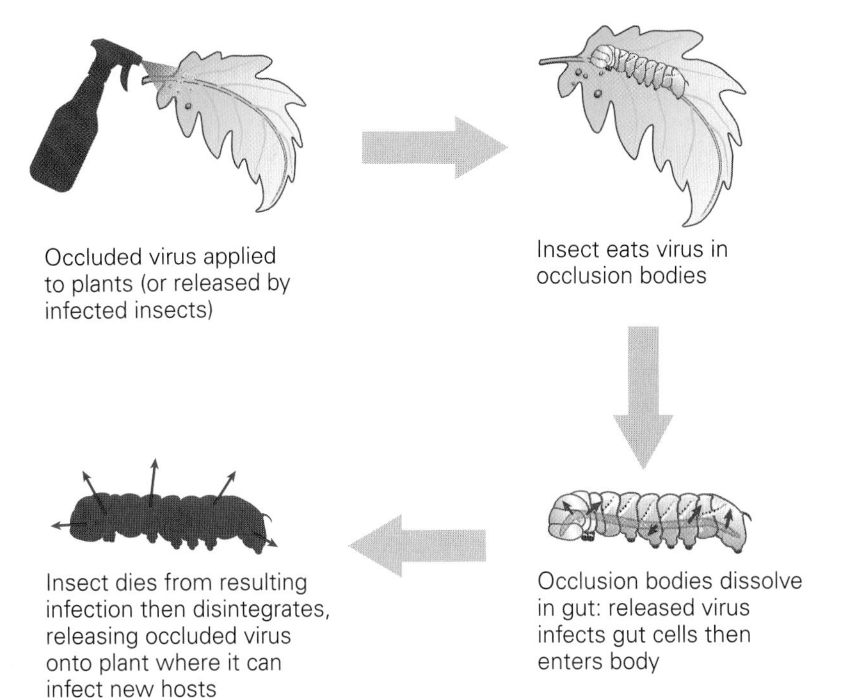

Figure 7.5
Insecticidal effect of generalized baculovirus infection. From, Harper D (2006) Biological control by microorganisms. In Encyclopedia of Life Sciences. With permission from John Wiley & Sons, Inc.

The entomopox viruses (family *Poxviridae*) and the cytoplasmic polyhedrosis viruses (family *Reoviridae*) also infect insects and produce OBs. However, neither of these has as yet been developed as a commercial biological control agent. Some work (including field trials) has taken place with other insect viruses, including the *Parvoviridae*, *Picornaviridae*, and *Tetraviridae*. Some of these approaches target urban rather than agricultural pests, including cockroaches, ants, and termites, but this is still at an early stage.

Viruses to control rabbits

It is not only insects that are controlled with viral agents. These have also provided the most successful approach to controlling the devastating numbers of European rabbits infesting much of Australia.

European rabbits were released into Australia in 1859 as part of a very unsuccessful attempt at improvement of the ecosystem by early settlers. With no effective predator and very limited competition, they rapidly became a highly destructive pest. Attempts to control them were unsuccessful, and the arrival of one potential biological control agent, the domestic cat, also proved an ecological disaster.

The myxoma virus is a member of the *Poxviridae*, and was originally observed as a cause of a mild skin disease in American rabbits (genus *Sylvilagus*) in Uruguay in the late nineteenth century. However, in European rabbits (genus *Oryctolagus*), it causes a hugely damaging systemic infection, apparently due to a different interaction with the immune system. With European rabbits between 90% and 100% are killed by the resultant infection, which is known as myxomatosis.

From 1938 onward it was evaluated as a possible control agent for introduced European rabbits in Australia. It was finally released in 1950 and proved highly effective, reducing the rabbit population by an estimated 500 million (around 85%) in two years. This provides an excellent example of the self-propagating nature of an effective biological agent.

However, in Australia, myxoma virus is spread by mosquitoes. During winter the lack of suitable insects to spread the disease meant that viruses that killed rabbits rapidly died out, while those that did not kill their host rapidly, permitting it to survive until the insect vector was again available, had an evolutionary advantage. In addition, small numbers of rabbits had some resistance to the disease, and these tended to be the ones that survived to breed. These two evolutionary pressures meant that, within a few years, rabbit numbers were rising once more. By 1957, only 25% of rabbits in Australia were killed by myxomatosis, and evidence has now appeared that the virus has established long-term infections in some rabbits. This is a clear example of antagonistic co-evolution, whereby pathogen and host co-evolve, changing the nature of their relationship. It also illustrates some of the limits of biological control in such systems.

The virus was introduced into Europe in 1952 in an attempt to control rabbits on an estate in France. As might have been expected (but apparently was not, at the time) it then spread across the continent (with the rabbit flea as its main vector) and is still active today.

Despite the evolution of resistance, myxoma virus still represented the most successful attempt to control rabbits in Australia. So, when a new and lethal rabbit disease (rabbit hemorrhagic disease, caused by a member of the *Caliciviridae*) was identified in China in 1984, there was pressure to repeat the experiment. The Australian government instituted a testing program on Wardang Island, three miles from the South Australian coast, to determine whether the virus would harm native Australian wildlife. However, during testing, the virus spread to the mainland which was perhaps not very surprising given the limited distance it had to cover. The cause of the escape was never confirmed, but evidence that it can be carried by flying insects provides one possible answer.

Within three months of the escape, 20 million rabbits had died. While the virus did not appear to harm native Australian species, hungry predators denied their usual food found alternatives from native wildlife. With the virus now released, the Australian government then undertook a program of controlled releases. The virus spread throughout the country, but while up to 90% of rabbits were eliminated in some dry areas, in other, wetter areas there was much less effect.

In 1997, the virus was introduced illegally into New Zealand, in what one commentator called "an act of biological warfare." Faced with the reality of the virus releases, the New Zealand government also authorized its use. However, the release had occurred at the wrong stage of the breeding cycle for maximum effect (since very young rabbits are unaffected) and rabbit numbers have now risen once again.

There have also been signs that, as with myxomatosis, virus and host are adapting to reduce the virulence of the circulating virus. Within three years of the accidental release of the virus on to the Australian mainland, there were calls for more effective versions to be developed.

Resistance

In contrast to the resistance to viruses developed by rabbits, resistance to viral insecticides in susceptible forms of the host insect is very rare, which may reflect the more limited (and typically non-adaptive) immune system present in insects. Many insects acquire increasing resistance as they age, with adult forms often very highly resistant. This process is known as developmental resistance and reflects changes in the maturing insect body; it is not transmitted to the next generation of larvae.

This development of resistance in insects is very limited compared with the rapid development of resistance to chemical pesticides, demonstrating a significant advantage for biological control.

Integrated pest management

The term **integrated pest management** (IPM) refers to the use of multiple, often low-potency, controls that together can reduce pest numbers to acceptable levels. Biological control agents are often used as part of an IPM strategy. This may include introducing or maintaining habitats for natural enemies, such as hedgerows or suitable plants, along with the reduction or elimination of plants or conditions that favor the pest organism. In such programs, predators may actually serve to spread infection—baculoviruses survive passage through the gut with high efficiency and can be spread in the droppings. Even at the microbiological level, it is possible to use multiple biological control agents simultaneously. For mosquitoes, for example, fungal and bacterial agents can be used together.

Chemical pesticides can also be used in combination with other approaches, including biological agents, provided that the biological agent itself is not harmed by the chemical treatments used. Since viruses are relatively unaffected by these chemical treatments they are well suited to such combined uses, and additive effects have been observed where both virus and chemical treatments exerted controlling effects.

A similar term, integrated pathogen management, has been suggested for the combined use of biological and other approaches in preventing and controlling infectious disease.

7.4 BACTERIOPHAGE THERAPY

Any cellular agent that causes human disease will be infected by its own pathogens, which can theoretically be developed as a control agent. Most progress in this area has been made with viruses that infect bacteria.

Bacteriophages

Bacteriophages are highly specific viruses that can target, infect, and (if correctly selected) destroy pathogenic bacteria. Antibacterial activity was first observed in the waters of the Ganges and Jumna rivers by Ernest Hankin in 1896. The causative agent was discovered independently in 1915 by Frederick Twort and in 1917 by Felix d'Herelle. It was d'Herelle who named them bacteriophages (devourers of bacteria) and who then expanded on his initial finding to establish many of the techniques that form the basis of modern virology.

Bacteriophages are believed to be the most numerous type of viruses accounting for the majority of the 10^{31} viruses present on Earth. They can be found at high concentrations in water, with over 10^8 per milliliter being recorded from some sources. More than 90% of characterized bacteriophages are classified in the order *Caudovirales* (**Figure 7.6**). These are the tailed bacteriophages, with a large double-stranded DNA genome in the range of 33,000 to 170,000 bp or even larger. Other families of bacteriophages also exist, with a range of morphologies, genome types, and genome sizes (**Table 7.7**).

From the 1940s onward, bacteriophages became one of the basic tools of molecular biology. Along with an understanding of the processes of the cell came a vastly increased understanding of the nature and activities of the bacteriophages themselves.

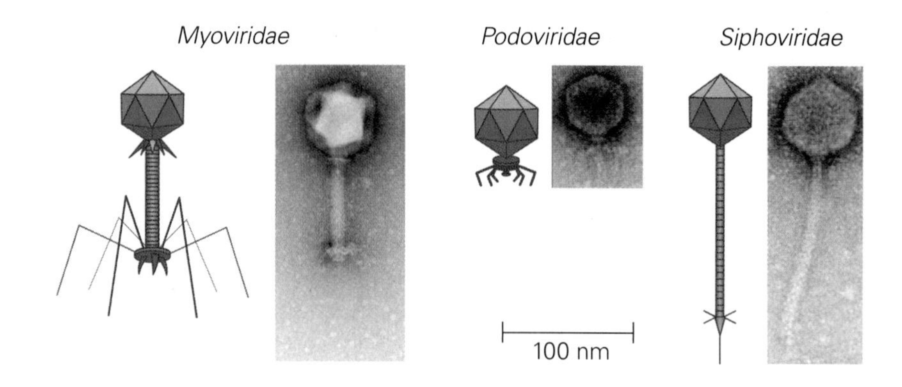

Figure 7.6
Tailed bacteriophages (*Caudovirales*).
Courtesy of Biocontrol Limited.

It soon became clear that very large numbers of bacteriophages existed, and that the vast majority were specific to a single host species (and indeed usually restricted to a limited range of strains within that species). While many bacteriophages produced rapid lysis of the host cell (**Figure 7.7**), others could integrate into the host chromosome, entering a latent state known in this context as lysogeny. The inserted bacteriophage DNA is then known as a prophage. Reactivation may occur in response to a variety of stimuli, and is directly analogous to the chromosomal insertion/reactivation cycle seen with the *Retroviridae* in eukaryotic cells.

The ability to enter the lysogenic state is associated with a range of characterized genetic functions within the bacteriophage genome including

Table 7.7 Families of bacteriophages

Virus family	Genome type	Genome size (kbp)	Structure	Example
Caudovirales				
Myoviridae	dsDNA	33.6–170	Non-enveloped, icosahedral head (50–110 nm, may be elongated) with long contractile tail	*Enterobacteria* phage T4
Podoviridae	dsDNA	40–42+	Non-enveloped, icosahedral head (60 nm) with short, non-contractile tail	*Enterobacteria* phage T7
Siphoviridae	dsDNA	48.5	Non-enveloped, icosahedral head (60 nm) with long, non-contractile tail	*Enterobacteria* phage λ
Other families				
Tectiviridae	dsDNA	147–157	Icosahedral, contains lipid, 63 nm with 20 nm spikes	*Enterobacteria* phage PRD1
Corticoviridae	dsDNA	9–10	Icosahedral, contains lipid, 60 nm+	*Pseudoalteromonas* phage PM2
Plasmaviridae	dsDNA	12	Enveloped, spherical/pleomorphic, 80 nm	*Acholeplasma* phage L2
Inoviridae	ssDNA	4.4–8.5	Non-enveloped, filamentous, 6–8 nm × 760–1950 nm	*Enterobacteria* phage M13
Microviridae	ssDNA	4.4–5.4	Non-enveloped, icosahedral, 25–27 nm	*Enterobacteria phage* ΦX174
Leviviridae	ssRNA	3.4–4.2	Non-enveloped, icosahedral, 26 nm	*Enterobacteria* phage MS2
Cystoviridae	dsRNA (segmented)	13.4 (3 segments)	Enveloped, spherical, 86 nm with 8 nm spikes	*Pseudomonas* phage Φ6

The *Caudovirales* account for more than 90% of all characterized bacteriophages. Other viruses (including members of the *Myoviridae* and the *Siphoviridae*) infect the Archaea.

a repressor of lytic gene function (that prevents killing of the host bacterial cell) and DNA integrases (that insert the viral DNA into the bacterial genome).

When lysogenic bacteriophages emerge from their latent state they may pick up and transfer bacterial DNA as part of this process. In some cases these may be associated with bacterial virulence, but this is not universal. The two cycles, lytic and lysogenic, are summarized in **Figure 7.8**.

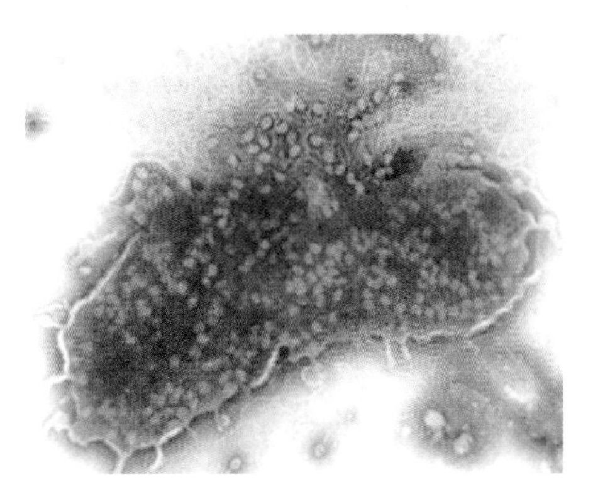

Figure 7.7
Bacteriophage lysis of the host cell. From Brown JC (2003) Virology. In Encyclopedia of Life Sciences. With permission from John Wiley & Sons, Inc.

Bacteriophages as therapeutic agents

Bacteriophages were discovered before effective antibiotics and so it was hoped that they would be able to control bacterial disease. In 1919, there were reports of the apparently successful treatment of typhoid in chickens and of dysentery in five humans. In 1921, bacteriophages were used against *Staphylococcus* in skin disease.

During the 1920s, both localized and large-scale experiments were undertaken in many countries, including the treatment of over a million patients in India. A wide range of commercial preparations were sold in Europe and the USA. The novel (1925) and film (1931) *Arrowsmith* presented a fictionalized account of their use.

Unfortunately, understanding of the nature of bacteriophages was extremely limited at that time. It was not until 1939 that use of early electron microscopes helped to settle an ongoing argument as to whether they were viruses or some form of chemical toxin. In consequence, much of the early work using bacteriophages was deeply flawed:

- Bacteriophages were used against diseases with no bacterial component (e.g. herpes, urticaria).
- It was assumed that bacteriophages were able to destroy a wide range of bacteria, whereas they are highly specific. They were thus used against inappropriate bacterial targets.
- Inappropriate growth conditions or preservatives were used that limited or prevented the inclusion of infectious bacteriophages.
- Methods of administration were used that inactivated any bacteriophages present (some new evidence shows this may not destroy them).

Figure 7.8
The lytic and lysogenic life cycles of a bacteriophage. Courtesy of Biocontrol Limited.

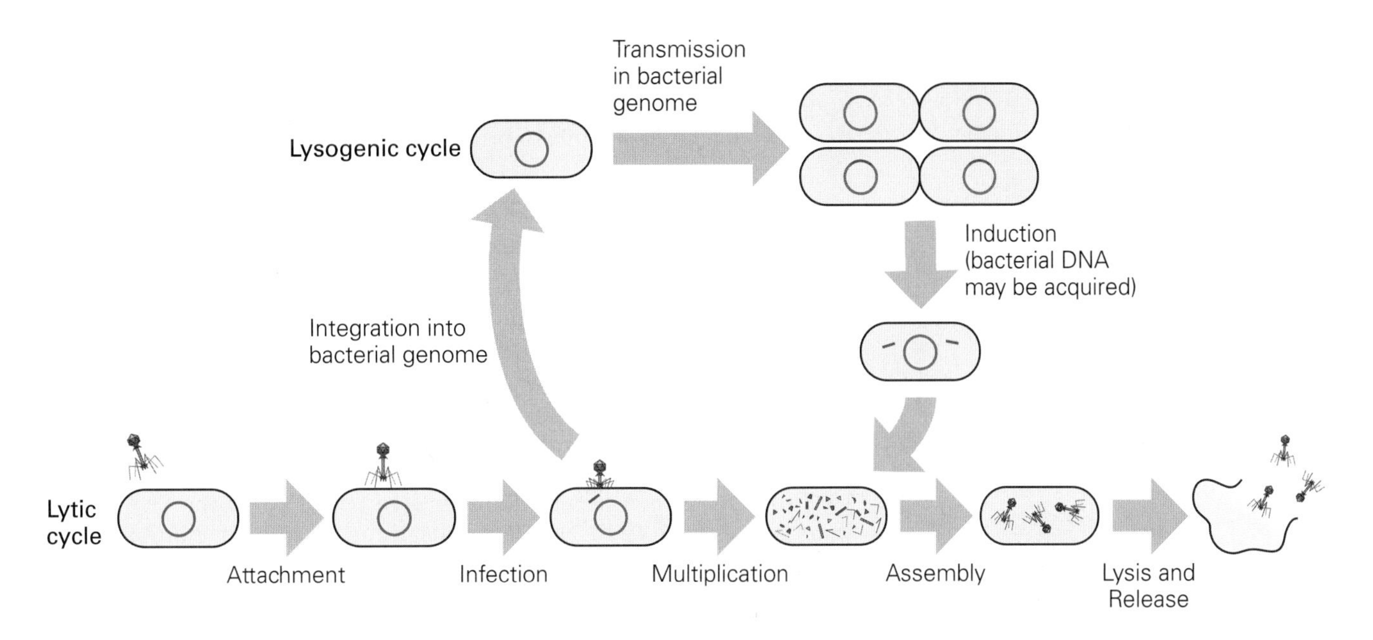

In 1934, the *Journal of the American Medical Association* published the results of a large study carried out by the US Council on Pharmacy and Chemistry. It concluded that, apart from a few restricted uses, there was no good evidence that bacteriophages actually worked in the therapeutic uses that had been evaluated. It further noted that most of the work carried out was deeply flawed, lacking proper controls, adequately purified therapeutic substance, or sufficient numbers of patients.

Despite this, the therapeutic use of bacteriophages continued through World War II. The German and Soviet armies used many bacteriophage preparations, notably against dysentery.

The antibiotic age

With the introduction of antibiotics, the popularity of bacteriophages declined. In 1932, Prontosil, the first sulfonamide drug, had been manufactured. In 1941, the first clinical use of penicillin showed the promise of this new class of drugs. Antibiotics began to appear in increasing numbers. They had broader activity, were simpler to use, and were supported by properly conducted clinical studies.

Some work with therapeutic bacteriophages continued, notably in Soviet Georgia, but the antibiotic era had arrived and was to last for over half a century. Work with bacteriophage therapeutics did continue in Western Europe, but on a very small scale, until the 1960s in France and even until the 1980s in Switzerland. However, the main locations for such work were the countries of the Soviet bloc.

In 1926, Georgiy Eliava, a bacteriologist from Soviet Georgia, visited Felix d'Herelle in Paris. On his return he began to work on bacteriophages. During the 1930s d'Herelle visited Georgia and started to work with Eliava to establish a centralized institute for bacteriophage research in Tbilisi, the capital of Soviet Georgia. This was derailed when Eliava was executed in 1937 during the Stalinist purges, but work continued and at its height the Institute produced millions of doses of a range of bacteriophage mixtures, with multiple targets and specific indications for their use. Alongside this, work also proceeded in Poland and Russia. Given the limited availability of antibiotics in the Soviet bloc, bacteriophage drugs became an accepted part of medical practice.

Meanwhile, in the West, the triumph of antibiotics continued. In 1969, US Surgeon General William Stewart was reported as saying in a report to the US Congress, "It is time to close the book on infectious diseases. The war against pestilence is over." He is on record as denying ever having said this, but it was a commonly held attitude at the time. However, antibiotic resistance was already a significant problem and was about to get a lot worse.

By the 1990s, rising numbers of cases were occurring where antibiotics could no longer cope. Resistance to all available antibiotics was by then apparent for a range of pathogenic bacteria, including methicillin-resistant *Staphylococcus aureus* (MRSA), vancomycin-resistant *Enterococcus* (VRE), and a range of others.

The situation was made worse by a lack of novel antibiotics. Drug companies had by this time moved away from research into antibiotics, where there were seen to be plenty of drugs available, into areas where the potential for long-term use of "lifestyle" drugs (such as those to lower cholesterol or to control heart disease) made them more profitable. In the five years from 1983 to 1987, fifteen new antibiotics were approved by the FDA. Twenty years later, from 2003 to 2007, it was just three. To add to the problem, even when a new type of antibiotic was finally developed, resistance appeared almost immediately.

Now, with resistance approaching crisis levels and few new drugs in the pipeline, interest turned once again to the potential of bacteriophages.

A renewal of interest in bacteriophages

In the 1980s, a British bacteriologist, William Smith, had conducted a series of experiments in both mice and farm animals. Bacteriophages were shown to be more effective than a variety of antibiotics in reducing the mortality of mice with generalized and intracerebral *Escherichia coli* infections, and bacteriophages were shown to be multiplying in the tissues. Very low doses of phage were found to be effective. Smith and his co-worker, Michael Huggins, evaluated the treatment of experimentally induced diarrhea in calves, piglets, and lambs due to several different strains of *E. coli*, and were able successfully to treat and to prevent infections, even when low doses of bacteriophages were used. However, some caution was warranted since these were experimental infections rather than trials against natural disease.

These demonstrations of the potential of bacteriophage therapy, combined with the growing urgency of the antibiotic crisis, led to renewed interest in this approach in the 1990s. The British clinician James Soothill showed that bacteriophages could prevent the destruction of skin grafts in guinea pigs and could protect mice against lethal levels of bacterial infection.

Combined with the continuing increase in antibiotic resistance, work showing that bacteriophages could be effective led several groups to look at the possibility of bringing therapeutic bacteriophage technology back from the former Soviet bloc, which was now accessible following the fall of the old Communist regimes several years earlier.

Unfortunately, despite the apparent promise of this technology, and despite widespread use including a very large body of work involving the treatment of many thousands of patients in local studies, the work which had been done was not enough to prove the case for bacteriophage therapy for Western use. Under the different pressures that applied in the Soviet era, particularly the desperate need for treatment options, this work had not involved the detailed documentation, procedures, and double-blind controls required for progress within a modern regulatory framework. The evidence was supportive, but not sufficient.

As a result, much of the initial excitement over access to former Soviet science in the mid-1990s proved unsupportable. While clinics in Poland and Georgia offered bacteriophage treatments to those motivated enough to travel to them for treatment, these could not be used in Western Europe or the United States. Those companies that managed to survive in the West generally did so on the basis of other kinds of applications, in particular the somewhat less complex areas of food and agricultural applications.

By 2007, five products were licensed in the United States. One was to prevent bacterial infections of tomatoes and peppers, two to destroy bacteria associated with food poisoning on food animals before slaughter, and two for the control of *Listeria monocytogenes* on ready-to-eat food. In this latter use, exposure of humans to the bacteriophages was accepted by the FDA—an important step forward.

Clinical work in humans was slower to develop, not least due to the relatively high cost of such work. The range of issues relating to the use of bacteriophages as antibiotics are summarized in **Box 7.1**.

Some groups have focused on investigating the use of genetically modified (GM) bacteriophages, often with enhanced abilities to destroy their bacterial target but with reduced ability to replicate—essentially a gene vector

Box 7.1 Issues relating to the use of bacteriophages as antibiotics

Advantages:

- Bacteriophages are highly specific, helping to avoid side effects. This is in line with an increasing trend in antibiotic usage to avoid drugs with broad specificity.
- The use of replicating bacteriophages also allows for the use of very low input levels that can multiply up at need. This can reduce both cost and toxicity.
- Bacteriophages are unaffected by antibiotic resistance, and there is evidence that they may be able to work synergistically with some conventional antibiotics.
- Bacteriophage preparations appear free of gross toxicity. Potential complications arising from the release of bacterial endotoxins from lysed cells (the Herxheimer effect, as seen with some antibiotics) do not seem to be an issue at the dosing levels now in use.
- Bacteriophages are relatively simple to manufacture (although the highly regulated manufacturing process required for clinical use adds to costs).
- Bacteriophages themselves may be able to adapt to counter bacterial resistance.

Disadvantages:

- High specificity means that it is important to identify the pathogenic bacteria present and to ensure that they are responsible for the disease. Multiple bacteriophages may be required to obtain useful levels of coverage, though this can in turn reduce the chance of resistance.
- Bacteriophages have limited suitability for systemic use since they can generate immune responses which can reduce efficacy. This can be avoided by use on infected sites on the body surface or oral administration. The potential for systemic use remains to be assessed in clinical trials.
- The mobilization of bacterial genetic material that may be able to moderate virulence is to be avoided. For this reason, only lytic (rather than lysogenic) bacteriophages are used, and regulators require careful monitoring of all forms of transduction (transfer of bacterial genes).

Regulatory and commercial issues:

- The attitudes of the EU and US regulators to this novel therapy have proven to be less of an issue than many had predicted, and it is clear that trials can be undertaken and completed successfully.
- Patentability of bacteriophages has been cited as an issue, since the basic technology is well established and there are a huge number of bacteriophages waiting to be isolated. However, monoclonal antibodies are now a commercial success despite occupying a very similar position.

approach. However, any use of GM technology brings with it very significant additional regulatory requirements.

In addition, many researchers consider the ability of bacteriophages to replicate and thus to expand their numbers at precisely those points where they are needed to be a major strength of the approach, and thus question the use of nonreplicating bacteriophages or bacteriophage-derived vectors (**Box 7.2**). Unlike gene therapy vectors, where replication appears to have the potential to cause problems, bacteriophages can only replicate in their prokaryotic target cells. To eukaryotic cells they are effectively inert, since

Table 7.8 Modern clinical trials of bacteriophage therapeutics

Phase	Location	Target	Condition
Phase 1, safety only			
Completed 1999	UK	*Enterococcus*	Gut infection
Completed 2005	Switzerland	*E.coli*	Diarrhea
Completed 2008	USA	*Pseudomonas*, MRSA, *E. coli*	Leg ulcers
Completed 2009	Belgium	*Pseudomonas*, MRSA	Burns
Phase 2, safety and efficacy			
Completed 2007	UK	*Pseudomonas*	Ear infection
Under way (with earlier Phase 1)	Bangladesh	*E. coli*	Diarrhea

Planned: MRSA (nasal carriage), *P. aeruginosa* (lung infections, ear infections, ulcers).

their controls for gene expression are specific for prokaryotic cells. While properly configured genes may be expressed if eukaryotic controlling elements are added, this does require the deliberate insertion of such controls from other sources.

It should be noted that all trials reported to date have used unmodified, replication-competent bacteriophages.

Into the clinic

The first modern clinical trial of a bacteriophage therapeutic was carried out in 1999 in London, though this used healthy volunteers without the targeted infection, and thus only addressed safety issues. The intended target was vancomycin-resistant *Enterococcus* infections. After this, additional safety trials were carried out (**Table 7.8**), but it was not until 2007 that the first modern clinical trial of the efficacy of a bacteriophage therapeutic was begun, targeting *Pseudomonas aeruginosa* infections of the ear. The trial completed in 2007 and larger, phase 3 trials are now planned. It will be necessary for these trials to be completed successfully and approved by the relevant regulators—the European Medicines Agency (EMA) in Europe and the Food and Drugs Administration (FDA) in the USA—before any bacteriophage therapeutic can proceed to market in these areas. However, it seems that progress is being made.

Box 7.2 Active and passive bacteriophage therapy

In the first phase 2 trial, completed in 2007 (against *Pseudomonas aeruginosa*), the input dose of bacteriophages was 600,000 infectious units, equating to only 2.4 ng of protein. This approach, known as *active bacteriophage therapy*, relies on bacteriophage multiplication to generate therapeutic doses of the bacteriophages against the far larger numbers of bacteria that are present.

Use of higher doses sufficient to produce antibacterial effects without relying on bacteriophage replication is known as *passive bacteriophage therapy*, and many approaches using genetically modified agents may use this since the modified bacteriophages are unable to replicate.

Key Concepts

- Viruses may be used in a variety of beneficial ways, including their use as model organisms in biological research.
- Viruses can be used in the delivery of therapeutic genes, known as gene therapy. Originally seen to have great potential for the control of inherited genetic diseases, progress in gene therapy has been much slower than was initially anticipated and its focus has switched to anti-cancer applications.
- Viruses may be used to treat cancers (virotherapy). Such treatment originally used the natural properties of specific viruses, but is now usually based on genetic modification and the use of viral and gene vectors.
- Biological control in agriculture only accounts for a small section of the pest control market. Viruses account for less than 0.1% of expenditure on agricultural pesticides, although this is expanding.
- The use of viruses of bacteria (bacteriophages) to control damaging bacterial infections following their discovery in the early part of the twentieth century was rapidly replaced by the use of chemical antibiotics, but recent developments are reviving interest in this approach.

DEPTH OF UNDERSTANDING QUESTIONS

Hints to the answers are given at http://www.garlandscience.com/viruses

Question 7.1: Why has gene therapy not delivered on its original promise for inherited genetic disease?

Question 7.2: Gendicine, a gene therapy product delivering a functional p53 gene into cancer cells, has been approved for use in China. Is it likely that the US and European regulators will now approve this approach?

Question 7.3: Why did farmers in New Zealand release rabbit Calicivirus, even though this has been classed as bioterrorism? What are the consequences of this act?

Question 7.4: Is it necessary to use genetic modification techniques to develop bacteriophage-based antibiotics?

FURTHER READING

Hacein-Bey-Abina S, Hauer J, Lim A et al. (2010) Efficacy of gene therapy for X-linked severe combined immunodeficiency. *N. Engl. J. Med.* 363, 355–364.

Harper DR (2006) Biological control by micro-organisms. In Encyclopedia of Life Sciences. John Wiley & Sons, Chichester. http://www.els.net/

Harper DR & Kutter E (2008) Therapeutic use of bacteriophages. In Encyclopedia of Life Sciences. John Wiley & Sons, Chichester. http://www.els.net/

Häusler T (2006) Viruses vs. Superbugs: A Solution to the Antibiotics Crisis? Macmillan, London.

Lacey LA & Kaya HK (eds) (2007) Field Manual of Techniques in Invertebrate Pathology: Application and Evaluation of Pathogens for Control of Insects and Other Invertebrate Pests. Springer, Netherlands.

Templeton NS (ed.) (2008) Gene and Cell Therapy: Therapeutic Mechanisms and Strategies, 3rd ed. CRC Press, Boca Raton.

Wright A, Hawkins CH, Änggård EE, Harper DR (2009) A controlled clinical trial of a therapeutic bacteriophage preparation in chronic otitis due to antibiotic-resistant Pseudomonas aeruginosa; a preliminary report of efficacy. *Clin. Otolaryngol.* 34, 349–357.

INTERNET RESOURCES

Much information on the internet is of variable quality. For validated information, PubMed (http://www.ncbi.nlm.nih.gov/pubmed/) is extremely useful.

Please note that URL addresses may change.

Biological control: a guide to natural enemies in North America. http://www.nysaes.cornell.edu/ent/biocontrol/

Gene Therapy Clinical Trials Worldwide, provided by the Journal of Gene Medicine. http://www.abedia.com/wiley/

Human Genome Project gene therapy resource. http://www.ornl.gov/sci/techresources/Human_Genome/medicine/genetherapy.shtml

CHAPTER 8

Emergence, Spread, and Extinction

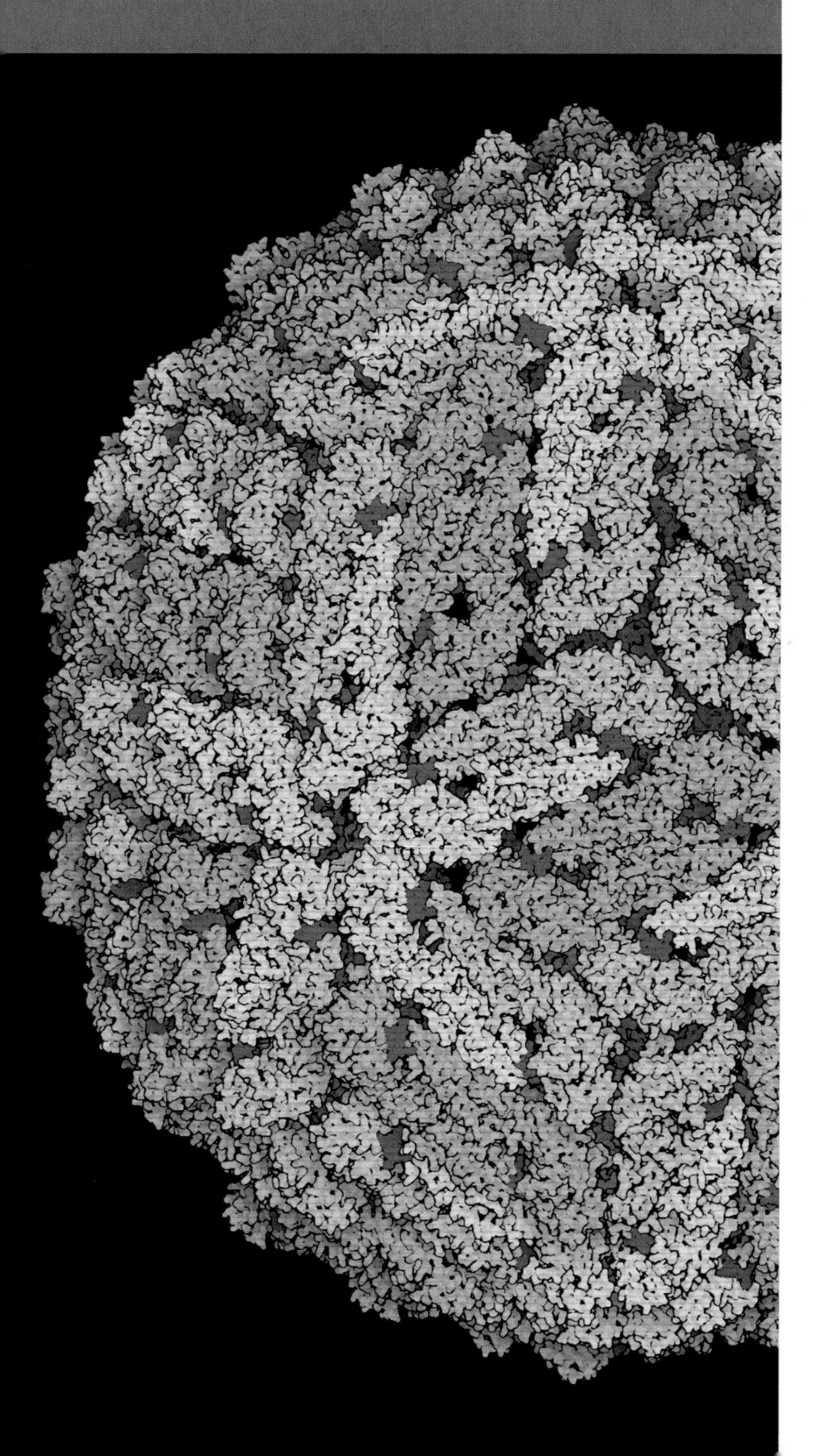

INTRODUCTION

Viruses all have an origin, typically emerging from a related agent, often from a different host species. If they can then spread, they may establish themselves in the new host and diverge further from their ancestor. If they cannot transmit themselves, they will become extinct in the new host unless continually replaced by cross-species infections. With modern technology, underpinned by an understanding of the nature and vulnerabilities of the virus in question, it is also possible to block transmission deliberately. The eventual outcome of this, as with smallpox (*Poxviridae*) and rinderpest (*Paramyxoviridae*), is to drive a virus into extinction.

About the chapter opener image
Dengue Virus
(Courtesy of the Research Collaboratory for Structural Bioinformatics Protein Data Bank and David S. Goodsell, The Scripps Research Institute, USA.)

The level of infection with a particular virus among members of its host population is never constant. In some locations or at some times, levels may be far higher than normal. We call these rises **emergences**, **epidemics**, or even **pandemics**, though only if the disease symptoms are apparent enough that we notice them. Similarly, a virus may drop below detectable levels, and sometimes can just vanish. However, within a complex environment, it is almost impossible to know just what is going on in every location and how complete or how permanent any extinction might be. For these reasons, a major requirement in understanding what is happening with any infection is effective monitoring of the levels and types of infections that are occurring.

8.1 SURVEILLANCE

Before understanding of the microbial origins of disease, monitoring the level of an infection was limited to observing its effects. With some diseases this was not difficult, since it is hard to miss the symptoms of a smallpox outbreak. However, even here there could be confusion between diseases with similar symptoms that result from different infections, as with the various types of pox. Smallpox is caused by a member of the *Poxviridae*, while the great pox is a name formerly used for the bacterial disease syphilis. To add to the confusion, chickenpox (probably from the old English word *tzuiken*, meaning itchy pox) is caused by a member of the *Herpesviridae*. Similarly, multiple viruses with different modes of transmission can cause a similar range of symptoms resulting from viral hepatitis.

With increased understanding of the nature and specific causes of diseases came the ability to monitor specific infectious agents. The late nineteenth century saw the establishment of a system of requiring reports to central authorities for diseases of particular concern. Key to this was the concept of **notifiable diseases** that, by law, had to be reported to the authorities.

The US Congress in 1878 authorized the Marine Hospital Service to collect reports on cholera, smallpox, plague, and yellow fever from diplomats overseas. These were to be used in designing quarantine measures to prevent spread of these diseases into the USA. Congressional funds were first provided for reports of these notifiable diseases a year later, and in 1893 reporting was expanded to include data from states and municipal authorities. Telegraphic alerts were required for specific diseases from 1912.

In the United Kingdom, diseases such as cholera, diphtheria, smallpox, and typhoid had to be reported in London from 1891, and in the rest of England and Wales from 1899. With the founding of the World Health Organization (WHO) in 1948, coordination of such reporting at the global level was established.

In the USA in 2009, a total of 21 viruses or groups of viruses cause notifiable diseases (see **Box 8.1**). While the more exotic viruses, such as West Nile virus and hantavirus pulmonary syndrome (HPS, *Bunyaviridae*) are to be expected, the list also includes some more familiar diseases, including chickenpox and mumps.

The list can expand or contract, but at any time it reflects those diseases that the health authorities consider it important to know about. There is also extensive monitoring of a far wider group of diseases at all levels from local to global. Additionally, from the earliest days of such **surveillance**, it was clear that surveillance overseas was and is an essential part of any such program.

One hundred and twenty years on from the first Congressional authorization of such reporting, the US Senate Foreign Operations Subcommittee noted in 1998 that "There is an urgent need ... to significantly augment

Box 8.1 Viruses causing nationally notifiable diseases in the USA 2009

- California serogroup virus disease
- Eastern equine encephalitis virus disease
- Powassan virus disease
- St. Louis encephalitis virus disease
- West Nile virus disease
- Western equine encephalitis virus disease
- Hantaviruses causing pulmonary syndrome
- Hepatitis A virus
- Hepatitis B virus
- Hepatitis C virus
- Human immunodeficiency virus
- Measles virus
- Mumps virus
- Influenza A virus (pediatric and novel forms)
- Poliovirus
- Rabies virus
- Rubella virus
- Severe Acute Respiratory Syndrome-associated Coronavirus disease
- Smallpox virus
- Varicella-zoster virus
- Yellow fever virus

Note: notifiable diseases of nonviral origin have been omitted from this list.

international surveillance and control mechanisms, and to strengthen the ability of developing countries, where deadly viruses often first gain a foothold, to protect and care for their people."

The process of identifying new or emerging diseases involves work at all levels, from the field investigator trying to find an outbreak, whose biggest problem may be where to find spare parts for his jeep, to the molecular biologist compiling sequence data who may know almost nothing about the source of the disease. It may also be heavily influenced by the political climate.

When a mistake is made, it is easy to apportion blame with the benefit of hindsight, but it is a lot more difficult to know where to use the necessarily limited resources available before an outbreak has occurred.

Two examples illustrate the opposing problems of too little and too much response.

Too little: hantavirus pulmonary syndrome

When an outbreak of an unexplained and deadly respiratory disease in the southwestern United States in 1993 was detected by alert epidemiologists, investigation by scientists at all levels from local to national soon identified the causative agent as a hantavirus (*Bunyaviridae*; discussed in more detail in Section 8.2). Hantaviruses were already known, but the Eurasian form of the disease (hemorrhagic fever with renal syndrome; HFRS) was distinctly different to this new American form, soon named hantavirus pulmonary syndrome (HPS).

Since the identification of the first hantavirus, Hantaan, during the Korean war the US Army had run a program investigating hantavirus disease, but budget cuts in 1991–1992 had greatly reduced this program. At the time, there was no sign that hantaviruses would cause any major problems in the USA, and cutting the budget seemed harmless.

Too much: swine flu

At the opposite extreme are the consequences of the 1976 Fort Dix influenza outbreak. The virus isolated in an outbreak at this military training facility appeared by all available tests to be derived from swine and closely related to the pandemic H1N1 1918–1919 "swine flu" (*Orthomyxoviridae*) which killed up to 50 million people (**Box 8.2**).

A massive public health program was begun to head off the expected and long-feared epidemic of swine flu, which some predictions, based on events in 1918–1919, suggested could kill a million Americans. Almost 50 million Americans were vaccinated, and huge efforts were made to ensure that there was sufficient vaccine for everyone. Had the epidemic developed as expected, this would have been hailed as a triumph of public health medicine. However, the virus proved to be poorly transmissible and caused only a very localized outbreak. In consequence, the main results of the vaccination program were substantial damage to the public perception of vaccination, together with a series of claims for damages due to the vaccine, most notably involving Guillain-Barre syndrome, a paralytic neurological condition. Settling the claims would eventually cost nearly US$100 million, almost as much as the vaccination program itself. The program is often described as "a fiasco."

Under different circumstances, canceling the hantavirus programs could have been harmless, and the 1976 swine flu vaccine could have saved millions of lives. While making such decisions is complex, and unlikely ever to be right all the time, the cost of the wrong decision can be very high indeed.

Box 8.2 On the naming of influenza

Influenza A viruses are subtyped by the nature of their surface glycoproteins; hemagglutinin (HA, or H) and neuraminidase (NA, or N). There are 16 H subtypes and 9 N subtypes. Each virus is named for the H and N on its surface. These may even be subdivided further, as with the $H_{sw}1N1$ virus that was responsible for the 1918–1919 Spanish influenza. The $_{sw}$ suffix refers to the apparent origin of this form of the H protein in viruses from pigs (swine). It is changes in these H and N proteins that occur in *antigenic shift*.

Influenza: from surveillance to pandemic

Influenza had not gone away. From the late 1990s, surveillance in Southeast Asia showed a worryingly high level of pathogenicity for H5N1 influenza (known as "bird flu") in humans.

Transfer appeared to require close contact with infected birds, but once the infection of a human occurred, mortality has exceeded 50% of confirmed cases: 262 deaths from 442 cases as at September 2009. Even if this figure is an overestimate (milder cases may be missed), this is still a terrifyingly high level of mortality, more than twenty times the overall mortality in the 1918–1919 influenza pandemic (although in some locations mortality then exceeded 70%). In consequence, surveillance was intense.

Years of intensive monitoring showed bird-to-human transmission in fifteen countries across Asia and Africa. However, despite all of this there were only a very few cases where human-to-human transmission was expected (between three and six cases appears to be the general consensus) and the evidence for H5N1 having pandemic potential is slight.

If it were to make the jump to effective human-to-human spread, it could in theory kill billions, but one aspect of the move between species is that virulence in the new host often declines. The chances of a fully virulent H5N1 influenza that could spread efficiently enough to cause a worldwide pandemic might be small, but they were enough to keep attention focused on bird flu until 2009.

The return of swine flu

A cluster of influenza cases in Mexico in April 2009 showed that the new pandemic was arriving. But it was not H5N1 bird flu. Instead, it was the most worrying kind of influenza, a new form of H1N1 influenza that appeared to be transmitted to humans from pigs—swine flu. Initial reports of mortality were very high—over 5% was reported (although missed milder cases were likely to reduce this number). Given the similarity to the swine-derived 1918 influenza, with what appeared to be an even higher mortality rate, that was enough to make people very worried indeed.

Surveillance measures moved into high gear, but the high levels of mortality that had been feared never really developed during 2009.

A full-scale pandemic was declared by the WHO on 11th June 2009. There was evidence that the new influenza was producing symptomatic infections in younger patients than was normal for the routine, "seasonal" influenzas. Some severe infections were seen, particularly in pregnant women. But the mortality rate as H1N1(2009) moved into its pandemic phase was considerably lower than had been feared (**Table 8.1**), and was broadly similar to that seen with circulating "seasonal" influenza types routinely present in the human population.

A spokesman for the WHO was reported as saying "It was believed that the next pandemic would be something like H5N1 bird flu, where you were seeing really high death rates, and so there were people who believed we might be in a kind of apocalyptic situation and what we're really seeing now with H1N1 is that in most cases the disease is self-limiting." In many quarters, public reaction to extreme pronouncements in some sections of the press followed by apparently mild disease was often annoyance rather than fear.

Perhaps reassuringly, studies determined that the genes in the 2009 swine flu differed from those in the 1918 variant, a genomic sequence for which was now available from samples taken from a cadaver buried in the permafrost and from stored tissue samples. In particular H1N1(2009) lacks a full-length

Table 8.1 Identified influenza pandemics*

Name	Date	Mortality rate	Deaths	Influenza subtype
Asiatic (Russian) Flu	1889–90	Unknown	1 million	H2N2?
Spanish Flu	1918–20	2–3%	50–70 million	H1N1
Asian Flu	1957–58	0.2%	1–2 million	H2N2
Hong Kong Flu	1968–69	0.2%	0.75–1 million	H3N2
Russian Flu**	1977	Very low	Very low	H1N1
Mexican Flu	2009–2010	0.03%***	18,500 confirmed, total likely to be less than 0.25 million	novel H1N1

* Pandemics before 1889–1890 not fully characterized; sporadic reports date back over 2000 years. ** Age restricted, may have been an artificial reappearance of a virus from 1950. *** Moderated by widespread use of vaccination and antiviral drugs as well as advanced medical care for severe cases.

PB1-F2 protein, which is associated with inflammation and apoptosis, that was present in the 1918 form. This lack is common in human influenza viruses seen in the last 60 years, but full-length PB1-F2 is still commonly seen in avian viruses. According to some analyses, the 2009 H1N1 virus seemed to be a triple reassortant, with genes from human, swine, and avian strains of influenza, with genes coming from both the Americas and Eurasia. It has also been suggested that pig-to-human transmission occurred in Asia, with an infected human introducing the virus to American pigs.

Tracing the origin of such infections is always difficult and rarely results in a clear progression. What does seem clear is that, as is often the case, Mexican flu was not originally Mexican (similarly, the 1918–1919 "Spanish" flu is thought to have begun in the United States, but received large-scale press coverage when it reached Spain, so it became known as Spanish flu).

However, the value of intensive surveillance was shown when, one day after the pandemic was declared, the first H1N1(2009) vaccine was produced. Barely three months later, the FDA approved four vaccines for use. So, before the virus moved into what could be a more severe second wave of infections, vaccines were tested, approved, and available for use.

There are other big differences to 1918, even if variations in the virus itself are not considered. In 1918, the world was just recovering from a debilitating war. Viruses were barely understood, and there were no vaccines or drugs to treat influenza. Perhaps as importantly, given the role of secondary bacterial infections in increasing mortality, no effective antibiotics were available.

In 2009, the virus was isolated, grown, and sequenced before the pandemic became established. Understanding of transmission and the processes of infection informed control efforts, but these still proved unable to contain the pandemic. However, two drugs (Tamiflu and Relenza) were available from the start of the pandemic, and four vaccines were approved for use within a few months, with millions of doses available ahead of the winter influenza season in the US and Europe.

Another difference is the increased level of general medical care. When a patient does experience serious consequences of infection, intensive care in hospital can support the body through to recovery. Of course, this requires enough beds to be available—which is not always the case. In July 2009, a 26-year-old Scottish woman had to be transferred to Sweden for extracorporeal membrane oxygenation, since all five beds available in the United Kingdom were in use. She has now recovered, but the case shows that the availability of the highly specialized facilities needed in such cases cannot always match the level of need.

There are also negative developments when 2009 is compared with 1918. Long-distance travel is a major route of virus spread and was much faster, easier, and more common in 2009 than in 1918. Similarly, intensive farming (suggested as involved in the April 2009 Mexican outbreak, but not shown to have any definite role) is also far more common, facilitating spread and amplification of diseases among livestock. Finally, there are many more humans to infect—over 6 billion people (with more than half living in densely populated cities), against 1.8 billion (with only one in six living in urban areas) in 1918.

However, the combination of improved nutrition, awareness of the nature of the disease, antiviral drugs, vaccines, and improved medical care means that the situation in 2009 was very different to that in 1918, even if the virus itself had been of similar virulence. And of course it was not.

The swine flu controversy

As the 2009–2010 pandemic unfolded it was clear that this was not the pandemic that everyone had feared. In fact it was soon evident that mortality rates were broadly comparable with seasonal influenza, which is present in the population every year (**Box 8.3**).

Human nature being what it is, this did not result in a universal sigh of relief. Rather, some groups began to see a conspiracy to inflate the seriousness of the 2009–2010 pandemic for the benefit of various groups, in particular the vaccine and drug manufacturers.

Of course, with the benefit of hindsight it is clear to all concerned that influenza A H1N1(2009) was very different to the (quite closely related) 1918 influenza virus. But in the early days of the pandemic, with worrying levels of mortality apparently seen in Mexico, choices had to be made.

What if 2009 had been a replay of 1918? Then, vaccines would have saved tens of millions of lives, and antiviral drugs many more. And the authorities would still have been blamed, but in that case for underestimating rather than overestimating what they needed to do.

The role of surveillance

It cannot be disputed that the detection and characterization of H1N1(2009) in Mexico in April 2009 allowed understanding (though not prevention) of the forming pandemic and the early development of vaccines. However, the publicity around this has also led to such high use of antiviral drugs that Tamiflu levels in Japanese rivers were high enough to cause concern.

However, it was also apparent that the virus had been circulating for months before it was first detected and did not produce the level of mortality that was feared. There is also the question of the value of the efforts made to monitor and control H5N1 influenza. If this did adapt to become a highly transmissible human infection, the consequences could be severe. But did focusing on bird flu prevent swine flu from being detected earlier? Certainly as late as early 2009, discussions of the next pandemic focused on H5N1. The head of the influenza division at the Centers for Disease Control and

Box 8.3 A second wave of influenza?

As the 2009 influenza pandemic developed, there was real concern that a second, more virulent wave of disease could emerge as the northern hemisphere countries moved into the traditional winter influenza season. After the appearance of a new form of influenza, the highest number of deaths tends to occur in a second wave some time later.

In the 1918–1919 influenza pandemic, a study in Copenhagen showed that 60% of deaths occurred in the second wave in late 1918 against 5% in the initial wave earlier in the year (35% occurred in a third wave in early 1919). This number is often used to support claims for a severe second wave, and in Copenhagen the actual mortality rate was 0.3% mortality in summer rising to 2.3% in the fall (the second wave)—nearly eight times higher.

Such observations were not universal, however. Analysis of the records for England and Wales for the same period indicates a drop from 0.27% to 0.1% mortality between fall and winter. It should also be remembered that diagnostic procedures at the time were predominantly symptom-based, with no procedures available to provide serological or virological confirmation.

Additionally, there is evidence that a milder first wave may provide immunity that can moderate the second wave.

With the 1968 H3N2 "Hong Kong" flu, 85% of deaths occurred in the second wave, in the winter of 1968–1969, against 15% in spring 1968, but the influenza itself was not a particularly severe form, killing 34,000 people in the USA and a million worldwide. This is a lot of deaths, but was broadly in line with death rates from circulating "seasonal" influenza in non-pandemic years. So, while the second wave accounted for an even higher percentage of deaths than in 1918, the actual mortality rate was not especially high and the number of deaths no greater than in many non-pandemic years.

Given the moderate nature of infections caused during 2009, increased severity of any second wave of influenza A H1N1(2009) was by no means certain as the northern hemisphere moved into winter and approached its annual flu season. Naturally there was considerable concern, and the highest monthly total was reached in December 2009, with 4400 laboratory-confirmed deaths (**Figure 1**). However, there was little sign of any increase in overall mortality and no "second wave" was observed.

The pandemic was formally declared at an end by the World Health Organization on 10th August 2010.

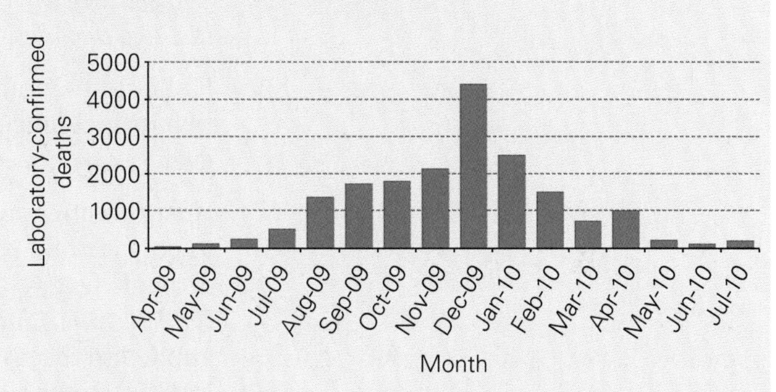

Figure 1
Mortality by month due to influenza A H1N1(2009).

Prevention (CDC) was quoted as saying "it was missed because flu surveillance—good and getting better in birds—was spotty at best in swine" (**Figure 8.1**).

It needs to be considered that surveillance can perhaps go too far, by focusing intensive efforts on the type of low-level event that has always been occurring, but has remained undetected until the advent of new diagnostic techniques.

In 1918, a few tens of cases of a new infection scattered in remote areas across the globe each year would be very unlikely to be noticed. Now they

Figure 8.1
Did surveillance for a possible H5N1 bird flu pandemic mask the emergence of H1N1 swine flu in 2009?

are seen, reported, and collated into a global picture. Certainly any one could lead to an epidemic, as with influenza H1N1 or Nipah virus spreading from pigs to humans. But there also exists the possibility that perhaps we are just seeing what has always been going on, rather than what will underlie the next major outbreak. If we detect too many such events, how do we know which are the important ones?

With a radio signal, too much amplification can result in the background noise swamping the signal. Great care needs to be taken to ensure that the same is not true for surveillance of potential infectious diseases.

8.2 EMERGENCE

An **emerging disease** has been defined by the CDC as:

"A disease of infectious origin with an incidence that has increased within the last two decades, or threatens to increase in the near future."

It should be noted that this does not require that the virus is new, simply that its incidence (the number of new infections) has increased. It does not even say by how much.

Unsurprisingly, while many diseases fit this definition, attention has tended to concentrate on viruses which cause high-profile diseases such as hemorrhagic fevers (e.g. Ebola, *Filoviridae*) or human immunodeficiency virus (HIV, *Retroviridae*). In general, the term emerging disease is most commonly applied to diseases that have historically been rarely seen (or identified). It is easier to see a change in a disease that has been restricted geographically or numerically. Indeed, even a major increase in an already common infection may be thought of as an epidemic or a pandemic (as with influenza H1N1 in 2009) rather than an emerging disease.

The example of adult chickenpox demonstrates emergence by a well-known and well-established infection. In England and Wales (Scotland compiles its statistics separately), both the overall number of cases and the number of deaths from chickenpox rose sharply between 1980 and the mid-1990s, the latter due to an increasing number of cases in adults, where the disease is more severe. Numbers dropped during the late 1990s for unknown reasons; the chickenpox vaccine introduced in the USA in 1995 was still not in routine use in the United Kingdom by 2009. Thus, after its apparent emergence, this particular manifestation of an otherwise common disease went into a decline.

It is clear that defining what constitutes an emerging disease is not simple, and that diseases increasing in one place may not be doing so elsewhere. It is also clear that the emergence of infections is not new. It has been occurring throughout history, and covers the whole range of infections.

There are a number of routes by which infections may emerge, as summarized in **Table 8.2**. It should be noted that these routes do not work in isolation. When considering the contribution of all of these factors to the emergence of diseases, it is useful to look at specific examples that illustrate the processes involved.

Zoonosis

A **zoonosis** is the transfer to humans of an infectious agent from an animal source in which the virus replicates and maintains itself (the reservoir host), and represents a major route for the introduction of novel pathogens. Examples of viral zoonoses are shown in **Table 8.3**.

Table 8.2 Possible routes for the appearance of novel virus infections

Route	Examples	Contributory factors
Zoonosis* (transfer from another species)	SARS coronavirus, nvCJD	Exposure to host species or vector, habitat destruction, movement of animals
Zoonosis via medical procedures	SV40, endogenous retroviruses	Animal (or animal cell)-derived medical products, xenotransplantation
Spread to a new geographical area	Smallpox, measles, West Nile virus	Population movements, trade routes, military expeditions, transfer of host or vector
Mutation (genetic alteration of existing virus)	Drug-resistant viruses, influenza antigenic drift	RNA genomes, inappropriate use of antiviral drugs
Recombination (exchange of genetic material with a virus or a host cell)	Influenza virus antigenic shift, cellular gene homologs in complex viruses	Segmented genomes, exposure to animal sources of related virus, collection of cellular genes by many viruses on an evolutionary time scale
Genetic manipulation	Virus vectors, recombinant baculovirus insecticides	GM medicines and pest control agents
Identification of preexisting but unknown virus	Hepatitis C virus, Sin Nombre hantavirus	Improved diagnostic techniques
Reappearance	Ebola, Marburg; potential exists for smallpox, SARS coronavirus	Re-infection of humans in absence of circulating infection; exhumations, excavations, illegal stocks
Deliberate release	Myxomatosis, smallpox	Pest control, biological weapons
Accidental release	Rabbit hemorrhagic disease, smallpox	Insufficient precautions

* Zoonotic spread forms part of the normal infectious process for many viral and other infectious agents, but may also introduce novel agents.

A true zoonosis is poorly transmissible between humans, since otherwise human-to-human transmission comes to dominate spread of the virus.

A common feature of viral zoonoses is that many are well adapted to coexistence with their normal host and cause a relatively mild disease. This is seen with the largely asymptomatic infections of simian immunodeficiency virus (SIV; *Retroviridae*) in apes and monkeys, which was the origin of HIV in humans (see below). It is also seen with hantaviruses in their rodent hosts, which can go on to cause lethal infections such as hantavirus pulmonary syndrome when they infect humans (see below). As a result, a virus may circulate unnoticed in the animal population. However, if the virus is also able to infect humans, in some cases the resultant disease may be much more severe.

Such a transfer requires contact between humans and animals, which may be increased by alterations in animal behavior or habitats caused by environmental changes. A vector species such as a mosquito or tick may also be involved in the transfer, either by passive transfer of the virus or with active replication (and thus amplification) in the vector species, as with yellow fever virus in *Aedes* mosquitoes.

An example of an infection where a vector species is involved is West Nile virus (*Flaviviridae*), which probably fits most peoples' image of an emerging virus. The virus is transferred to humans, usually by *Culex* mosquitoes (**Figure 8.2**), from infected birds. Most infections of humans are asymptomatic, with about 20% leading to febrile illness (West Nile fever). Less than 1% lead to neuroinvasive disease, but it is this potentially lethal infection which makes West Nile so feared. Humans and other mammals appear to

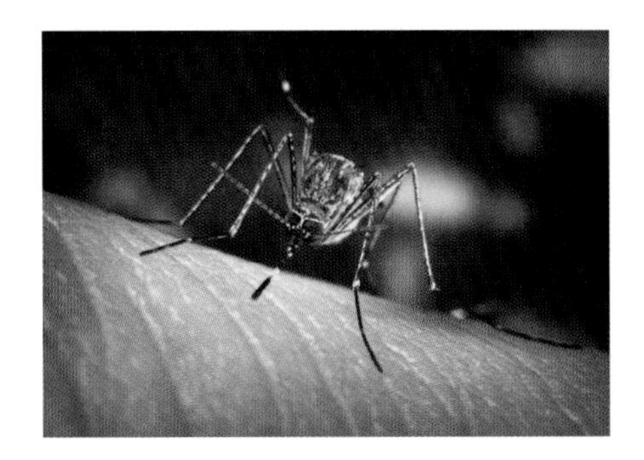

Figure 8.2
***Culex tarsalis*, the main vector for West Nile virus (*Flaviviridae*) in the United States.** From CDC Public Health Image Library (http://phil.cdc.gov/).

Table 8.3 Examples of viral zoonoses

Disease	Virus family	Frequency	Distribution	Reservoir host	Source of human infection	Direct human-to-human spread
Yellow fever	*Flaviviridae*	Endemic	Tropical	Monkeys	Vector (mosquito bite; arbovirus)	Not seen
West Nile	*Flaviviridae*	Endemic	Worldwide	Birds	Vector (mosquito bite; arbovirus)	Not seen
Colorado tick fever	*Reoviridae*	Endemic	North America	Rodents	Vector (tick bite; arbovirus)	Extremely rare (blood transfusion)
Dengue fever	*Flaviviridae*	Endemic	Tropical	Monkeys	Vector (mosquito bite; arbovirus)	Not seen
Lassa fever	*Arenaviridae*	Endemic	West Africa	Rats	Rats (urine, feces)	Common (requires close contact)
Rabies	*Rhabdoviridae*	Endemic	Most areas	Mammals	Mammals (saliva)	Extremely rare (organ transplantation)
Hantavirus diseases	*Bunyaviridae*	Common	HFRS (Eurasia), HPS (Americas)	Rodents	Rodents (urine, feces)	Extremely rare (one case of Andes virus HPS)
Filovirus hemorrhagic fevers (Ebola, Marburg)	*Filoviridae*	Rare	Africa	Bats	Contact with infected animals	Common
Nipah virus disease	*Paramyxoviridae*	Rare	Asia	Bats	Pigs (body fluids)	Common
Hendra virus disease	*Paramyxoviridae*	Very rare	Australia	Bats	Horses (body fluids)	Not seen
Herpes B virus disease	*Herpesviridae*	Extremely rare	Limited	Macaque monkeys	Macaques (bite)	Extremely rare (one case)
Severe acute respiratory syndrome (SARS)	*Coronaviridae*	Extinct	(Asia/worldwide)	Bats?	Bats, possibly animals in markets, including civets	Common

be "dead-end hosts" (see Section 8.3), unable to produce enough virus to transmit the infection back to the vector species.

First observed in 1937 in Uganda, West Nile virus was introduced into the United States in 1999. Since then there have been almost 30,000 cases with 1139 deaths (**Figure 8.3**). However, numbers of cases and of deaths have declined sharply since a peak in 2003, and the number of states reporting cases has remained relatively constant. This raises the question, when does a virus cease to emerge?

Sin Nombre and Hantaan viruses

Another route of emergence is the recognition of a long-established but unnoticed disease. A good example of this is hantavirus pulmonary syndrome, which also provides a good example of a direct zoonosis—one which does not involve an intermediate vector species.

In May 1993, a cluster of cases of a severe respiratory infection was noted in the Four Corners area of the USA, where Arizona, Colorado, New Mexico, and Utah meet. While the initial numbers were relatively small (12 deaths from 24 cases in the first half of 1993), mortality was extremely high, and

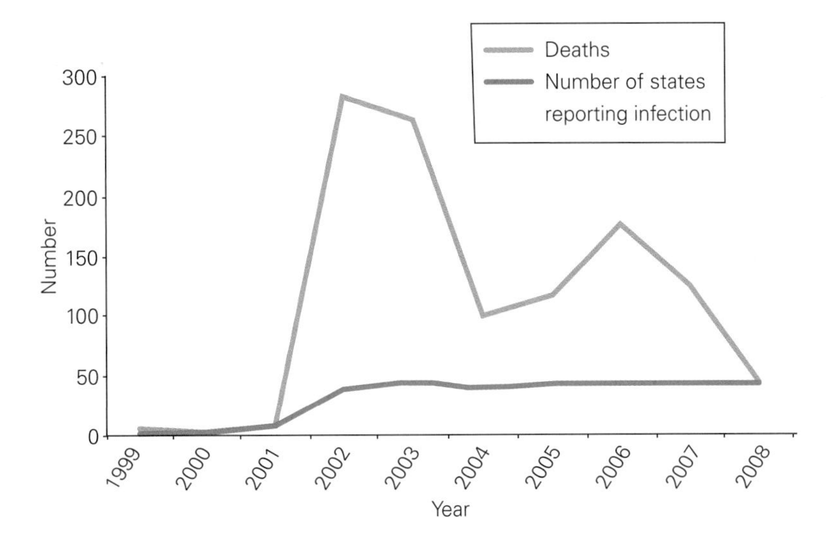

Figure 8.3
West Nile virus infections in the United States.

would later rise to 80% in the initial patient group. The occurrence of multiple deaths in a relatively small area and in otherwise healthy individuals resulted in an urgent need to find the agent responsible. By early June 1993, it was determined that this was a hantavirus. The first of this group (genus *Hantavirus*, family *Bunyaviridae*) to be identified was the Hantaan virus, noted during the Korean War as the cause of hemorrhagic fever with renal syndrome (HFRS) in United Nations troops. Further studies showed Hantaan-like HFRS disease in many countries throughout Asia and Europe caused by a range of viruses.

The new virus was named Sin Nombre ("No Name") hantavirus since naming a lethal disease after a specific location can be both misleading when it occurs elsewhere and unpopular with the inhabitants of the area. The disease that it was causing was clearly different from HFRS, but knowledge from the Korean outbreak was extremely valuable. It had been established that Hantaan was transmitted to humans by inhaling materials contaminated with the urine or feces of the reservoir host, the Asian striped field mouse (*Apodemus agrarius*).

In the case of Sin Nombre, the animal host was quickly identified as the deer mouse (*Peromyscus maniculatus*; **Figure 8.4**). The ending of a drought appears to have increased the mouse population. As a result, people were increasingly exposed to their droppings and urine, and the disease, now termed hantavirus pulmonary syndrome (HPS), emerged. In 1993 a total of 48 cases were seen, of which 21 were fatal. But there have not been as many cases in any year since, with numbers ranging from 8 in 2001 to 45 in 2000. The mortality rate remains around 35%. Historical surveys have identified cases back to 1959. So, is hantavirus pulmonary syndrome an emerging disease?

Hantaviruses illustrate many of the elements contributing to a zoonotic virus infection. Hantaan and Sin Nombre cause only inapparent disease in the animal host, in which they maintain a high level of infection, and which acts as a reservoir for human disease. They are introduced into the human population anew each time by contact with the natural host. They can then cause severe disease in humans, but are not transmissible (only one case is known, of the South American Andes virus).

Despite this, Sin Nombre has been responsible for significant human disease in a highly developed country with excellent health surveillance, without being identified until the Four Corners outbreak brought it to the attention of the epidemiological community. Clearly there are likely to be many more such infections awaiting identification.

Figure 8.4
The deer mouse (*Peromyscus maniculatus*), the animal host of Sin Nombre hantavirus (*Bunyaviridae*). From CDC Public Health Image Library (http://phil.cdc.gov/).

Ebola and Marburg

While hantaviruses make the jump to humans many times each year but are then very unlikely to spread (the only case to date did not spread from the second human host), the hemorrhagic fever viruses Ebola and Marburg (*Filoviridae*) can spread between humans, but not efficiently enough to sustain themselves without further transfer from their reservoir hosts. Thus, humans are a dead-end host for these viruses. Each zoonotic infection appears to involve a transfer from asymptomatically infected bats (most likely the cave-roosting Egyptian fruit bat *Rousettus aegyptiacus*) into primates, where it produces symptoms similar to humans, and then transfer on into humans as a result of hunting. Once established within the human population, limited spread occurs causing a localized outbreak. The resultant disease is typically a severe hemorrhagic fever, involving destruction of the blood vessels and many other organs, with 23–90% mortality. The infection is controlled by isolation or barrier nursing (using microbiological containment to prevent transfer), and the outbreak then stops. A significant source of transmission between humans seems to be nursing care where facilities for barrier nursing are not available. Once the outbreak is controlled, a new outbreak awaits a new crossover into the human host.

HIV

Finally, HIV infection provides a good example of a zoonosis where efficient human-to-human spread takes over from zoonotic infections. HIV type 1 (HIV-1; *Retroviridae*) is found around the world and is the main cause of the AIDS pandemic. This appears to have originated from a simian immunodeficiency virus (SIV; *Retroviridae*) infection in an African chimpanzee that was transmitted to humans around 1929, following transmission into chimpanzees from monkeys thousands of years earlier. Clearly, even if HIV infection was originally a zoonosis, it is now a transmissible disease within the human population and is capable of sustaining itself without further reintroductions.

However, two further zoonotic events are believed to have introduced the related HIV-2 virus from a different species (sooty mangabeys) in the 1940s. HIV-2 is mainly restricted to West Africa and is generally associated with a lower level of disease. In 2009, a third type of HIV was identified. While classified as HIV-1 it appears to resemble a form of SIV seen in gorillas rather than chimpanzees and may well be the result of another zoonotic event. It would appear that SIV is capable of making more than one jump to humans.

There appear to have been only a maximum of four identified zoonotic events, and yet 33 million people are now infected. However, while the number of people living with the infection (prevalence) does still appear to be rising slowly in absolute terms, this reflects rising population generally, and the percentage of the global population living with HIV now appears to be declining from a peak of almost 6% in 2000, falling to less than 5% in 2007 (**Figure 8.5**). The incidence (the number of people newly infected) also now appears to be falling slightly from a 2005 peak of 2.2 million, although it is estimated that 2.1 million people became infected during 2007. At the same time, advances in antiviral therapy (see Chapter 6) have converted HIV infection from an inescapably terminal condition into one where it is possible to expect a long and relatively healthy life, at least where the patient can afford the necessary range of drugs to control the progression of disease. These drugs save lives, but unlike a vaccine they do not contain the infection. In cold epidemiological terms, it can be argued that they actually have the reverse effect since they keep infectious individuals alive and apparently healthy, thus increasing the opportunities for spread of the virus.

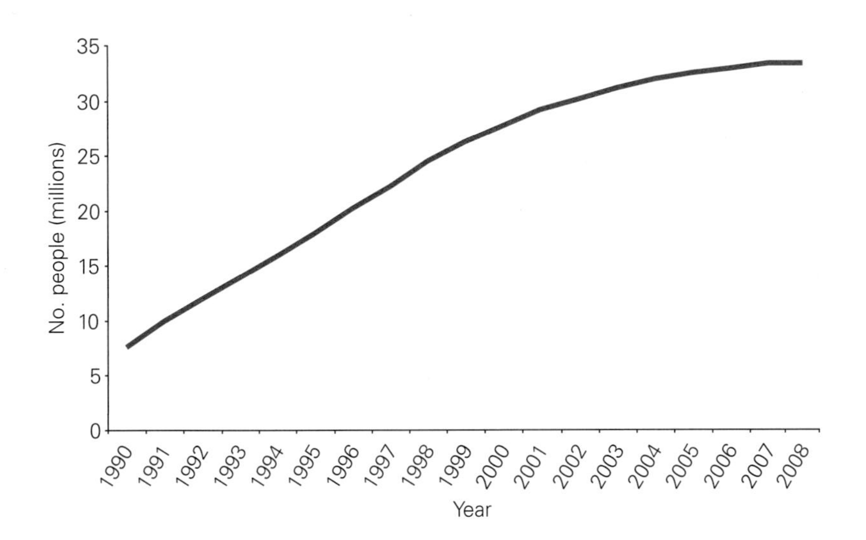

Figure 8.5
Total number of people living with HIV/AIDS. Data from http://www.avert.org/worldstats.htm

Severity of zoonotic infections

One question that must be asked is: why is a zoonotic infection often more severe than in its original host? Although many zoonoses do not appear to cause significant disease (and may as a result not be noticed), it is clear that some cause severe infections in the new host while being relatively harmless in the normal host. There are alternative explanations for this.

The classical theory states that it is in the interest of a virus not to kill or immobilize the host too quickly, since this reduces the chances for spread of the virus. Over time, both virus and host adapt so that the virus produces less severe disease in its normal host. The moderation of virulence of myxomatosis in the Australian rabbit population (see Chapter 7) illustrates such adaptation. However, when a novel host is infected, there has been no chance for adaptation to a less virulent form, and severe disease may result, at least initially. More recent theories suggest that this is an oversimplification, and that such benign equilibrium is subject to disruption by more virulent agents and by methods of transmission that are relatively unaffected or may even be favored by the sickness of the host (spread by arthropod vectors, or virus-induced diarrhea as a way of spreading rotaviruses, for example). However, it should be noted that myxomatosis is primarily transmitted by arthropods, but still provides a classical example of benign adaptation.

A more extreme form of the benign adaptation hypothesis has also been suggested (the "virus X" theory) in which viruses may act as a symbiote (providing mutual benefit), providing defense against related species with which the host comes into contact, assisting the normal host to win out in competition with the sickened opponent species. Such effects are even observed within one species, as with the effects of smallpox in the European conquest of the Americas (see below). While unlikely to be a factor in all cases, this could play a role under certain circumstances.

As with all biological systems, it is likely that no single theory will explain all of the observed events. What is clear is that zoonoses represent a major source of severe infections in humans.

Zoonosis through medical procedures

Any procedure where material is introduced into the body carries with it a very real risk of infection. Most infections arising from this source involve transmission from humans to humans, and can include infections arising from surgery, blood transfusions, or re-use of needles. Since these routes allow direct introduction of agents into the body they allow for very high-efficiency transmission.

However, for such a procedure to introduce a novel virus, the material that is introduced will need to have come from a nonhuman source. Two such routes are vaccination and xenotransplantation.

With vaccination, a useful example is provided by simian virus 40 (SV40; *Polyomaviridae*). This is a naturally occurring macaque virus which was present in early polio vaccines due to their growth in rhesus monkey kidney cells (see Chapter 5). SV40 was not inactivated by the treatments used to inactivate the poliovirus (*Picornaviridae*) in some early batches of the inactivated polio vaccine, and it is a highly transforming virus in some systems. It is also believed to have been present in some live polio vaccines from that time.

It is estimated that 10–30 million Americans (of the 98 million vaccinated) may have received SV40-contaminated vaccine in the period 1955–1963. Given the properties of SV40, there was a great deal of concern about this. However, long-term monitoring has shown no SV40-associated disease in recipients of contaminated vaccine. Though SV40 DNA has been found in some cancers, in particular those of the brain, no human pathology has been definitively established as resulting from SV40 infection from polio vaccine in over 50 years. Following multiple studies the US National Cancer Institute concluded in 2005 that "although SV40 causes cancer in laboratory animals, substantial epidemiological evidence has accumulated to indicate that SV40 likely does not cause cancer in humans."

Viral contamination of such a widely used vaccine indicated the need for great care in the use of animal cells for vaccine preparation. Advances in testing, particularly at the nucleic acid level, have contributed to a much greater ability to prevent such incidents, and the cell lines used are now very closely monitored by the regulatory authorities.

While it is possible to perform an extensive range of testing and purification procedures on vaccines, there are some medical procedures where this is not possible. One such would be **xenotransplantation**, where organs from animals could be adapted for use in replacing failed human organs. Rigorous testing or cleaning of these organs to prevent virus transmission might well compromise their function.

While initial attention focused on the use of primate organs, notably those from the baboon, concerns about the ethics of using primates and the apparent ability of baboon viruses to infect human cells caused interest to switch to transgenic pigs expressing human cellular marker proteins in order to minimize rejection. Pigs are more readily available than baboons and their organs are of a similar size to those of humans.

The pig is known to transmit viruses to humans, including Nipah virus (*Paramyxoviridae*) and swine flu. Pigs bred as transplant donors would, of course, be bred under gnotobiotic conditions (as sterile as possible and with any infectious agents identified), but as the FDA notes, "new infectious agents may not be readily identifiable with current techniques." It was only in 1998 that Nipah virus was identified as a result of a transfer from pigs to humans that caused an outbreak of lethal disease. In addition, multiple replication-competent porcine endogenous retroviruses (PERVs) have been found in the pig genome which are able to infect human cells. Attempts to remove these are complicated by the presence of multiple copies of the genomes in question, although there is some evidence that production of germ-line cells without these PERV sequences may be possible.

The supply of human donor organs is limited, and potential recipients often see xenotransplantation as a personal issue, with the possibility of future, unknown infections outweighed by the near certainty of death without a donor organ which may not be available from a human source. However, epidemiologists need to be concerned about the possibility of introducing potential zoonotic infections directly into patients who are immunosuppressed to prevent rejection of the implant. Entirely new infections could then be amplified and introduced into the human population as a whole.

The WHO has stated that "no nation should undertake any xenotransplantation in humans without an appropriate regulatory framework and surveillance," while the Australian National Health and Medical Research Council has gone further in agreeing a long-term halt to such work. Certainly the commercial enthusiasm for such work in the 1990s has faded in the face of these challenges.

Geographical contact

In a predominantly agricultural society, with little travel between regions, which existed for most of recorded human history, it was possible for viruses to exist in isolated areas of infection. However, as society developed, trade routes were established that allowed for the transmission of infections between such areas. As more and more people traveled, such transmission became more common along the trade routes. Military expeditions can also be responsible for spreading infection. It has been suggested that the ability of the Spanish conquistadores under Cortes to conquer the complex and populous Aztec nation was assisted by the passing of European viral infections (including smallpox and measles) to the totally non-immune native population. Certainly the death toll from such introduced diseases was very high, with some areas being effectively depopulated.

While HIV appears zoonotic in origin, it also seems that the virus was present in restricted areas of central Africa for approximately 50 years. Given the lack of epidemiological data on such areas, the virus would not have come to the attention of the medical community while it was restricted to this region. The opening of central African regions to commerce, notably by the building of the road system known as the Kinshasa Highway across the regions, is likely to have played a major role in spread of the virus, culminating in transmission to urban areas and then by air travel into other countries where highly promiscuous and high-risk sexual activities among "amplifier" populations combined with injecting drug abuse accelerated its conversion into a pandemic.

Sometimes geographical contact can alter the pattern of an existing disease. Dengue virus (*Flaviviridae*) is a major example of a zoonosis transmitted by mosquitoes, with between 50 and 100 million new cases each year. It is present in many tropical and subtropical areas worldwide. In humans, exposure to one of the four serotypes (serological variants) of Dengue virus can result in Dengue (breakbone) fever, characterized by fever and muscle and joint aches. Following infection, the patient is then immune to that particular serotype, but immunity to the other serotypes is transient. This can, however, make subsequent infections more severe, increasing the risk of the life-threatening complications of Dengue hemorrhagic fever (DHF) or Dengue shock syndrome (DSS). This is because the nonprotective immunity resulting from exposure to the earlier serotype actually appears to enhance the subsequent infection. Clearly, development of the more severe forms (DHF or DSS) is influenced by the presence of multiple serotypes of the virus in the same area, which allows for sequential infections. Until World War II the different serotypes were generally restricted to different geographical areas. It is thought that the population movements (civilian and military) and loss of effective hygiene during World War II were responsible for spreading these previously isolated serotypes among the human population, allowing for infection by multiple serotypes and consequently increasing the severity of the disease.

Population movements into different rural areas provide an important route for the introduction of zoonotic viruses. A virus circulating in one species will only be able to cross into a new host species if the two come into contact. Settlers in new areas may come into increased contact with potential animal hosts or carriers. In addition, the resultant disruption to the life of the animal hosts may also play a role. If a virus is circulating efficiently in a large animal population, there is little selective pressure for the virus to evolve the ability to infect another host species, particularly if this incurs a decrease in the ability to replicate in the normal host. However, if the numbers of the normal host are reduced (e.g. by habitat destruction), circulation of the virus will be disrupted and a virus that can infect another host may be at a selective advantage, even if it has lost some ability to replicate in the normal host.

It should be noted that it is not just human population movements that can introduce diseases. International travel is usually thought of as a human phenomenon, but movement of agricultural and experimental animals is also a significant factor. Ebola Reston strain was introduced to America in a shipment of monkeys from the Philippines. The monkeys died of the infection, but fortunately this particular strain proved not to cause human disease. Spread of animal viruses is in some ways easier to control. Transfer of animals is generally more regulated than human travel, and it is possible to introduce programs of testing, quarantine, and (on occasion) mass slaughter that would be unacceptable with humans. In the early history of the HIV pandemic, many countries (including some with significant levels of infection already present within their borders) introduced requirements similar to the milder forms of animal disease regulation, in that they required that travelers be certified as HIV-negative before entering the country. Even these mild controls proved very unpopular, and they were generally short-lived.

With international travel at its present level, few areas remain where isolated diseases can circulate. Despite this, some concerns about this remain. For example, in the late 1990s attention focused on American troops returning from military training exercises in areas of Australia where the mosquito-borne Ross River virus (*Togaviridae*) is endemic and is also a notifiable disease. It had already spread into the South Pacific islands of Fiji, New Caledonia, Samoa, and the Cook Islands, where tens of thousands of cases of polyarthritis resulted in 1979–1980. The virus can be spread between humans by mosquitoes. There was serious concern that this disease could be introduced into America, and extensive control measures (aimed particularly at vector control) were taken before, during, and after the exercise to prevent this. Despite these, at least one case was identified (and contained) in the United States following the exercise.

An extreme form of the "geographical contact" route has been suggested based on the identification of organic materials in cometary matter. The suggestion that "viruses from outer space" invade this planet is extremely hypothetical to say the least. However, influenza was given its name in Renaissance Italy as it was thought that the disease was influenced by the stars.

Mutation

While all viruses must have evolved by mutation, it is difficult for us to observe the effects of this process in the limited time that we have been able to study viruses. Historically, there are reports of apparently novel diseases, but the reliability of such observations is doubtful given the rudimentary state of medical knowledge at the time. In addition, the records used to support such findings are fragmentary and cover very limited areas, so it is entirely possible that a similar disease was active, but unrecorded, in other areas. More recently, there is evidence of new diseases arising, for example the appearance of rabbit hemorrhagic disease calicivirus (see Chapter 7) in China in the early 1980s (it was identified in 1984), followed by rapid and devastating spread. The virus is thought to have arisen from a precursor that, while highly infectious, did not produce clinical disease.

A more relevant example in humans is the appearance of drug-resistant variants of almost all viruses that are treated with antiviral drugs (see Chapter 6). Even in a single host, mutation during normal replication results in the presence of a population of viruses with differing genetic and physical properties, rather than a single type. Some of these mutants may be at a disadvantage normally, but will be preferred when the antiviral drug to which they are resistant is used. The use of the drug will then provide selection pressure to fix the mutation in the population. The problem of drug resistance is particularly marked in bacteria, where the genes responsible may be transferred on extrachromosomal genetic elements (plasmids), even between different species. Drug-resistant bacteria such as methicillin-resistant *Staphylococcus aureus* (MRSA) or multidrug-resistant Gram-negative pathogens such as *Pseudomonas aeruginosa* and *Acinetobacter baumannii* are often referred to as emerging diseases in their own right, and drug-resistant viruses may yet achieve this dubious status. It is to be hoped that lack of equivalent resistance active-transfer mechanisms in viruses and the judicious use of antiviral drugs will slow the appearance of viral equivalents. However, there are already reports of forms of HIV resistant to multiple drugs of the same general type, and the rapid mutation of viruses makes this likely to become more of a problem in future.

Another example of ongoing virus mutation is *antigenic drift* in influenza virus (see Chapter 1). However, the mutant forms produced are not easily resolved from the parental strains, and are not thought of as new or emerging viruses.

Mutation of viruses is the key to escaping immune surveillance, as well as resisting the effects of antiviral drugs. As stated above, few of these viruses are sufficiently different to be classified as emerging diseases. Despite this, mutation does contribute to emerging diseases. For example, hepatitis C (see below) is showing a dramatic increase in frequency at least in part due to the extreme antigenic variation of the virus, the result of a high mutation rate.

Recombination

While it is possible for any virus to undergo recombination, where genetic material is exchanged with related viral or cellular sequences, this is most evident in influenza virus, where the RNA genome is present in multiple different segments which can be exchanged without any need to cut and splice nucleic acids, making the process far easier and more frequent.

It is known that influenza, as well as undergoing antigenic drift by mutations of the surface glycoproteins hemagglutinin and neuraminidase (see Box 8.2), can also undergo sudden changes of these proteins. This results from the exchange of genome segments with mammalian or avian influenza viruses during mixed infections, often as a result of the close proximity of humans, farm animals, and birds (all of which can act as hosts), and is known as *antigenic shift* (see Chapter 1). Immunity to the original human virus then provides at best partial protection against the shifted virus, which may thus be able to cause the worldwide outbreak known as a pandemic.

While all viruses can undergo recombination, the presence of a segmented genome makes this more likely since no molecular changes are required.

There are many other viruses with multiple genome segments, including the *Reoviridae* and the *Bunyaviridae*. The former include rotaviruses, which are a major cause of infant mortality in the developing world, while the latter include the hantaviruses discussed above.

Genetic manipulation

With the routine use of recombinant DNA technology, it is now theoretically possible to produce a virus with almost any desired characteristics. Experience with biological pesticides such as baculoviruses expressing insect-specific toxins suggests that such approaches are quite possible.

One example of a genetically modified virus already in use is a vaccinia (*Poxviridae*) recombinant expressing the G glycoprotein of rabies virus within the vaccinia thymidine kinase gene. Using this location for inserted DNA reduces the virulence of the vaccinia vector. The vaccine is used for preventing rabies in wild animal populations. It is given inside edible bait, and is thought to enter the bloodstream via cuts in the mouth. The recombinant vaccine has been in use in the field since 1987, and although a case of human infection resulting from handling of bait was reported in 2001, it appears to be both safe and effective in controlling rabies in wild fox and racoon populations.

Identification

While many hundreds of viruses infecting humans are now known, it is likely that thousands more remain to be discovered. Some may cause serious but rare diseases, and the identification of the agent responsible may await recognition of the disease itself, as with hantavirus pulmonary syndrome (see above). Many will cause only mild disease (or no disease at all), since those which cause obvious symptoms tend to be the first to be noticed. However, even in recent years there has been a range of well-known diseases for which a cause had not been found. Molecular techniques have greatly extended our ability to determine the cause of such a disease, and in many cases the agents responsible are now being identified, as the following two cases illustrate.

Hepatitis A virus (*Picornaviridae*) and hepatitis B virus (*Hepadnaviridae*) had been identified by the 1970s, but it became clear that there was also a form of post-transfusion and community hepatitis that was not caused by either of these agents. Identification of the cause of this non-A, non-B hepatitis (NANBH) was not possible at the time, and was hampered for many years thereafter by difficulty in growing any causative agent in culture. However, infectivity could be passed between chimpanzees, and in a procedure harking back to the earliest days of virology infectious material was found to pass 50–80 nm filters, indicating that a small virus was responsible. Between 1982 and 1988, a random-primed complementary DNA library was constructed from plasma containing the NANBH agent, and an expressed protein (later found to code for an immunodominant epitope within the viral nonstructural protein 4) was identified which reacted with plasma from a patient with NANB hepatitis. This allowed identification of the role of this agent in NANBH, and permitted diagnostic tests to be developed that allowed donated blood to be screened for this agent. In subsequent work, culture methods were developed, and hepatitis C virus (HCV) was seen. This has been classified as a member of the *Flaviviridae* and is a major cause of human disease, with an estimated 170 million carriers worldwide. Prevention of infection by screening of donated blood is the main control method, although treatment of infections with ribavirin and pegylated interferon is also used. Specific drugs and vaccine approaches are in development, although these are challenged by the high level of variability of the virus. None of this would have been possible without the initial identification of the agent.

Human herpesvirus 8 (HHV-8; *Herpesviridae*) was identified in 1994. It is implicated in Kaposi's sarcoma (KS) (predominantly in AIDS patients) as well as with other malignant conditions and is now more commonly known as Kaposi's sarcoma-associated herpesvirus (KSV). The viral genome was detected by a technique known as representational difference analysis. DNA from a KS lesion was combined with DNA from a sample taken from elsewhere on the same patient. The DNA that did not react, and was therefore present only in the KS lesion, was examined, and sequences were found which were clearly those of a member of the *Gammaherpesvirinae*, related to the oncogenic virus of squirrel monkeys, Herpesvirus saimiri, and more distantly related to Epstein-Barr virus (EBV; human herpesvirus 4), which is linked with a variety of human cancers (see Chapter 1). KS was traditionally a rare and relatively benign cancer, and was first described in 1872. However, both the incidence and the severity of the disease increased due to more aggressive infections in patients with immunosuppression caused by HIV, illustrating another route for disease emergence.

Clearly, identification of a preexisting virus does not make it either new or emerging, but it does allow studies to identify the role, incidence, and prevalence of the virus in question. With both of the viruses discussed above, such studies showed that the virus was both a major health problem and had also shown increases in incidence during the two decades before its discovery, classifying both of the newly identified viruses as causes of an emerging disease.

Reappearance

Clearly, for a virus to reappear, it must have disappeared in the first place, and few viruses that we know about meet this requirement. One such virus is smallpox. When smallpox was first eliminated (see Chapter 5), monkeypox virus was a particular concern. It is a close relative of smallpox, and observed rises in transmission are thought to be due to decreased levels of smallpox vaccination. However, monkeypox in humans remains to date a limited, non-epidemic disease. Initial concerns that monkeypox was actually a form of smallpox were resolved when full genetic sequences of the two viruses were obtained, and it became clear that there were substantial differences at the genetic level.

Release of smallpox viruses from biowarfare stocks remains a real concern as noted below, but another potential route does remain. Since poxviruses are very stable, there is still some concern that opening of ancient graves, either accidentally through building works in long-established cities or deliberate archeological investigations, could result in release of live virus. While the virus appears to lose at least some activity within months in warm conditions, it is stabilized within scab tissue from victims of the disease and at colder temperatures. Morphologically intact virus has been observed from mummified tissue 500 years old, although the virus was not viable. However, there is sufficient concern about survival of smallpox virus in scab tissue within closed burial caskets in cool climates that vaccination of workers coming into contact with such tissues has been proposed.

One other candidate for reappearance is the SARS coronavirus (*Coronaviridae*). This was an entirely new infection, first seen in November 2002. Early cases were poorly reported but a case in an international traveler in February 2003 brought it to the attention of the world. Air travel was to prove a major route of international spread. By July 2003, 8098 cases had caused 774 deaths in fourteen countries across three continents. A second, laboratory-associated outbreak around the Chinese National Institute of

Figure 8.6
The greater horseshoe bat (*Rhinolophus ferrumequinum*), one of the species implicated as a reservoir host for SARS-like coronaviruses. Drawing by Friedrich Specht, from Brehms Tierleben, Small Edition 1927.

Virology in Beijing in March and April 2004 caused eight cases, with one death. A previously unknown coronavirus was identified as the cause of severe acute respiratory syndrome (SARS) in April 2003. SARS was contained by rigorous isolation and quarantine, barrier nursing, and hygienic measures. Investigations of the source indicated palm civets at markets in rural China, but these were found not to be the reservoir host. They may be an intermediate host, permitting amplification and onward spread to humans, but analysis indicates that infection may have been passed on to them from humans, who became infected directly from the reservoir host. As with Hendra and Nipah viruses (*Paramyxoviridae*) and Ebola and Marburg viruses, the reservoir host seems to be bats, in this case Chinese horseshoe bats of the genus *Rhinolophus* (**Figure 8.6**). These bats are hosts to a range of related coronaviruses resembling the SARS agent, although these viruses appear to use a different cellular receptor. It is presumably an unusual change in receptor specificity that produced SARS, and if this has happened once it can of course happen again, meaning that the potential for another transfer to humans and thus the reappearance of a SARS-like infection is very real.

Deliberate release

It is sadly true that deliberate release of pathogenic organisms has been used against humans as well as in agricultural control (see Chapter 7). While some cases were the unplanned result of contact between disparate groups (see above), deliberate releases have also occurred. These are rarely battlefield uses (**biological weapons**), since the relatively slow effects of biological agents mitigate against their use where rapid incapacitation of an armed enemy is required.

From the siege of Kaffa in 1346 to the siege of Tallinn in 1710, the use of plague-infected corpses to carry infection into a besieged city has a long but rather dubious history, since the actual effect of such attacks is not clear. It is very likely that the same rats that carried the fleas that act as plague vectors crossed freely between besieger and besieged, carrying infection by more mundane means. This has not stopped the Mongol use of this tactic at Kaffa being considered as the origin of the Black Death.

Anthrax has also been developed as a biological weapon, from the British delivery trials on Gruinard Island in 1942, to the leak from a Soviet weapons plant in 1979, and the anthrax bioterrorism letters in the United States in 2001.

Extensive use of bacterial agents was also undertaken by the infamous Unit 731 in China during World War II, with claims that 400,000 casualties resulted from their activities.

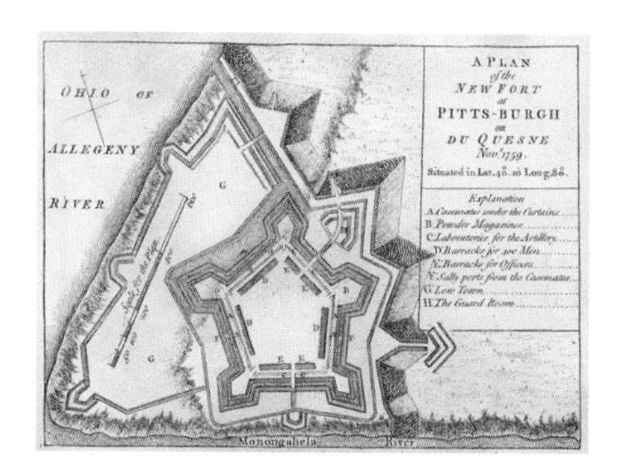

Figure 8.7
Map of Fort Pitt (Pennsylvania). Published in 1765, drawn by John Rocque.

Use of viral agents has been less common, but smallpox in particular has been used in this way, with the supply of contaminated blankets to Native American tribes in the early years of the colonization of America. In June 1763, there was an outbreak of smallpox at Fort Pitt, Pennsylvania (**Figure 8.7**). The fort was under siege by the Shawnee, Delaware, and Mingo tribes. In July, following correspondence between Colonel Bouquet, the commander of the fort, and Lord Jeffrey Amherst, the British military commander in North America, it was decided to use smallpox-contaminated blankets to pass the disease to the besiegers (**Figure 8.8**). It is hard to be sure whether the attempt succeeded, since smallpox was already causing disease in the area. However, the smallpox epidemic then spread across the continent, killing many thousands of Native Americans from at least six tribes, as well as substantial numbers of colonists. Unlike biological controls in agriculture (see Chapter 7), biological weapons are unable to recognize the difference between their human user and their human target.

There are now extensive international treaties in place to prevent the development or use of biological weapons by governments, most notably the 1972 Convention on the Prohibition of the Development, Production and Stockpiling of Bacteriological (Biological) and Toxin Weapons and on their Destruction. This was the first multilateral disarmament treaty to ban an entire class of weapons, and came into force on 26th March 1975. Of course, it is only as effective as any other international treaty, and did not prevent the 1979 anthrax leak from a bioweapons plant in Sverdlovsk which killed approximately 100 people.

Research on both bacterial and viral agents was carried out by all sides in the Cold War, but it is the statistics for Soviet production which appear genuinely frightening. It is impossible to be certain of the absolute truth in such a murky area, but Alibek and Handelman claim that at its peak, Soviet annual production included 150 tons of Venezuelan equine encephalitis (*Togaviridae*), 100 tons of smallpox, and 250 tons of Marburg virus as well as thousands of tons of bacteriological agents. Even more worryingly, as reported in *The New Yorker* in 1999, rumors continue to surface that while scientists debated the destruction of the last remaining, tightly controlled stocks of smallpox in Atlanta and Novosibirsk, more than 20 tons of weaponized smallpox remained from these programs. To put this in context, the infectious dose for smallpox is around a hundred viruses. A single ton of pure smallpox virus contains approximately ten billion virus particles for every person on the planet.

Now that it is possible to synthesize whole viral and bacterial genomes, even destruction of stocks of viruses does not eliminate the possibility of their use. Fortunately, however, such use is not simple. As with the Aum Shinrikyo attack on the Tokyo subway in 1995 using Sarin nerve gas demonstrated, errors in formulation and delivery can hugely reduce the efficacy of a weapon. The situation with biological agents is even more complex, since they are typically far less stable than a chemical agent. For example, as published in *The New Yorker* in 1999, there are claims that the Soviet Union had SS-18 intercontinental missiles ready to load with smallpox, each able to deliver an 8.8 ton payload. It is undoubtedly possible, but it would take a great deal of work to make an intercontinental ballistic missile into a reliable and efficient delivery system for a biological agent that would need to be viable and infectious when it is released. The damaging effects of the extreme temperatures that would be encountered during extra-atmospheric transit and re-entry would be only one of many problems.

Against this, a single case of smallpox can lead to a widespread outbreak, and in a nonvaccinated population this would have the potential to cause a pandemic. In Yugoslavia in 1972, one infected traveler caused an outbreak that took eighteen million doses of vaccine to stop. It is for this reason that stocks of smallpox vaccine, which had dropped to very low levels in the 1990s, have been built up once again to the level where protective vaccination could be offered to whole populations.

Although no virus-based biological weapon has been used since 1763, the possibility of release by terrorists or other sources is considered realistic enough to justify extensive and complex measures to protect against such an event.

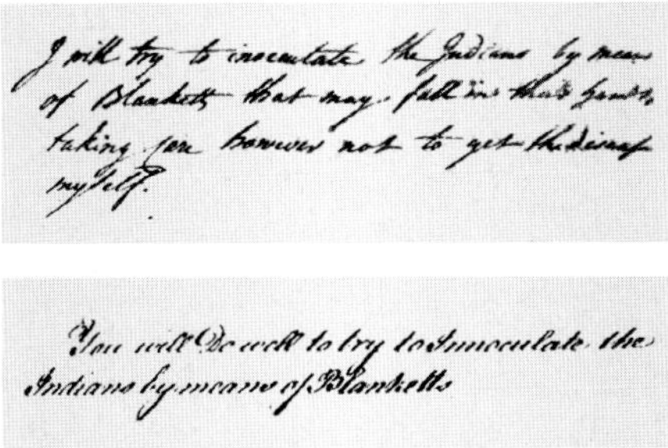

Figure 8.8
Correspondence between Colonel Henry Bouquet and Lord Jeffrey Amherst. Colonel Bouquet writes: "I will try to inocculate the Indians by means of Blanketts that may fall in their hands, taking care however not to get the disease myself." Lord Amherst replies: "You will Do well to try to Innoculate the Indians by means of Blanketts". From, http://www.nativeweb.org/pages/legal/amherst/lord_jeff.html. Courtesy of Peter d'Errico, University of Massachusetts, USA.

Accidental release

A good example of accidental release is provided by the escape of rabbit hemorrhagic disease calicivirus from the Wardang Island facility in South Australia, as detailed in Chapter 7. This led to a nationwide epidemic, which was then extended by authorization of additional releases.

A rather different example is provided by the release of smallpox from an experimental facility at the University of Birmingham, England in 1978, a year after the disease had officially been eliminated. Following leakage of smallpox virus into a ventilation system Janet Parker, a medical photographer working on the floor above the laboratory, was infected and then became the last person to die from smallpox. Her mother also became infected, but survived. The outbreak was contained to just these two cases, but led to major changes in how work with such agents is authorized and controlled in the United Kingdom.

Based on genetic similarities, there has also been speculation that the 1977 influenza A H1N1 outbreak (which began in China but did not quite reach pandemic levels) was actually the result of accidental release of an isolate preserved from the 1950s. Interestingly, immunity to this virus may actually help to moderate the effects of infection with influenza A H1N1(2009), the new pandemic strain. This may be an example of an unplanned beneficial effect.

Factors favoring the emergence of a disease

Increasing levels of immunosuppressive procedures, infections, and other agents are a cause for concern. Immunosuppression, whether as a medical procedure or due to other effects such as HIV infection, could allow novel mutants or poorly replicating zoonoses to establish and transmit to other humans.

Even if a virus establishes itself within the human population, it may remain restricted to a few hosts or one area. Examples of this include Dengue fever prior to World War II, and HIV in rural areas of Africa before transmission to urban areas and into "amplifier" populations where more intense and high-risk sexual activities combined with injecting drug abuse converted it into a pandemic.

The roles of some relevant factors in disease emergence, both in initial infection and in subsequent transmission, are summarized in **Table 8.4**. These factors rarely operate in isolation, for example war increases poverty, reduces hygiene, and can alter the pattern of sexual activity in an area. Specific examples of viral disease amplification in the human population include:

- The endemic status of hepatitis A (*Picornaviridae*) and hepatitis E (*Hepeviridae*) in urban areas with poor sanitation where fecal contamination of drinking water is unavoidable.
- The rise in almost all blood-borne diseases, notably hepatitis B (*Hepadnaviridae*) and hepatitis C (*Flaviviridae*) resulting from use of shared needles in injecting drug abuse, combined with the poverty and poor hygiene of many drug abusers and possible direct immunosuppressive effects of some drugs of abuse. Re-use of needles has also been important in the spread of HIV worldwide.
- The explosive increase in HIV infection resulting from extreme promiscuity and specific practices (notably, unprotected receptive anal sexual intercourse) in homosexual populations in the early stages of the HIV pandemic. In counterpoint to this, population movements within

Table 8.4 Contributory factors to the emergence of virus infections

Factor	Consequences	Examples
Poverty	Poor nutrition	Rotavirus diarrhea, measles, hepatitis E
	Lack of clean water	
	Lack of medical care	
	Poor hygiene (exposure to disease vectors)	
	Immunosuppression from other infections	
Urbanization	Close proximity of infected hosts and humans	Rotavirus diarrhea, measles, hepatitis E
	Poor hygiene (exposure to disease vectors)	
	Poor housing	
	Lack of clean water	
	Relaxation of social codes	
Habitat destruction or invasion	Increased contact with host and vector species	Filovirus disease, hantaviruses (HFRS, HPS)
War	Disruption of medical services and disease control	Rotavirus diarrhea, measles, hepatitis E
	Movement of refugees and troops	
	Breakdown of social codes	
	Overcrowding, poor nutrition, and poor hygiene in refugee camps	
	Increased poverty	
Intensive farming	Amplification of diseases in livestock	nvCJD, Nipah virus, influenza
Irrigation	Creation of breeding pools for vector mosquitoes	Arboviral diseases
Population and material movements	Spread of previously localized diseases via carriers or vectors	West Nile virus, HIV, (smallpox)
Multiple-partner sexual activity	Transmission of sexually transmitted or blood-borne diseases (risk levels greatly increased by extreme practices)	HIV
Drug injection (non-medical)	Transmission of blood-borne agents	HIV, hepatitis C
	Immunosuppression from other infections	
Medical procedures	Transmission of blood-borne agents by re-use of equipment	HIV, hepatitis C, nvCJD
	Transmission in blood, blood products, or organs	
Immunosuppressive conditions	Emergence of rare or normally nonpathogenic agents as causes of disease	HHV-8 (KSV)

Africa, in particular the displacement of male workers from their families for long periods, has led to a far greater problem of heterosexual AIDS in such countries, with prostitutes as major vectors.

- The increase in the number of infections in Marburg and Ebola virus hemorrhagic fever outbreaks resulting from centralized care of victims in hospitals with very limited facilities. The lack of barrier nursing and the necessary re-use of medical equipment in such hospitals are thought to be a significant route of spread and since carers in such hospitals do not have access to effective isolation procedures, they may become infected and can then further spread the virus.

It can be seen that while diseases must first be introduced into the human population, a combination of other factors is required for them to become established and transmitted within that population. It is probable that many agents with the potential to cause significant human disease have failed at this stage (as with influenza A H5N1), and the eradication of the risks outlined in Table 8.4 plays an important role in disease control.

8.3 SPREAD

The key to the future of any infection is the efficiency with which it can transmit itself to new hosts.

Modes of transmission

There are a large number of routes by which viruses may transfer to new hosts (**Table 8.5**). It is important to bear in mind that these are not exclusive, and many viruses are transferred by several of the routes indicated. For example, HIV may be transmitted sexually, in blood or blood products, or vertically from mother to child. With influenza, the virus may be transmitted by fresh or dried aerosol droplets, and contamination of **fomites** (inanimate objects capable of carrying infectious organisms) such as surfaces, doorknobs, or taps may also transmit the virus. If an infected person touches their nose or mouth and then handles the fomite, such transfer can be direct, without involving an aerosol stage.

The efficiency of each route of transmission is highly variable. The vast majority of viruses shed as aerosols or into sewage will not encounter a suitable host. This is particularly true of the fecal–oral route in developed societies where sewage disposal is managed so as to avoid such transmission. In contrast, direct transmission can be highly efficient, but requires much closer contact. The most intimate form of contact is direct transfer of blood or organs, which can initiate an infection with even poorly transmissible viruses. It is this that underlies concerns about *xenotransplantation* (see Section 8.2).

It is also apparent that some modes of transmission are more prolific than others. Although it is very wasteful of individual viruses, the sheer numbers produced mean that aerosol spread in dried droplets can be highly efficient and can spread a pandemic round the world in months, as with influenza A HIN1(2009). Spread by other routes can also be efficient, but may take longer as with HIV-1; it took well over 50 years from the original transfer from chimpanzees to humans to establish a worldwide pandemic.

Even within a particular mode of transmission, individual viruses will show differing levels of efficiency. Hepatitis B virus (*Hepadnaviridae*) is more efficiently spread in blood than is HIV, for example, due to the higher stability of the virus. This is reflected in the prevalence of the virus (the number of humans infected). So far HIV has infected about 60 million people in total. For hepatitis B, the best estimate is that two billion people have been infected, of whom 350 million remain chronic carriers of the disease.

It is also important to remember that efficient killing of the host is not in the interest of the virus (**Box 8.4**). Of the two billion humans infected with hepatitis B virus, 600,000 a year die, a mortality rate of 0.03%. Six hundred times that number are chronically infected, acting as incubators for the next generation of the virus. By contrast, Ebola virus kills most of those infected fast and with very obvious symptoms, limiting its spread.

Table 8.5 Modes of virus transmission

Route	Contact required	Requirements	Examples	Prevention
Airborne droplets	Direct exposure (short-range transmission in fresh droplet aerosol, or on contaminated fomites*)	Release of virus into airways	Influenza, rhinoviruses, varicella-zoster	Capture of aerosol (tissues)
Droplet nuclei (dried droplets)	Minimal (droplet nuclei stable in air, enter airways with high efficiency, stable on fomites*)	Release of virus into airways; infectivity survives drying	Influenza, rhinoviruses	Capture of aerosol (tissues)
Fecal–oral	None (viruses enter drinking water supply or other environmental source)	Excretion of virus in feces or urine; mixture of sewage with food or water	Hepatitis A, hepatitis E, poliovirus, rotavirus	Separation of sewage from water supplies or food; hygienic food preparation; hand washing
Sexual	Intimate exposure	Presence of virus in body fluids	Hepatitis B, HIV, HSV-2, papillomaviruses	Barrier methods (condoms), sexual continence
Oral transmission	Direct exposure (oral contact, possibly by fomites*)	Presence of virus in saliva or on mouthparts	Cytomegalovirus, Epstein-Barr virus, HSV-1	Limitation of contact
Direct contact	Direct exposure, including contact with body, body fluids, and contaminated fomites*	Presence of virus on body surface or in relevant body fluids	Filoviruses, papillomaviruses	Limitation of contact, barrier nursing
Blood or organ transfer	Entry of blood or tissue (shared needles, blood transfusions, organ transplants)	Presence of infectious virus in transferred material	Hepatitis B, hepatitis C, HIV; cytomegalovirus in organs	Sterilization or single use of needles and medical equipment, testing of blood or organs before use
Vertical transmission	Transfer from mother to child (*in utero*; during birth; by breast milk)	Respectively: ability to cross placenta; presence in body fluids or on female genital tract; presence in breast milk	Respectively: varicella-zoster, rubella, HIV; HSV-2, HIV; HIV	Reduction in virus load (using antiviral drugs) or avoidance of exposure (e.g. birth by Caesarian section)
Vector-borne	Transfer by vector species (mosquitoes, ticks, others) often involving replication and amplification in the vector	Ability to replicate in or survive on the vector	Arboviruses (e.g. dengue fever, Colorado tick fever, West Nile, yellow fever)	Control or avoidance of vector species

* Fomites are inanimate objects capable of carrying infectious organisms. Their precise nature varies with the infection, but can include door handles, food preparation equipment, mops, needles, toothbrushes, towels, and many others.

Transmission by arthropods

Transmission by biting insects and ticks (both belonging to the phylum *Arthropoda*) has long been a major route of virus infection. Viruses transmitted in this way are given the functional classification of **arboviruses** (from ***ar***thropod ***bo***rne ***viruses***).

The virus is acquired when the vector feeds on the blood of the host. While it is possible for the virus to be transferred passively if a susceptible human is bitten soon afterward, more commonly the virus will replicate in the vector species, typically in the cells lining the gut and in the salivary glands. Amplification in this location ensures that an infected arthropod vector will amplify the virus and that virus will be available to infect a new host when they are bitten in turn.

Box 8.4 An ideal virus?

It is possible to imagine an ideal virus—at least from the point of view of the virus. Such a virus would produce no or minimal symptoms, potentially even benefiting its host in some ways, thus allowing the host to act as an efficient vehicle for its spread. It would then be shed continuously over the lifetime of the infected host, by an efficient route.

Some viruses, such as herpes simplex virus type 1 (HSV-1; *Herpesviridae*), approach this ideal. Around 75% of adults show evidence of infection, and the virus then becomes latent in the spinal ganglia, maintaining a long-term infection with repeated shedding from cold sores at the site of the initial infection. The virus has even adapted so that peak shedding of infectious virus occurs before the major symptoms of the cold sore appear, facilitating spread since the potential new host cannot see any signs of infection.

An even closer approach to the hypothetical perfect virus is seen in endogenous retroviruses (see Chapter 1). These are maintained stably within the host genome. While the mode of spread (vertical transmission) is slow, over time an individual variant may become present in billions of hosts. This can, however, represent an evolutionary dead end, since many such viruses lose the ability to transmit to a new host in any other way. Thus, they last just as long as their host species, and no longer. It is not yet known whether dinosaurs had endogenous viruses in their genomes, but if they did then extinction for one meant extinction for the other.

The main type of arthropod vector is the mosquito, but ticks and biting flies can also transmit specific arboviruses. Two main types of mosquito that act as arboviral vectors are from the genus *Culex* and the genus *Aedes*. For example, among the *Flaviviridae*, *Culex tarsalis* (Figure 8.2) is the main vector for West Nile virus, while the established urban vector for yellow fever virus and the Dengue viruses is *Aedes aegypti* (**Figure 8.9A**). However, other mosquitoes may be involved, in particular *Aedes albopictus*, the aggressive and urban-dwelling Asian "tiger mosquito" (**Figure 8.9B**). When *Aedes albopictus* was introduced to the Americas in 1985 in water contained in a shipment of used car tires this enhanced significantly the potential for the transmission of many arboviruses.

There are over 500 identified arbovirus infections, some examples of which are shown in **Table 8.6**. Many of these can be classified as emerging diseases. For many arboviruses, the virus may be passed to subsequent generations of the vector species in the eggs (vertical transmission) and it may also overwinter in the arthropod host. In many cases, such as Crimean-Congo hemorrhagic fever (*Bunyaviridae*), the arthropod vector is itself an important reservoir of infection.

Arthropods are also capable of transferring bacterial and protozoal infections including bubonic plague (fleas) and malaria (mosquitoes), and arthropod-borne infections are a major risk to the health of humans and livestock worldwide.

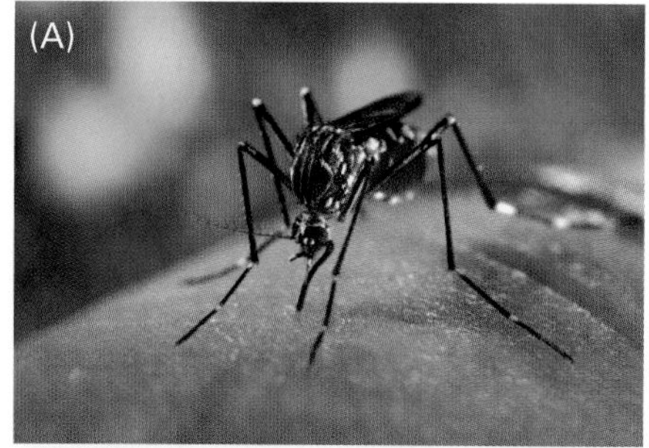

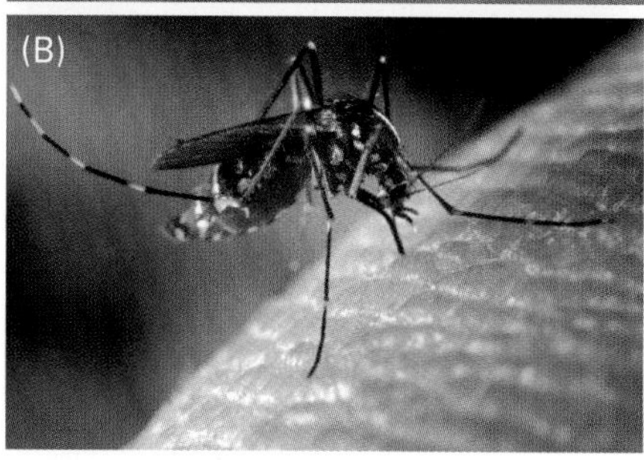

Figure 8.9
Mosquito vectors of the genus *Aedes*.
(A) *Aedes aegypti*; the main vector for yellow fever and Dengue fever. (B) The Asian tiger mosquito, *Aedes albopictus*, which can act as vector for a range of arboviruses.

Sylvatic and urban cycles

The maintenance of arboviruses typically involves cycling between their reservoir host and a mosquito in the **sylvatic cycle**, defined as affecting only wild animals. When a human receives the virus this can either be as a dead-end host, unable to transmit the virus further (**Figure 8.10A**) or as a host able to amplify the virus and to initiate further transfer, potentially beginning an "**urban**" **amplification and infection cycle** with humans as another reservoir host (**Figure 8.10B**). The importance of the two cycles varies with different viruses.

For many arboviruses including, for example, La Crosse encephalitis (*Bunyaviridae*), Japanese encephalitis, and (usually) West Nile virus (*Flaviviridae*), the human host is a dead end. The virus relies on its reservoir host to maintain itself, and the human is irrelevant to the life cycle of the virus.

For yellow fever (*Flaviviridae*), the sylvatic cycle is well established and, while the virus can establish an urban cycle, this is not common in the Americas. For other arboviruses, including Chikungunya and Ross River

Table 8.6 Examples of arboviruses infecting humans

Family	Virus	Areas affected	Vector type	Reservoir host	Transmission from humans?
Bunyaviridae	La Crosse encephalitis	North America	Mosquito	Chipmunks, squirrels, other small mammals	No
Bunyaviridae	Crimean-Congo hemorrhagic fever	Africa, Asia, Europe	Ticks, flies	Livestock, mammals, ticks (birds)	No (except by blood transfer)
Flaviviridae	Dengue virus (types 1–4)	Tropical and subtropical areas worldwide	Mosquitoes	Humans (primates)	Yes (primary route)
Flaviviridae	Tick-borne encephalitis virus	Europe, Asia	Tick (and milk from infected livestock)	Small mammals	No
Flaviviridae	Japanese encephalitis virus	Southern and Eastern Asia	Mosquitoes	Birds, pigs	No
Flaviviridae	West Nile virus	Worldwide	Mosquitoes (ticks?)	Birds (multiple species)	Rare (blood and organ transfer, other routes may exist)
Flaviviridae	Yellow fever virus	Africa, South America	Mosquitoes	Monkeys	Yes
Reoviridae	Colorado tick fever	Western North America	Ticks	Rodents	No (except by blood transfer)
Togaviridae	Chikungunya virus	Africa, Southern Asia	Mosquitoes	Primates	Yes
Togaviridae	Eastern equine encephalitis	North and South America	Mosquitoes	Birds	No
Togaviridae	Ross River virus	Australasia, Pacific	Mosquitoes	Marsupials, bats	Yes
Togaviridae	Venezuelan equine encephalitis	Rats, opossums (small mammals)	Mosquitoes	Birds and rodents	Possible

viruses (*Togaviridae*), both sylvatic and urban cycles exist side by side and both are important to maintenance of the virus. Finally, with Dengue fever the importance of the sylvatic cycle in human infections is not certain and infection is maintained predominantly by the urban cycle. Thus, the urban cycle is more important for human disease in this particular case.

Control of arbovirus infections

While arboviruses are difficult to control since their vectors actively go looking for new hosts, they are also susceptible to control at two levels. Options for controlling the virus itself are limited: vaccines exist for a few, such as yellow fever, Japanese encephalitis, and tick-borne encephalitis. Specific antiviral drugs or other therapeutics are not (yet) available. However, unlike many other viruses it is also possible to seek to control the vector. This may be by:

- The use of mosquito nets impregnated with the insecticide permethrin
- Removal of breeding habitats, in particular static water
- The use of chemical insecticides such as temephos (in the USA) or dichlorodiphenyltrichloroethane (DDT) (banned in many developed countries but still in use elsewhere)

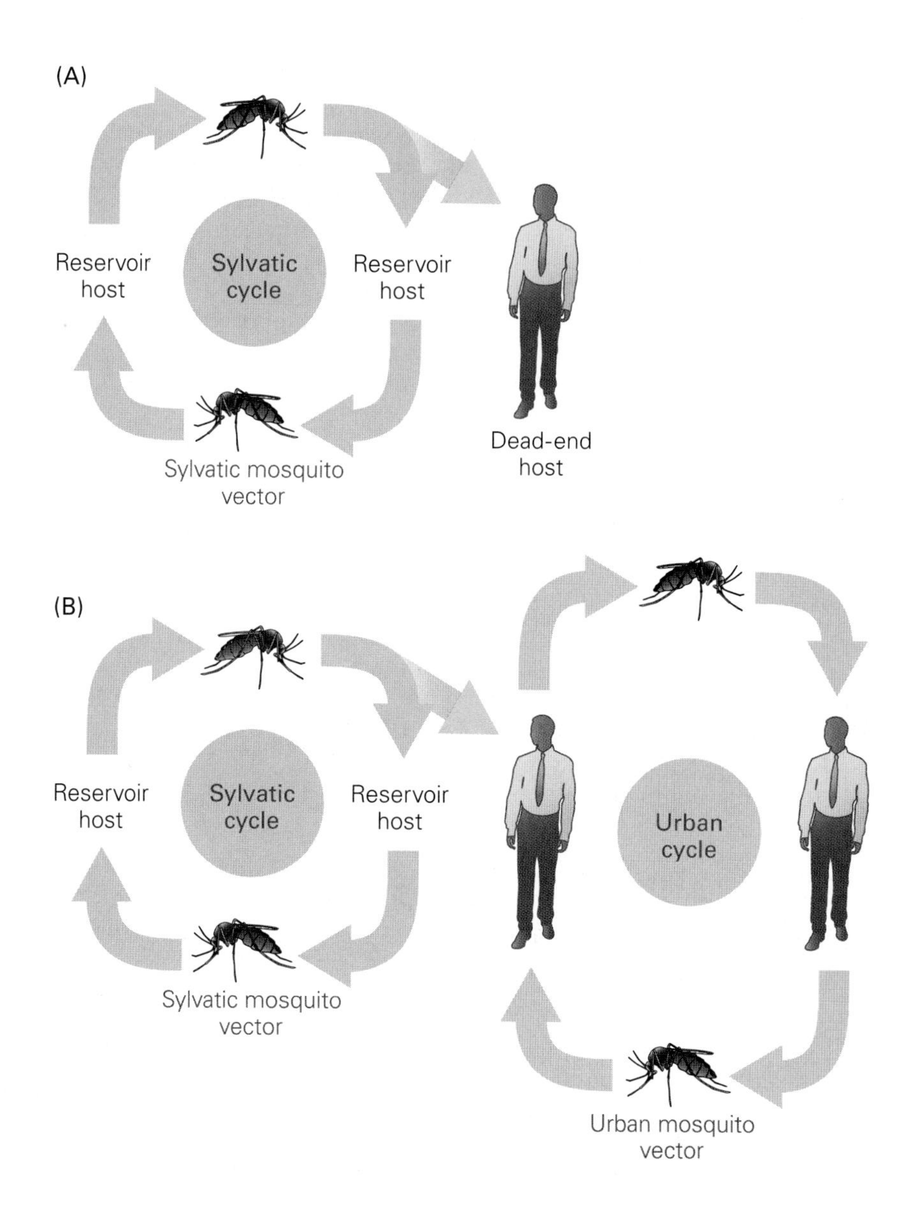

Figure 8.10
Urban and sylvatic cycles of arboviruses. The sylvatic cycle involves circulation and amplification with a wild animal host and a mosquito vector. Transfer to humans usually takes place in the sylvatic setting. (A) The infected human host may then act as a dead-end host, unable to transmit the virus further. (B) Alternatively, the human host may act as a source of virus to begin the urban cycle, with amplification by vector-mediated transfer to new hosts. Amplification occurs by multiplication in both host and vector.

- Use of larval growth inhibitors such as methoprene
- Use of biological control agents including *Bacillus thuringiensis Israeliensis* (BTI), predator fish (*Gambusia*), or crustaceans (*Mesocyclops*)

However, despite the potential for control at multiple levels, arbovirus infections are a major and enduring health problem in many areas of the world.

8.4 EXTINCTION

By many estimates, 99.9% of all species that have ever existed are now extinct. Since viruses appear to be ubiquitous and to have been around for billions of years, it is a reasonable surmise that many species of virus are also extinct. However, the definition of species in the virus world is of limited value, and the viruses existing today are likely to have derived from many forms that are no longer in existence.

Within the period of less than a hundred years when mankind has understood what a virus actually is and has been able to observe and identify them, the number of viruses that have become extinct is, unsurprisingly, limited.

Local disappearances are observed routinely. Following each outbreak, Ebola and Marburg viruses disappear back to their reservoir host, but reappearances are both expected and frequent. Measles virus (*Paramyxoviridae*) infection generates protective immunity, but the virus itself is antigenically locked, unable to change its coat as does influenza since the major antigenic regions are essential for infectivity and cannot be varied easily. As a result, it needs new, naive hosts to infect. In any isolated population, enough children need to be born to keep this number high enough for the virus to circulate. The population level needed to support measles appears to be approximately 500,000 people. In any smaller population without regular mixing with outsiders, the virus eliminates itself. Of course, this is rarely an issue with the current level of travel and population movement, but it does provide a useful example of a mechanism of purely local "extinction."

True worldwide disappearance of a virus is much rarer. Three examples are available that will serve to illustrate how such extinction may occur.

First, smallpox was very good at sustaining itself in the human population. It was only contained and eventually eliminated by direct human intervention with the use of an effective vaccine, as outlined in Chapter 5. For all of the deep debate about eliminating the few remaining stocks it now appears that the virus may not be restricted to the two official depositories. However, for now the virus is no longer circulating and is thus functionally extinct. There is at least a realistic possibility that it may be followed by poliovirus and measles virus, both again due to the use of effective vaccines as outlined in Chapter 5.

Second, SARS coronavirus appeared in November 2002 and caused a worldwide outbreak through to July 2003 (with a laboratory-based outbreak approximately a year later). It was contained by the use of isolation and quarantine along with effective barrier nursing and is no longer circulating. However, a spectrum of closely related viruses have been observed in the bat reservoir host, and the possibility for further transfer of a closely related virus from this reservoir remains very real.

Finally, pandemic influenza viruses appear to settle down to cause recurring, seasonal influenza once the multiple waves of the initial pandemic have passed. However, they are then usually displaced when a new pandemic strain appears. The nature of circulating influenza after the 1918–1919 pandemic is much less well defined than with later pandemic strains, but it is clear that the 1918 H1N1 strain is no longer in circulation. The cycle of pandemic–seasonal–replaced then happened for the 1957–1963 H2N2 influenza A (Asian flu), which was replaced by the 1968–1969 H3N2 influenza A (Hong Kong flu). This then circulated alone until 1977, and thereafter until 2009 it circulated alongside the H1N1 influenza A that arose from the more limited 1977 outbreak (Russian flu). This led to an unusual situation with both subtypes (H1N1 and H3N2) circulating as seasonal influenzas. The new pandemic H1N1(2009) strain now appears to have substantially replaced both preexisting seasonal strains.

This replacement even of successful pandemic strains meant that when scientists wanted to examine the 1918 strain, it was necessary to collect samples from a frozen cadaver of a casualty of the pandemic, or to use fixed tissue specimens from the period. This has been done, and the sequence of the 1918 virus is known. If the sequence is available it could, in theory at least, be produced at any time.

In a time when any virus genome sequence, if known, may be produced from the basic building blocks of DNA and then brought to life in a suitable host cell, no virus can truly become extinct. Knowledge may be forgotten or concealed, even in a digital age, but it cannot be unlearned.

Key Concepts

- Levels of virus infection are never static; they increase and decrease, new viruses appear and old ones disappear (often temporarily). All of this is part of normal variation.
- Surveillance of trends in virus infection using the best available technology can provide valuable advance warning of infections, but there is also the risk of generating so much data that important events can be hidden in the detail.
- An emerging disease has been defined by the Centers for Disease Control and Prevention (CDC) as: *"A disease of infectious origin with an incidence that has increased within the last two decades, or threatens to increase in the near future."*
- Many routes exist for novel viruses to infect humans, including transfer from other species (zoonosis), geographical spread, genetic variation, and release by human agency. Existing infections may be newly identified and viruses known previously may reappear.
- Viruses that cause serious disease are often the focus of attention, but these may be poorly adapted to their host. It is not usually an evolutionary benefit for a virus to kill or incapacitate its host too fast.
- Once a virus is established in the human population, the method and efficiency of its transmission will determine its future. If it can spread efficiently it will establish itself and diverge from its original source, often moderating its virulence in the process. If it cannot it will die out unless continually reintroduced from another source.
- Viruses can spread by a wide variety of routes, from respiratory aerosol to direct organ transfer. Arthropod vectors are an important method of virus spread.
- As with all forms of life, viruses are subject to extinction. But when its genome sequence is known and the technology to produce it exists, it is not really possible for a virus to disappear completely.

DEPTH OF UNDERSTANDING QUESTIONS

Hints to the answers are given at http://www.garlandscience.com/viruses

Question 8.1: Why and how do viruses cross between species?

Question 8.2: Why was attention focused on the H5N1 bird influenza, right up until the 2009 H1N1 swine influenza was already on its way to pandemic status?

Question 8.3: Why do some viruses damage an old host less than a new one when they cross between species?

Question 8.4: If aerosol spread in dried droplets is so efficient as a method of transmission, why don't all viruses use it?

Question 8.5: Why are arboviruses so common?

FURTHER READING

Alibek K & Handelman S (1999) Biohazard: The Chilling True Story of the Largest Covert Biological Weapons Program in the World–Told from Inside by the Man Who Ran It. Random House, London.

Belshe RB (2005) The origins of pandemic influenza--lessons from the 1918 virus. *N. Engl. J. Med.* 353, 2209–2211.

Garrett L (1994) The Coming Plague. Penguin Books, New York.

Harper DR & Meyer AS (1999) Of Mice, Men and Microbes. Academic Press, San Diego.

Houghton M (2009) The long and winding road leading to the identification of the hepatitis C virus. *J. Hepatol.* 51, 939–948.

Miller J, Engelberg S & Broad W (2001) Germs: Biological Weapons and America's Secret War. Simon and Schuster, New York.

Miller MA, Viboud C, Balinska M & Simonsen L (2009) The signature features of influenza pandemics--implications for policy. *N. Engl. J. Med.* 360, 2595–2598.

Preston R (1999) The demon in the freezer. In The New Yorker, July 12, 1999, pp 44–61. (Available online at http://cryptome.org/smallpox-wmd.htm)

Trevejo RT & Eidson M (2008) Zoonosis update: West Nile virus. *J. Am. Vet. Med. Assoc.* 232, 1302–1309.

Wang LF & Eaton BT (2007) Bats, civets and the emergence of SARS. *Curr. Top. Microbiol. Immunol.* 315, 325–344.

INTERNET RESOURCES

Much information on the internet is of variable quality. For validated information, PubMed (http://www.ncbi.nlm.nih.gov/pubmed/) is extremely useful.

Please note that URL addresses may change.

2007 AIDS Epidemic Update, World Health Organization. http://data.unaids.org/pub/EPISlides/2007/2007_epiupdate_en.pdf

Center for Biologic Counterterrorism and Emerging Diseases CBC-ED. http://www.bepast.org/

Emerging Infectious Diseases Online. http://www.cdc.gov/ncidod/eid/index.htm

ProMED mail. Program for Monitoring Emerging Infectious Diseases Email discussion forum. http://www.promedmail.org/

World Health Organization disease outbreak news. http://www.who.int/csr/don/en/index.html

CHAPTER 9
Viruses, Vectors, and Genomics

INTRODUCTION

DNA was known as early as 1869, when Swiss physician Friedrich Miescher observed a material derived from cell nuclei in pus from discarded surgical bandages, which he then called nuclein. In 1919, Levene identified the chemical groups present and proposed a phosphate-linked nucleotide structure. However, despite all of this work, it was still widely accepted that protein was the most likely candidate for the genetic material, until Avery and co-workers reported in 1944 that DNA was the "transforming principle" that could alter the properties of bacterial cells. The work of Hershey and Chase in 1952 further confirmed this using radiolabels specific for nucleic acids and proteins.

About the chapter opener image
Bacteriophage ϕX174
(Courtesy of the Research Collaboratory for Structural Bioinformatics Protein Data Bank and David S. Goodsell, The Scripps Research Institute, USA.)

Box 9.1 The basics of genetic manipulation

Bacterial cells have a system of recognizing and defending against foreign (often viral) DNA. This is based on specific patterns of methylation of their own DNA and the ability to recognize and digest invading DNA that lacks this modification, a process known as restriction.

The digestion process is mediated by *restriction endonucleases*, enzymes that can destroy the DNA molecule by cutting it at precise sequences within the molecule (**Figure 1**). The resultant pieces often have short, single-stranded regions known as sticky ends at the site of the cut. These can then allow molecules with identical sticky ends (typically those cut with the same enzyme) to adhere by base pairing. Bacterial ligase enzymes can then join-up the pieces.

The ability to cut at specific sites and then to join-up the fragments is the basic technology that resulted in the techniques of genetic manipulation.

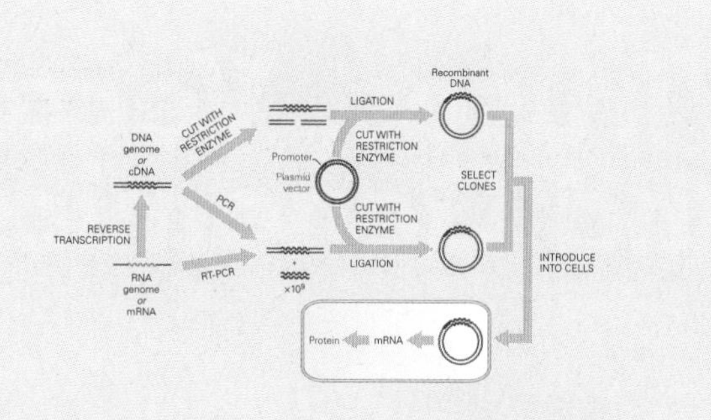

Figure 1
Recognition sites for three commonly used restriction enzymes. These type II enzymes both recognize and cut the DNA at very specific sequences, which are often palindromic, reading the same in both directions.

Any direct manipulation of the genetic material had to await the identification of the nature of that material, and a vital step in this occurred in 1953 when Watson, Crick, Wilkins, and Franklin characterized the double-helical structure of DNA. This led to an understanding of how, using specific base pairing, DNA could pass its precise sequence of bases on to new DNA molecules during replication—the true basis of genetic inheritance.

Recognition of the nature and structure of the DNA molecule underlay the understanding of the mechanisms and processes of the regulation and expression of genetic information. Following on from this, the identification of bacterial enzyme systems that were responsible for cutting, moving, and adapting bacterial DNA began to provide the tools that would allow the direct manipulation of the genetic material (**Box 9.1**). Using these methods, in 1973 Cohen and Boyer successfully transferred a section of foreign DNA into an *Escherichia coli* bacterium.

The way was now open to one of the most controversial sciences of the twentieth century. Variously termed **genetic engineering** and **genetic manipulation**, it has expanded our ability to adapt life to the needs of humanity at the most basic level. But while the technology has moved ahead rapidly, public acceptance has been much slower, especially in areas where the technology is perceived to benefit companies rather than individuals. It is a challenge for the twenty-first century to find responsible and acceptable ways of using this technology and to make a reasoned understanding of the risks and benefits clear to all.

9.1 GENETIC MANIPULATION

With knowledge of the structure of the genetic material, it became possible to manipulate and modify this directly, rather than by the less precise techniques of selective breeding that had been practiced since the earliest days of agriculture. Viruses have played a central role in the development of the basic approaches of genetic manipulation. Although synthetic techniques using manufactured DNA (see Section 9.3) are now widely used, the basic approaches of genetic manipulation are still applicable. These are:

- **Gene cloning**: the construction of a library containing the DNA of interest for expression, probing, or other uses

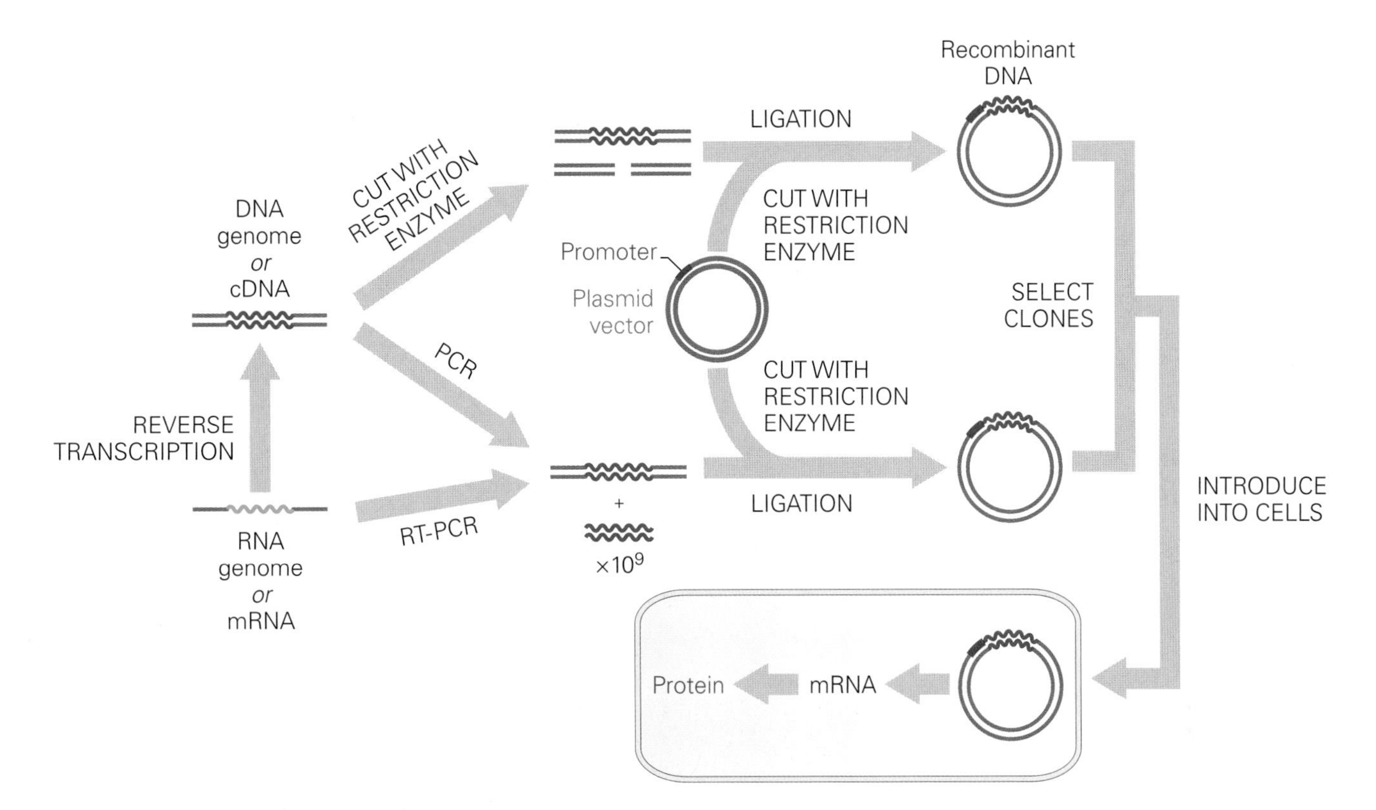

Figure 9.1
Basic principles of cloning and expression. DNA containing the gene of interest (wavy line) is cut into fragments with restriction enzymes. These fragments are then mixed with similarly cut vector DNA and ligated to form recombinant DNA. The transformed vectors are allowed to replicate and then clones containing the gene of interest are selected and introduced into cells for expression. Alternatively, the gene of interest can be amplified by polymerase chain reaction (PCR), or reverse transcriptase PCR (RT-PCR) in the case of RNA, in which case, joining to the vector is achieved by either blunt end ligation or restriction enzyme digestion of sites in the PCR primer followed by ligation.

- **Gene expression**: using the cloned DNA to produce proteins or other products
- **Gene probing**: using the sequence-specific binding properties of the DNA (or RNA) to detect matching sequences in any of a wide range of hybridization applications (see Chapter 10)

The basic process of cloning and expressing genes is illustrated in **Figure 9.1**.

Cloning

In the early days of genetic manipulation, the first step in generating useful material was typically the use of **restriction enzymes** to digest the genome of interest, generating a library of essentially random fragments of DNA. This was followed by insertion into a plasmid **vector** to allow the inserted gene(s) to be maintained and replicated within a suitable (usually prokaryotic) host cell. This was then used to produce the cloned DNA.

A plasmid is a covalently closed circular DNA which can replicate within a host cell, while a plasmid vector is a hybrid DNA molecule derived from but often much smaller than a naturally occurring plasmid, which is able to replicate and which sometimes is also designed to drive the expression of DNA inserted into it. While plasmids are both relatively simple to use and readily available, they are only capable of carrying limited sizes of inserts, typically up to 20 kilobase pairs (kbp). To clone larger inserts, a variety of other systems may be used, many of which contain virus-derived elements (**Table 9.1**).

The introduction of DNA into cells is referred to as *transformation* in bacterial or plant cells and as **transfection** in animal cells (since the term transformation already has another meaning in this setting, as noted in Section 1.5). As well as physical methods including chemical, electrical, liposomal, or even magnetic approaches, a viral vector can be used. With such a vector, the DNA is packaged into a virus particle and introduced into the cell by infection with the resulting hybrid virus, a process referred to as

Table 9.1 Cloning vectors

Vector	Nature	Maximum carrying capacity
Plasmid	Derived from self-replicating bacterial plasmids	Small (up to 20 kilobase pairs)
Cosmid	Hybrid plasmid containing bacteriophage lambda sequences, able to package into bacteriophage capsids	Medium (37–52 kilobases)
Phagemid	Vector based on the filamentous F1 bacteriophage	Small (ssDNA; up to 6 kilobases)
Fosmid	Vector based on the *E.coli* F plasmid (maintained at very low numbers per cell)	Medium (as cosmid)
Bacteriophage	Genome of a bacterial virus, with insert	Small to medium (as virus)
Virus	Genome of a virus of eukaryotes, with insert	Small to medium (4–25 kilobase pairs)
Bacterial artificial chromosome	Large DNA based on the F plasmid	Large to very large (> 700 kilobase pairs)
Yeast artificial chromosome	Large eukaryotic (yeast) DNA	Large to very large (> 3000 kilobase pairs)

transduction. It should be noted that transduction is actually a natural process mediated by many viruses of bacteria (bacteriophages), in particular those which can integrate into the host chromosome (lysogenic bacteriophages).

Once the library of DNA fragments was established, regions of interest could be identified by a variety of means, mainly those based around specific binding of nucleic acid probes of known sequence to regions with appropriate complementary sequences in the sample (*hybridization*, see Section 10.5).

Expression

Expression of the gene of interest may involve subcloning into a suitable expression vector, a plasmid that is capable of both replication and expression in the system used.

A wide variety of systems have been used to allow the expression of foreign genes. The promoter that drives expression of the inserted DNA will reflect the host cell type, with the most obvious types being those active in prokaryotic or in eukaryotic cells. Given that DNA is DNA, wherever it is located, some expression vectors (shuttle vectors) can contain the multiple sequences necessary for expression in both types of cell (see **Figure 9.2**). There are several key elements:

- *Promoter*: to drive expression of inserted genes by interaction with transcription factors and RNA polymerases.
- *Cloning* (*or polycloning*) *site*: a region of synthetic DNA containing one or more sites that can be cut by different restriction enzymes allowing DNA to be inserted.

- *Reporter gene*: the expression of which can be detected by a simple assay system, positioned such that its transcription is disrupted when DNA is inserted into the cloning site, thus confirming that the plasmid now contains additional DNA. Examples include the *lacZ* gene (for prokaryotic systems) and the green fluorescent protein (GFP) for eukaryotic systems.
- *Selectable gene*: imparts a selectable characteristic such as antibiotic resistance to ensure that only cells containing the plasmid will grow under the conditions used. Examples include ampicillin (for prokaryotic systems), neomycin (for eukaryotic systems), and zeocin, which functions for both systems.
- *Origin of replication*: regulates the replication rate of the plasmid. Many plasmids now also contain an origin of replication from filamentous bacteriophages (f1) to allow mutagenesis of the (single-stranded) DNA in this system.

Additional features are also necessary for efficient expression in eukaryotic systems. These include a polyadenylation site adjacent to and downstream from the inserted gene, which is required to produce functional eukaryotic mRNA and which may also control splicing of the mRNA produced. Other regions of DNA which bind specific eukaryotic transcription factors and transcriptional activators are typically included to enhance transcription.

One useful addition that can be used in both prokaryotic and eukaryotic systems is the inclusion of a synthetic DNA sequence coding for a region of consecutive histidine residues in frame with the protein expressed, along with a proteinase site to allow their removal. This sequence binds the recombinant protein to a nickel affinity column, allowing rapid purification, followed by proteolytic removal of the histidine tag.

Post-translational processing

While expression in prokaryotes is relatively straightforward and easy to scale up to commercial levels, expression in eukaryotic cells is far more complex. Proteins produced in prokaryotic cells may have the correct amino acid sequence but it is important to remember that the amino acid chain (polypeptide) alone does not constitute a protein. Both during and after translation, the amino acid chain undergoes a very wide range of other chemical modifications, ranging from the attachment of small chemical

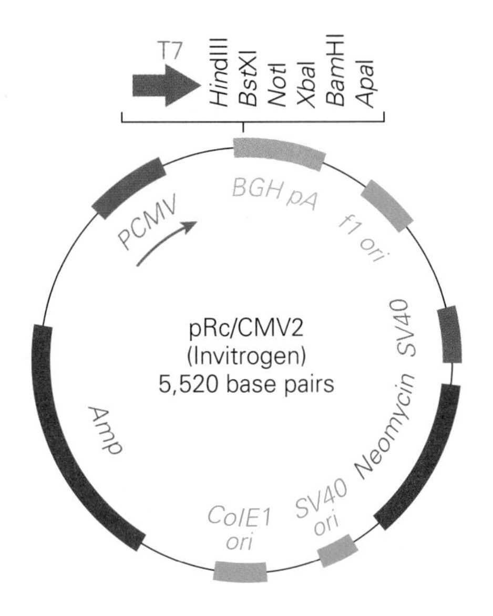

	Eukaryotic	Prokaryotic
Promoters	PCMV SV40	T7
Origins of replication	SV40 ori	ColE1 ori f1 ori
Selectable genes	Amp	Neomycin
Polyadenylation site	BGH pA	
Polycloning site Restriction enzymes that cut at this site are listed		

Figure 9.2
A prokaryotic/eukaryotic expression vector. The stable eukaryotic expression vector pRc/CMV2 (Invitrogen) is designed for replication and expression in both prokaryotic and eukaryotic systems. Note that this plasmid does not contain a reporter gene. The filamentous bacteriophage origin of replication (f1 ori) is used for mutagenesis work.

groups such as phosphate or sulfate up to the attachment of complex sugar chains, which can be bigger than the polypeptide itself (**Table 9.2**). In addition, the environment of the translated protein may also be important for correct folding. The amount and type of post-translational processing of a protein is partially determined by signals in the amino acid chain, but also varies with the type of cell, with cells from more complex organisms typically producing more complex modifications (**Table 9.3**). These modifications can have major effects on the structure and function of the protein. Cleavage or folding of the protein can also modify its function, as illustrated by findings concerning the PrPSc protein of scrapie (see Chapter 2).

Table 9.2 From DNA to a mature protein

Stage	Process	Examples of indicator sequence motifs
Genome (DNA) sequence		
Transcription ↓	Gene finding	TAC (transcribes to AUG) start codon, open reading frame of suitable length before stop codon, promoter sequences
	Intron identification	GT–consensus sequence–AG motif
	RNA editing	A6G3 sequence mediates insertion of additional Gs into mRNA (*Paramyxoviridae*)
mRNA sequence		
Translation ↓	Alternate start sites	Internal ribosomal entry sites (IRES; *Picornaviridae*) and translational initiation at internal AUG (and ACG) codons (*Paramyxoviridae*)
Polypeptide sequence		
Post-translational modifications ↓	Folding	Simple structures such as alpha helices and beta-pleated sheets may be predicted. More complex folding may be assisted by other (chaperone) proteins
	Cleavage	The wide range of proteinase enzymes prevent identification of general consensus sequences, although those for proteinases of known type (e.g. viral proteinases) may be identified
	Covalent binding	Mediated by cysteine residues (disulfide bonds)
	Glycosylation	N-linked glycosylation at asparagine-X-serine/threonine sequences (X = any amino acid except proline); O-linked glycosylation occurs at serine or threonine; mannose attachment at serine/threonine or tryptophan. The structure of the attached sugar groups may vary widely
	Fatty acylation	Attachment of myristic acid at N-terminal amino acid (usually glycine) during translation or at other glycine residues post-translationally; reversible attachment of palmitic or other long-chain fatty acid to cysteine residues
	Phosphorylation	Reversible attachment of phosphate groups to serine or threonine
	Sulfation	Attachment to tyrosine residues
	Other modifications	Acetylation, alkylation, citrullination, glycosylphosphatidylinisotol (GPI) anchors, isoprenylation, linkage to other proteins or structures; many others
Mature protein		

Note that similar residues may be used for a variety of modifications, e.g. cysteine for disulfide bonds or fatty acid attachment, serine or threonine for glycosylation or phosphorylation. The presence of a consensus sequence does not indicate that the modification is always present.

Table 9.3 Equivalence of post-translational processing in protein production systems as compared to mammalian systems

Examples of modifications	System			
	Bacteria	Yeast	Baculovirus / insect cells	Mammalian cells
Protein folding	+	+	++	+++
Proteolysis	+	+	++	+++
Fatty acylation	±	++	++	+++
Phosphorylation	+	++	+++	+++
Glycosylation	±*	+*	++*	+++
Secretion	±**	++	++	+++
Function	±	++	++	+++

* Nature of attached sugar chains is different to that in mammalian cells.
** Signals differ from those in eukaryotic cells; proteins may aggregate and show poor solubility.

Post-translational processing is very different in prokaryotic and eukaryotic cells, with the result that the proteins produced in prokaryotic systems can be very different from authentic proteins produced when viruses infect eukaryotic cells (Table 9.3 and **Table 9.4**). These differences can result in poor solubility and in reduction or loss of immunogenicity or function. Where it is important to obtain proteins with properties similar to those produced by the virus in its natural host (for example in vaccine formulation), it is often necessary to express the gene of interest in a eukaryotic system. Such work often uses **yeast expression systems**, which are less demanding than some other eukaryotic expression systems.

Virus-based systems for expression in eukaryotic cells

Examples of widely used prokaryotic and eukaryotic virus expression systems are shown in **Table 9.5**. As can be seen, different systems are suited to different applications. For example, for production of high levels of proteins in a eukaryotic system, placing an inserted gene under the control of the highly efficient baculovirus polyhedrin promoter can result in the inserted gene accounting for up to 30% of cellular protein production. Baculoviruses (*Baculoviridae*) infect insects, and the polyhedrin gene is hyperexpressed late in infection to produce enough polyhedrin protein to form a larger

Table 9.4 Protein production systems

System	Authenticity (for mammalian material)	Effort required	Scale-up potential	Maximum level of expression (% of cell protein production)
Bacteria	±	+	+++	30
Yeast	++	++	+++	1–5
Baculovirus	++	+++	++	10–30*
Mammalian cells	+++	+++	+	< 1–10*

* Maximum yield restricted to specific systems.
Cell-free systems derived from the above may also be used.

Table 9.5 Examples of viral vector systems

Bacteriophage lambda	
Host cells	Prokaryotic (*Escherichia coli*)
Virus used	*Enterobacteria* phage λ (*Siphoviridae*)
Uses	Cloning, prokaryotic expression, genomic libraries, experimental use as a vaccine vector
Advantages	Very wide range of well-characterized systems available, large inserts possible (up to 52 kbp in cosmid systems)
Disadvantages	Overall size must resemble that of normal viral genome, limited splicing and protein processing
Bacteriophage M13	
Host cells	Prokaryotic (*Escherichia coli* expressing sex pili)
Virus used	*Enterobacteria* phage M13 (*Inoviridae*)
Uses	Based on single-stranded nature of M13 genome: site-directed mutagenesis, DNA probes
Advantages	Easily purified single-stranded genome, suitable for mutagenesis, wide range of well-characterized systems, maintained as plasmid in host cells
Disadvantages	Limited insert size (1–3 kb), DNA rearrangements may occur, DNA orientation may be fixed, limits on splicing and protein processing
Baculovirus	
Host cells	Eukaryotic (insect)
Major viruses used	*Autographa californica* multiple nuclear polyhedrosis virus (AcMNPV), *Bombyx mori* (silkworm) nuclear polyhedrosis virus (BmNPV)
Uses	Recombinant protein production, vaccine vector under evaluation
Advantages	Nonpathogenic for humans, high levels of protein production possible, can be grown in caterpillars, inserts possible up to 15 kbp
Disadvantages	Different post-translational processing in insect cells, fusion proteins may be preferred, splicing may differ, possible insolubility of product
Adenovirus	
Host cells	Eukaryotic (specific types of human cells)
Major viruses used	Usually adenovirus type 5, others may be used (*Adenoviridae*)
Uses	Vaccine and gene therapy vectors, cancer therapies, recombinant protein production, production of transformed cell lines, infecting cells for immunological assay
Advantages	Oral or nasal delivery possible, authentic post-translational processing, efficient nuclear entry, high levels of expression possible, specialized vectors available, insert size up to 8 kbp (36 kbp in some systems)
Disadvantages	May be cytotoxic, immunity to adenovirus may prevent use, narrow host range with some types
Adeno-associated virus	
Host cells	Eukaryotic (human/mammalian) plus helper virus if replication required
Major viruses used	Adeno-associated virus (*Parvoviridae*)
Uses	Gene therapy and vaccine vectors
Advantages	Wide host range, nonpathogenic if helper virus not present, infects a broad range of cell types, easy to manipulate ssDNA genome, low immunogenicity, can produce long-lasting expression (vectors necessarily unable to replicate), efficient integration into host genome at defined site
Disadvantages	Limited insert size (5 kb), high levels of preexisting immunity

Table 9.5 *cont.*

Herpesvirus	
Host cells	Eukaryotic (wide range of vertebrate cells)
Viruses used	HSV-1 (HHV-1), Epstein-Barr virus (HHV-4) (*Herpesviridae*)
Uses	Neural targeting, vaccine vectors, recombinant protein production, cancer therapies, amplicon and episomal expression vectors
Advantages	Well-characterized, large viruses, wide choice of insertion sites, inserts up to 10 kb (larger in amplicon or episomal vectors—up to 150 kbp)
Disadvantages	Pathogenic for humans, cytotoxic, concerns over latency, may transform cells, limited availability
Poxvirus	
Host cells	Eukaryotic (human/mammalian)
Viruses used	Vaccinia (*Poxviridae*)
Uses	Vaccine vector, gene transfer, cancer therapies, recombinant protein production, infecting cells for immunological assay
Advantages	Wide host range, wide choice of insertion sites, large inserts possible (25 kbp), some systems allow high-level expression, wide availability
Disadvantages	May be pathogenic for humans, risk of early termination with some inserts, introns problematical, lack of nuclear stage in replication, involvement of virus-specific pathways
Retrovirus	
Host cells	Eukaryotic (human/mammalian)
Viruses used	Moloney murine leukemia virus, *Lentiviruses* (*Retroviridae*)
Uses	Gene therapy vectors, high-efficiency gene transfer and integration
Advantages	High efficiency of gene transfer, efficient integration into host genome, multiple systems available
Disadvantages	Concerns over safety and oncogenicity (including leukemia induction in clinical trials), integration at variable sites, limited insert size (8–10 kbp maximum), requirement for actively dividing cells (except *Lentiviruses*)
RNA viruses	
Host cells	Eukaryotic (human/mammalian/other)
Viruses used	Multiple (*Coronaviridae, Paramyxoviridae, Picornaviridae, Reoviridae, Rhabdoviridae, Togaviridae*)
Uses	Recombinant protein production, neural targeting, cancer therapies
Advantages	Capability to target specific cell types, high levels of protein production
Disadvantages	Small genomes restrict insert size, high mutation rate from RNA genome

protective capsule around the virus. Using the polyhedrin promoter to drive the expression of an inserted gene takes advantage of this natural hyperexpression.

Since many viruses are required to produce large quantities of their proteins in a relatively short time, many viral genes are expressed at very high levels. In addition, the factors controlling virus gene expression have been widely studied. As a result, many of the controlling elements in eukaryotic expression vectors are derived from viruses. For example, sequences from SV40 (*Polyomaviridae*) or cytomegalovirus (*Herpesviridae*) are widely used.

Some promoters have high efficiency in order to ensure a high level of transcription independent of the cell type, while others are active only in certain cells or under specific conditions, allowing expression of the inserted gene to be controlled. Examples of efficient promoters include the CMV immediate-early promoter or the Rous sarcoma virus LTR promoter, both of which are highly active in many cell types. An example of a conditional promoter is the metallothionein promoter, which can be induced by heavy metals such as cadmium.

Unlike prokaryotic genes, eukaryotic genes are usually broken up by introns, regions of untranslated DNA that are removed from the pre-mRNA by splicing prior to translation. Viral genes generally have far fewer true introns than cellular genes, probably due both to limitations of space in the viral genome and to the involvement of viral factors which exploit this difference to enhance viral mRNA translation over that of cellular transcripts. Some viral vectors (see below and Table 9.5) may not be able to express intron-containing genes. However, spliced viral mRNAs are not uncommon, and specific splicing signals may be required.

The requirement for splicing of genes containing introns can be avoided by cloning a DNA copy (a cDNA) of an mRNA, which will already have had any introns removed. However, the lack of introns in such a clone may result in altered transport within the cell and in inhibition of translation.

Viral vectors

While many vectors are based on genetic elements that can replicate within a host cell (plasmids or episomes) and may include some elements of viral DNA, viruses themselves have been used for this purpose for a number of reasons. These include the presence of high-efficiency promoters and also the ease with which a virus can introduce foreign nucleic acid into cells. While systems for the introduction of naked nucleic acid into cells are well defined, they can be difficult to use and of very low efficiency with some cell types. Viruses, optimized for this purpose by millions of years of evolution, provide a useful (if sometimes complex) alternative.

Viral vectors in common use are summarized in Table 9.5.

Some viruses, usually those with larger genomes such as the tailed bacteriophages (for prokaryotes) or the *Herpesviridae* or the *Poxviridae* (for eukaryotes), may be capable of independent replication even when large amounts of foreign DNA are inserted into their genome. In contrast, viruses with small genomes may require replacement of essential parts of their genome in order to carry a useful amount of foreign DNA. A good example of this is those retroviruses where replacement of essential viral genes with cellular DNA occurs during natural infections, and in which the insertion of cloned DNA usually requires the loss of essential genes. Viruses with inserts are maintained using another source to supply the missing functions. This can be a complete, co-infecting 'helper' virus, a modified helper virus that can supply the missing functions but not replicate itself (a suicide virus), or viral genes stably expressed in the cell used to culture the virus.

Since virus replication often damages or kills infected cells, many approaches use defective viruses, which in some systems contain little viral nucleic acid other than packaging signals.

Use of viral vectors for vaccine production is an important aspect of this work, and is covered in more detail in Chapter 5, where the effect of folding and structure on protein immunogenicity is also reviewed.

9.2 SEQUENCING

Matching the development of an understanding of the nature of and the ability to manipulate the genetic material is the need to understand what it says and what it means.

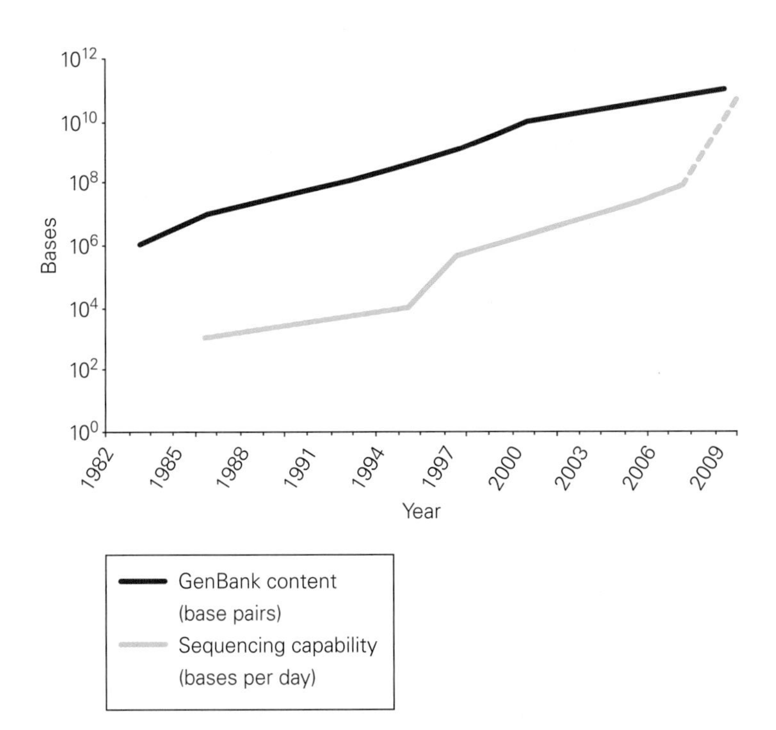

Figure 9.3
Sequence data and sequencing power.

Huge amounts of sequence data are now available, and more is becoming available at an ever-increasing rate (**Figure 9.3**). However, it is important to remember that while a genome sequence can usually be used to predict the order and type of amino acids in the polypeptides that can be produced from it, the nature of any final protein product is by no means a simple or even predictable extension from the genetic sequence. There are a large number of intermediate steps (see Table 9.2).

During the early part of the 1970s, a number of highly labor-intensive techniques were developed for identifying the order of bases in a molecule of DNA or RNA. Using such techniques, Fiers and co-workers reported the sequence of the 3659 bases of the RNA making up the genome of the MS2 bacteriophage in 1976. This was the first whole genome sequence to be characterized.

By 1975, the English scientist Frederick Sanger had developed a much more rapid method of sequencing DNA, the dideoxy method, which his group used to provide the first sequence for a DNA genome. The sequence of the 5386 base single-stranded DNA genome of the bacteriophage ΦX174 was reported in 1977. Alternative techniques from this period came, in time, to be replaced by improved methods derived from the Sanger technique.

In time, automated systems became available for these techniques, increasing output up to a million bases per day. These, in turn, have been replaced by new, massively parallel sequencing methods with far higher levels of output. These use highly sensitive detection systems with fluorescent tags to report the release of light resulting from the addition of individual bases to forming DNA molecules (**Figure 9.4**). While each new DNA molecule is small, powerful software is used to assemble these into a full sequence. This is the foundation of the science of **genomics**, the study of whole genomes of organisms.

The most prominent such method is referred to as 454 sequencing, after the 454 Life Sciences company that developed the technique. With more than 400,000 reactions occurring simultaneously, these systems can assemble sequence data for up to several hundred million bases per run, and their capability continues to expand. Even faster systems are now becoming available.

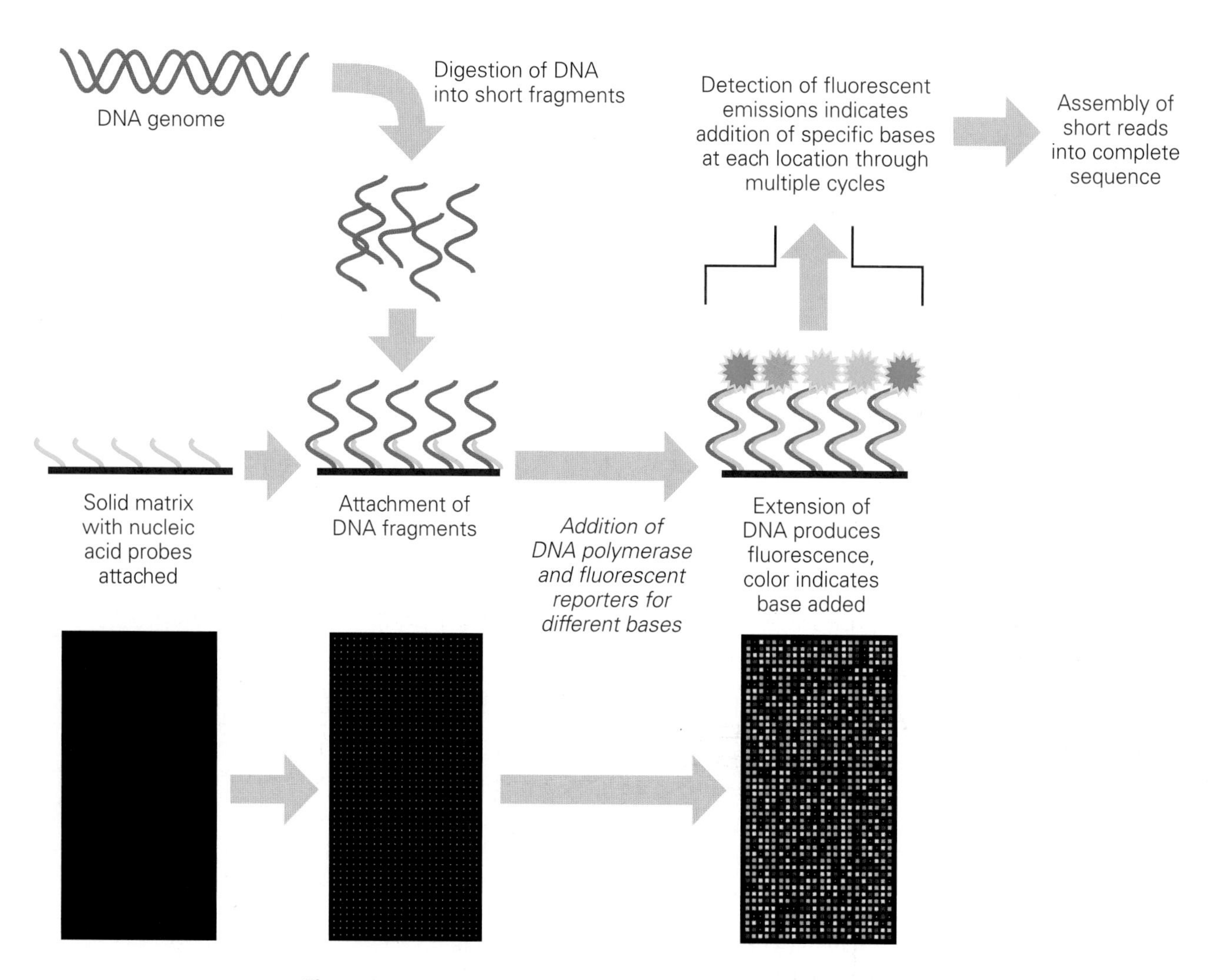

Figure 9.4
Sequencing by synthesis. The use of a different fluorescent probe for each base (G, C, A, T) allows the sequence to be 'read' as synthesis proceeds.

Metagenomics

A side effect of the use of a massively parallel approach is that DNA molecules present at very low levels (down to one in a few hundred thousand copies or even lower in some systems) in the sample will also be sequenced—an effect referred to as "depth." This allows the identification of sequences present at low levels, such as low-abundancy mRNAs, rare mutations, or the range of viral strains present in a specimen. However, this same sensitivity can also cause problems with low-level contamination similar to those seen with the polymerase chain reaction (PCR, see Chapter 10).

When applied to mixed samples, an approach referred to as **metagenomics**, such sequencing can be used to determine the range of viral types present in an environmental sample or in a clinical specimen. A metagenomic approach can provide useful information, for example with specimens from a patient with a highly variable virus such as HIV.

However, the limitations of the sampling procedure used do need to be considered. A small sample will not usually be fully representative of the body (or of the environment) as a whole, especially if a virus is present at high levels in one specific location which is not that from which the sample is taken.

Virus genome sequencing

As noted above, the genomes of viruses were the first to be sequenced, due to their small size. Despite this, the pr eukaryotic cellular genomes. With some viruses, such as hepatitis B, overlapping genes mean that the amount of coding capacity that is used for proteins exceeds 100% of the genome size. In comparison, less than 2% of the human genome seems to code for proteins.

Even the small absolute size of many viral genomes is no longer always helpful and they are actually too small for many of the sequencing systems now in development. It is an unavoidable consequence of the relatively small size of virus genomes that they benefit far less from the development of the latest ultra-high-throughput systems. The often-cited lower cost of these systems may not apply for short viral genomes, since the actual cost per run can be high, and if the full capacity of the system is not used the cost per base can rise very significantly. It is possible to counter this by loading multiple samples in a single well, but this can cause problems if viruses are closely related, with the possibility of a single consensus sequence being generated for multiple viruses. Tagging of individual sequences may now be used to permit this. However, older systems may still be used to analyze viral genomes and other short nucleic acids.

As of 2009, the GenBank DNA sequence database (Figure 9.3) contained approximately 2000 complete viral genome sequences (including at least one representative of every virus family known to infect humans), and many more partial sequences. However, given that the estimated number of viruses on the planet is greater than 10,000,000,000,000,000,000,000,000,000,000 (10^{31}) it is clear that we have a long way to go before we can claim to have characterized a significant percentage of viruses.

As more and more sequences become available, understanding of virus function at this most basic level is increasing, with viruses previously considered to be closely related being identified as significantly different (as with hepatitis E virus, which has been moved from the *Caliciviridae* to its own family, the *Hepeviridae*) and with relationships becoming apparent that would not otherwise be detected (for example, similarities between the *Herpesviridae* and the tailed DNA bacteriophages).

As with many systems studied, work with (relatively) simpler viral systems is likely to underpin understanding of more complex systems.

Junk or not junk?

As noted, viruses have very dense coding within their genomes (sometimes exceeding 100% of the genome by the use of overlapping genes), but the human genome does not. Less than 2% of the human genome codes for proteins, prompting some observers to call the rest "junk DNA."

As is often the case, this simple and dismissive term reflects a limited understanding of a complex situation. DNA specifying many controlling factors as well as a range of non-protein effector molecules lies in these "junk" regions. These latter include the functional RNAs of the cell (transfer and ribosomal RNAs) as well as a range of small RNAs such as those of the RNAi system. Some of the DNA is likely to represent genuine "junk," such as defunct retroviral elements (though even these may have a role, for example in moderating tolerance of fetal tissue during pregnancy), but much of it will not.

What is extremely clear is that the simple ACGT sequences of the DNA do not tell us what the products of those genes will do. It was a belief that this might somehow be the case that underlay many of the overly optimistic expectations for rapid commercial results from early work in genomics.

9.3 SYNTHESIS AND AMPLIFICATION

With understanding of the basic nature of the genetic material coupled with advances in synthetic technology (see Chapter 10), it soon became possible to produce short oligonucleotides to order by chemical synthesis. Simultaneous advances in sequencing methods provided the necessary information on exactly what to produce.

Under the correct conditions, such oligonucleotides have sequence-specific binding properties, allowing their use as probes for the detection of specific nucleic acid sequences, or for initiating DNA amplification in PCR and related techniques (see Section 10.5). Thus, cloned probes were usually no longer needed where a nucleic acid sequence was available.

PCR and its related and derivative techniques have produced as much of a revolution in genetic manipulation as they have in diagnostics and detection. Prior to the development of DNA amplification systems, it was necessary to purify and extract a DNA of interest as outlined above. Often this needed to be further purified by the removal of flanking DNA sequences before it could be used. Specific issues with individual DNA molecules could make cloning using such methods extremely difficult.

With PCR, so long as enough is known of the sequence to design short oligonucleotide primers to flank the region of interest, a simple reaction process can produce billions of copies of that specific region within hours. By including restriction sites in the primers, these can then be cut and used to insert the amplified DNA directly into any desired vector for further amplification or for other uses. It is, of course, important to use a high-fidelity polymerase (see Section 10.5) in such reactions to ensure that the product DNA is as accurate a copy as possible.

By 2002, synthetic technology had advanced to the point where it was possible to produce a whole 7741-base cDNA coding for the poliovirus (RNA) genome from single nucleotide building blocks. This could then produce infectious poliovirus when incubated in a cell-free cytoplasmic extract from suitable host cells. As the authors noted, referring to the 1991 paper by Molla et al. that first showed synthesis of poliovirus in cell-free systems, life can be expressed as a molecular formula. In the case of poliovirus, this was expressed as $C_{332,652}H_{492,388}N_{98,245}O_{131,196}P_{7501}S_{2340}$. Equally, the complete works of Shakespeare are just a collection of 26 different letters—the arrangement is everything, and that was what was now becoming possible.

Needless to say, progress continued, and by 2008 the 582,970 base pairs of the DNA chromosome of *Mycoplasma genitalium* could be synthesized, allowing the creation of what has been termed "artificial life"—although in order for the genes so created to be expressed and replicated it is still necessary to insert this DNA into a host cell. Given this, it can be argued that the synthetic DNA is in fact acting as a giant virus.

From the above it can be seen that, so long as the DNA sequence is known, it is now possible to bypass traditional techniques entirely. With the synthesis of a virus genome reported in 2002 and that of a bacterial genome reported in 2008, larger and more complex genomes are certain to follow. This, of course, raises concerns since the genome sequences of many pathogenic organisms are readily available and the potential exists to produce them using standard techniques. Ways to resolve the many aspects of this issue are now under active discussion.

In addition to the above, there are a number of more esoteric uses of DNA, for example its use as a structural material in nanotechnological applications.

9.4 PERSONALIZED MEDICINE

Personalized medicine was one of the great early promises of genomics. Using genomic information it was to be possible to look at the genes of the individual patient, and to determine by that analysis not only whether they have any genes associated with disease, but which drugs would work best for that patient, and which might not work at all.

Certainly, it has been known for a very long time that the responses of individual patients to specific medicines are highly variable. This is also true of viruses. For example, some forms of the HLA B35 antigen carried by people of European origin are associated with more severe forms of hantavirus disease (see Section 4.7), while mutations in the CCR5 protein that acts as a second receptor for the human immunodeficiency virus (see Chapter 3) can delay the development of disease. Thus, as with a wide range of diseases it is quite possible that genomic data for individuals could inform them of their susceptibility to some (but by no means all) virus infections.

It is clear that genomics and its related sciences, necessarily matched with and dependent on developments in computing power and in bioinformatics to handle the huge quantities of data generated, will provide the basis for much of future medicine. It is perhaps unsurprising that this fusion of biotechnology and information technology holds such promise, but it will need a great deal of work to realize it in full.

Key Concepts

- Work with viruses underlies most of modern molecular biology and is still an important part of such work.
- Viral promoters and regulatory sequences are commonly used, and viral vectors provide valuable systems for genetic manipulation.
- Early procedures and technologies for cell-based cloning and dideoxy sequencing have been supplemented by more powerful approaches, such as the direct synthesis of genes of interest rather than their isolation and extraction.
- The increasing capacity of sequencing technologies is underpinning a huge expansion in knowledge at this most fundamental level.
- While it is now possible to generate artificial life at both the viral and the cellular level, it is essential to balance this and other such technologies with public acceptance and understanding.

DEPTH OF UNDERSTANDING QUESTIONS

Hints to the answers are given at http://www.garlandscience.com/viruses

Question 9.1: Why are viruses used as vectors in genetic manipulation?

Question 9.2: Why does genome sequence data not tell us all we need to know about a particular virus?

Question 9.3: Should sequence data for dangerous viruses be published?

FURTHER READING

Cello J, Paul AV & Wimmer E (2002) Chemical synthesis of poliovirus cDNA: generation of infectious virus in the absence of natural template. *Science* 297, 1016–1018.

Lodge J, Lund P & Minchin S (2007) Gene Cloning: Principles and Applications. Taylor & Francis, New York.

Molla A, Paul AV & Wimmer E (1991) Cell-free, de novo synthesis of poliovirus. *Science* 254, 1647–1651.

Pevsner J (2009) Bioinformatics and Functional Genomics. Wiley-Blackwell, New Jersey.

Shen Y & Post L (2007) Viral vectors and their applications. In Fields Virology, 5th ed. (DM Knipe, PM Howley eds). Lippincott Williams & Wilkins, Philadelphia.

Wooley JC, Godzik A & Friedberg I (2010) A primer on metagenomics. *PLoS Comput. Biol.* 6, e1000667.

INTERNET RESOURCES

Much information on the internet is of variable quality. For validated information, PubMed (http://www.ncbi.nlm.nih.gov/pubmed/) is extremely useful.

Please note that URL addresses may change.

Entrez protein database. http://www.ncbi.nlm.nih.gov/sites/entrez?db=protein

GenBank sequence database. http://www.ncbi.nlm.nih.gov/Genbank/index.html

RCSB Protein Data Bank. http://www.rcsb.org/pdb/home/home.do

The EMBL Nucleotide Sequence Database. http://www.ebi.ac.uk/embl/

CHAPTER 10
Culture, Detection, and Diagnosis

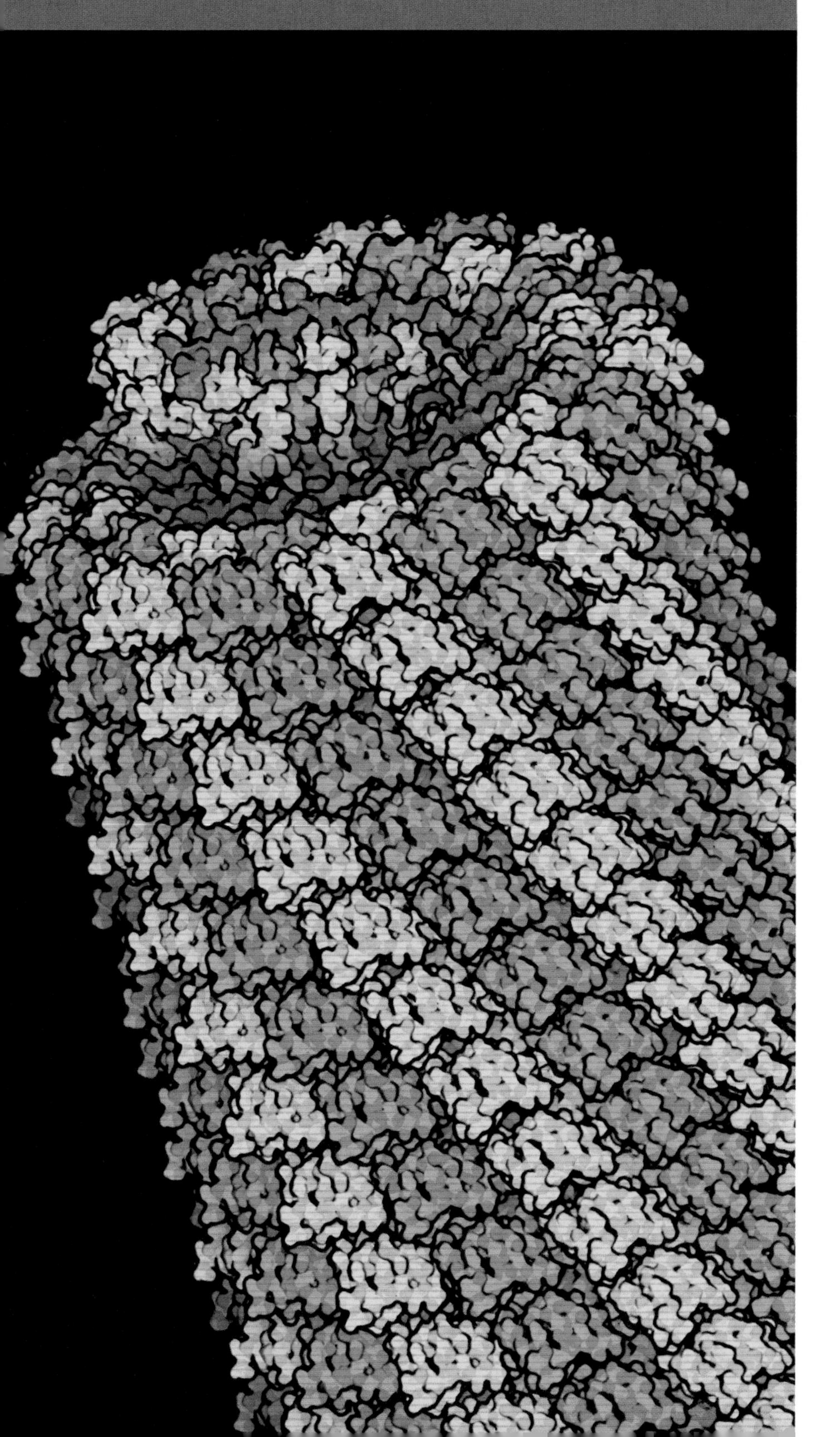

INTRODUCTION

The ability to detect viruses has always evolved alongside the increasing understanding of the underlying processes of disease and, more recently, has undergone radical changes reflecting the development of molecular biological techniques. In particular, the ability to detect specific proteins by the use of monoclonal antibodies and the ability to detect and even amplify nucleic acids by the use of sequence-specific probes have revolutionized diagnostic virology.

About the chapter opener image
Tobacco Mosaic Virus
(Courtesy of the Research Collaboratory for Structural Bioinformatics Protein Data Bank and David S. Goodsell, The Scripps Research Institute, USA.)

Box 10.1 A hepatitis alphabet

The effects of classifying viral disease due to symptoms are still with us in the varying hepatitis viruses. All are associated with damaging liver infection (hepatitis) with its associated symptoms of jaundice, abdominal pain, nausea, *et seq*. Five hepatitis viruses (A–E) have been classified and two more (F and G) have also been suggested. But despite the similarity in names, the hepatitis viruses are almost completely unrelated.

Hepatitis A has a positive-sense RNA genome and is a member of the *Picornaviridae*.

Hepatitis B has a DNA genome (but replicates via an RNA intermediate) and is a member of the *Hepadnaviridae*.

Hepatitis C virus is a negative-sense RNA genome and member of the *Flaviviridae*.

Hepatitis D is not actually a virus, but most closely resembles the viroids of plants.

Hepatitis E has a positive-sense RNA genome, but has a family all to itself, the *Hepeviridae*.

The earliest method of identifying a virus was by the observation of the clinical effects of the infection. This led to some odd situations, for example the classification of smallpox (*Poxviridae*), chickenpox (*Herpesviridae*), and the great pox (bacterial syphilis) as a related group of diseases due to similarities in their symptoms. The hepatitis viruses share similar symptoms but are, in fact, only distantly related (**Box 10.1**).

As described in Box 1.1, the existence of viruses was first recognized late in the nineteenth century, based on their ability to pass through filters that would stop bacteria. But even the expanding germ theory of disease did not explain viruses, since no "little animalcules" could be seen to explain the nature of the agent. Robert Hooke's enthusiastic statement made in 1665 about the power of the light microscope that "by the help of microscopes, there is nothing too small, as to escape our inquiry" was proven to be rather too optimistic. Given this inability to understand their nature, viruses were often thought of as toxins or a *contagium vivum fluidum*, rather than infectious agents. It was not until the development of the far more powerful electron microscope in the 1930s that the nature of these elusive agents could be confirmed.

Electron microscopy (**Figure 10.1**) formed the basis of virus classification (see Chapter 2) but has largely been superseded as a method of detection by molecular techniques. The history of this technique and also more recent refinements are described in Section 10.1.

The culture of viruses (Section 10.3) was for a long time the gold standard for virus detection but was slow and needed a high level of expertise. Again, molecular techniques, sometimes in conjunction with virus culture, have taken over as methods of choice.

The broad range of diagnostic techniques in use is summarized in **Table 10.1**, moving from those used to establish diagnostic virology through to more modern molecular techniques.

10.1 ELECTRON MICROSCOPY

With a few exceptions (and consequently fuzzy images) at the very top of the size range, viruses cannot be visualized with the light microscope. As outlined in Chapter 1, the development of electron microscopy allowed some of the earliest identification of viral agents. By the early 1940s Helmut Ruska, the pioneer of this instrument, had published a range of reports on the visualization of viruses. In 1943 he proposed that viruses should be classified by their morphology (thus by a physical property of the virus itself) rather than by the nature of their host or of the disease syndromes associated with them. This was the first step toward classifying viruses by their nature rather than their effects, and elements of this system are still in use in today's taxonomical system (see Figures 1.3 and 2.5) alongside genetic and other information.

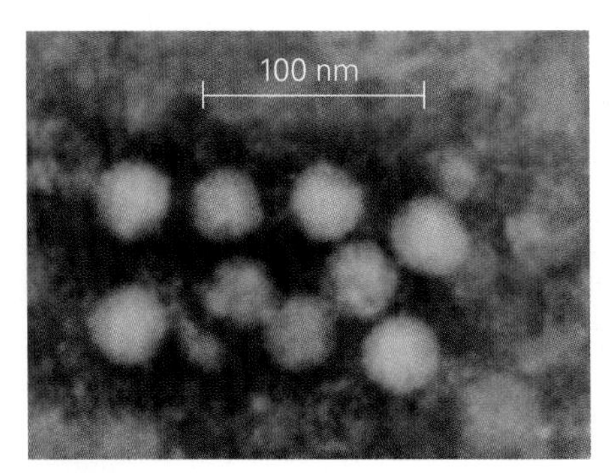

Figure 10.1
Otherwise unidentifiable "small round viruses" observed in stool specimen by electron microscopy.

Diagnostic electron microscopy

Despite its limitations as a diagnostic tool (Table 10.1), electron microscopy is still in routine use for virus diagnosis, although because of the small volumes that are examined the procedure is limited to specimens where high levels of viruses are present, or to where these can be concentrated. Normally the transmission electron microscope is used, along with heavy metal stains such as uranyl acetate or phosphotungstic acid to enhance the image, but other instruments such as the scanning electron microscope may sometimes be used. It is possible to amplify viruses by infection of cultured cells as was done with the initial identification of the SARS

Table 10.1 Commonly used techniques in virus diagnosis

Technique	Specimen types	Advantages	Disadvantages
Antibody titration (ELISA, RIA, IF, CFT, Western blot, many other techniques)	Serum, CSF	Rapid, applicable to almost all viruses, inexpensive, can detect prior (cleared) infection or demonstrate markers of immunity—detection of IgM indicates recent infection	Identifies only immune response to virus infection
Virus culture	Blood (buffy coat), CSF, stool, vesicle fluid, BAL, NPA, throat washings, urine, swabs	Detects multiple viruses (depending on host cell used), traditionally the "gold standard" for virus detection	Can be very slow, needs expert technicians, relatively expensive, specimen transport conditions are critical, only detects infectious virus, many viruses will not grow in cell culture
Accelerated virus culture (centrifugation onto cell sheet)	As virus culture	As virus culture, but faster	As virus culture, but additional labor and equipment required
Culture plus immunodetection or nucleic acid detection	As virus culture	As virus culture, but much faster and highly specific without need for expert observation	As virus culture, but additional labor and equipment required along with specific immunological reagents or nucleic acid probes
Culture plus reporter cell lines	As virus culture	As virus culture, but much faster and highly specific without need for expert observation	As virus culture, but requires specific genetically modified cell lines (limited range available)
Electron microscopy	Vesicle fluid, stool, cultured virus, skin scrapings, (urine if concentrated)	Rapid, can identify general type of virus	Only usable where virus is present at high concentrations, cannot discriminate between morphologically similar viruses or within virus families, expensive equipment, requirement for expert technicians
Electron microscopy plus immunodetection or hybridization	As electron microscopy	As electron microscopy, but can also discriminate between similar viruses if reagents available	Only usable where virus is present at high concentrations, need for specific immunological reagents, expensive equipment, requirement for expert technicians
Cytology	Biopsy or tissue specimens, vesicle fluid, skin scrapings, blood (buffy coat)	Rapid, can show infected cell type (often by use of specific stains), demonstrates active infection	Relatively insensitive, labor-intensive, not applicable to many viruses, requires expert technician
Immunocytochemistry or immunofluorescent assay	As cytology	Rapid, can show infected cell type, applicable to many viruses	Labor-intensive, requires specific antisera, may require specialized equipment (immunofluorescence)
Direct detection of nucleic acid	Biopsy specimens, vesicle fluid, stool, cultured virus, skin scrapings, (urine if concentrated)	Rapid, can show infected cell type, applicable to many viruses, may be highly sensitive (branched DNA)	Labor-intensive, may require specific (labeled) nucleic acid probe derived from sequence data, may require specialized equipment
Nucleic acid amplification	Any specimen	Rapid, extremely sensitive, applicable to many viruses	Labor-intensive and technically complex, requires expensive equipment, requires specific nucleic acid probe derived from sequence data, good controls needed to avoid false positives, may detect insignificant viruses, RNA detection may require RT step (for RNA viruses)
Real-time PCR	Any specimen	Rapid, extremely sensitive, applicable to many viruses, quantitative, reduced risk of contamination	Labor-intensive and technically very complex, requires expensive equipment, requires specific nucleic acid probe derived from sequence data, controls needed to avoid false positives, may detect insignificant viruses, RNA detection requires RT step

Buffy coat is white blood cell layer from centrifuged blood specimens. BAL, bronchoalveolar lavage; CFT, complement fixation test; CSF, cerebrospinal fluid; ELISA, enzyme-linked immunosorbent assay; IF, immunofluorescence; ISH, *in situ* hybridization; NASBA, nucleic acid sequence-based amplification; NPA, nasopharyngeal aspirate; PCR, polymerase chain reaction; RIA, radioimmunoassay; RT, reverse transcriptase.

coronavirus by electron microscopy, but this requires live virus and permissive cells, adding the limitations of culture-based systems to those of electron microscopy.

One primary use is the examination of stool specimens, where large number of unculturable and otherwise unidentifiable viruses may be seen (see Figure 10.1). Even today's improved versions of the electron microscope may be unable to resolve sufficient morphological characteristics to differentiate these viruses. However, the electron microscope does have unique abilities, and is able to identify viruses without any prior knowledge of their structure or characteristics.

A commonly used enhancement of electron microscopy is the use of monoclonal antibodies tagged with small gold particles that show up as black dots on the electron micrograph since they block the electron beam. By their specificity for protein targets at the molecular level, the correct antibodies can allow the typing of morphologically identical viruses, bypassing this limitation of basic electron microscopy. However, this does require monoclonal antibodies for the target to be available, so cannot be used to identify novel agents, and an alternative approach using gold-labeled nucleic acid probes has been developed.

10.2 CYTOLOGY

Cytology is the direct examination of cells within clinical specimens. In some (but by no means all) cases, virus infections can produce characteristic effects that can be seen using a light microscope. The use of specific stains may assist in identifying these. Examples include the staining of acetone-fixed cells from lesions to identify the multinucleate giant cells formed by herpesvirus-induced fusion (**Figure 10.2A**), or the characteristic inclusions formed by some viruses within the host cell (**Figure 10.2B**).

It is also possible to enhance detection by the use of monoclonal antibodies (immunocytochemistry) or nucleic acid probes to detect viral proteins or nucleic acids, respectively. The initial probe may be labeled to allow detection, or a second-stage binding may be used allowing for signal

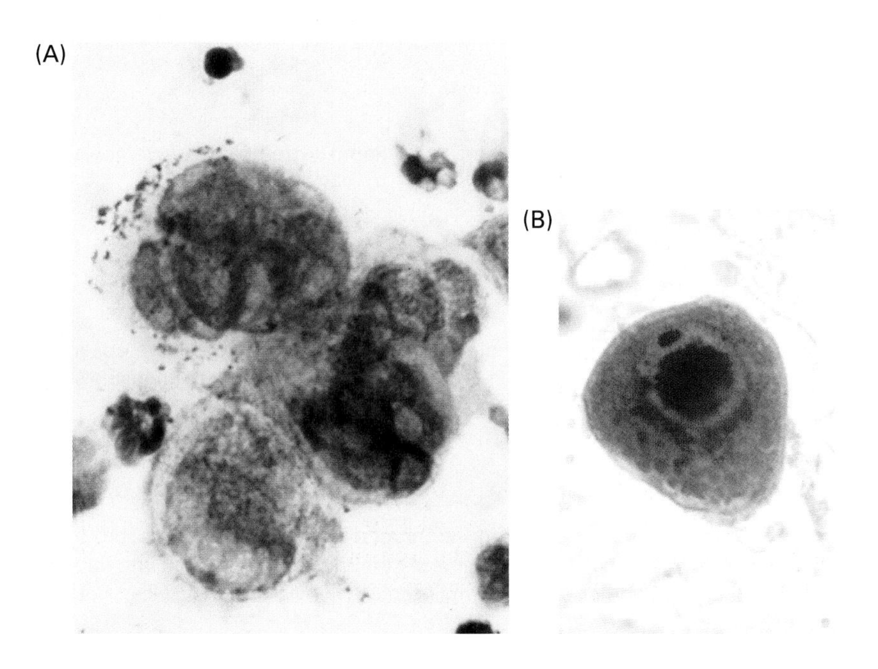

Figure 10.2
Cytology. (A) Tzanck smear (stained, acetone-treated cells) of herpes simplex virus—infected cells from a genital lesion, showing multinucleate giant cell (syncytium). (B) "Owl eye" intranuclear inclusion in a cell infected with cytomegalovirus. From, CDC Public Health Image Library (http://phil.cdc.gov/).

(A) Reporter binding systems

Target

DIRECT
(one stage)

INDIRECT
(two stage)

Second-stage
binding allows
amplification

Reporter

Reporter

(B) Reporter detection systems

Chromogenic	Luminescent	Fluorescent	Radiometric
Detected by microscopy or automated reader	Detected by film exposure or automated reader	Detected by UV microscopy or automated reader	Detected by film exposure or automated reader

Colorless substrate / Colored product / Enzyme / Probe

LIGHT / Substrate / Luminescent product / Enzyme / Probe

UV / Fluorescence / Fluorophore / Fluorescence / Probe

Radiation / Probe

Figure 10.3
Reporter systems for use in detection of bound antibodies or nucleic acids.
(A) Reporter binding systems. For proteins, reporters may be by antibodies or another protein affinity system (e.g. avidin/biotin). Similar approaches are used for nucleic acids, where reporters may be sequence-specific or employ an affinity system.
(B) Reporter detection systems.
UV, ultraviolet.

amplification. Commonly used systems are shown in **Figure 10.3**, and an example of immunofluorescent detection of a virus in **Figure 10.4**. All may be detected directly (often using a film for luminescent or radiometric systems). It should be noted that radiometric detection is far less widely used than it was previously due to the development of alternative detection systems that do not have the cost and safety penalties of working with radioisotopes.

An increasingly common variation on the use of fluorescent probes is the expression of fluorescent proteins such as the green fluorescent protein of the jellyfish *Aequorea victoria* under the control of a relevant promoter. This can mark cells where this promoter is in use, for example by an infecting virus.

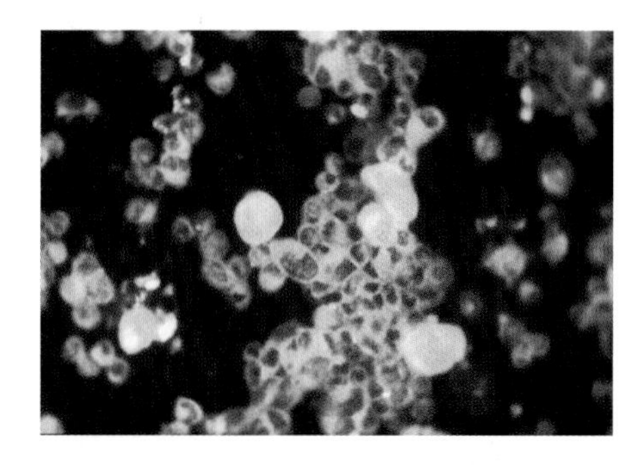

Figure 10.4
Immunofluorescent detection of cytomegalovirus antigens. From CDC Public Health Image Library (http://phil.cdc.gov/).

10.3 VIRUS CULTURE

For many years, growth of a virus from a specimen taken from a patient represented the "gold standard" of diagnostic virology. In part, this arose from **Koch's postulates**, which outlined the requirements to conclude that a microorganism had a causative role in an observed disease:

1. The microorganism must be found in abundance in all organisms suffering from the disease, but should not be found in healthy animals.
2. The microorganism must be isolated from a diseased organism and grown in pure culture.
3. The cultured microorganism should cause disease when introduced into a healthy organism.
4. The microorganism must be re-isolated from the inoculated, diseased experimental host and identified as being identical to the original specific causative agent.

At the time that Koch and Loeffler first defined the postulates (1884–1890), no virus of any type could be cultured. Koch himself also recognized that postulate #1 did not apply in the case of asymptomatic carriers of cholera, and we now realize that there are specific cases where each postulate is invalid.

Indeed, in some cases very few of the postulates apply. For example, hantavirus pulmonary syndrome (HPS) is caused by several different hantaviruses (see Chapter 8). The actual pulmonary disease occurs due to a massive immune response, typically occurring after the virus is cleared from the lungs (breaching postulate #1). In addition, disease only appears to result from exposure in about 10% of cases, with the risk of disease being related to the HLA type of the human host (breaching postulate #3). Sin Nombre virus, the first identified virus associated with HPS, remains very difficult to culture (postulate #2 looking problematical) and was identified entirely by molecular means.

Over one hundred years after the postulates were first proposed, it is now clear that they do not allow sufficiently for the variability inherent in biological systems. Despite this, for many years there was a strong tendency to regard the ability to isolate and grow a virus as the key step in identification.

It is now recognized that many viruses are difficult to culture, and indeed that many cannot be cultured at all with current techniques, even when they are observed in association with disease. In some cases, this is due to requirements for specific cell types, such as papillomaviruses only growing in differentiated keratinocytes, which are extremely difficult to culture. In other cases, there is a need for specific additions to the culture media, as with astroviruses and many orthomyxoviruses which require an active proteinase to be present in the medium. This then prevents the use of serum (which contains proteinase inhibitors) as a growth supplement, making cell culture much more complex, although other supplements such as allantoic fluid from embryonated eggs may be used instead.

Traditionally, it has been impossible to work out what specific growth requirements might apply for a virus that has not been cultured, except by extension from what works for similar viruses. However, with an increasing number of viruses from hepatitis C onward being identified by molecular biology, this may no longer always be the case since interpretation of the genomic data may in time indicate some specific growth requirements. However, it is a very long way from a nucleic acid sequence to even a single functional protein, and this is more a potential than an actual approach at present.

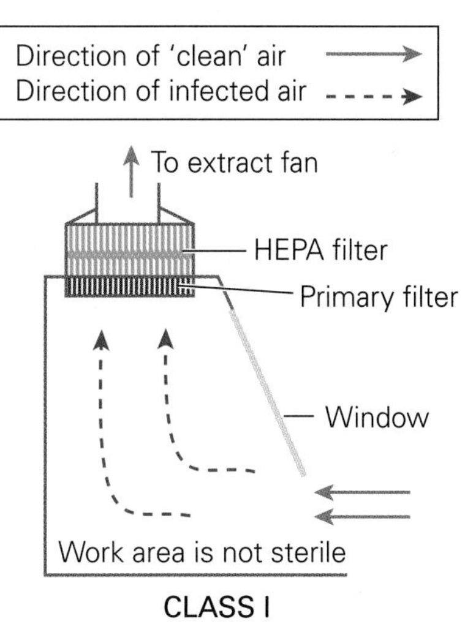

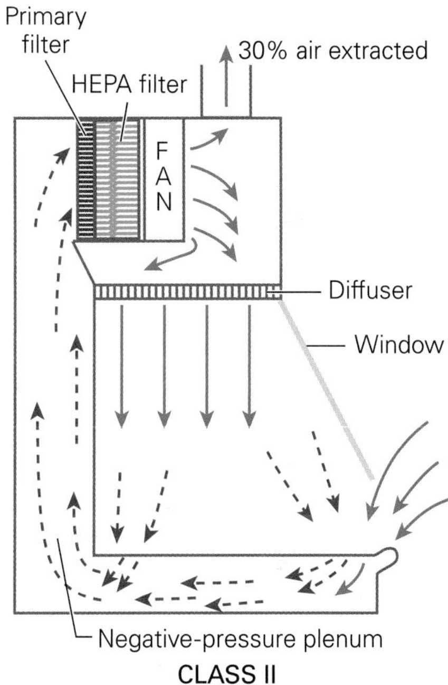

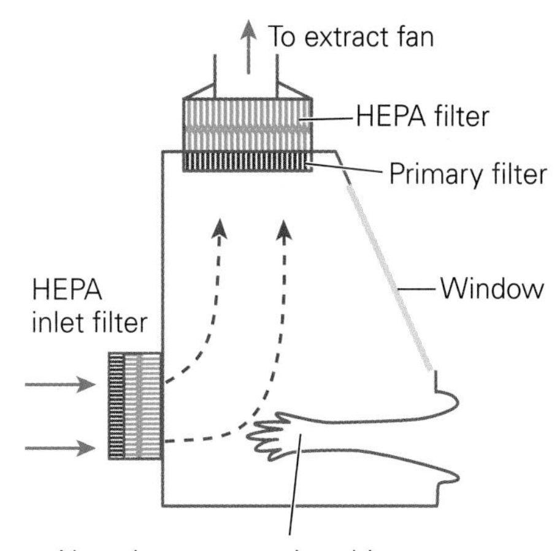

Figure 10.5
Types of biological safety cabinet. From, Collins CH (1988) Laboratory-Acquired Infections, 2nd ed. With permission from Elsevier. HEPA, high-efficiency particulate air.

Containment

When working with live viruses, it is necessary to consider their possible effects on health and the environment, and to work with them under appropriate levels of biosafety and with appropriate equipment, including appropriate biosafety cabinets where required (**Figure 10.5**).

There are four general levels.

Biosafety level 1: for work involving well-characterized agents not known to consistently cause disease in healthy adult humans, and of minimal potential hazard to laboratory personnel and the environment. Much of the work may be done on the open bench.

Biosafety level 2: for work involving agents of moderate potential hazard to personnel (and to which they may have already been exposed in the community) and the environment. It includes various viruses that cause only mild disease to humans, or are difficult to contract in the laboratory setting, such as hepatitis A, some forms of influenza, many herpesviruses, mumps, and measles. Genetically modified organisms are often given a base level 2 classification to limit the release of modified organisms into the environment. Laboratories have restricted access and special facilities for disinfection and disposal of equipment. Work likely to generate aerosols is carried out in biological safety cabinets.

Biosafety level 3: for work with infectious agents which may cause serious or potentially lethal diseases as a result of exposure by the inhalation route. Viruses worked with under category 3 include HIV, encephalitis viruses, and yellow fever virus. Laboratories have double entry doors and are sealed to prevent fluid leaks. They have an inflow of air to prevent release of aerosols. Very strict safety precautions are applied and any activity likely to generate aerosols is carried out in a biological safety cabinet. Waste material is routinely sterilized by autoclaving.

Biosafety level 4: this is the highest level, and very few such laboratories exist. All precautions are additional to those used in biosafety level 3. Biosafety level 4 is for work with dangerous and exotic agents that pose a high individual risk of aerosol-transmitted laboratory infections, and which cause severe to fatal disease in humans for which vaccines or other treatments are not available, such as Marburg and Ebola viruses and Lassa fever virus. In the United States, sealed hazardous materials (Hazmat) protective suits are used with a self-contained oxygen supply. In the United Kingdom a different approach is taken, and all material is handled inside sealed class III biological safety cabinets. The entrance and exit of a level 4 laboratory will contain multiple sterilizing procedures and barriers and the facility itself will often be in a specialized, isolated location.

Effects on economically or ecologically important animals or plants are also considered when assigning viruses and other agents to appropriate biosafety levels.

Routine diagnostic virology typically takes place within biosafety level 2.

IMPORTANT: Please note that the above is for information purposes only and does not constitute a full summary of appropriate procedures and safety measures. Before carrying out any such work the reader is referred to specific advice from local authorities such as the Centers for Disease Control and Prevention (USA) or the Health Protection Agency (UK) and to appropriate statutory regulatory bodies.

Table 10.2 Sources of cultured cells

Type of culture	Source	Comments
Primary	Cells taken directly from host organism	Usually mixed culture, may show wide variation. Adherent[1] and usually contact inhibited.[2] May be required for some virus types
Secondary	Primary cultures	Generated by passaging[3] from primary cultures, less heterogeneous than primary cultures. Adherent and usually contact inhibited.
Diploid	Usually from banks of cells held by specialist units and/or collections	Similar to secondary cultures, normal chromosome numbers. Adherent[1] and usually contact inhibited.[2] Most will only sustain a limited number of passages before becoming senescent (approximately 50 for human cells)
Continuous (transformed)	Cancer cells, cells transformed by viruses or chemical agents. May be new isolates or from collections	May have unusual chromosome numbers, may not show contact inhibition[2] and overgrow each other, may grow in suspension (non-adherent), may require fewer complex growth factors, do not usually become senescent (unlimited passages)
Hematopoietic	Blood cells	Grow in suspension, usually primary cells. Required for some virus types

[1] Adherent: grow by adhering to solid surface (flask, tube, or bioreactor). [2] Contact inhibited: grow until touching other cells then cease, forming a sheet one-cell thick (a monolayer). [3] Passaging: dissolution of cell sheet using proteinases and chelators, dilution in fresh medium, and seeding into new culture vessels.

Growing and counting viruses

In order to grow viruses, permissive host cells are required. The basic sources of cells for culture are shown in **Table 10.2**. While cell cultures were originally (and may still be) derived directly from clinical specimens or from animal sources, the need for standardization (particularly in a diagnostic setting) means that most will be obtained from collections such as the American Type Culture Collection (ATCC) in the USA and the Health Protection Agency Culture Collections in Europe. These culture collections will provide details of the necessary culture conditions and types of media that should be used. Similarly, viruses provided by such culture collections will be supplied with details of suitable host cells and culture conditions in order to allow productive virus infections (**Box 10.2**).

As well as its role in virus diagnosis, it should be remembered that virus culture is the basic technology underlying most scientific work with viruses. It is necessary both to grow viruses in order to be able to work with them, and to be able to study their effects on their host cells.

It is also widely used to enumerate the amount of virus present. While many techniques may be used for this, including particle counts in the electron microscope, protein or nucleic acid quantitation, or assay of virus functions such as hemagglutination (see below), the basic technique for counting infectious virus particles has been and remains the plaque assay (see **Box 10.3** for details).

Box 10.2 The basics of cell culture

In all cases, cells must be grown aseptically, often with the use of antibiotics to prevent bacterial contamination (although other agents such as mycoplasmas and fungi will not be inhibited). Cells are grown in flasks or tubes (bioreactors are not usually used in the diagnostic laboratory) bathed in a nutrient medium at an appropriate temperature. This is typically 32 to 37°C for mammalian cell cultures. The basic medium will contain nutrients, salts, and usually a pH indicator. A very wide range of media are available and are suitable for specific types of cells. Despite many efforts to use fully defined media, biological supplements such as calf serum or the more expensive fetal calf serum are often used to enable cell growth. Light is usually kept to a minimum, and in many cases the pH of the medium is stabilized by the use of an atmosphere supplemented to 5% carbon dioxide, although buffering systems are available to avoid the need for this.

Diagnostic virus culture

Until the advent of monoclonal antibodies and nucleic acid probes, virus culture was the core technique of the diagnostic virology laboratory. However, many viruses will not grow at all in such systems. Among those that do grow, some produce observable effects in hours or days, but others can take weeks to produce apparent effects in culture, leading to the often-repeated claim that the virology lab report would often arrive after the patient had gone home or had died.

Box 10.3 Plaque assay—the counting of viruses

Where a host cell is killed by virus infection, the sequential process of infection, cell death, release of virus, infection of neighboring cells, cell death, release, in a continuing cycle will result in a zone of dead cells and often a physical hole in the cell monolayer. This is known as a plaque.

To count infectious viruses, varying dilutions of a virus preparation are allowed to infect cultured cell monolayers (sometimes called cell sheets) and these are overlaid with an appropriate medium (often thickened with a gelling agent such as agar or the more refined agarose in order to localize virus spread). These are then incubated for an appropriate length of time (since viruses kill cells at differing rates).

The formation of plaques can be observed through a microscope, and at an appropriate time point (when the plaque is large enough to be visible but has not yet merged with neighboring plaques) the medium is removed and the cell monolayer stained with a vital stain that shows only live cells. These stains, such as neutral red or crystal violet, will produce a characteristic perforated appearance (see **Figure 1**).

The individual plaques are then counted and, by cross-referencing with the dilution used for each monolayer, the number of infectious virus units (often termed plaque-forming units, or PFU) in the original preparation can be calculated.

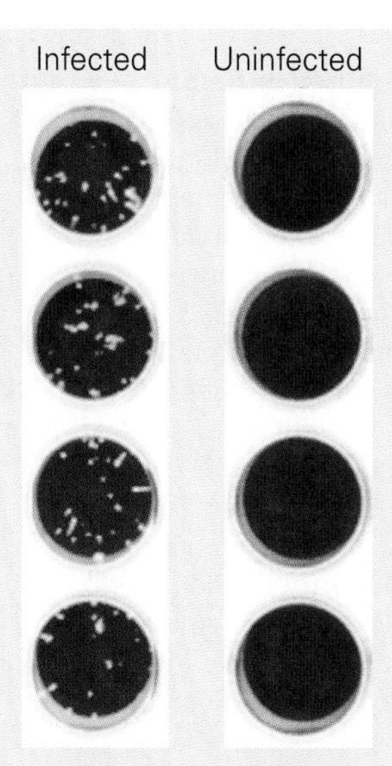

Figure 1
Plaque-forming unit (PFU) assay. Plaque formation by herpes simplex virus in Vero monkey kidney cells stained with a vital stain (crystal violet). Plaques formed by destruction of the monolayer are the clear areas in the infected wells. Courtesy of S. Argent, Department of Virology, Barts and The London School of Medicine and Dentistry, London. From, Harper D (2001) Viral culture methodologies. In Encyclopedia of Life Sciences. With permission from John Wiley & Sons, Inc.

Another concern was that the virus in the sample had to be viable when it reached the diagnostic laboratory. In large hospitals with virology laboratories on site, this was not usually a problem. However, for smaller facilities where transport off-site was required, great care had to be taken to ensure that viruses in the specimen would not deteriorate and thus become undetectable. This often necessitated special procedures, such as the provision of specific types of swab and their immersion in specialized "viral transport medium" as well as accelerated and/or temperature-controlled delivery of certain types of specimen.

In order to carry out diagnostic virus culture, specimens taken from patients are inoculated into cell cultures that are known to be able to support the growth of viruses thought likely to be present. A swab from a suspected shingles vesicle might be inoculated into human fibroblast cultures, for example, while a specimen thought to contain influenza virus might be put into canine kidney cells. None of these cell types are those infected naturally, and in the case of the virus responsible for shingles, varicella-zoster virus (VZV; *Herpesviridae*), it will actually grow very inefficiently in such cultures—but it will grow. Some specimen types are by their nature usually sterile, for example cerebrospinal fluid. Others are definitely not—stool specimens will contain both bacteria and viruses and will need both clarification (often by centrifugation) and the use of antibiotics to prevent growth of bacteria.

Traditionally, the effect of a virus on cell culture was monitored at frequent intervals by a skilled technician who would observe characteristic changes in the cells, usually in the form of cell killing (cytopathic effect), detachment from the growth substrate (rounding), the presence of reflective inclusion bodies within the cells, or the formation of giant cells (syncytia) caused

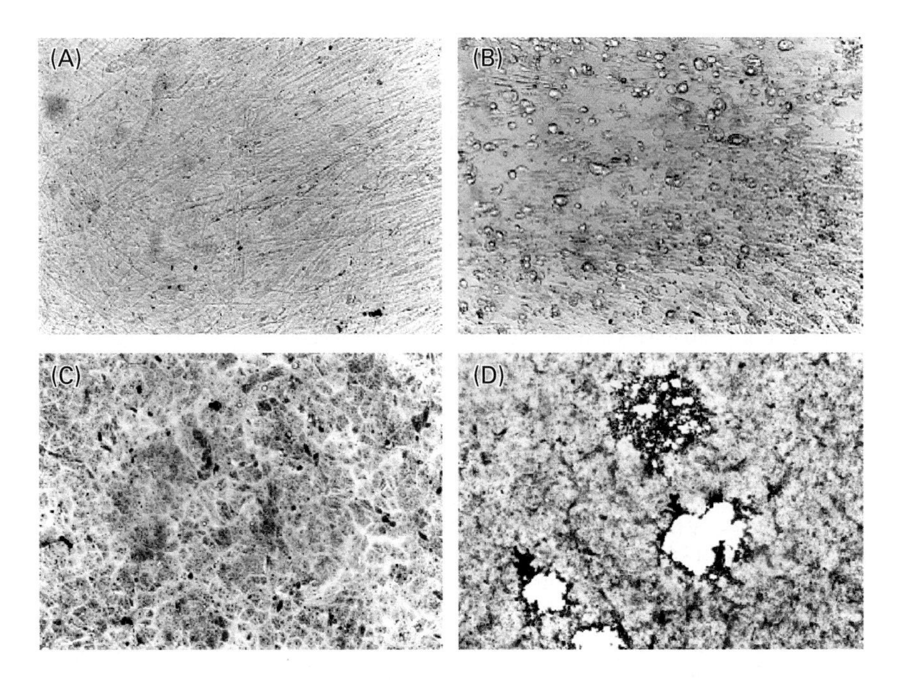

Figure 10.6
Cytopathic effect caused by herpes simplex virus in different cell types. (A) Uninfected human embryo lung (HEL) cells. (B) HEL cells infected with herpes simplex virus (HSV) showing rounding of cells. (C) Uninfected Vero monkey kidney cells. (D) Vero cells infected with HSV, showing formation of plaque by destruction of cell monolayer (see Box 10.3). Figure also shows different appearance of fibroblastic (HEL) and epithelial (Vero) cell types. Courtesy of S Argent, Department of Virology, Barts and The London School of Medicine and Dentistry, London. From, Harper D (2001) Viral culture methodologies. In Encyclopedia of Life Sciences. With permission from John Wiley & Sons, Inc.

by virus-induced fusion (see Figure 10.2 and **Figure 10.6**). These changes could sometimes be made easier to detect by staining, but once a culture is stained the cells are killed and the virus is unable to grow further. Thus, the level of skill required of the technician carrying out the work was high.

For some viruses, innate properties of the infection could be used to speed up detection. With many of the *Paramyxoviridae* (e.g. mumps, parainfluenza) and the *Orthomyxoviridae* (influenza), expression of the viral hemagglutinin (the viral glycoprotein that is named for its ability to bind red blood cells) on the surface of infected cells could be detected by washing the infected cell monolayer with red blood cells then gently rinsing them. The red blood cells adhere to cells with viral hemagglutinin on the surface, producing distinctive pale-red areas. A related technique, the hemagglutination assay, relies on the ability of multivalent hemagglutinin on virus surfaces to form complexes with red blood cells and keep them from settling out of suspension. However, this is normally used to determine the amount of virus produced from culture systems (titration) rather than diagnosis.

Bearing in mind the need for skilled observation, along with the extended time that could be required for these assays [up to several weeks for cytomegalovirus (human herpesvirus 5; *Herpesviridae*) in standard cultures, for example], it is unsurprising that once rapid diagnostic techniques using immunodetection (with monoclonal antibodies) or hybridization (with nucleic acid probes) became available, these were taken up avidly by the vast majority of diagnostic virology laboratories. Such approaches also permitted a far higher level of automation, with consequently reduced costs. The techniques used are similar to those used in cytology (see Figure 10.3).

Alongside detection systems which can reveal viral products in cells long before visible changes occur, efforts were made to speed up the processes of infection. Alongside routine optimization of media and cell types, one approach was to centrifuge the virus-containing sample onto the cells. This approach, known as shell vial culture, can result in significant reductions in the time required for the effects of virus infection to become apparent. When combined with immunodetection or *hybridization* (see Section 10.5), the time required can be cut dramatically. In the case of cytomegalovirus, this could be reduced from several weeks to less than a day.

Another approach is the use of genetically engineered cells expressing reporter genes under the control of viral promoters. One such is the oddly named ELVIS (enzyme-linked virus-inducible system) cell line which contains the β-galactosidase gene from the bacterium *Escherichia coli* under the control of the herpes simplex virus (HSV) U_L39 gene promoter. When these cells are infected with HSV, the gene is expressed and the colorless substrate X-gal is converted to a blue product, marking the infected cells and allowing an extremely rapid identification of the infecting virus. This technique does however use proprietary, genetically modified cells, which can limit its applicability in some laboratories.

10.4 SEROLOGICAL AND IMMUNOLOGICAL ASSAYS

A huge range of immunological markers relating to virus infection can be identified, from skin-test estimation of the allergic response to precise quantitation of the cytokine response. Many have a role in virology, but the main element of the immune response used in such investigations is the production of virus-specific antibodies.

Immunoassays are one of the oldest techniques of diagnostic virology, with detection of antibodies in patient sera being indicative of past (IgG) or recent (IgM) infection with the agent in question as well as of possible immunity. The nature and role of antibodies is covered in Chapter 4, and the range of such tests is illustrated in **Table 10.3**. Viral antigens may also be detected using similar approaches.

The first test to identify antiviral antibodies was the **complement fixation test**, which came into use in 1929. For many years detection of antibody was dependent upon functional activities such as hemagglutination inhibition or particle agglutination (red blood cells or latex particles) as well as the fixation of complement.

This was followed by the development of plate-based assays using viral antigens to capture antibodies from clinical samples, or antibodies to capture viral antigens. The use of polyclonal antisera in such systems results in high levels of background binding and very noisy results. The development of *monoclonal antibodies* (see Chapter 6) made such assays far more useful and supported their widespread adoption.

The production of test antigens for diagnostics has also advanced, based on molecular techniques that can allow antigens to be produced cheaply and efficiently even if a virus is troublesome or impossible to grow in culture. The systems used to prepare and evaluate such antigens are substantially as those for vaccine components (see Chapters 5 and 6), with similar concerns and capabilities.

10.5 NUCLEIC ACID DETECTION AND AMPLIFICATION

Nucleic acid detection systems can identify latent or inactive virus, since the presence of protein products is not required. Even without amplification (see below), such systems can be of comparable or greater sensitivity compared to immunological detection systems.

For the direct detection of viral nucleic acids from clinical specimens, **electrophoresis** can be used to separate nucleic acid fragments on the basis of size and charge (**Figure 10.7**). Where used, this method is often employed to type viruses, either directly as in the case of dsRNA genome segments

Table 10.3 Assays used with antibody-containing clinical samples

Assay	Method	Advantages	Disadvantages
Binding assays (colorimetric, fluorescent, luminescent, radiometric)	Viral antigens attached to a microplate are used to capture antibodies, which are then detected by a second, reporter-bearing antibody	Fast, simple format. Use of correct reporter antibodies can identify types of antibody. Easily automated	Does not identify targets within bound antigen
Hemagglutination inhibition	Measures ability of antibody in sample to interfere with virus-induced agglutination of red blood cells	Fast, simple format	Restricted to hemagglutinating viruses. Limited information on antibody type. Difficult to automate. Not now in widespread use
Complement fixation test	Measures the ability of antiviral antibodies to fix complement, preventing lysis of indicator red blood cells	Fast, simple format	Limited information on antibody type. Difficult to automate. Not now in widespread use
Latex agglutination	Viral antigen bound to particles is agglutinated by antibodies in the sample	Fast, simple format	Does not identify targets within bound antigen. Limited information on antibody type. Quite difficult to automate
Immunoblotting	Proteins separated on a polyacrylamide gel are transferred onto a membrane then reacted with the clinical sample. Bound antibodies are detected by a second, reporter-bearing antibody	Identifies antibodies binding to individual antigens	Complex, antigen is highly denatured and many antibodies will not bind. Very difficult to automate
Radioimmune precipitation	Radiolabeled viral proteins in solution are reacted with the clinical sample, then large antibody-binding proteins are used to precipitate the complexes which are analyzed by electrophoresis	Identifies antibodies binding to individual antigens in a non-denatured state	Very complex, requires radiolabel. Extremely difficult to automate, not used diagnostically
Neutralization assay	Detects ability of antibodies to prevent virus infection or to activate complement-mediated viral inactivation	Detects functional antiviral activity	Extremely complex. Extremely difficult to automate, not used diagnostically

Samples may include serum (from clotted blood), plasma (from unclotted blood), cerebrospinal fluid, or material from any other location where detectable levels of antibody may be present.

purified from rotaviruses (from stool specimens), or by the assay of restriction fragment length polymorphisms (RFLPs) in the genomes of cultured virus. This technique is usually used for viruses with DNA genomes, although DNA copies (cDNAs) of RNA genomes can be analyzed. Restriction endonuclease enzymes are used to cut the DNA at specific sites. The frequency and locations of the recognition sites for each restriction endonuclease in the DNA determines the number and size of DNA fragments generated, which can then be separated by electrophoresis. This method is also referred to as restriction endonuclease analysis (REA). RFLP analysis has been used to type strains of many viruses, and may represent the most convenient means of discriminating between strains of some viruses. However, diagnostic assay of clinical specimens usually requires the additional sensitivity of hybridization- or amplification-based systems (see below).

Hybridization

Many nucleic acid–based detection systems exploit the ability of single-stranded nucleic acids that are complementary in sequence to form

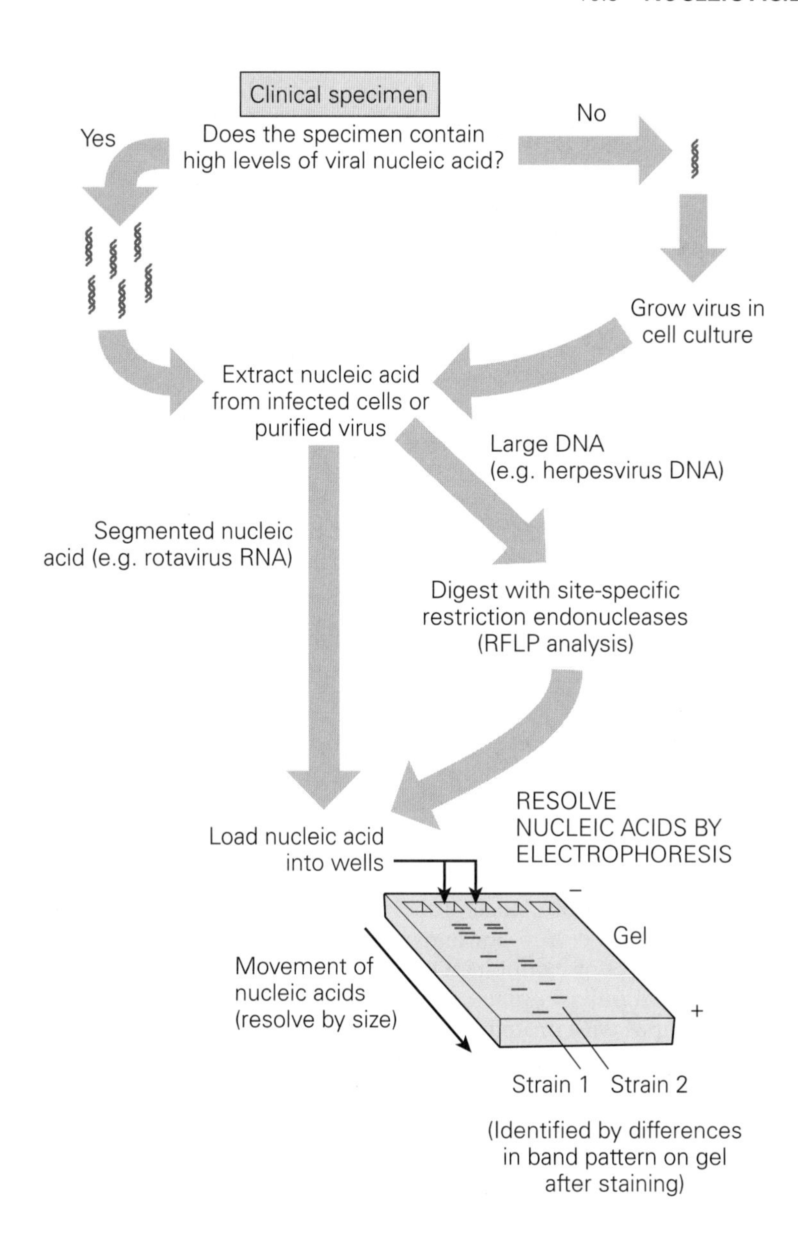

Figure 10.7
Direct detection of viral nucleic acid by electrophoresis. Type of gel used may differ. dsRNA (rotavirus) uses polyacrylamide gels similar to those used for protein separations, while DNA uses agarose gels, with a varying current applied to resolve larger DNA molecules.

double-stranded molecules by base pairing, a process known as **hybridization**. Short, synthetic nucleic acid probes are often used to detect matching sequences in the target nucleic acid.

In DNA, cytosine bases in one strand pair with guanine in the other, and adenine pairs with thymine (or uracil (U) in RNA or in some synthetic DNAs). The cytosine–guanine (GC) pair has three hydrogen bonds compared with the two of an adenine–thymine (AT) pair, and so forms a more stable bond. Thus, more energy (heat or chemical) is needed to disrupt GC bonds than AT bonds. This specificity can be exploited both to separate double-stranded target nucleic acids to enable testing, and to control the rate of hybridization during testing.

Hybridization of target DNA to a complementary probe nucleic acid will be weak if only a few bases match, and far stronger if there is a complete match—every C with its G, every A with its T (or U). So, if the conditions selected for rebinding are relatively harsh (raised temperature or disruptive chemical conditions) only good matches will pair up. The situation is similar to the use of high- and low-affinity antibodies, and allows selection of the best matches.

Hybridization forms the basis of a large number of techniques used in the detection, diagnosis, and analysis of viruses. These include *in situ* hybridization, dot-blot hybridization, DNA microarrays, and nucleic acid amplification.

The nature of the viral genetic material must be considered, since many viruses have RNA genomes, and do not have any DNA stages in their life cycle, necessitating the detection of RNA, often by its conversion into complementary DNA (cDNA).

Probes were originally DNA purified from a biological system containing the desired sequence such as a plasmid containing cloned viral genes. However, more commonly, short probe oligonucleotides (typically up to 50 nucleotides in length) are now synthesized directly from nucleotide monomers. These are often RNA rather than DNA, since nucleases are available which allow the digestion of single-stranded RNA, and thus the simple removal of unbound probe material. Once bound, the probe is detected as described in Figure 10.3B. As with antibody-mediated detection, radiometric systems are now used much less frequently.

Uses include dot blotting, *in situ* hydridization, and branched-chain detection systems.

In **dot blotting**, extracted nucleic acids are attached to a membrane. They are then exposed to the probe before rinsing to remove unbound material and blocking of unused binding sites. An appropriate detection system is then used. This approach is suitable for quantitation of the target nucleic acid, but is relatively insensitive and is only applicable where high levels of viral nucleic acids are available, for example from cultured virus.

For ***in situ* hybridization**, nucleic acid probes are used directly on cells from specimens or from culture systems. These are usually attached to a solid matrix, and fixed (for example, with paraformaldehyde or glutaraldehyde) to hold the specimen in place. They are then treated to permeabilize the cell (for example, with proteinase K or acetone). Probes are then added, followed by rinsing to remove unbound material. The resultant image is similar to that seen with immunofluorescence techniques (Figure 10.4), allowing localization of the target nucleic acid within the specimen and even within individual cells, as well as some level of quantitation.

A more sensitive form of hybridization detection, the **branched DNA system**, uses a probe that has multiple binding sites for a second, reporter probe, enhancing the resulting signal. This is often mediated by the binding of a series of oligonucleotide probes that bridge from the bound, sequence-specific probe to a much larger number of reporter probes. This is broadly similar to the use of indirect systems (Figure 10.3A) such as peroxidase–antiperoxidase (PAP) complexes in protein detection where repeated binding of enzyme–antibody complexes is used to amplify the signal.

Nucleic acid amplification

Amplification-based techniques differ from hybridization-based detection systems in that probes are modified (usually by extension) following binding to the target nucleic acid sequence. This process is then repeated in a series of cycles, typically allowing all nucleic acids produced in one reaction to function as templates for the next, giving a chain reaction that is broadly similar to the amplification produced during a virus infection. Over time and with repeated cycles, this results in an exponential increase in the amount of target-complementary nucleic acid.

With the potential for doubling the amount of nucleic acid product in each cycle (subject to availability of reagents and appropriate binding), each ten cycles can theoretically produce a thousandfold increase, with a 40-cycle reaction thus capable of amplifying the original target a trillion times in a single assay. As with hybridization-based systems, the specificity of the reaction comes from the selective nature of the binding of the short synthetic nucleic acid probes, referred to in this setting as primers, to the target nucleic acid. The large number of copies produced makes detection relatively simple.

The polymerase chain reaction

PCR was the first amplification-based system to be developed, and was developed by Dr Kary Mullis in 1984–1985 while working for the Cetus Corporation in California. As a result, the basic technique of PCR has, unlike most diagnostic techniques, been protected by patent almost since its inception. These patents have been held worldwide by Hoffmann-La Roche since 1992, meaning that the use of the system needs a licence from the company, although some basic patents expired in the United States in 2005.

One effect of this position is that many PCR assays are kit-based, and a wide range of viruses may be detected in this way including retroviruses, herpesviruses, and a variety of hepatitis viruses.

The basic principles of PCR are summarized in **Box 10.4**. However, it should be noted that the conditions (e.g. temperatures, times, ionic type, and concentration present) need to be optimized for each reaction.

In the earliest PCR work, the polymerase used was a fragment of the *Escherichia coli* DNA polymerase (the Klenow fragment) which was destroyed by heat and so needed to be replenished for each cycle. It was only with the use of heat-resistant DNA polymerase isolated from the thermophilic bacterium *Thermophilus aquaticus* (Taq polymerase) that PCR became a practical technique, since one addition of the polymerase at the start of the

Box 10.4 The polymerase chain reaction

An initial heating period (1–9 minutes at 94–96°C) denatures the target DNA and separates the strands. The core cycle of heat–anneal–extend then begins.

Heat (94–96°C for 20–30 s) denatures the strands

Anneal (50–65°C for 20–40 s) allows short oligonucleotide primers (red bars in **Figure 1**) to bind to complementary sequences flanking the target DNA to be amplified (blue)

Extend (70–74°C for 45 s) DNA synthesis occurs from the bound primers

The cycle is repeated 20–35 or more times before a final elongation step of 5–15 minutes at the extension temperature. If working at maximum efficiency and if there is no limitation of reagents, the amount of product will double with every cycle.

All of the above is performed in an electronically controlled thermal cycler, which will then hold the products at a lowered temperature (4–15°C) until they are collected.

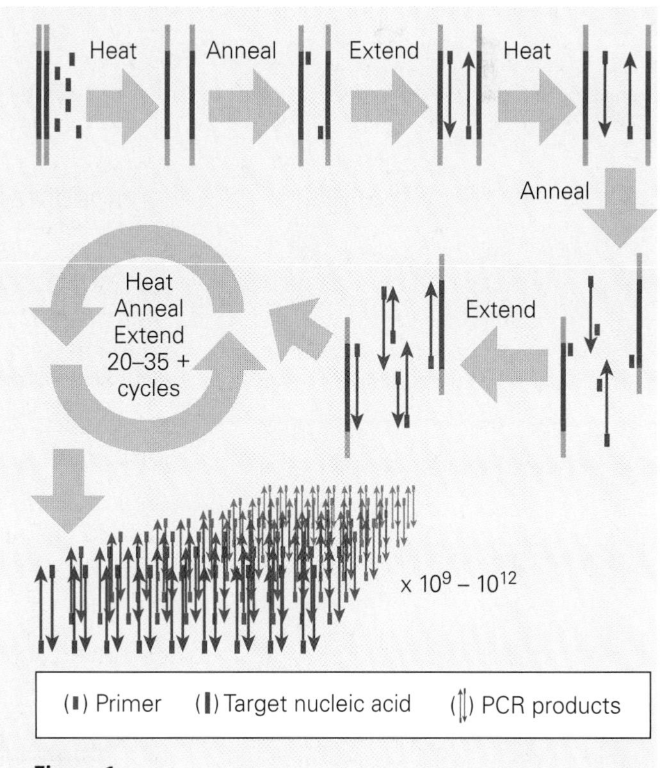

Figure 1
The polymerase chain reaction.

reaction is all that is required. A cloned form of the enzyme (Amplitaq®) further reduced the cost of the technique. This was combined with the use of electronically controlled heating/cooling blocks (thermal cyclers) able to hold a large number of tubes under precisely controlled temperatures, making the procedure both convenient and reproducible.

Taq itself lacks a proofreading (checking and exonuclease) activity, and so it cannot correct errors in the synthesized DNA. The resultant low fidelity (high error rate) can be a problem, particularly when using PCR to generate DNA for cloning. A range of other heat-resistant polymerases are now available, some of which are able to check and remove incorrectly incorporated bases, with a resultant tenfold or more decrease in the errors in product DNA and a consequent increase in maximum product size. Examples include Pfu, Vent™, and Deep Vent™, which also offer greater thermostability, making them more suitable for high melting temperature GC-rich regions. Others offer other desirable properties such as an inherent reverse transcriptase activity (e.g. Tth), allowing the use of RNA as a starting material without additional reverse transcriptase enzymes.

In some applications, mixtures of polymerases are used to optimize yields.

Alternative amplification-based detection systems

Driven by the commercial protection around the original PCR system, alternative systems were developed by other companies.

The **ligase chain reaction** (used in the proprietary LCx system) relied on linking two primers directly, rather than synthesizing DNA between them, but appears no longer to be in routine use.

Nucleic acid sequence-based amplification (**NASBA**), also known as *self-sustained sequence replication* (*3SR*) or *transcription-based amplification* (*TBA*), is shown in **Figure 10.8**, and allows direct detection of RNA without a

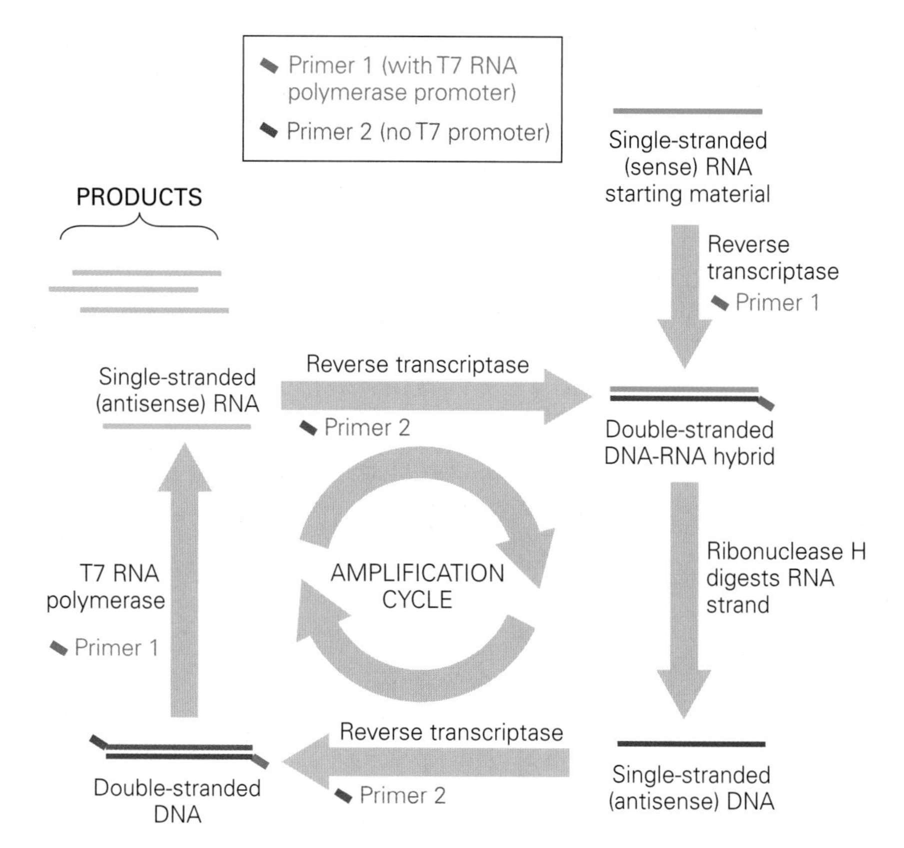

Figure 10.8
Nucleic acid sequence-based amplification (NASBA). A variant of the system with T7 promoters on both primers can produce both sense and antisense RNAs.

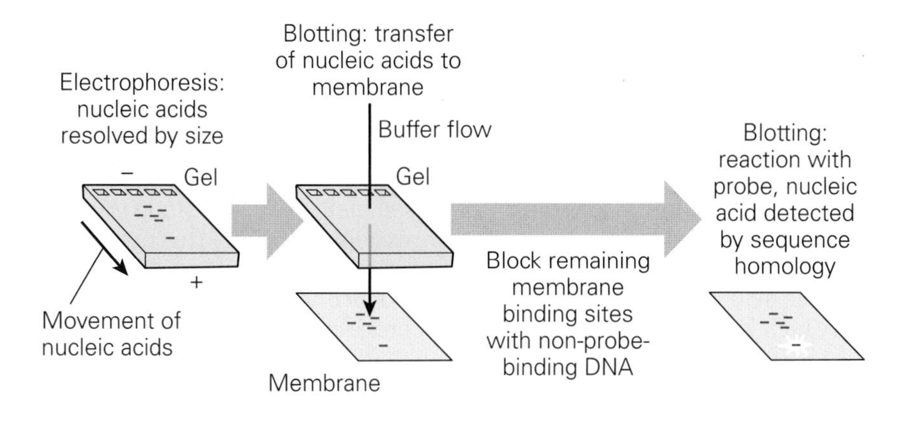

Figure 10.9
Electrophoresis and Southern blotting. DNA fragments separated by size using electrically induced migration through a gel matrix (electrophoresis) are transferred (blotted) onto a membrane which is then reacted with a probe to detect the target DNA. Similar methods exist for RNA (Northern blotting) and proteins (Western blotting, where a current is applied to remove the proteins from the gel matrix). Despite the now-accepted "geographical" naming system, Southern blotting was actually named for its inventor, Ed Southern.

reverse transcriptase step, and diagnostic kits for use with viruses are available, in particular for viruses with RNA genomes (SARS coronavirus, influenza). Unlike PCR, NASBA is isothermal, requiring no heating or cooling. It is typically carried out for 90 to 120 minutes, at 41°C.

Detection of amplified nucleic acids

The original method for the detection of amplified DNA from PCR was on the basis of size by electrophoresis. This is labor-intensive and provides no information on the PCR product other than its size(s). As a result, it was often combined with **Southern blotting**, where the resolved nucleic acids were transferred to a membrane and reacted with nucleic acid probes specific for the desired PCR product (**Figure 10.9**). This was and is very labor-intensive and time-consuming, and not well suited to diagnostic use.

Subsequently, microplate systems based on specific hybridization were developed which avoided the need for complex and time-consuming electrophoretic procedures (**Figure 10.10**), and were thus more suited both to diagnostic use and to automation.

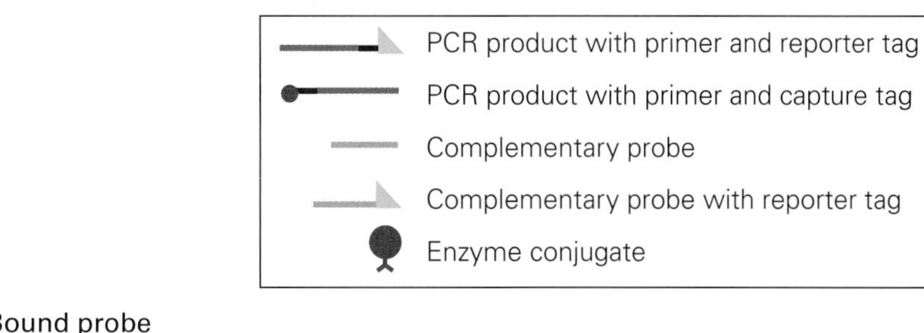

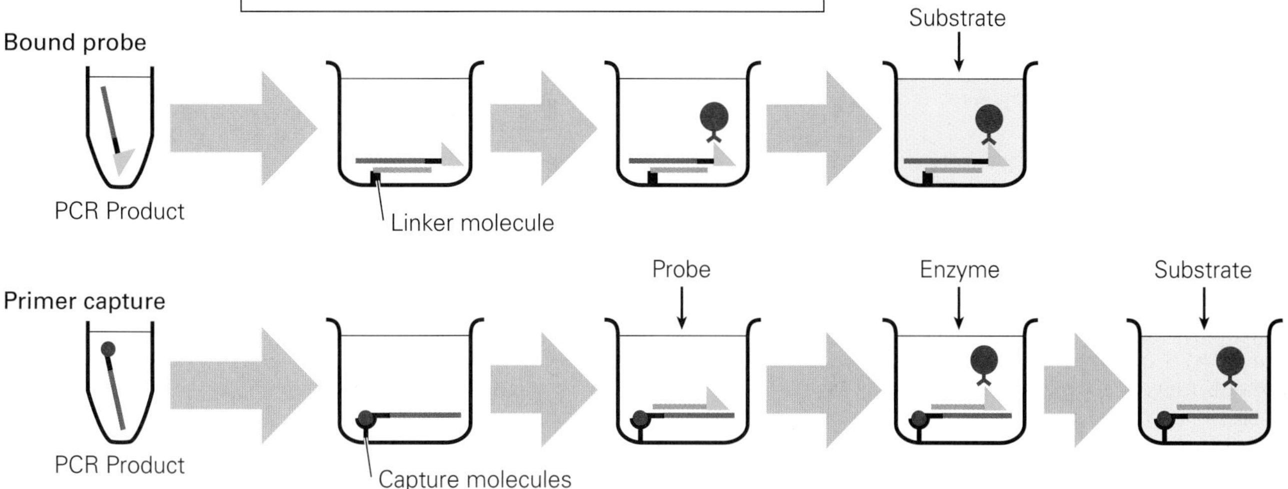

Figure 10.10
Microplate hybridization for detection of PCR product. The illustration shown here uses a reporter system that binds to an enzyme when the PCR product complements the probe sequence. A color change is seen when the enzyme substrate is supplied; if the PCR product does not complement the probe sequence, then no color develops. Alternatively, fluorescent or luminescent reporter systems may be used.

Problems of amplification-based nucleic acid detection systems

All amplification-based systems can provide exquisitely sensitive diagnostic systems that can detect latent or inactive virus, on occasion down to single molecules of DNA. However, in consequence they may detect insignificant levels of virus. Additionally, most PCR-based methods require targets with a known nucleic acid sequence, and all are extremely sensitive to contamination. With their extreme sensitivity, even very low levels of contaminating nucleic acid can give false-positive results since they will be able to reach very high levels during the amplification process.

Methods of avoiding contamination when using PCR include the inclusion of good controls, and strict separation of reagent preparation, test setup, and test analysis. Fresh sterile reagents and equipment should be used, and ultraviolet irradiation of work surfaces and areas may be considered.

Use of deoxyuridine triphosphate rather than deoxythymidine triphosphate as a DNA precursor, combined with a uracil-specific degrading enzyme (uracil N-glycosylase, UNG, sold as Amperase®), can prevent contamination by carryover of PCR product from previous reactions. Any DNA produced by the previous PCR will contain uracil rather than thymine and will be degraded, while the UNG itself is inactivated by the heating step before DNA is produced in the new PCR reaction.

In order to address such concerns, multiple developments were made from the original reaction system. Some of these are summarized in **Table 10.4**.

The most important of these for diagnostic purposes is **quantitative real-time PCR**, which answers many of the concerns raised with the basic PCR process.

Table 10.4 Enhancements of the polymerase chain reaction

Name	Method	Advantages	Disadvantages
Reverse transcriptase PCR (RT-PCR)	Initial reverse transcriptase step copies RNA to DNA (may use Tth polymerase with innate RT activity)	Allows direct detection of RNA	Lower sensitivity
Multiplex PCR	Uses multiple primer sets in one tube	Allows simultaneous assay for multiple targets, which can be useful in diagnosis	Primer sets must be specific and be optimized to work in similar conditions, multiple bands can be confusing
Touchdown PCR	Lowers annealing temperature from 3–5°C above optimum to 3–5°C below optimum as cycles progress	Gives highest-affinity primers a "head start," thus reducing contamination	Requires specialized equipment
Hot-start PCR	Prevents enzymes from becoming active until high temperatures are attained (may use modified or conjugated proteins, or wax barriers)	Reduces background arising from nonspecific products formed before operating temperatures attained	Requires specialized reagents
Nested PCR	Sequential PCR reactions with second primer set detecting product of first reaction	Added sensitivity, added specificity	Increased risk of contamination during set-up of second reaction
In situ PCR (IS-PCR)	PCR on fixed cells with microscopic detection of product	Shows location of target within cells	Complex, expensive
Quantitative real-time PCR (RT-PCR*, RQ-PCR, QRT-PCR, RTQ-PCR)	Release of reporter after each extension cycle	Provides quantitation of target, real-time monitoring reduces issues with contamination	Complex, expensive

* Not to be confused with reverse transcriptase PCR. Many other variant forms of PCR exist, but are of limited relevance to virology and to diagnostics.

In a normal PCR assay, while targets present at high levels are amplified faster, during the reaction limitations on raw material mean that all targets that are amplified can eventually reach similar levels, making the system essentially non-quantitative—it shows that a target is present, but not how much there is.

Multiple approaches have been taken to allowing quantitation of the target. Some of the earliest used a series of reactions carried out in parallel with a modified target, generating a ladder showing how much target would be produced from each, against which the reaction products from the test PCR could be compared. However, this approach was extremely complex and could be of limited accuracy. More recently, systems have been developed that allow real-time monitoring of the amount of product after each (heat–anneal–extend) cycle by the detection of a (usually fluorescent) reporter that is released during production of the PCR product. This often involves removal from the influence of a fluorescence quencher attached to the same primer, either by unfolding of a hairpin that holds the two together (Molecular Beacon system) or by degradation of an oligonucleotide (Taqman® system). **Figure 10.11** shows the approach used in the Taqman® system, one of the earliest such systems to have been developed.

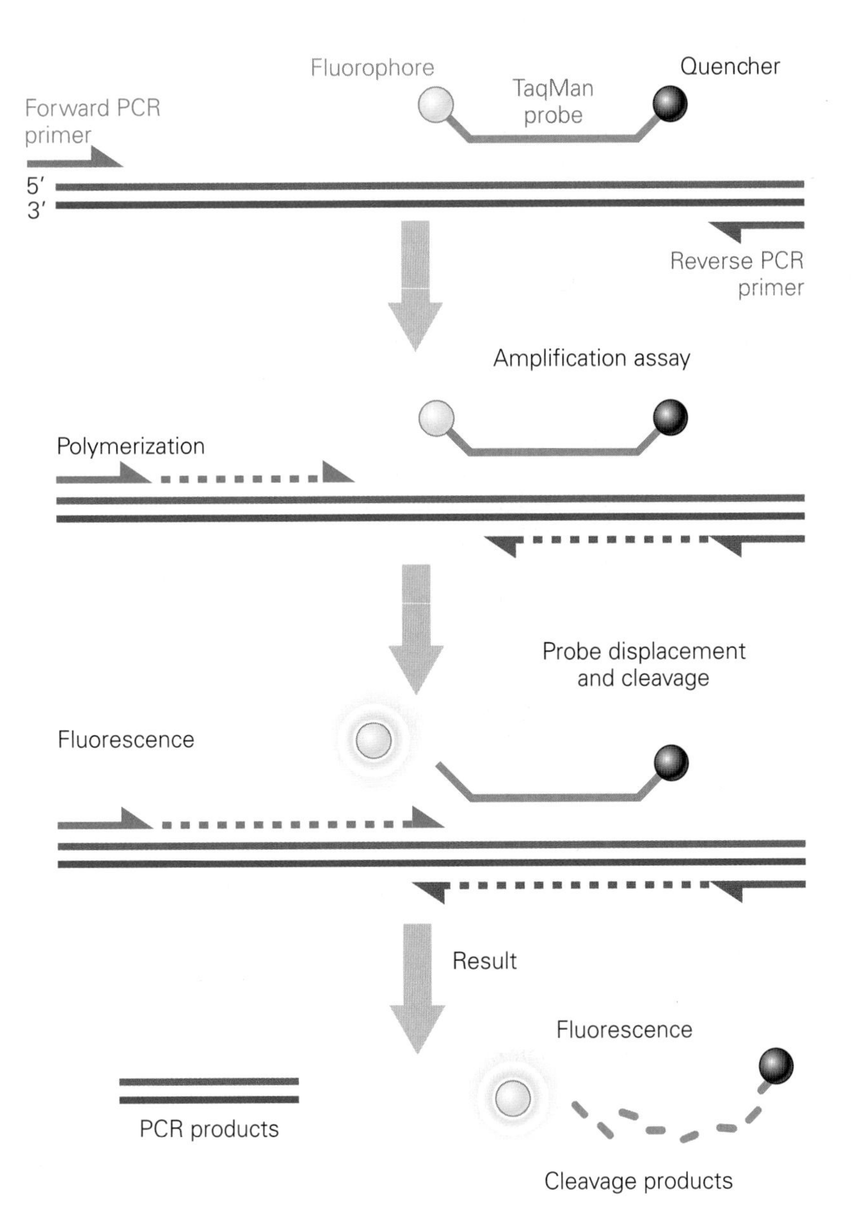

Figure 10.11
The Taqman® fluorescent reporter system.

Such approaches allow the appearance of PCR product to be monitored after each cycle, rather than only when a final, plateau level of product has been reached, and are thus able to deliver information on the amount of target DNA present in the sample. Rather confusingly, this method is often known as real-time or RT-PCR, leading to confusion with the reverse transcriptase or RT-PCR system used to detect RNA. Other names including the letter Q (for quantitative) such as RQ-PCR, QRT-PCR, or RTQ-PCR have been suggested to avoid this.

Microarrays

DNA **microarrays** are simply very large numbers of synthetic oligonucleotides (oligos) fixed to a solid matrix by any of a range of techniques, including some derived from inkjet printers. Microarrays bearing up to 50,000 oligos are in routine use, and some commercial systems can now have millions of different oligos present.

When probed with RNA or DNA along with a suitable (usually fluorescent or luminescent) reporter, binding can be detected for each oligonucleotide (**Figure 10.12**). By selecting these to match to known mRNA sequences, transcriptional activity can be monitored. Alternatively, gene polymorphisms can be analyzed in detail.

In virology, as well as transcriptional analysis, it is possible to use oligos matching known or potential subspecies of a highly variable virus such as HIV. By this method, the presence of a very large range of viral variants can be assayed. In a similar approach, microarrays have also been developed for epidemiological (and even diagnostic) studies of influenza viruses.

It should be noted that microarray technology is not restricted to nucleic acids. Protein microarrays assaying immunological or other protein-binding interactions are in widespread use, as are antibody microarrays, where the antibodies themselves are bound to the matrix. Other approaches use carbohydrates or even tissue sections in similar arrangements.

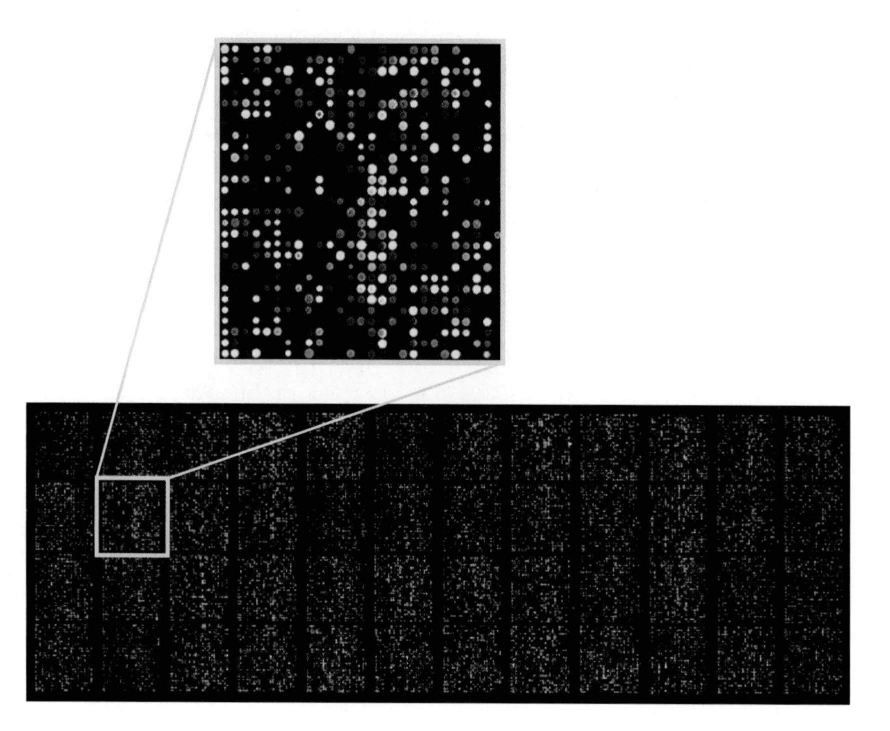

Figure 10.12
A DNA microarray using fluorescence detection.

High-throughput sequencing

One interesting area where rapid development is under way is the application of sequence data to diagnostic virology. Using the latest massively parallel sequencing-by-synthesis approaches (see Chapter 9) it is possible to obtain sequence data for multiple viruses and viral variants present in clinical specimens.

Combining advances in high-throughput sequencing with the use of advanced bioinformatic systems has the capability to provide previously unthinkable levels of information. The challenge with such systems is to sort the useful data from the mass of other information.

While this approach remains complex, expensive, and outside the scope of the typical diagnostic laboratory as of 2010, this is already changing.

10.6 FUTURE DEVELOPMENTS IN DIAGNOSTIC VIROLOGY

It is very clear that nucleic acid detection systems, particularly those based on amplification, are now a vital part of virus diagnosis. However, antibody and antigen assays, often based on the use of monoclonal antibodies and cloned proteins, continue to play a prominent role.

The trend away from in-house assays, toward the use of commercially produced assay kits, has continued apace. Such kits are evaluated and certified by the producing company, and their use simplifies quality control. Kits may also be useful where facilities and expertise for assay development are not available, extending the range of services that can be offered by the nonspecialist laboratory. They are also far better suited to use with automated systems.

Automated analytical systems represent another strong element in modern diagnostics. The newest machines are complex and flexible, and are capable of handling multiple assays simultaneously while performing all stages of testing from specimen preparation to issuing of test results. However, the tests that can be used with such a system are of course limited to those provided by the manufacturer, which can be restrictive, particularly where the laboratory has specialized requirements. In addition, such machines can be extremely expensive, although this is often moderated by potential savings in staff costs and by package deals linked to purchases of assay kits. Indeed, the latter may offset much of the initial cost, although long-term contracts may be required. Specimen turn-round times may be shorter than with manual testing, and out of working hours testing may be simpler to arrange.

It is also clear that molecular techniques can support investigations even where the causative viruses are unknown. In some of the earliest such work, hepatitis C virus (*Flaviviridae*) was identified entirely by such methods, years before culture of the virus became possible. Molecular techniques have already identified other novel viruses, including human herpesvirus 8 (*Herpesviridae*), linked to Kaposi's sarcoma, and Sin Nombre hantavirus (*Bunyaviridae*), a major cause of the lethal hantavirus pulmonary syndrome.

Although great care needs to be taken with some of the more optimistic claims arising from genomics, transcriptomics, proteomics, and others, it is clear that molecular virology has as yet demonstrated only a small fraction of its value in the detection and diagnosis of viruses.

Key Concepts

- Diagnosis of viruses has moved from observation of symptoms to understanding of the nature of viruses, and now to molecular analysis at the genomic level. However, clinical information still plays a vital role in determining what viruses to look for, and which may have a role in pathology.
- Electron microscopy, while useful in classifying viruses, is now less used in diagnosis, mainly due to the need for very high levels of the target virus along with expensive equipment and skilled operators.
- Microscopic examination of patient specimens can provide useful information as many viruses produce characteristic changes in infected cells. Use of specific antibodies and nucleic acid probes can enhance such detection.
- Virus culture, long-regarded as the "gold standard" of diagnosis, has now been supplemented by rapid diagnostic techniques which can detect viral products far more rapidly than traditional systems relying on cell killing. However, many viruses cannot be cultured even with the most advanced techniques.
- The immune response to virus infection generates high levels of antibodies, which can be used to confirm infection. Levels of specific types of antibody can be assayed in order to provide evidence for when the infection occurred.
- Nucleic acid technology in general and amplification-based systems [the polymerase chain reaction (PCR) and its derivatives] in particular have brought immense changes to many aspects of virology and provided huge increases in the sensitivity and range of assays.
- PCR and other amplification-based systems have matured from challenging and sometimes fraught experimental procedures to a routine approach supported by a wide range of established procedures, reagents, and kits.
- Microarrays that can detect the levels of a broad range of nucleic acids and high-throughput nucleic acid sequencing are already establishing themselves as diagnostic tools in specialist applications. Their use will continue to expand as costs decline and power of the systems used increases.
- Recent years have seen a profound movement from skilled technicians observing viruses to automated, kit-based analytical systems.
- Molecular technology has revolutionized virus diagnosis, and will continue to do so.

DEPTH OF UNDERSTANDING QUESTIONS

Hints to the answers are given at http://www.garlandscience.com/viruses

Question 10.1: What are the problems with a symptom-based classification of viral agents?

Question 10.2: Why can't we grow any virus that we find?

Question 10.3: What are the main concerns and advantages in the use of nucleic acid amplification-based diagnosis?

Question 10.4: Why is automation such a strong trend?

FURTHER READING

Harper DR (ed.) (1993) Virology Labfax. BIOS Scientific Publishers, Oxford.

Harper DR (2001) Viral culture methodologies. In Encyclopedia of Life Sciences. John Wiley & Sons, Chichester. http://www.els.net

Kudesia G & Wreghitt T (2009) Clinical and Diagnostic Virology (Cambridge Clinical Guides). Cambridge University Press, Cambridge.

Monis PT & Giglio S (2006) Nucleic acid amplification-based techniques for pathogen detection and identification. *Infect. Genet. Evol.* 6, 2–12.

Mullis KB (1990) The unusual origin of the polymerase chain reaction. *Sci. Am.* 262, 56–65.

Schmidt NJ, Lennette DA, Lennette ET et al. (eds) (1995) Diagnostic Procedures for Viral, Rickettsial and Chlamydial Infections, 7th ed. American Public Health Association, Washington, DC.

Storch GA (2005) Diagnostic virology. In Virology, 5th ed. (BN Fields, PM Howley eds). Lippincott Williams & Wilkins, Philadelphia.

Zuckerman M (2009) Pathogenic viruses: clinical detection. In Encyclopedia of Life Sciences. John Wiley & Sons, Chichester. http://www.els.net

INTERNET RESOURCES

Much information on the internet is of variable quality. For validated information, PubMed (http://www.ncbi.nlm.nih.gov/pubmed/) is extremely useful.

Please note that URL addresses may change.

All the Virology on the WWW. http://www.virology.net/

American Type Culture Collection. http://www.atcc.org

Health Protection Agency Culture Collections. http://www.hpacultures.org.uk/

PCR. A review of PCR and polymerase chain reaction. http://www.horizonpress.com/pcr/

Appendix I
Virus Replication Strategies

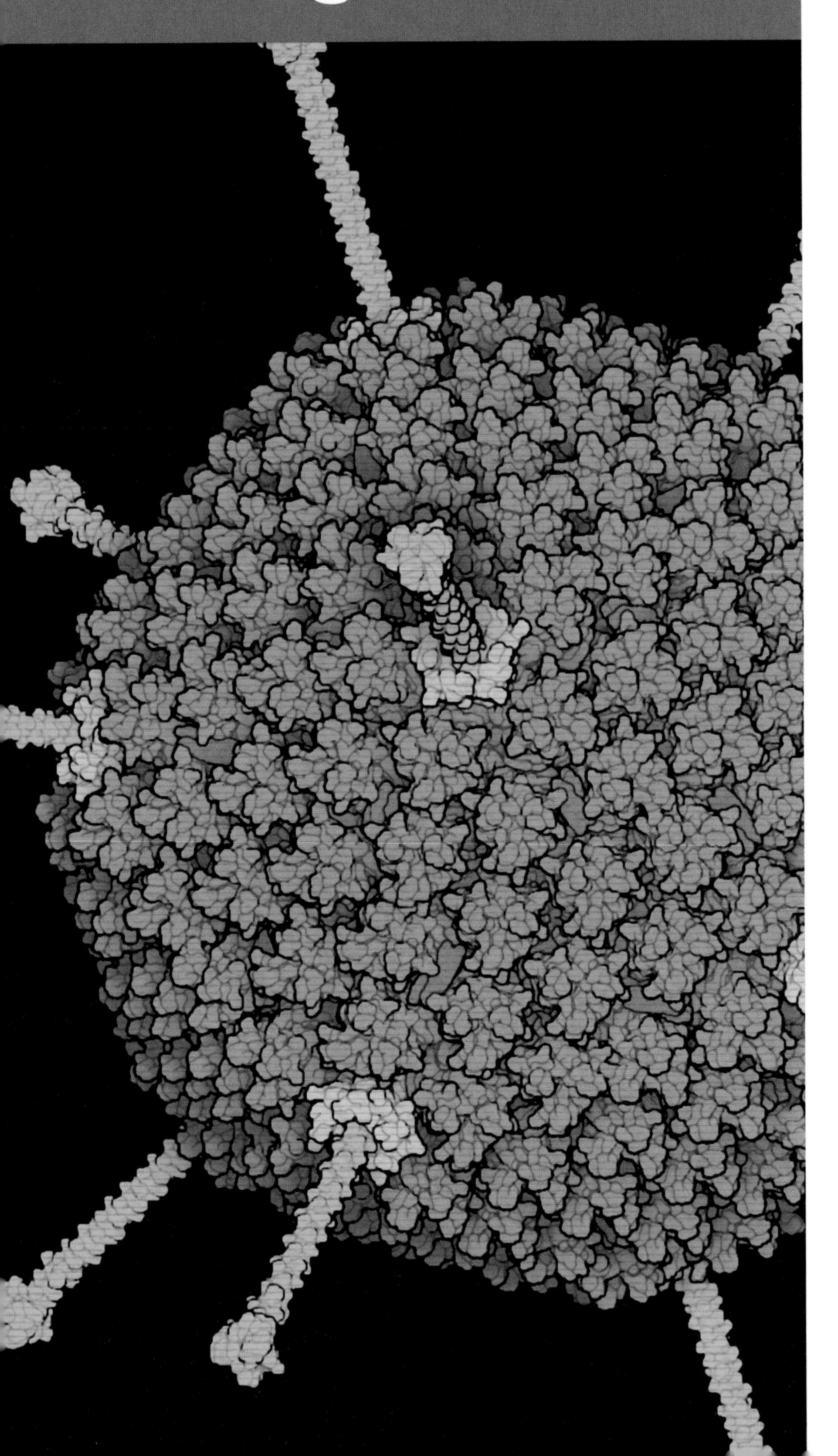

VIRUS REPLICATION STRATEGIES AND CHARACTERISTICS BY FAMILY

The following section provides a brief outline of the replication strategies used by each of the major virus families known to infect humans. The amount of information that can be presented is of necessity limited, and the following points should be noted.

About the image
Adenovirus
(Courtesy of the Research Collaboratory for Structural Bioinformatics Protein Data Bank and David S. Goodsell, The Scripps Research Institute, USA.)

The level of detail that is known for the replication strategies of individual virus families varies widely, for example, from the *Picobirnaviridae*, where most aspects of infection remain hypothetical, to the very high level of understanding of the much more complex *Herpesviridae*. In many cases, only a very brief outline of the processes of infection can be given. The effects of infection on the host cell are highly complex and are covered in Chapter 1 except where these relate directly to the major processes of virus replication (as with cap stealing by the *Bunyaviridae* and the *Orthomyxoviridae*).

Unless otherwise stated, virus diagrams are drawn to a scale of 2 cm to 100 nm.

Members refers only to species as defined by the International Committee on the Taxonomy of Viruses (ICTV). Many more unassigned viruses exist, at varying levels of characterization.

Infecting humans includes those viruses where such infections appear to be common enough to justify defining the virus as a (pathogenic) infection of humans rather than a rare occurrence with very limited epidemiological significance. However, it should be noted that alongside those viruses where humans appear to be the only natural host (such as hepatitis B virus and many of the *Herpesviridae*), many viruses that infect humans may do so as part of a multi-host life cycle (such as the arboviruses, that replicate in both humans and, typically, insect hosts) or as a dead-end zoonotic infection (such as Sin Nombre hantavirus).

Vaccines and antiviral drugs—only those that are in clinical use, although the level of such use may vary widely and may only apply to a limited number of viruses within the family. Brackets indicate restricted or discontinued availability, or availability for only a few of the viruses within the family.

Mimivirus

Mimivirus has not yet been assigned to a virus family and is primarily a virus of algae, though there have been suggestions of involvement in viral pneumonias in humans. Given the huge genome of this virus, replication is complex, but is broadly in line with other nucleocytoplasmic large DNA viruses (NCLDVs), such as the *Poxviridae*.

Deltavirus

Replication and pathogenesis of hepatitis D (the delta agent; the sole agent in the genus *Deltavirus*) is viroid-like in nature, although with the additional production of two proteins that form a nucleocapsid-like structure with the 1700-base genomic RNA. It is dependent on hepatitis B virus infecting the same cell to provide essential functions for its replication, including the proteins that form the outer shell of the infectious particle. Replication of the delta agent is covered in Section 2.4.

Virus images on the following pages were provided by:

Diagrams: Cull P (ed.) (1990) The Sourcebook of Medical Illustration, DA Information Services.

Micrographs: Dr Ian Chrystie, Department of Virology, St. Thomas' Hospital, London, UK; From, Harper DR (ed.) (1993) Virology Labfax. BIOS Scientific Publishers, Oxford (*Paramyxoviridae, Caliciviridae, Coronaviridae, Togaviridae, Herpesviridae, Polyomaviridae, Hepadnaviridae*); CDC Public Health Image Library (http://phil.cdc.gov/) (*Adenoviridae, Arenaviridae, Bunyaviridae, Filoviridae, Flaviviridae, Orthomyxoviridae, Parvoviridae, Picornaviridae, Poxviridae, Retroviridae, Rhabdoviridae*); Graham Colm under Creative Commons Attribution 3.0 License (*Reoviridae*).

Adenoviridae

Family	Genome type	Genome size (kbp)	Envelope	Capsid	Size (nm)	Virion proteins
Adenoviridae	dsDNA	35.8–36.2	No	Icosahedral	80–110	10

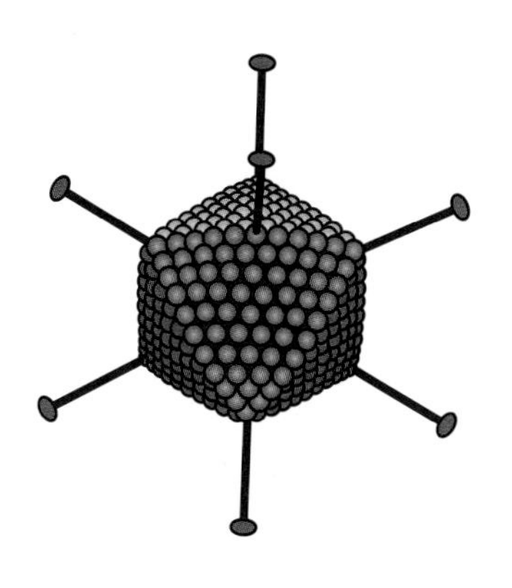

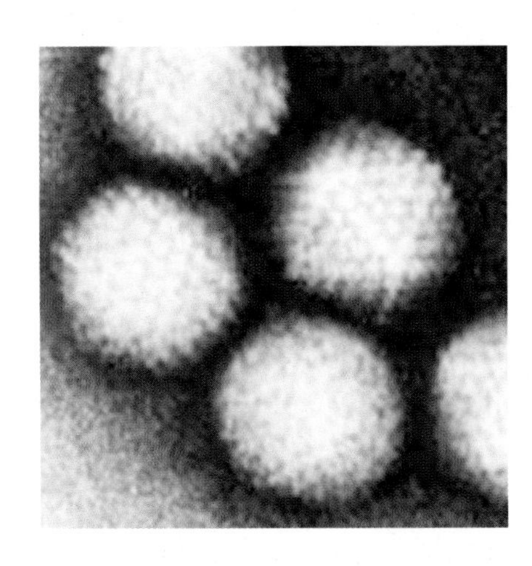

Examples of human pathogens

Human adenoviruses A–F

Members	Main host types	Infecting humans	Example of human disease	Vaccines	Antiviral drugs
31	Vertebrates: mammals (primates, humans), birds, reptiles, fish	6	Common cold	(Yes)	No

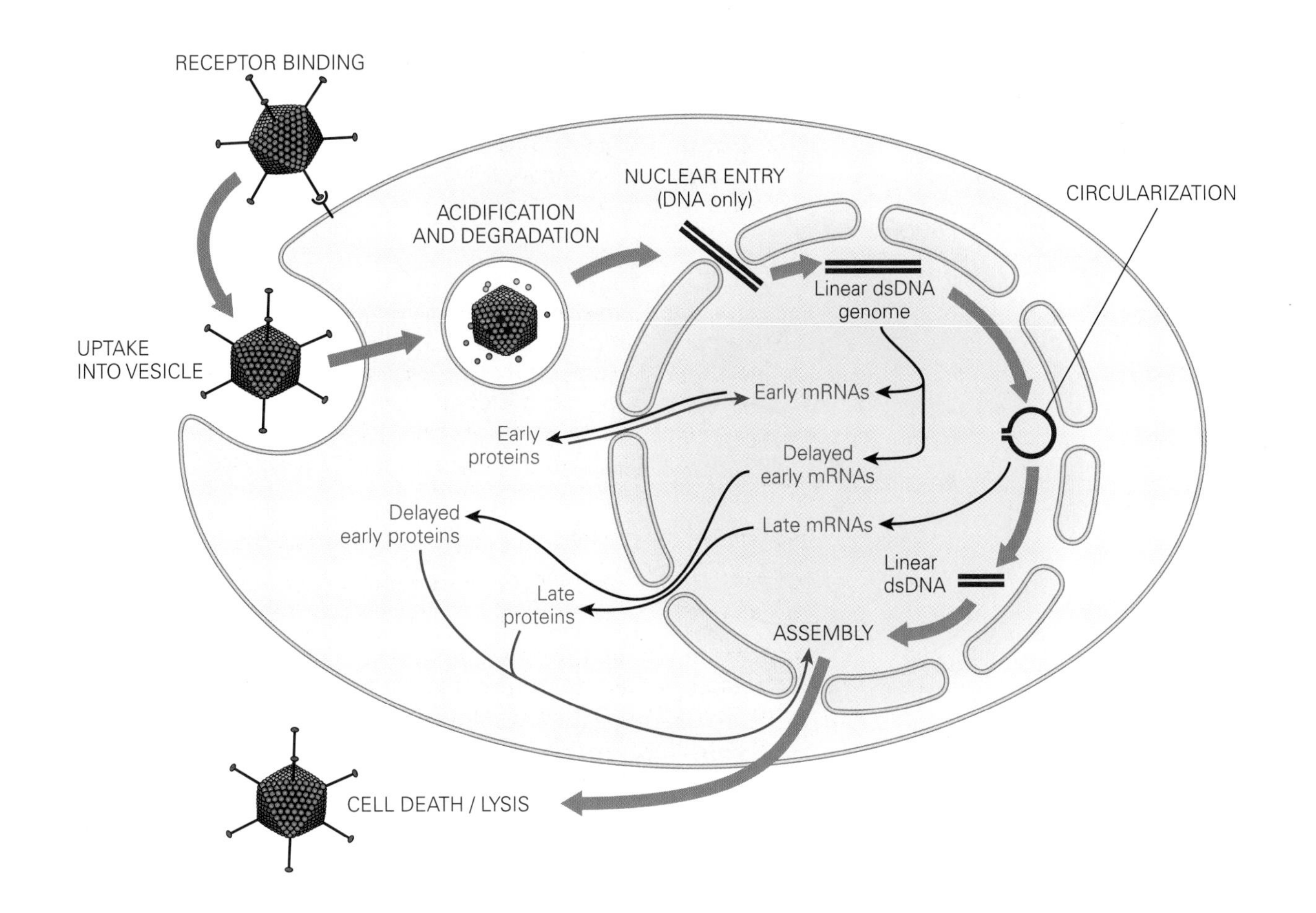

Arenaviridae

Family	Genome type	Genome size (kb)	Envelope	Capsid	Size (nm)	Virion proteins
Arenaviridae	Segmented ssRNA (ambisense)	11	Yes	Filamentous × 2	110–130	5

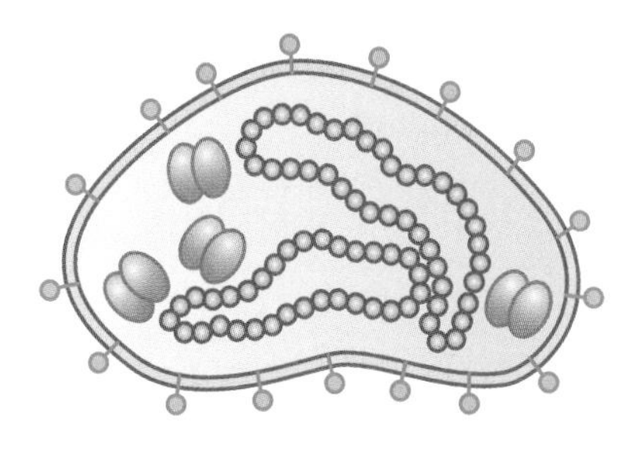

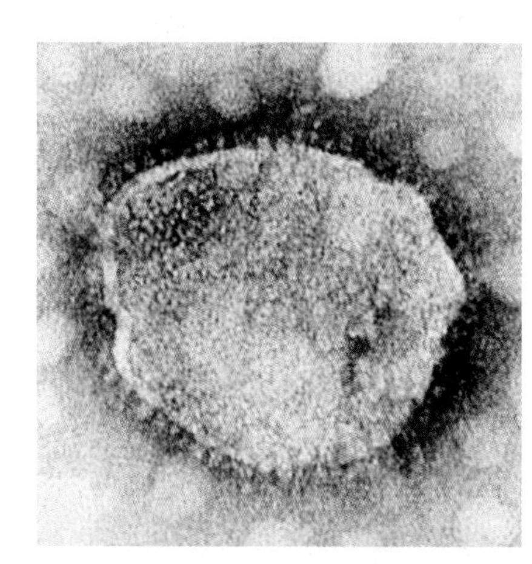

Examples of human pathogens

Lassa fever; lymphocytic choriomeningitis; Argentine, Bolivian, and Venezuelan hemorrhagic fevers

Members	Main host types	Infecting humans	Example of human disease	Vaccines	Antiviral drugs
22	Mammals (humans), arthropods (arboviruses)	Variable (7+)	Meningitis, hemorrhagic disease	No	Yes

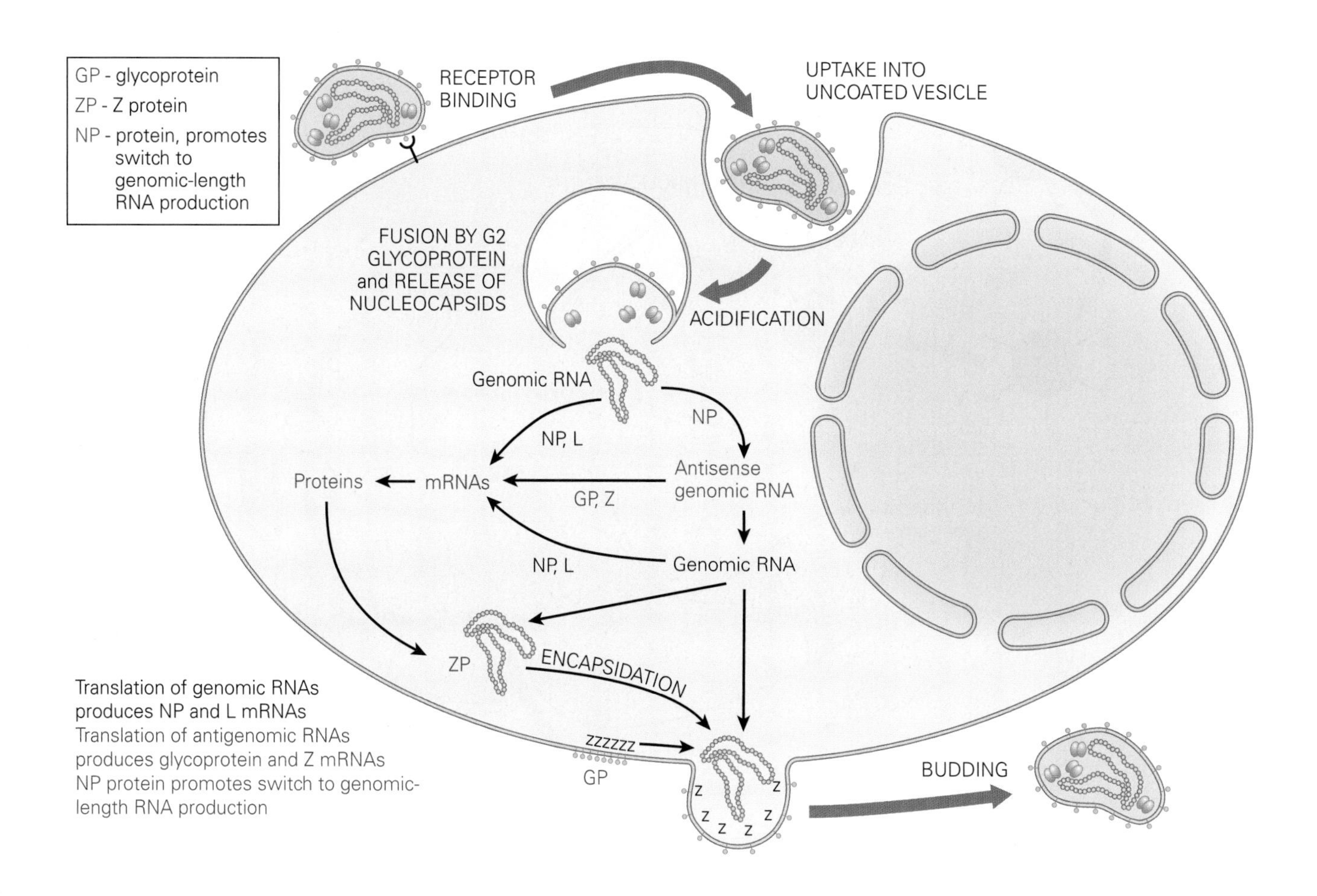

Astroviridae

Family	Genome type	Genome size (kb)	Envelope	Capsid	Size (nm)	Virion proteins
Astroviridae	ssRNA (+)	6.8–7.9	No	Icosahedral	27–30	3

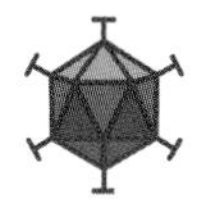

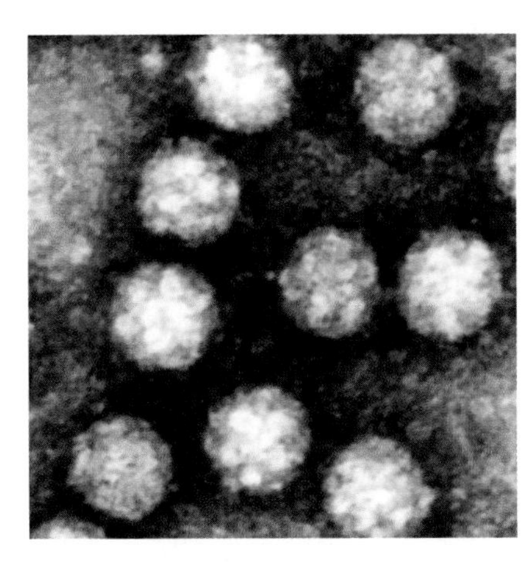

Examples of human pathogens

Human astrovirus

Members	Main host types	Infecting humans	Example of human disease	Vaccines	Antiviral drugs
9	Mammals (many, including primates, humans), birds	1	Gastroenteritis	No	No

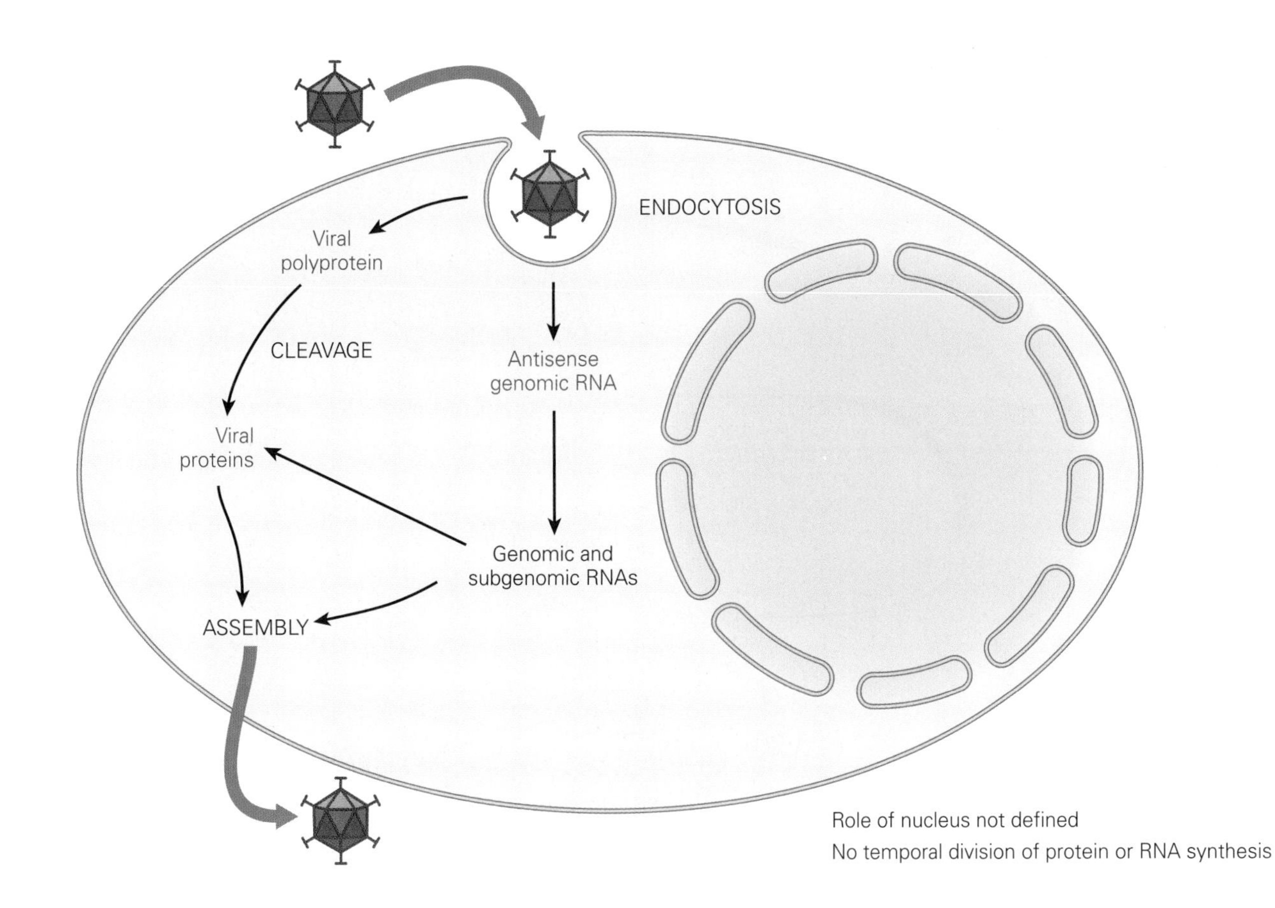

Role of nucleus not defined
No temporal division of protein or RNA synthesis

Bornaviridae

Family	Genome type	Genome size (kb)	Envelope	Capsid	Size (nm)	Virion proteins
Bornaviridae	ssRNA (–)	8.9	Yes	Crescentlike	80–100	5–7

Examples of human pathogens

(Borna disease virus; link to psychiatric illnesses in humans not yet confirmed)

Members	Main host types	Infecting humans	Example of human disease	Vaccines	Antiviral drugs
1	Mammals (many, including primates, humans), birds	1?	Neurological disease?	No	(Yes)

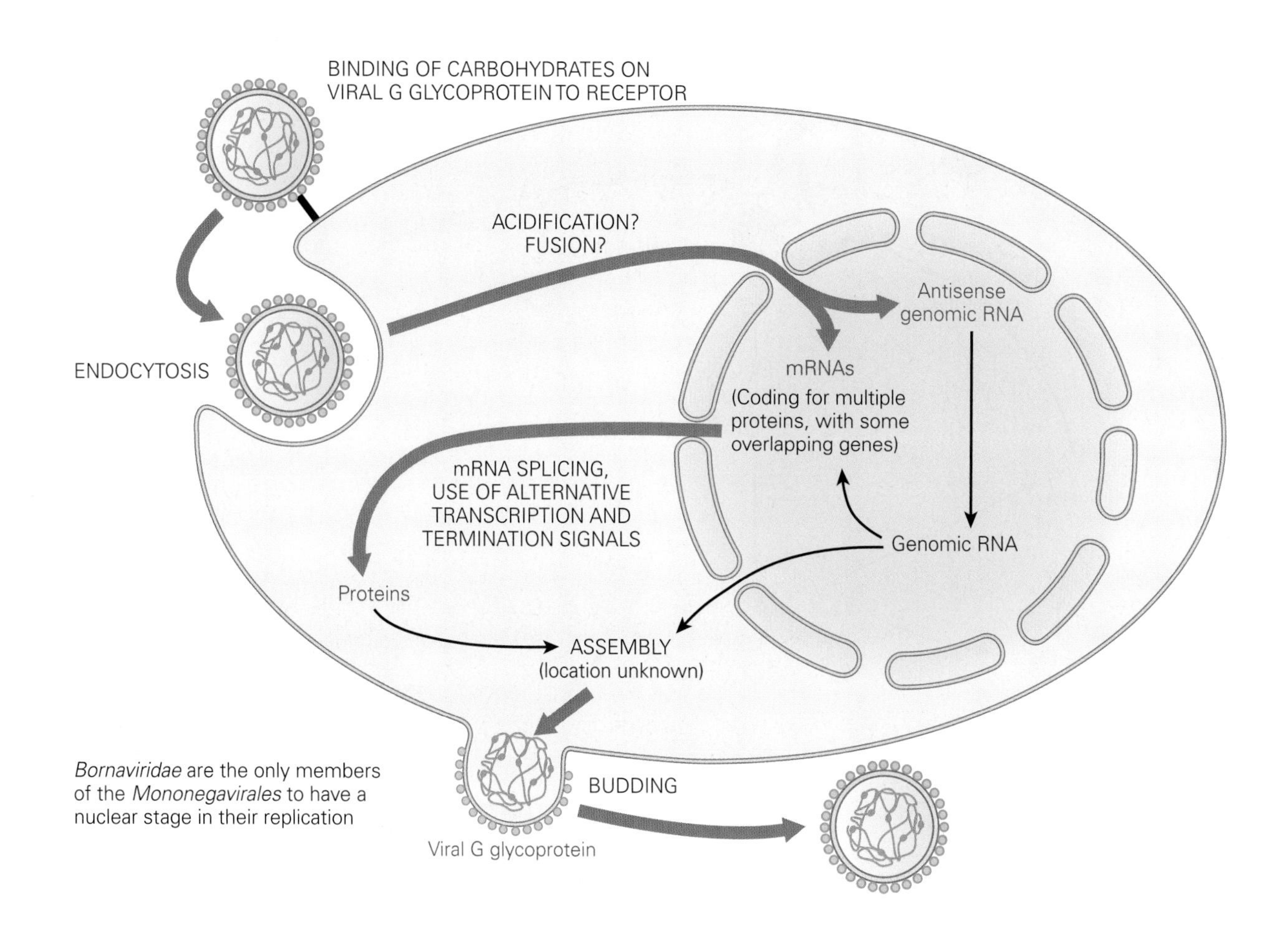

Bornaviridae are the only members of the *Mononegavirales* to have a nuclear stage in their replication

Bunyaviridae

Family	Genome type	Genome size (kb)	Envelope	Capsid	Size (nm)	Virion proteins
Bunyaviridae	segmented ssRNA (ambisense)	6.3–12	Yes	Helical (circular) × 3	80–120	4

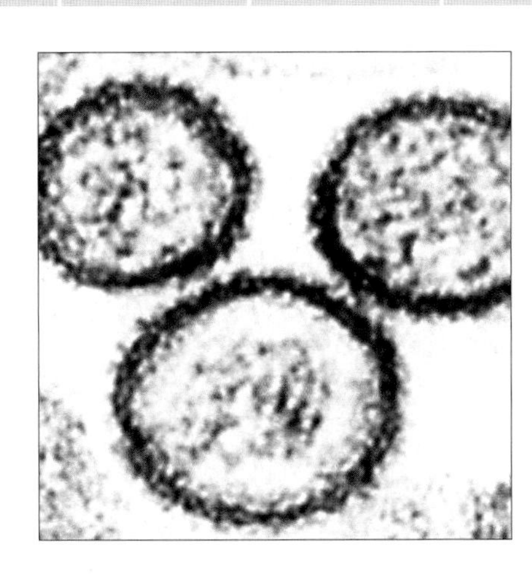

Examples of human pathogens

Bunyamwera, California Encephalitis, Hantaan, La Crosse, Puumala, Rift Valley Fever, Sandfly Fever, Sin Nombre, Seoul, Uukuniemi, and many others

Members	Main host types	Infecting humans	Example of human disease	Vaccines	Antiviral drugs
104	Mammals (humans), insects, plants (arboviruses)	Variable	Hemorrhagic disease	(Yes)	(Yes)

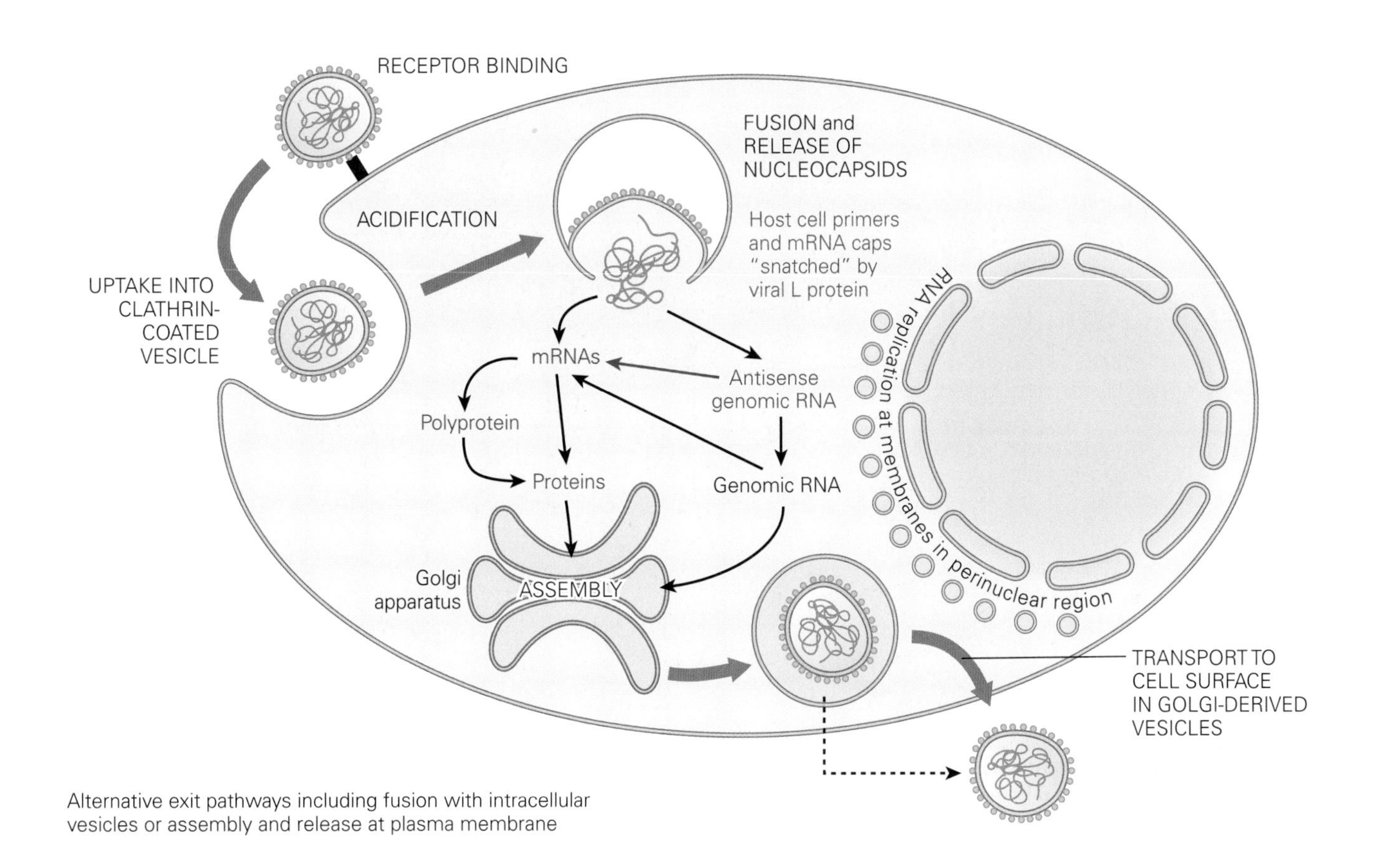

Alternative exit pathways including fusion with intracellular vesicles or assembly and release at plasma membrane

Caliciviridae

Family	Genome type	Genome size (kb)	Envelope	Capsid	Size (nm)	Virion proteins
Caliciviridae	ssRNA (+)	7.4–8.3	No	Icosahedral	35–39	1–2

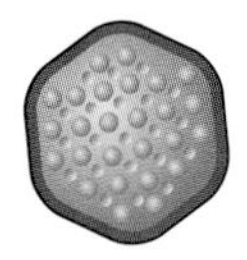

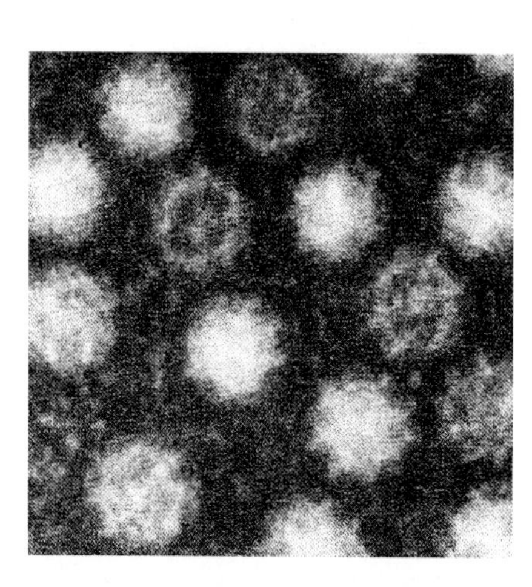

Examples of human pathogens

Norwalk virus, Sapporo virus

Members	Main host types	Infecting humans	Example of human disease	Vaccines	Antiviral drugs
6	Mammals (primates, humans), birds, reptiles, fish	2	Gastrointestinal illness	No	No

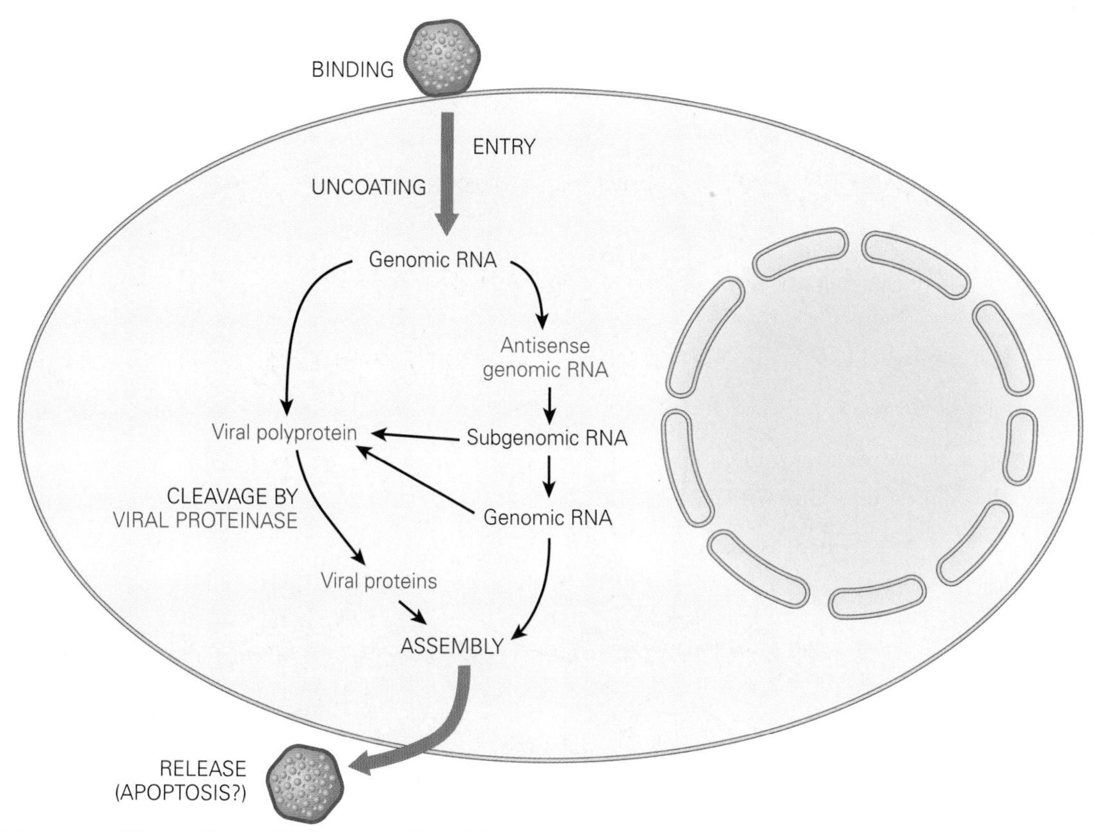

Hepatitis E virus was until recently classified as a member of the *Caliciviridae*. Owing mainly to differences in genome organization, it has now been classified as the only known member of the family *Hepeviridae*. Very little is known of its method of replication, and what is known suggests a strategy similar to that of the *Caliciviridae*, including the presence of a viral proteinase.

Coronaviridae

Family	Genome type	Genome size (kb)	Envelope	Capsid	Size (nm)	Virion proteins
Coronaviridae	ssRNA (+)	25–33	Yes	Elongated helical	120–160	5

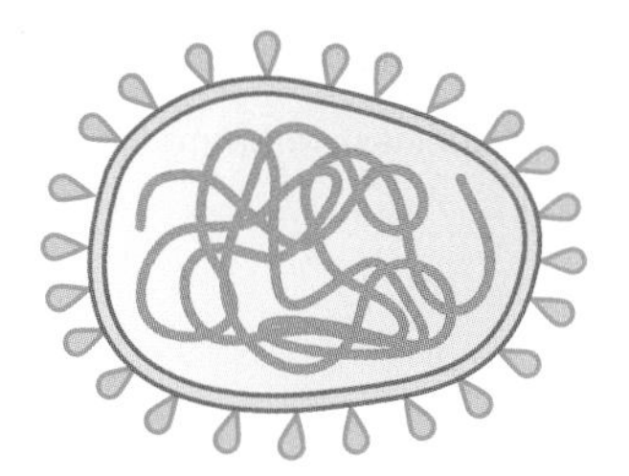

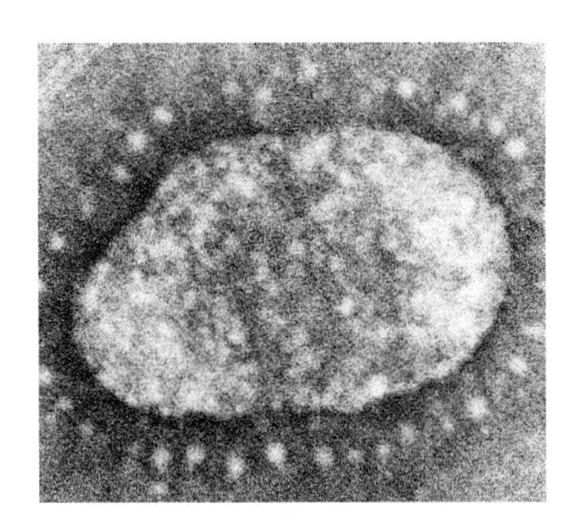

Examples of human pathogens

Human coronavirus 229E, Human coronavirus OC43, Human coronavirus HKU3, Human coronavirus NL43, (SARS coronavirus)

Members	Main host types	Infecting humans	Example of human disease	Vaccines	Antiviral drugs
20	Mammals (humans), birds	7	Common cold	No	No

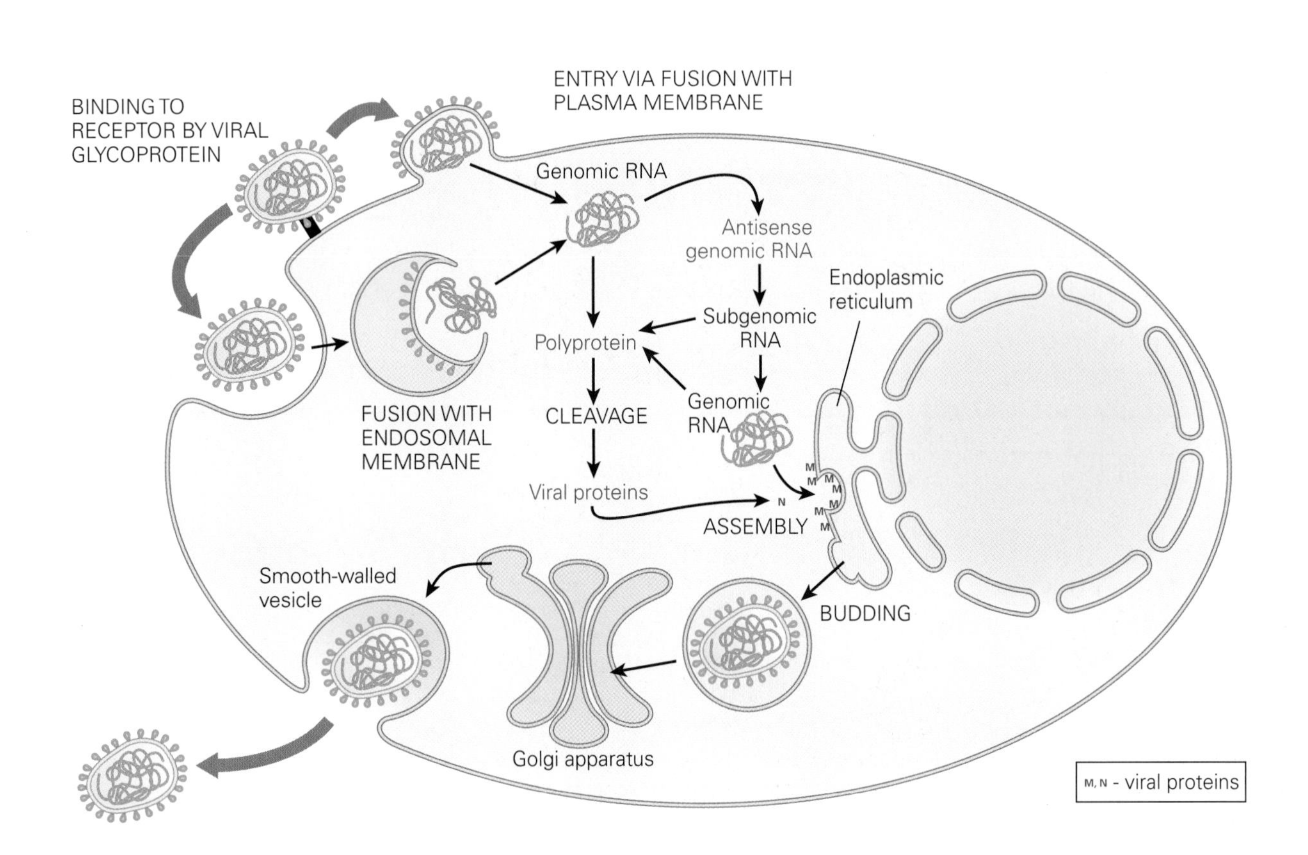

Filoviridae

Family	Genome type	Genome size (kb)	Envelope	Capsid	Size (nm)	Virion proteins
Filoviridae	ssRNA (–)	18.9–19	Yes	Elongated helical	790–1400 × 80	5

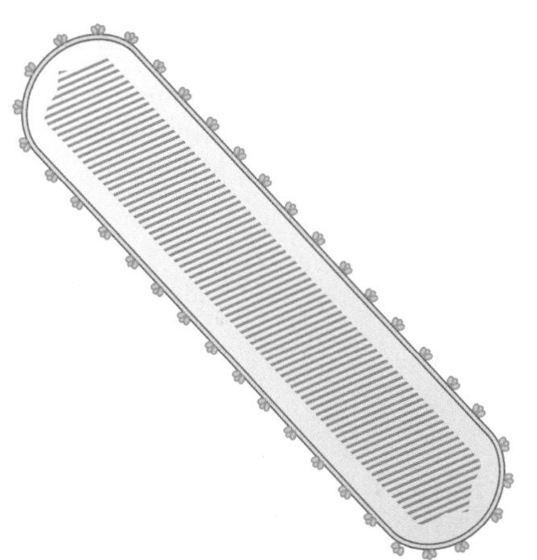

Examples of human pathogens

Ebola virus (4 species), Marburg virus (1 species)

Members	Main host types	Infecting humans	Example of human disease	Vaccines	Antiviral drugs
5	Primates (humans)	5	Hemorrhagic fever	No	No

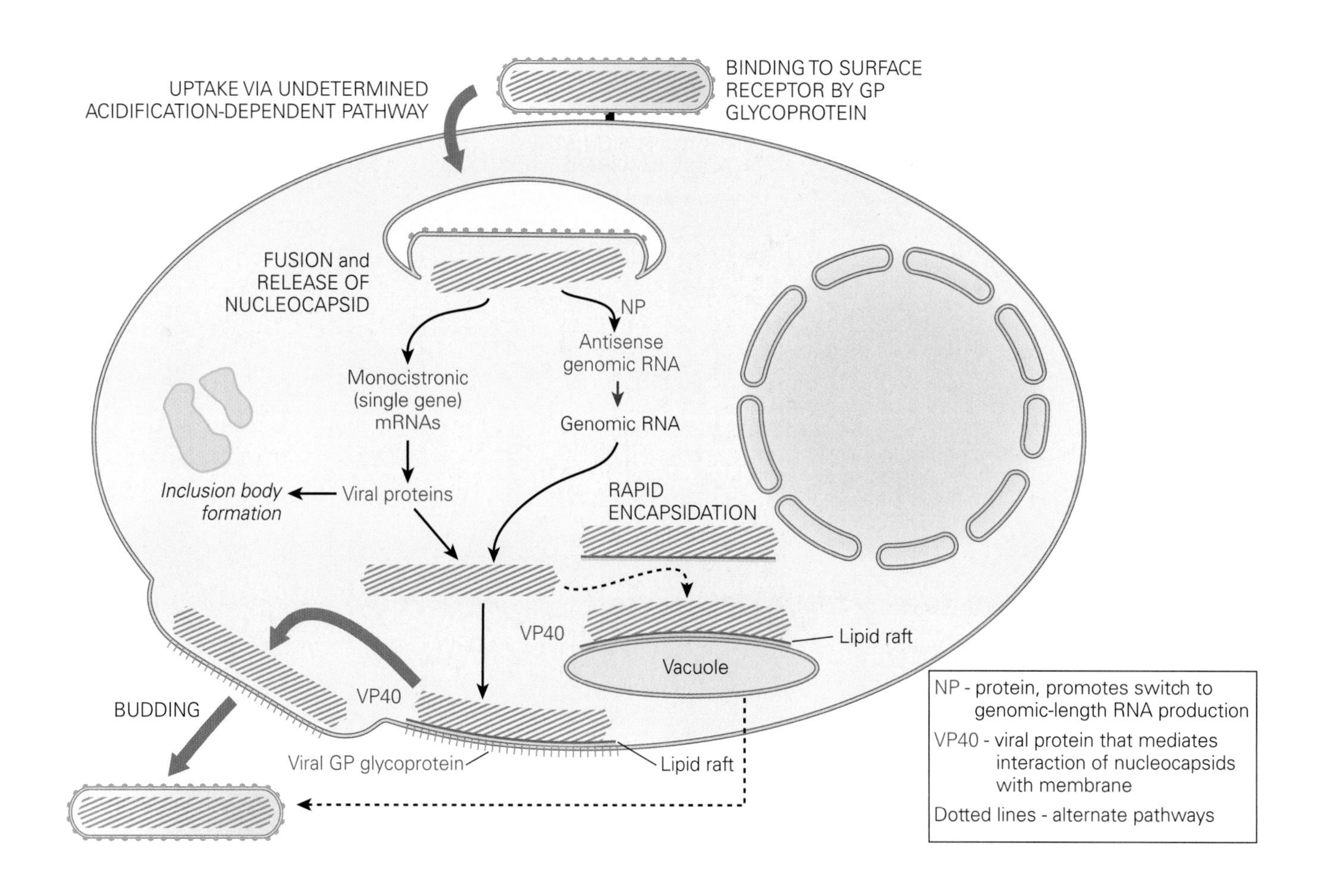

Flaviviridae

Family	Genome type	Genome size (kb)	Envelope	Capsid	Size (nm)	Virion proteins
Flaviviridae	ssRNA (+)	9.5–12.5	Yes	Polyhedral	40–60	3–4

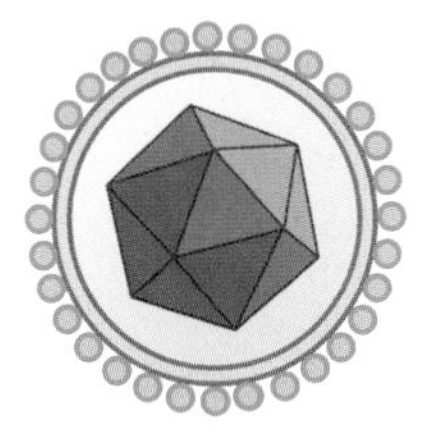

The diagram is drawn at double size for illustration purposes.

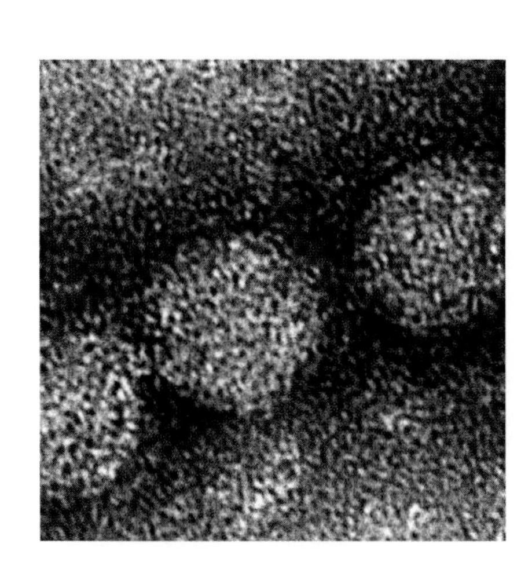

Examples of human pathogens

Yellow fever virus, Dengue viruses types 1–4, Japanese encephalitis virus, hepatitis C virus, others

Members	Main host types	Infecting humans	Example of human disease	Vaccines	Antiviral drugs
58	Mammals (humans), birds, arthropods (arboviruses)	Variable	Yellow fever, hepatitis	(Yes)	(Yes)

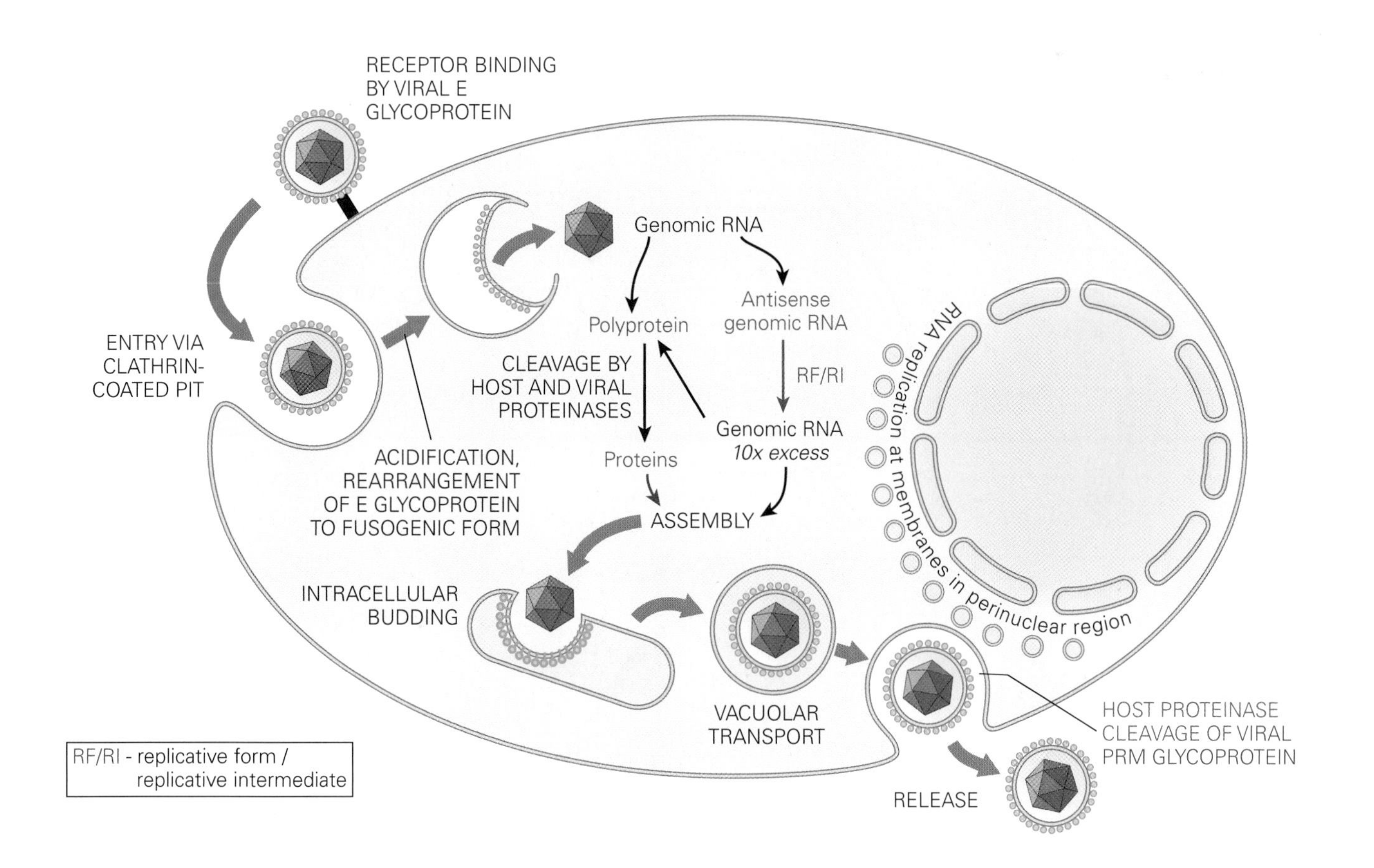

Hepadnaviridae

Family	Genome type	Genome size (kbp)	Envelope	Capsid	Size (nm)	Virion proteins
Hepadnaviridae	Partially dsDNA (via RNA)	3.0–3.3	Yes	Icosahedral	40–48	4

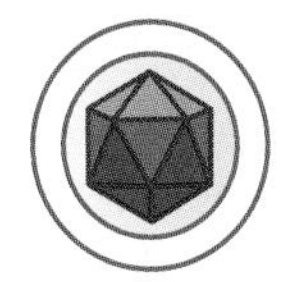

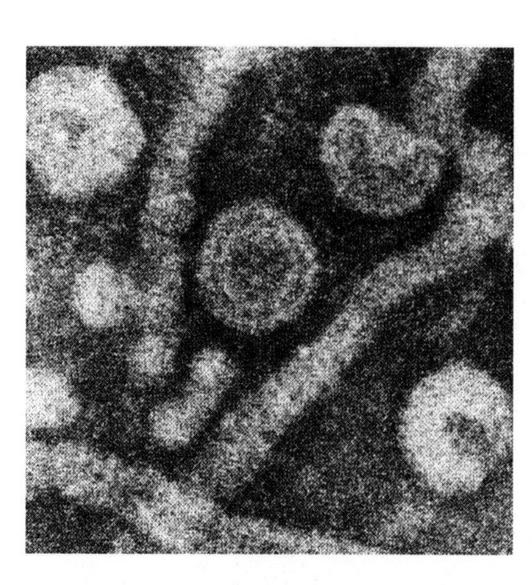

Examples of human pathogens

Hepatitis B virus

Members	Main host types	Infecting humans	Example of human disease	Vaccines	Antiviral drugs
6	Mammals (humans), birds	1	Hepatitis	Yes	Yes

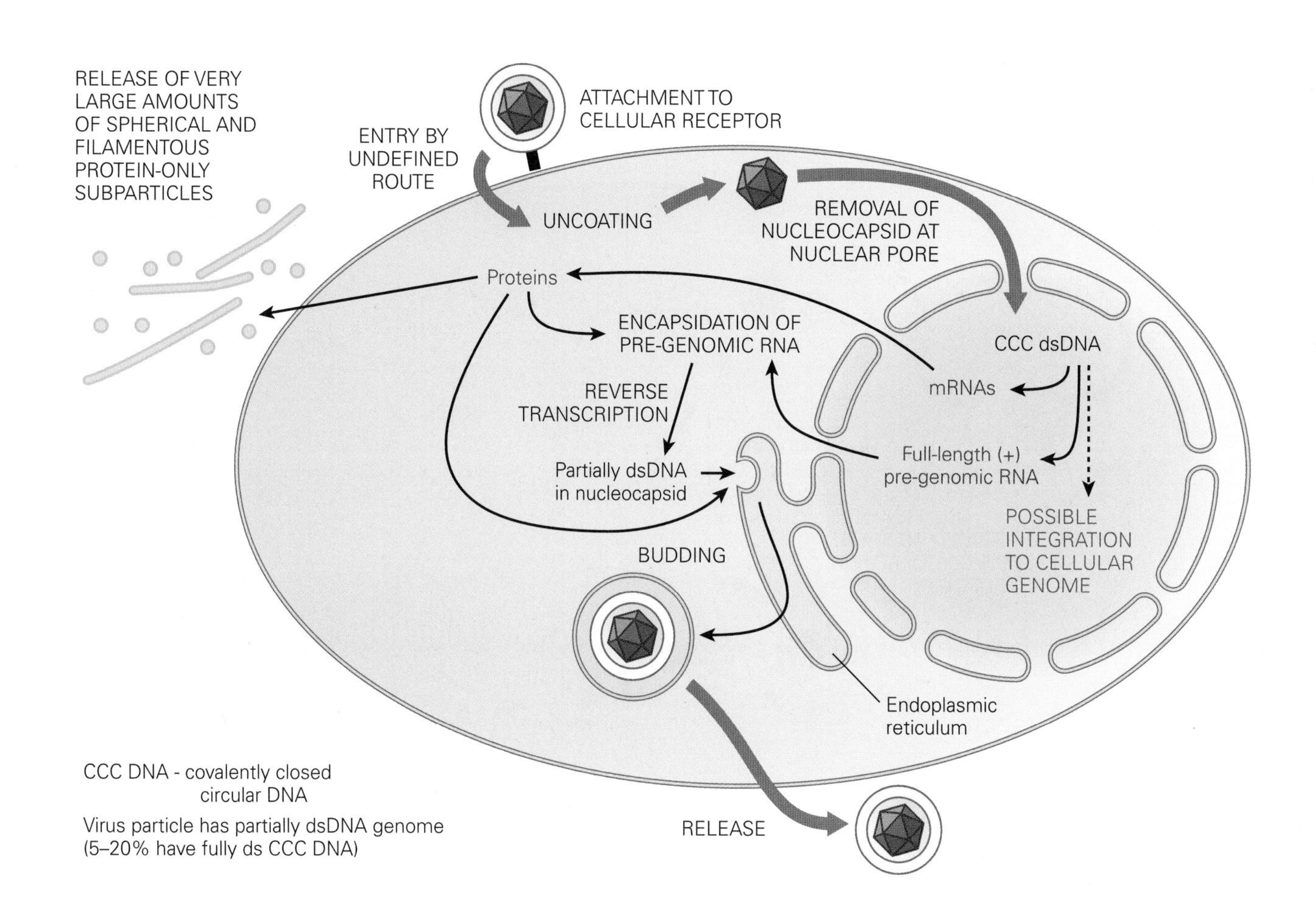

Herpesviridae

Family	Genome type	Genome size (kbp)	Envelope	Capsid	Size (nm)	Virion proteins
Herpesviridae	dsDNA	120–260	Yes	Icosahedral	120–200	24–71

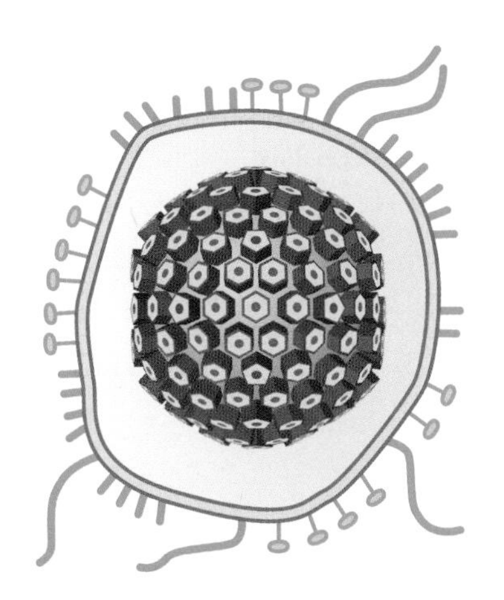

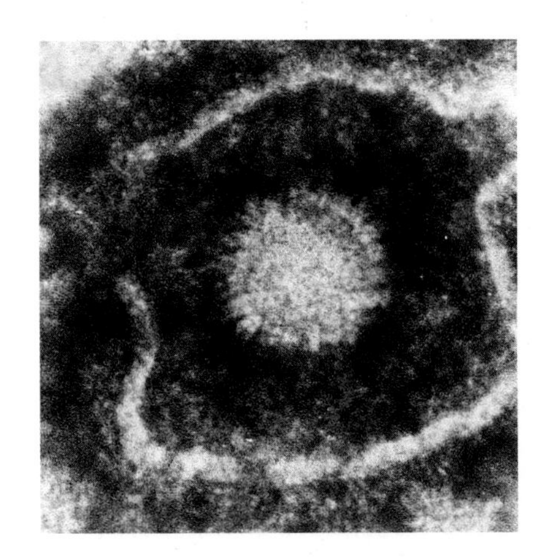

Examples of human pathogens

Herpes simplex viruses 1 and 2 (HHV-1, HHV-2), Varicella-zoster virus (HHV-3), Epstein-Barr virus (HHV-4), Human cytomegalovirus (HHV-5), HHV-6, HHV-7, HHV-8 (Kaposi's sarcoma-associated herpesvirus)

HHV = Human herpes virus

Members	Main host types	Infecting humans	Example of human disease	Vaccines	Antiviral drugs
66	Mammals (humans), birds, fish	8	Cold sores	(Yes)	Yes

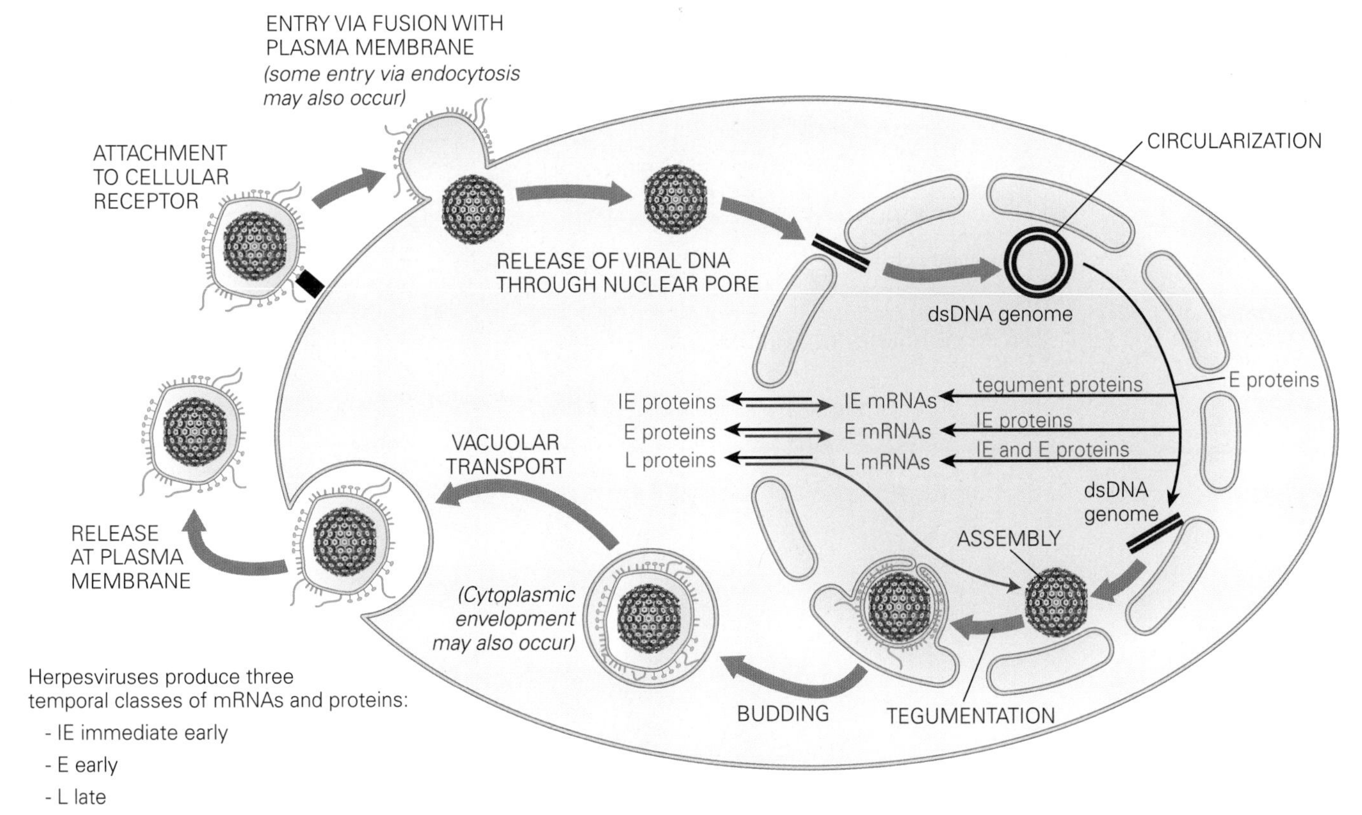

Herpesviruses produce three temporal classes of mRNAs and proteins:

- IE immediate early
- E early
- L late

The *tegument* is a layer between the nucleocapsid and envelope, containing a range of proteins concerned with regulation of viral and cellular metabolism

Orthomyxoviridae

Family	Genome type	Genome size (kb)	Envelope	Capsid	Size (nm)	Virion proteins
Orthomyxoviridae	Segmented ssRNA (–)	10–14.6	Yes	Helical	80–120	7

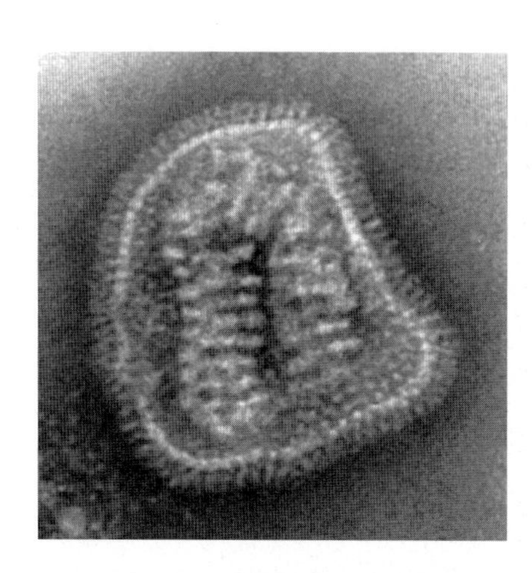

Examples of human pathogens

Influenza viruses A, B, and C; Thogoto and Dhori viruses (encephalitis)

Members	Main host types	Infecting humans	Example of human disease	Vaccines	Antiviral drugs
6	Mammals (humans), birds, arthropods, fish	3 + 2	Influenza (encephalitis—Thogoto and Dhori viruses)	Yes	Yes

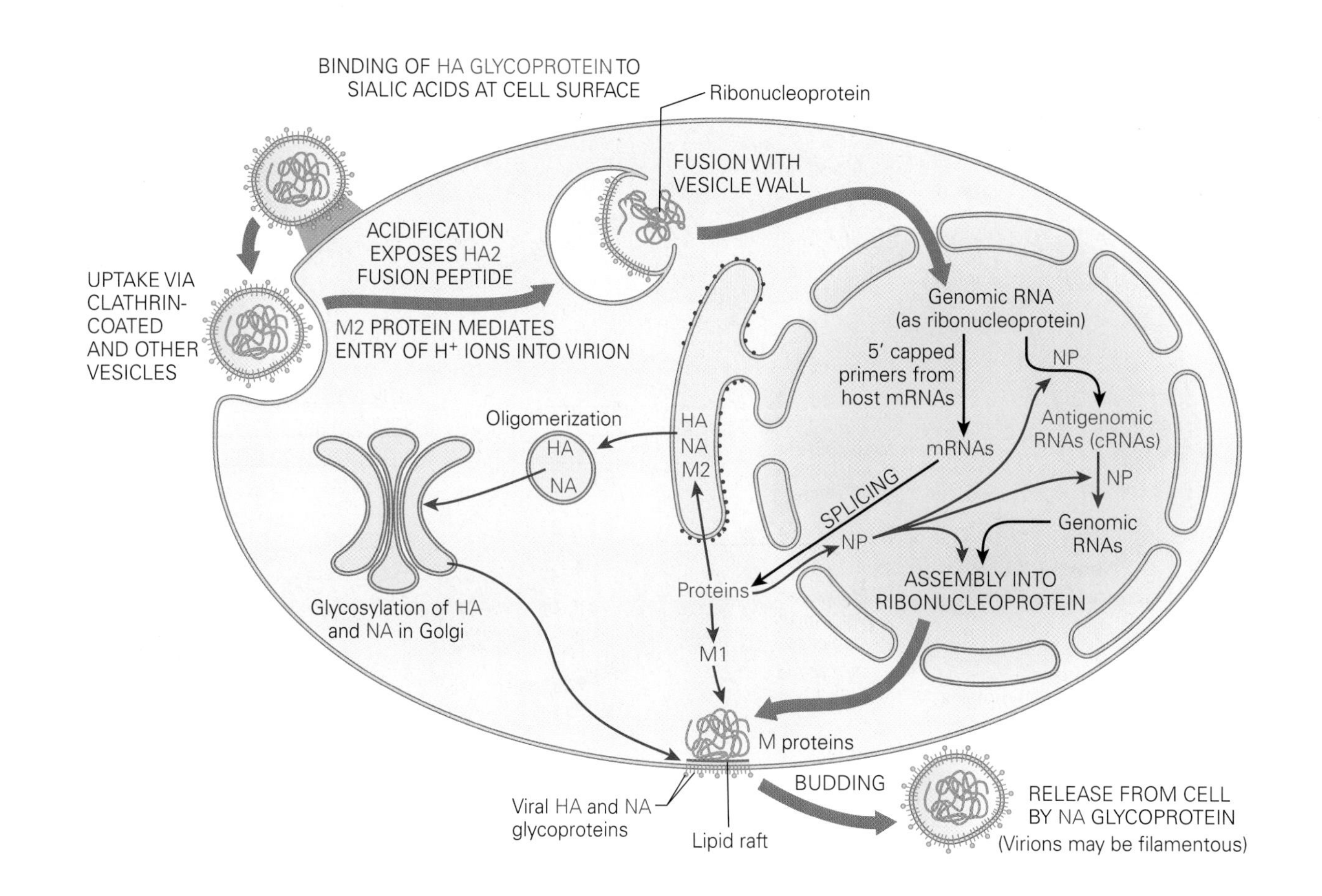

Papillomaviridae

Family	Genome type	Genome size (kbp)	Envelope	Capsid	Size (nm)	Virion proteins
Papillomaviridae	dsDNA	8	No	Icosahedral	52–55	2 + cellular histones

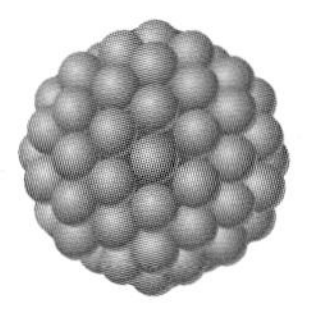

The diagram is drawn at triple size for illustration purposes.

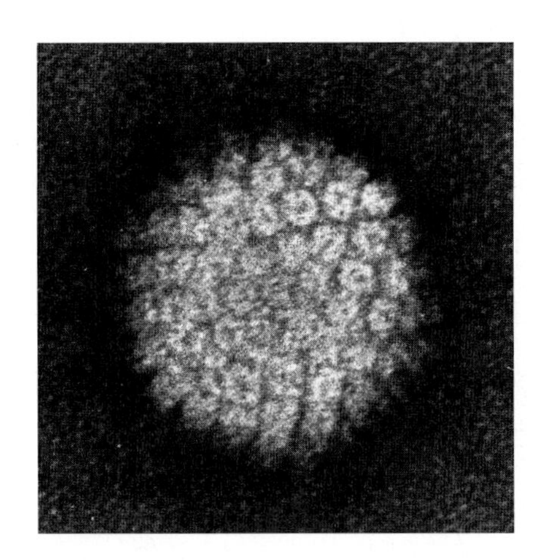

Examples of human pathogens

26 types of human papillomavirus have been assigned to species, many more have been identified. A subset of these (most notably types 16 and 18) are associated with cervical cancer

Members	Main host types	Infecting humans	Example of human disease	Vaccines	Antiviral drugs
44	Mammals (humans)	26	Warts, cervical cancer	Yes	Yes

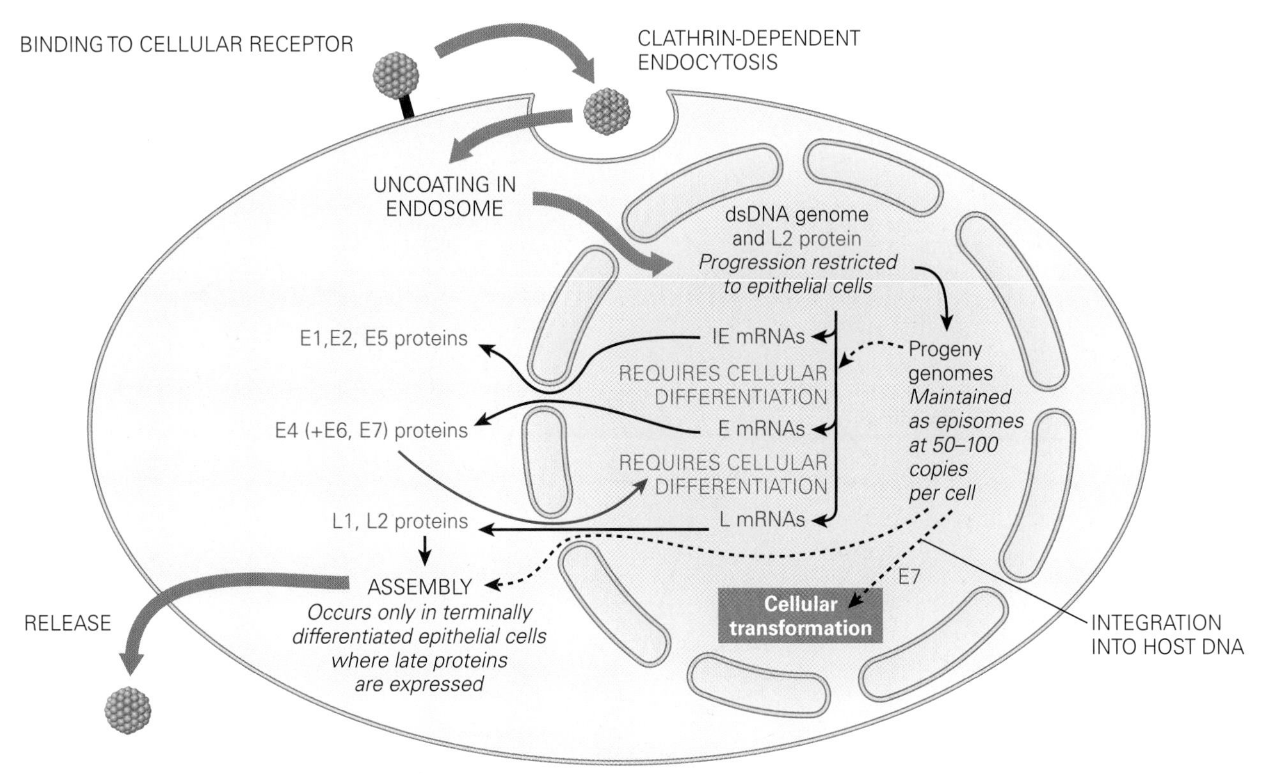

Papillomaviruses produce three temporal classes of mRNAs and proteins: immediate early (IE), early (E), and late (L). Production of the E and L classes occurs only in terminally differentiated cells

Paramyxoviridae

Family	Genome type	Genome size (kb)	Envelope	Capsid	Size (nm)	Virion proteins
Paramyxoviridae	ssRNA (–)	15.2–15.9	Yes	Elongated helical	150–200	6–7

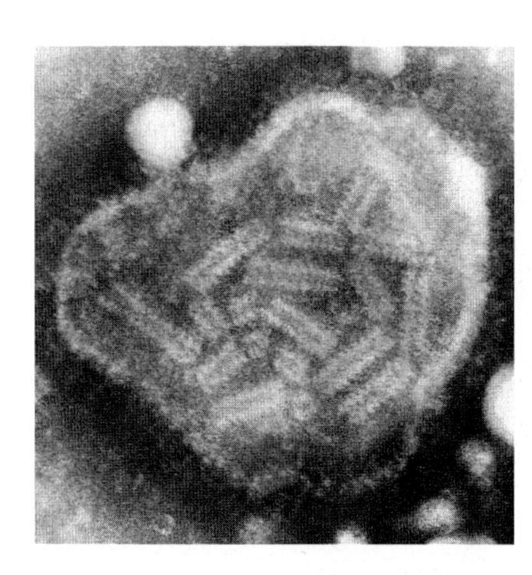

Examples of human pathogens

Measles virus, mumps virus, human parainfluenzaviruses 1–4, respiratory syncytial virus, human metapneumovirus, Hendra virus, Nipah virus

Members	Main host types	Infecting humans	Examples of human disease	Vaccines	Antiviral drugs
34	Mammals (humans), birds	10	Measles, mumps	Yes	(Yes)

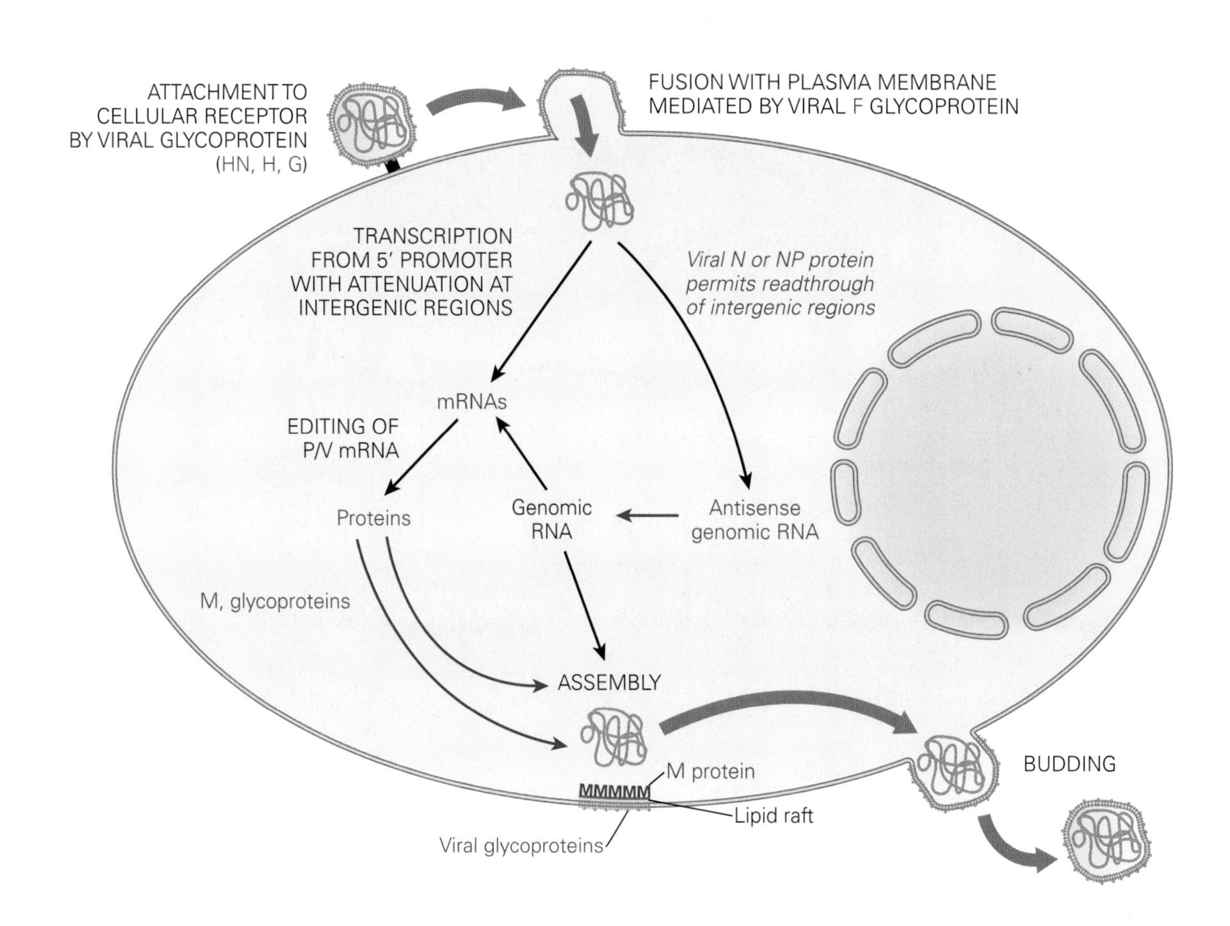

Parvoviridae

Family	Genome type	Genome size (kb)	Envelope	Capsid	Size (nm)	Virion proteins
Parvoviridae	ssDNA (–)	5	No	Icosahedral	18–26	2–3

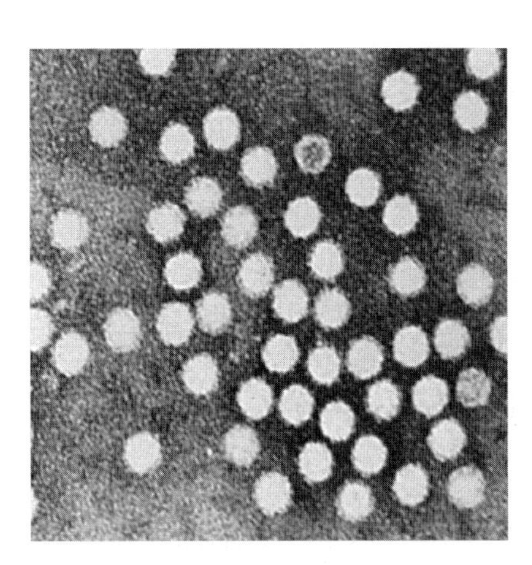

Examples of human pathogens

Human parvovirus B19

Members	Main host types	Infecting humans	Examples of human disease	Vaccines	Antiviral drugs
37	Mammals (humans), birds, insects	6	Anemia	No	No

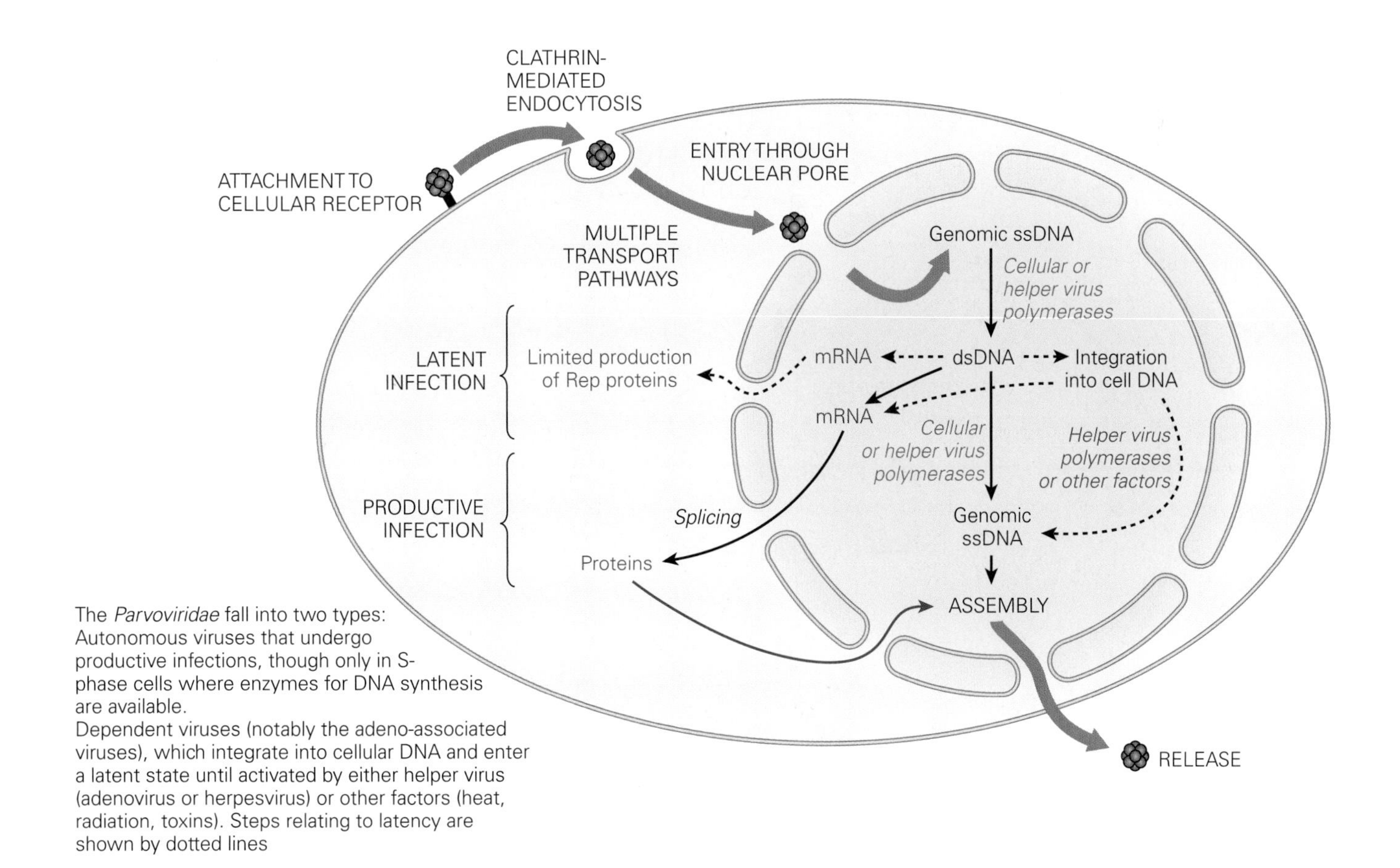

The *Parvoviridae* fall into two types:
Autonomous viruses that undergo productive infections, though only in S-phase cells where enzymes for DNA synthesis are available.
Dependent viruses (notably the adeno-associated viruses), which integrate into cellular DNA and enter a latent state until activated by either helper virus (adenovirus or herpesvirus) or other factors (heat, radiation, toxins). Steps relating to latency are shown by dotted lines

Picobirnaviridae

Family	Genome type	Genome size (kbp)	Envelope	Capsid	Size (nm)	Virion proteins
Picobirnaviridae	segmented dsRNA	4	No	Icosahedral	35–40	4?

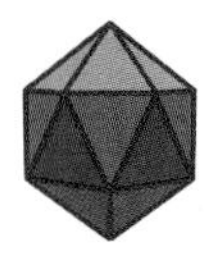

Examples of human pathogens

Human picobirnavirus

Members	Main host types	Infecting humans	Examples of human disease	Vaccines	Antiviral drugs
2	Mammals (humans)	1	Gastrointestinal illness	No	No

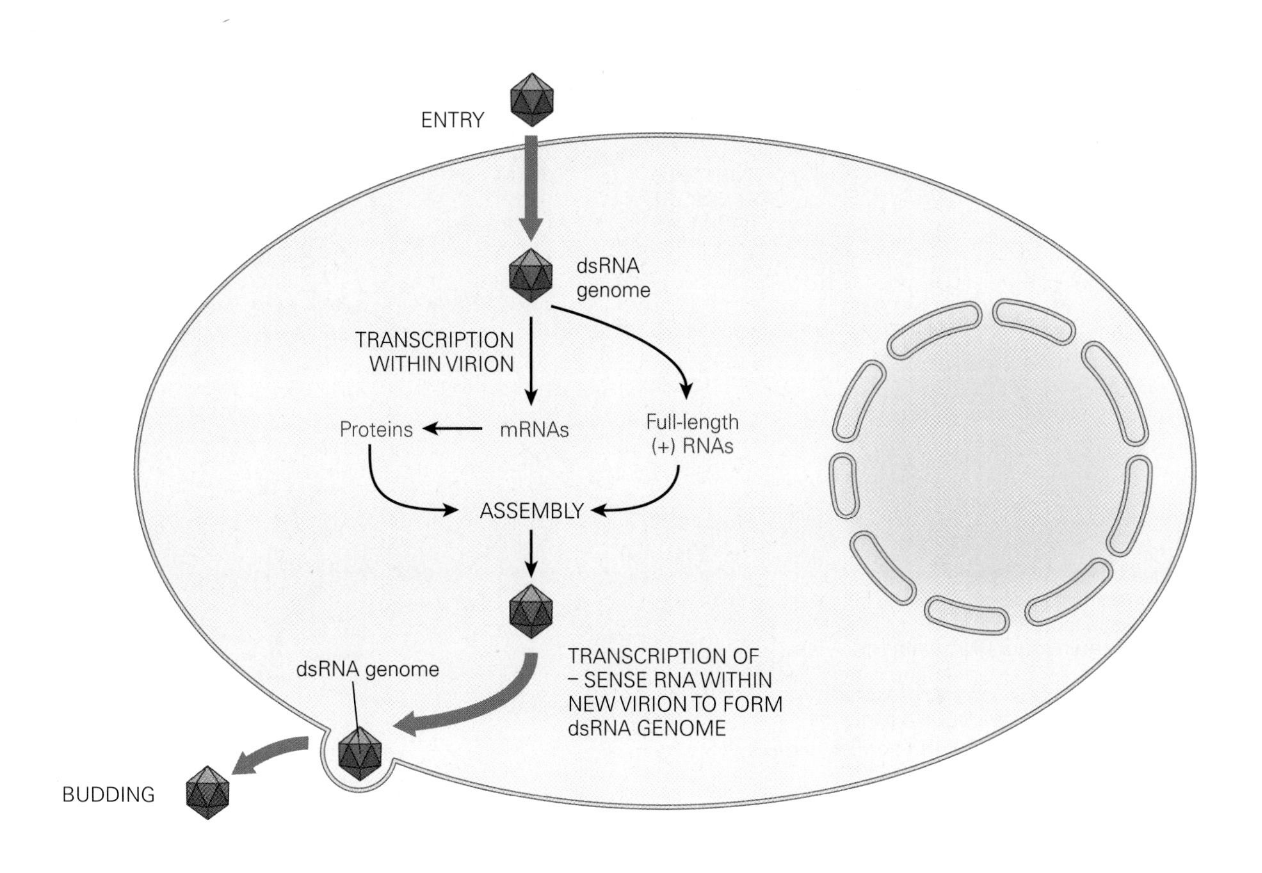

Picornaviridae

Family	Genome type	Genome size (kb)	Envelope	Capsid	Size (nm)	Virion proteins
Picornaviridae	ssRNA (+)	7–8.5	No	Icosahedral	27–30	5

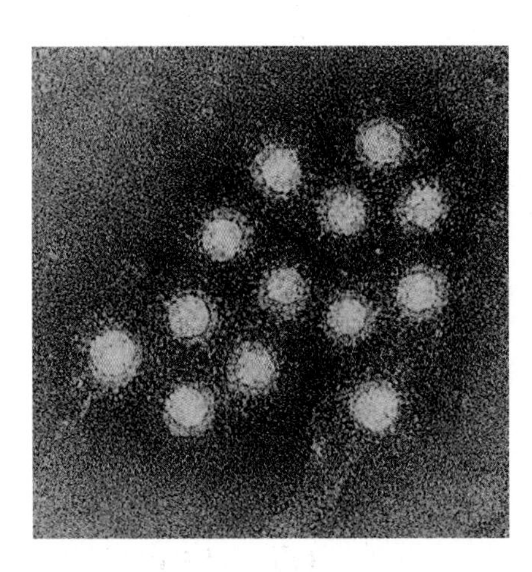

Examples of human pathogens

Polio virus types 1–3, hepatitis A virus, human rhinoviruses, human coxsackieviruses, human echoviruses, human parechoviruses, human enteroviruses, Vilyuisk encephalitis virus

Members	Main host types	Infecting humans	Examples of human disease	Vaccines	Antiviral drugs
22	Mammals (humans), birds	11–13	Polio, meningitis	(Yes*)	No

* Polioviruses 1–3 (trivalent), hepatitis A.

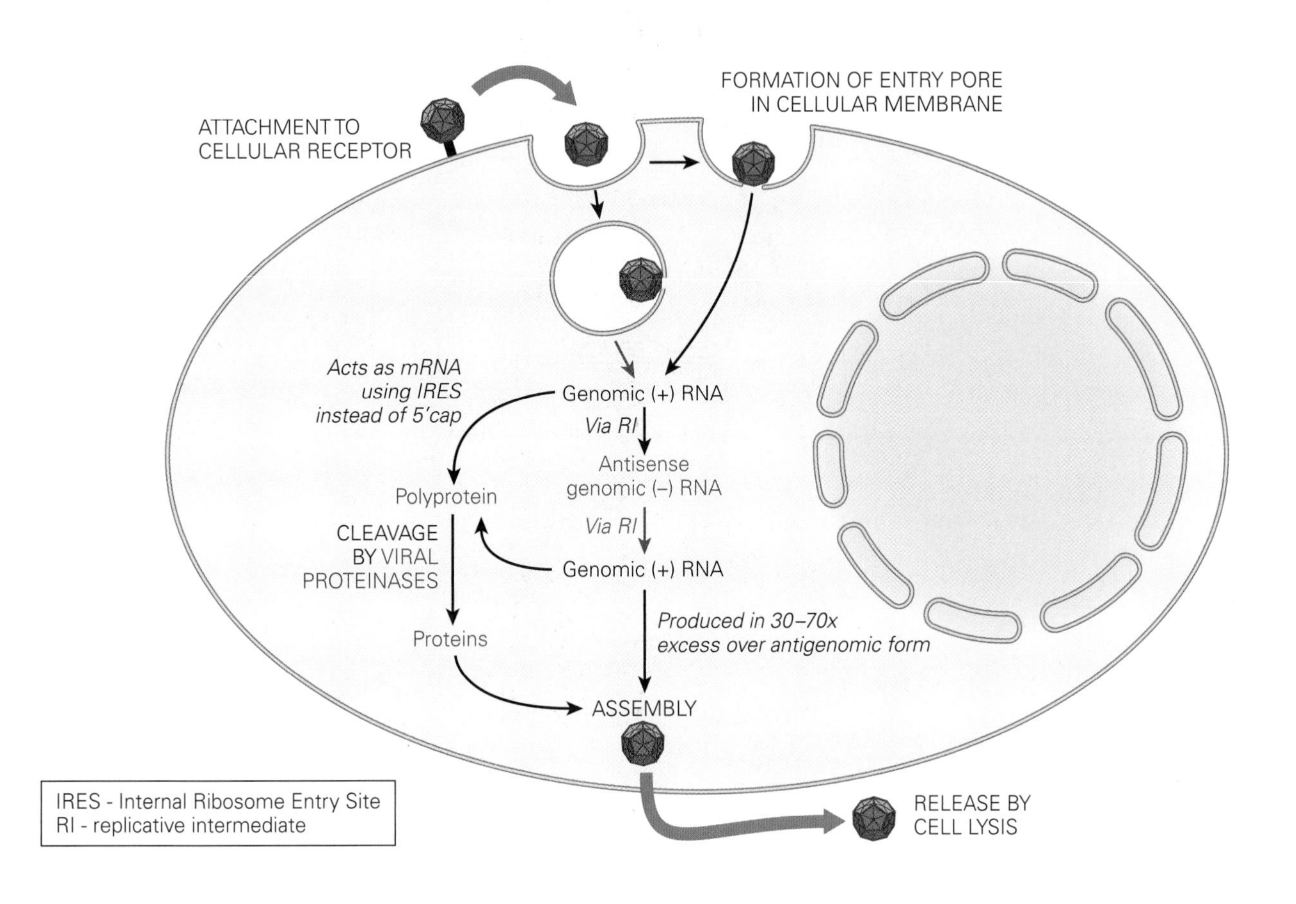

Polyomaviridae

Family	Genome type	Genome size (kbp)	Envelope	Capsid	Size (nm)	Virion proteins
Polyomaviridae	dsDNA	5	No	Icosahedral	40–55	3 + cellular histones

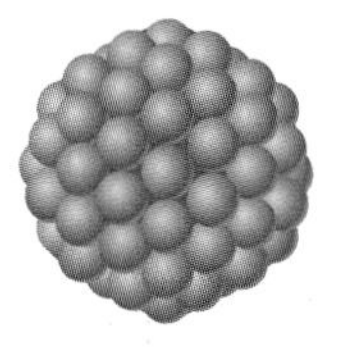

The diagram is drawn at triple size for illustration purposes.

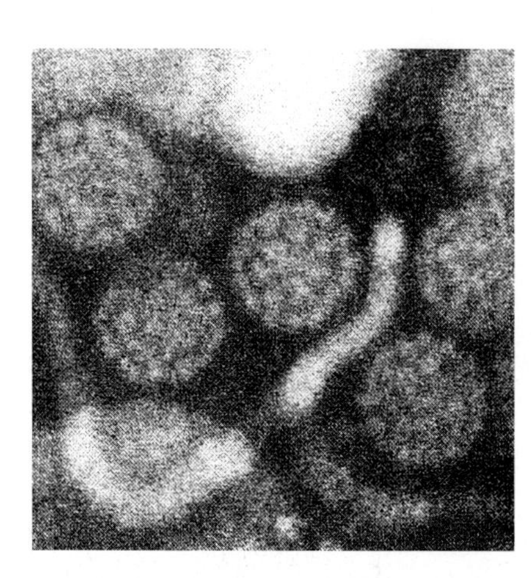

Examples of human pathogens

JC virus, BK virus

Members	Main host types	Infecting humans	Example of human disease	Vaccines	Antiviral drugs
13	Mammals (humans), bird	2	Progressive multifocal leukoencephalopathy; malignancies?	No	No

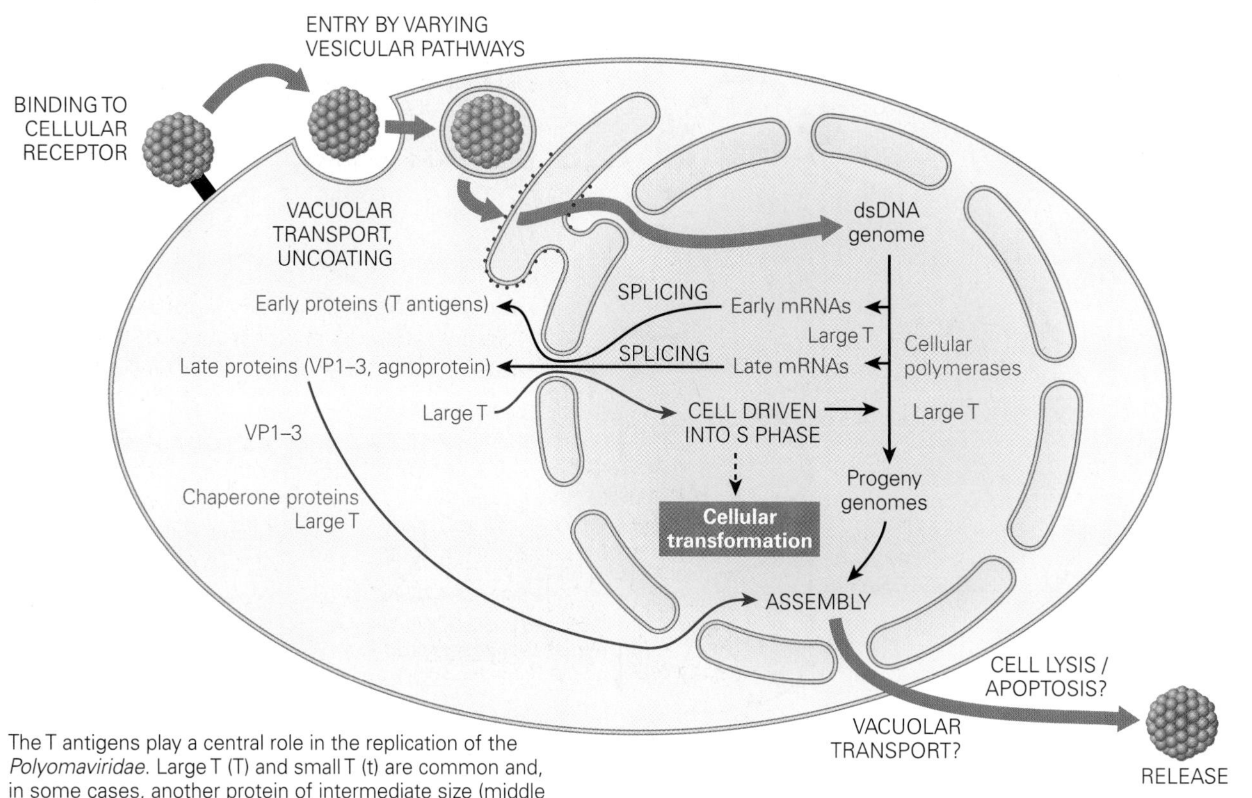

The T antigens play a central role in the replication of the *Polyomaviridae*. Large T (T) and small T (t) are common and, in some cases, another protein of intermediate size (middle T) is produced. Large T antigen is involved in many of the processes of virus infection as shown above

Poxviridae

Family	Genome type	Genome size (kbp)	Envelope	Capsid	Size (nm)	Virion proteins
Poxviridae	dsDNA	130–375	Yes	Complex	140–260 or 140–260 × 220–450	75+

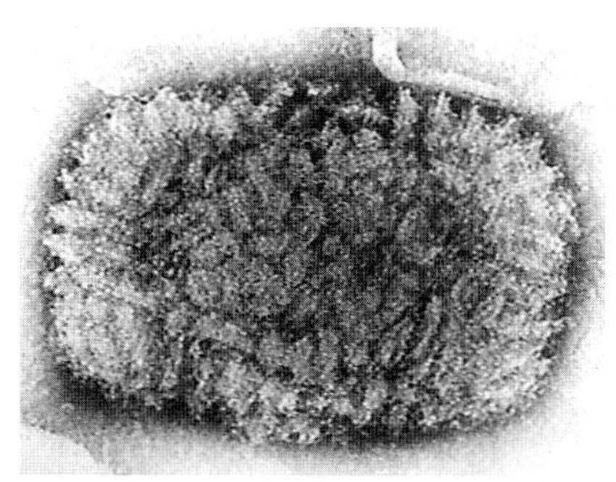

Examples of human pathogens

Orf virus, molluscum contagiosum virus, (smallpox virus)

Image shows dried "mulberry" form. Dried images showing internal structure (see Figure 1.5) are referred to as "capsular" form. Hydrated samples show smooth surface. An additional envelope may also be present.

Members	Main host types	Infecting humans	Examples of human disease	Vaccines	Antiviral drugs
62	Mammals (humans), birds, insects	6	Orf, molluscum contagiosum, (smallpox)	(Yes)	No

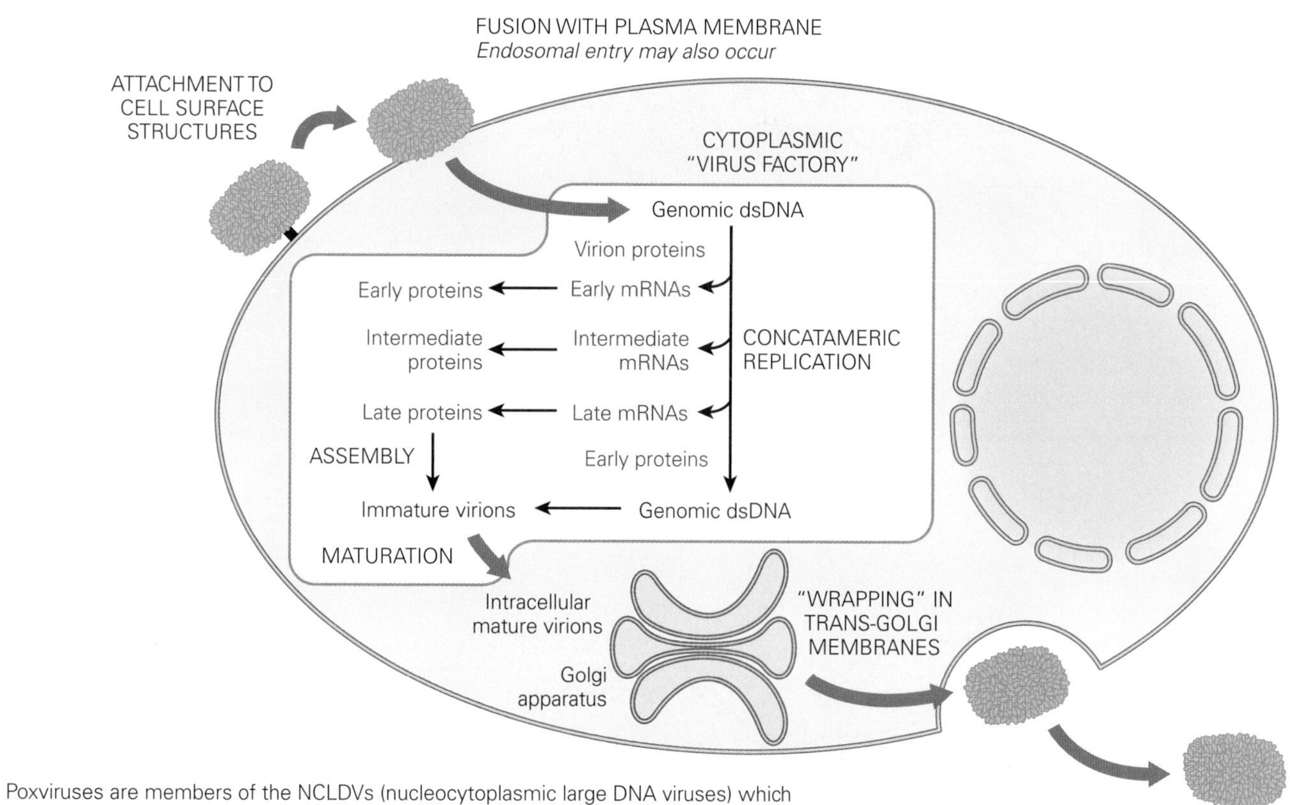

Poxviruses are members of the NCLDVs (nucleocytoplasmic large DNA viruses) which share a range of genes and which replicate in the cytoplasm, independently of the cell nucleus. Other members are not known to infect humans (except possibly *Mimivirus*)

Reoviridae

Family	Genome type	Genome size (kbp)	Envelope	Capsid	Size (nm)	Virion proteins
Reoviridae	dsRNA	18.2–30.5	No	Icosahedral	60–80	10–12

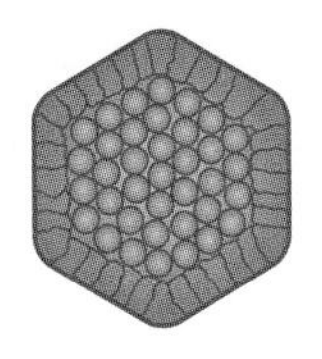

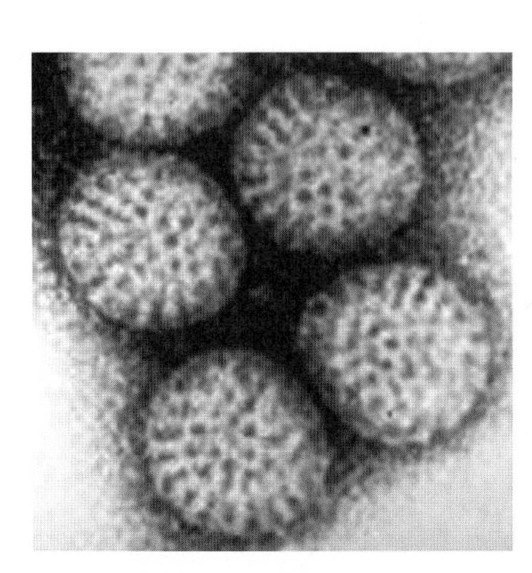

Examples of human pathogens

Human rotaviruses A–E, Colorado tick fever virus

Members	Main host types	Infecting humans	Examples of human disease	Vaccines	Antiviral drugs
79	Mammals (humans), birds, reptiles, fish, molluscs, arthropods, insects, plants, fungi	13+	Diarrhea	(Yes)	No

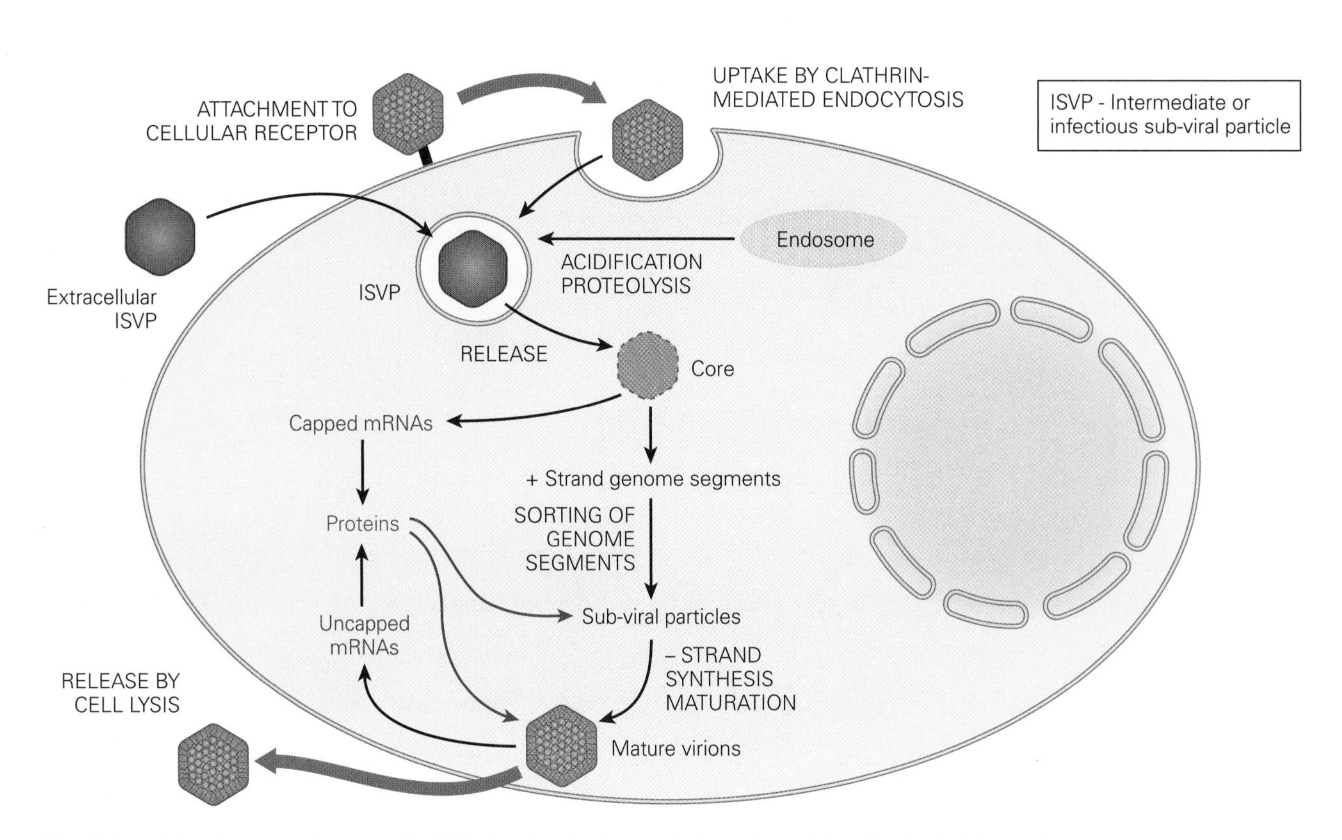

Virus-induced switch to use of uncapped mRNAs late in infection results in preferential synthesis of viral proteins

Retroviridae

Family	Genome type	Genome size (kb)	Envelope	Capsid	Size (nm)	Virion proteins
Retroviridae	diploid RNA (via dsDNA)	7–11	Yes	Spherical or pleomorphic	80–100	3–9

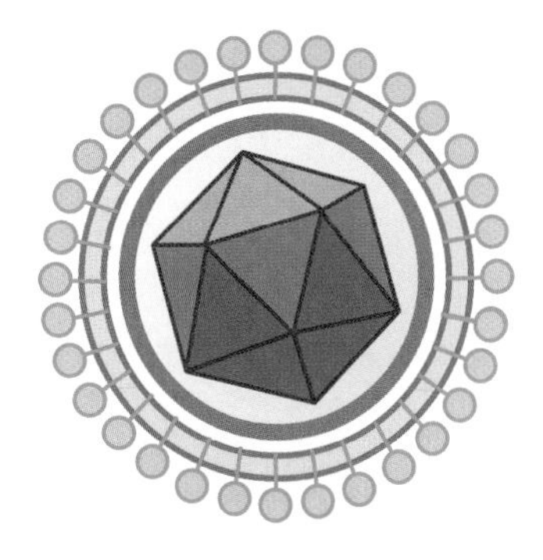

The diagram is drawn at double size for illustration purposes.

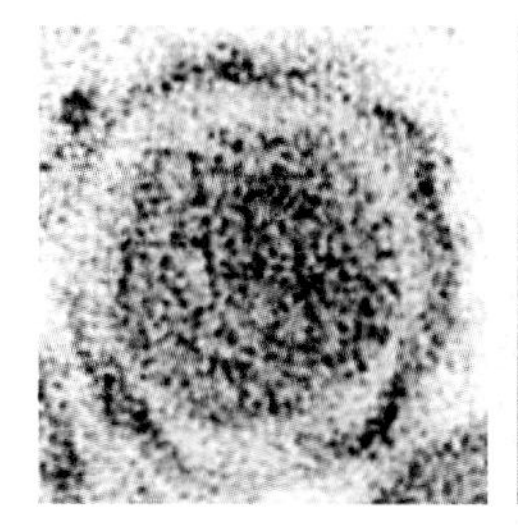
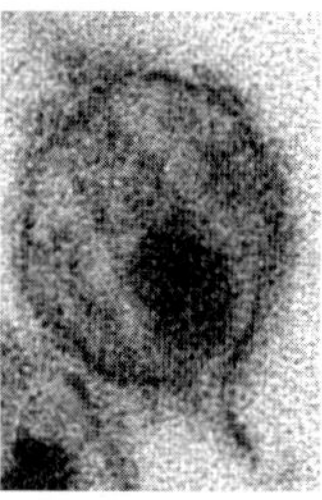

HTLV and HIV (showing distinctive conical nucleocapsid)

Examples of human pathogens

Human immunodeficiency virus (HIV), Human T-cell leukemia viruses (HTLV) types 1 and 2

Members	Main host types	Infecting humans	Examples of human disease	Vaccines	Antiviral drugs
53	Mammals (humans), birds, reptiles	4+	AIDS	No	Yes

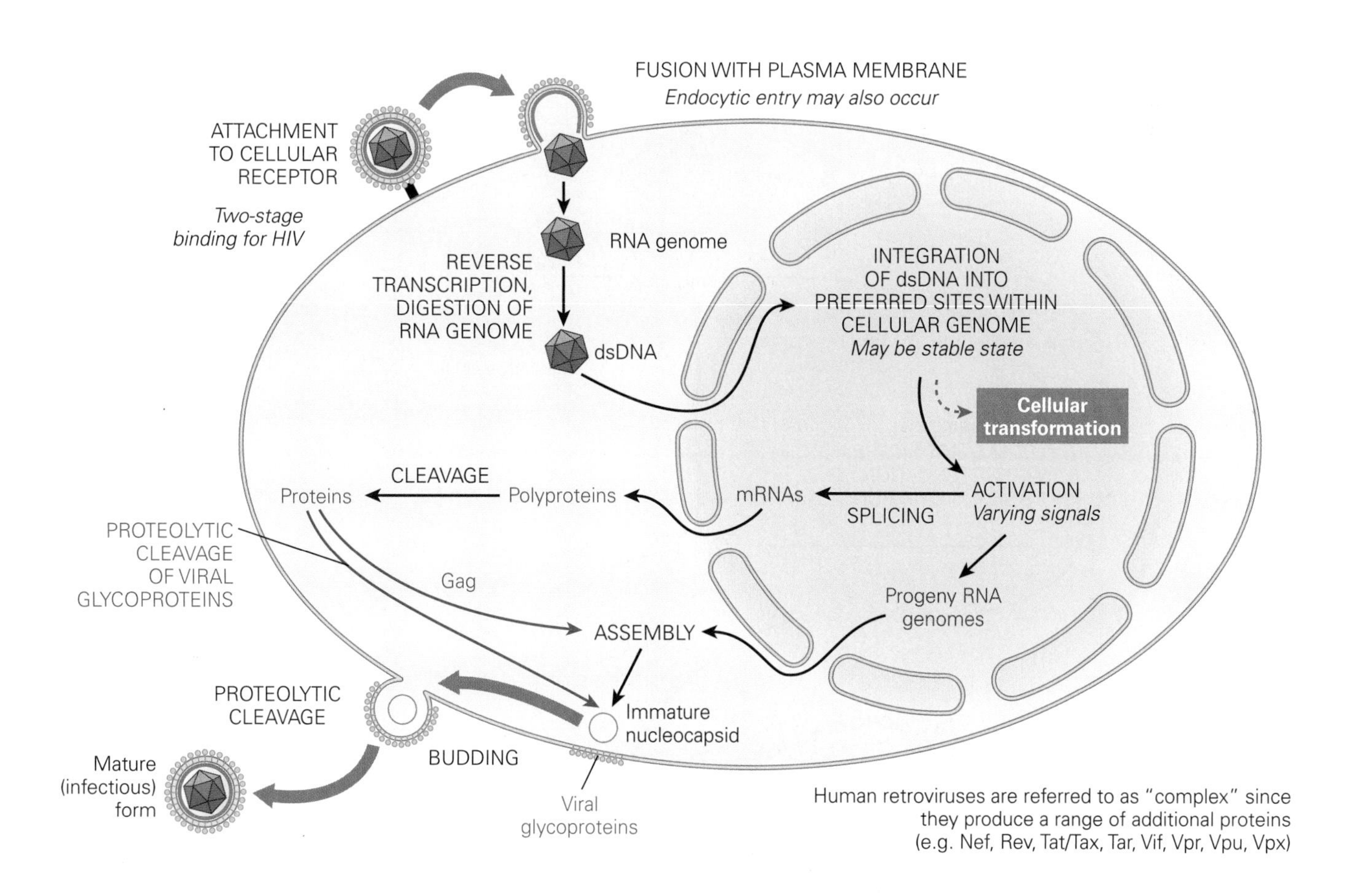

Human retroviruses are referred to as "complex" since they produce a range of additional proteins (e.g. Nef, Rev, Tat/Tax, Tar, Vif, Vpr, Vpu, Vpx)

Rhabdoviridae

Family	Genome type	Genome size (kb)	Envelope	Capsid	Size (nm)	Virion proteins
Rhabdoviridae	ssRNA (–)	11–15	Yes	Bullet-shaped	45–100 × 100–430	5–11

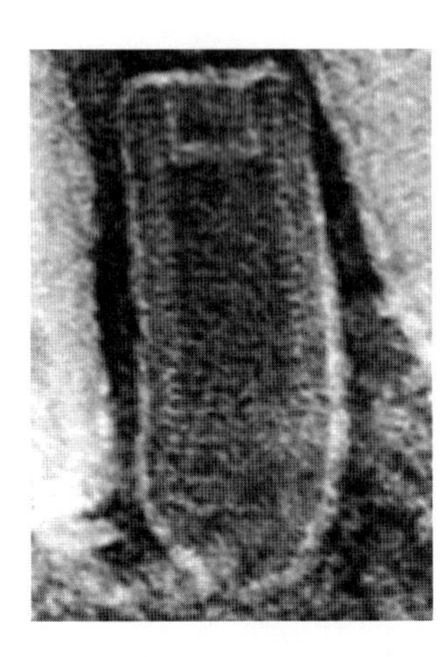

Examples of human pathogens

Rabies virus

Members	Main host types	Infecting humans	Examples of human disease	Vaccines	Antiviral drugs
38	Mammals (humans), fish, plants	4+	Rabies	Yes	No

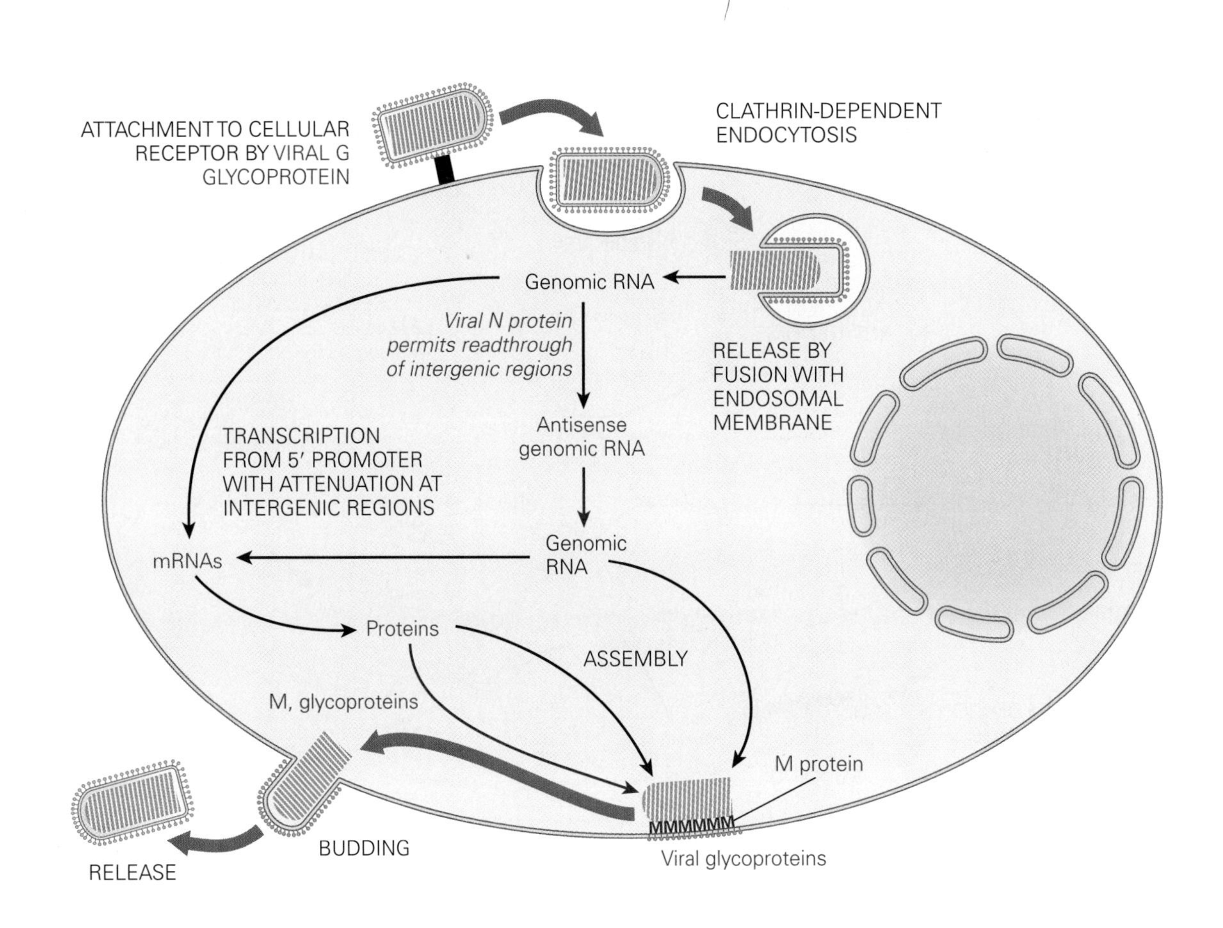

Togaviridae

Family	Genome type	Genome size (kb)	Envelope	Capsid	Size (nm)	Virion proteins
Togaviridae	ssRNA (+)	9.7–11.8	Yes	Spherical or pleomorphic	70	5–7

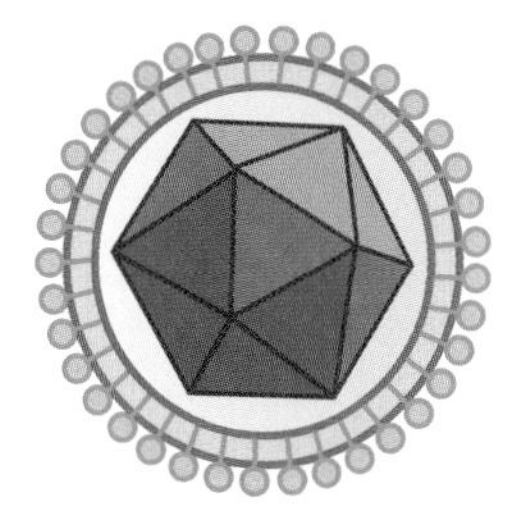

The diagram is drawn at double size for illustration purposes.

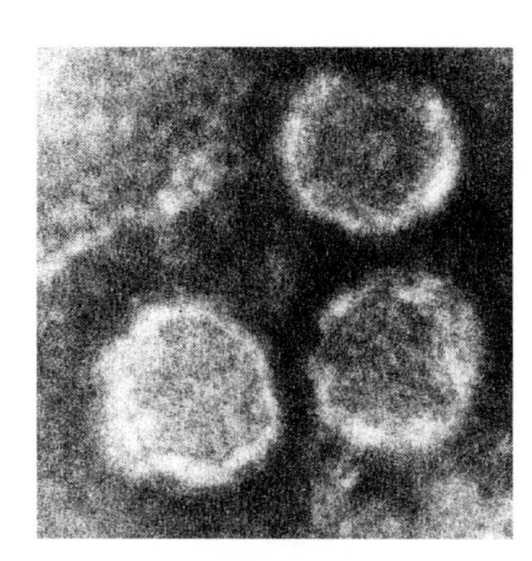

Examples of human pathogens

Rubella virus, Sindbis virus, eastern, western, and Venezuelan equine encephalitis, Ross River virus, Chikungunya virus, O'nyong-nyong virus

Members	Main host types	Infecting humans	Examples of human disease	Vaccines	Antiviral drugs
40	Mammals (humans), birds, arthropods (arboviruses)	13+	Rubella	(Yes)	No

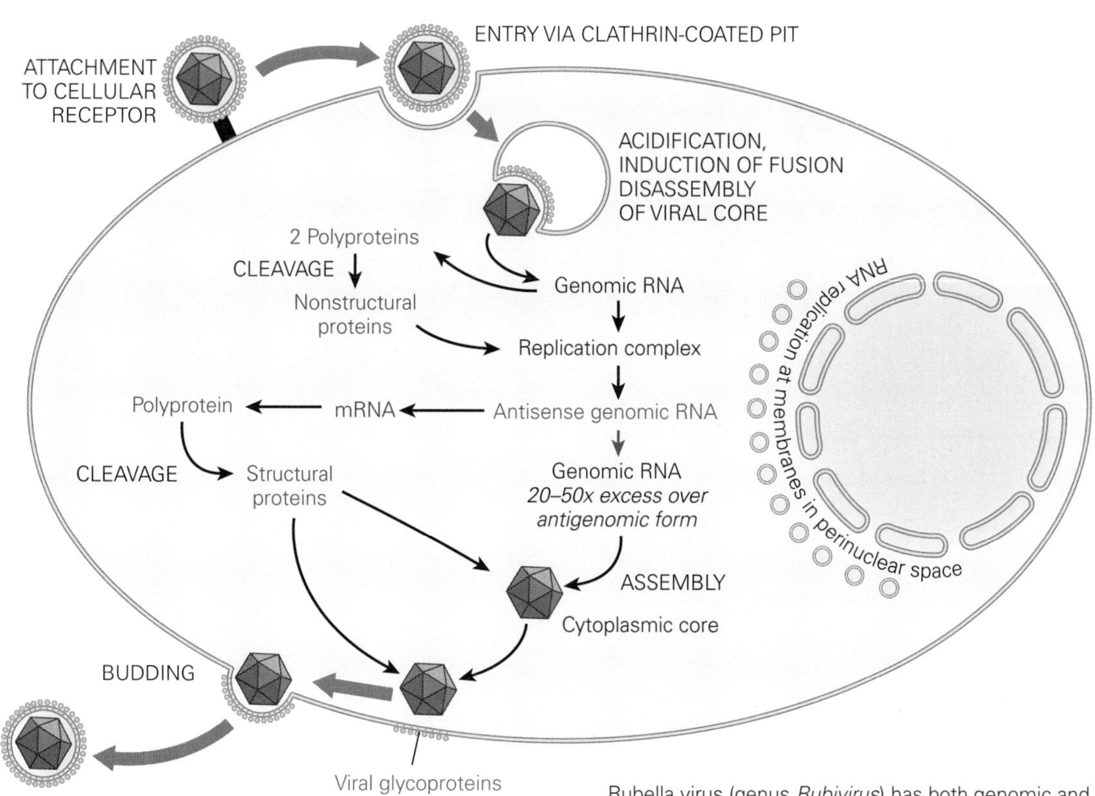

Rubella virus (genus *Rubivirus*) has both genomic and replicative differences to other members of the *Togaviridae*. One example of this is that Rubella virus buds into the Golgi apparatus as well as from the plasma membrane

Appendix II
Current Antiviral Drugs

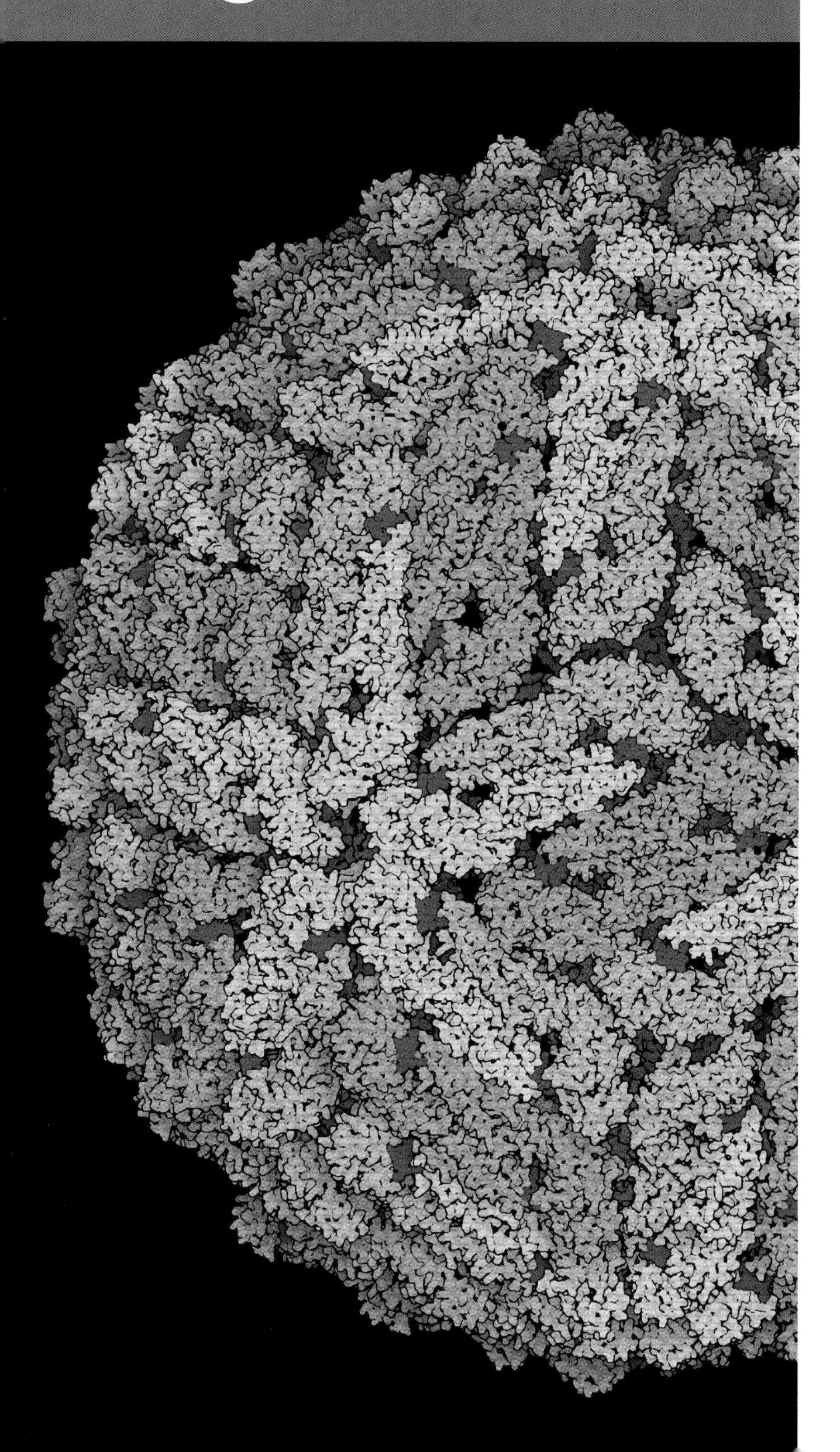

Note that while this table is as extensive as was possible at the time of writing, it is not possible to provide details of all antiviral drugs that may be available worldwide since approval status and/or availability varies between countries. Some of the drugs listed (e.g. brivudine, inosine pranobex, rimantadine, and tromantadine) may be available in a limited range of countries.

About the image
Dengue Virus
(Courtesy of the Research Collaboratory for Structural Bioinformatics Protein Data Bank and David S. Goodsell, The Scripps Research Institute, USA.)

Drug name (Trade name**)	Chemical name*	Virus target(s)	Mode of action	Route of admini-stration	Structure
Abacavir (Ziagen)	[(1*R*)-4-[2-amino-6-(cyclopropylamino) purin-9-yl]-1-cyclopent-2-enyl] methanol	Human immuno-deficiency virus (*Retroviridae*)	Nucleoside analog reverse transcriptase inhibitor (chain terminator)	Oral	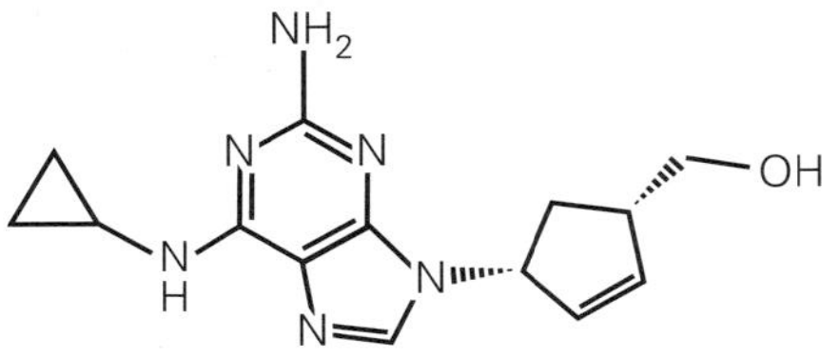
Aciclovir (Zovirax, Zovir)	2-amino-9-(2-hydroxyethoxy-methyl)-3,9-dihydro-6*H*-purin-6-one	Herpes simplex virus, varicella-zoster virus (*Herpesviridae*)	DNA polymerase inhibitor (nucleoside analog, chain terminator)	Oral, topical, intravenous	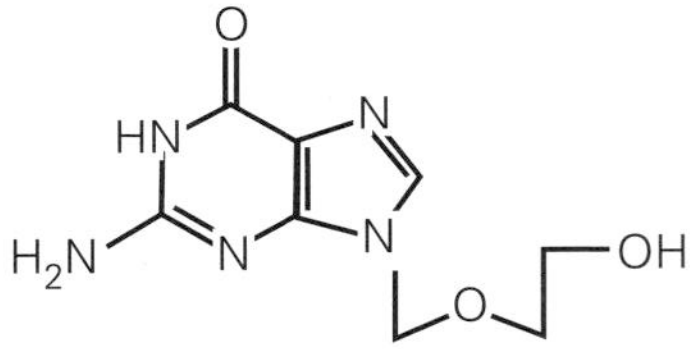
Adefovir dipivoxil (Preveon, Hepsera)	{[2-(6-amino-9*H*-purin-9-yl)ethoxy]methyl} phosphonic acid	Hepatitis B virus (*Hepadna-viridae*)	Reverse transcriptase inhibitor (nucleotide analog, chain terminator)	Oral	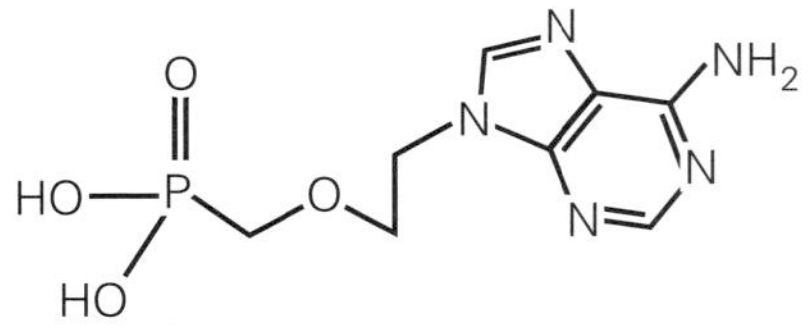
Amantadine (Symmetrel)	Adamantan-1-amine	Influenza A virus (*Orthomyxo-viridae*)	Ion channel inhibitor, interferes with viral penetration and uncoating, may interfere with assembly	Oral	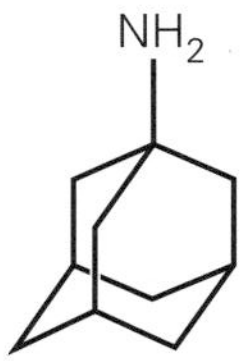
Amprenavir (Agenerase)	Tetrahydrofuran-3-yl[3-[(4-aminophenyl) sulfonyl-(2-methylpropyl) amino]-1-benzyl-2-hydroxy-propyl] aminomethanoate	Human immuno-deficiency virus (*Retroviridae*)	Protease inhibitor, replaced by fosamprenavir	Oral	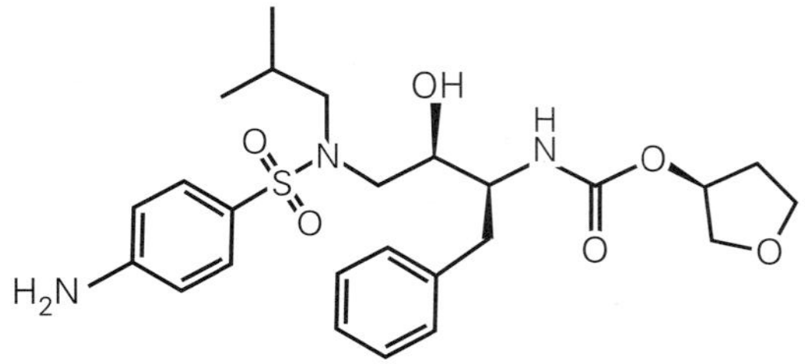

Atazanavir (Reyataz)	Methyl N-[(1*S*)-1-[[[(2*S*,3*S*)-2-hydroxy-3-[[(2*S*)-2- (methoxy-carbonylamino)-3,3-dimethyl-butanoyl] amino]-4-phenyl-butyl]-[(4-pyridin-2-ylphenyl)methyl] amino] carbamoyl]-2,2-dimethyl-propyl] carbamate	Human immuno-deficiency virus (*Retroviridae*)	Protease inhibitor	Oral	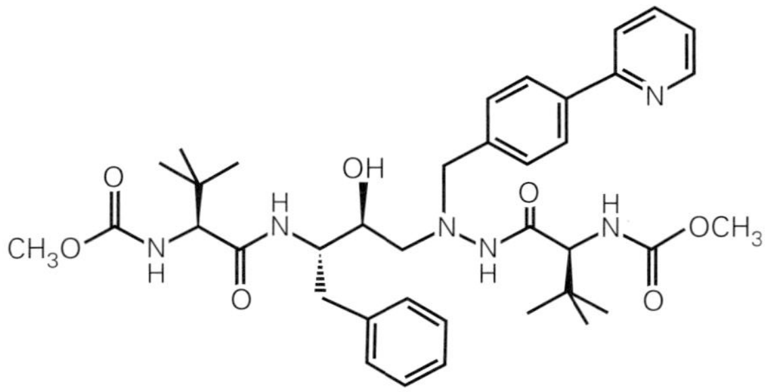
Brivudine (Helpin, Zerpex, Zonavir, Zostex)	5-[(*E*)-2-bromoethenyl]-1-4*S*,5*R*)-4-hydroxy-5-hydroxymethyl) oxolan-2-yl]pyrimidine-2,4-dione	Varicella-zoster virus (*Herpesviridae*)	DNA polymerase inhibitor (nucleoside analog)	Oral	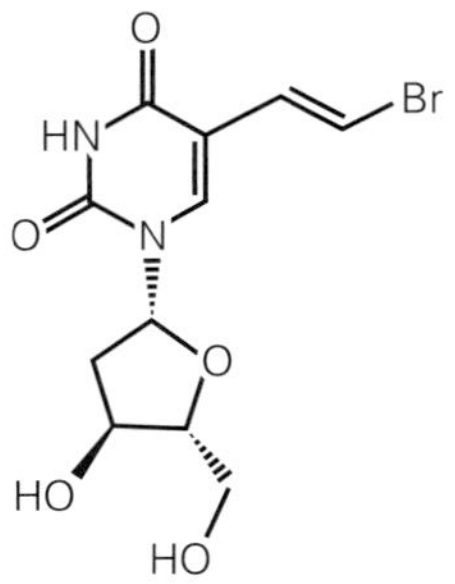
Cidofovir (Vistide)	[1-(4-amino-2-oxo-pyrimidin-1-yl)-3-hydroxy-propan-2-yl] oxymethyl-phosphonic acid	Cytomegalovirus (*Herpesviridae*)	DNA polymerase inhibitor (nucleotide analog)	Intravenous	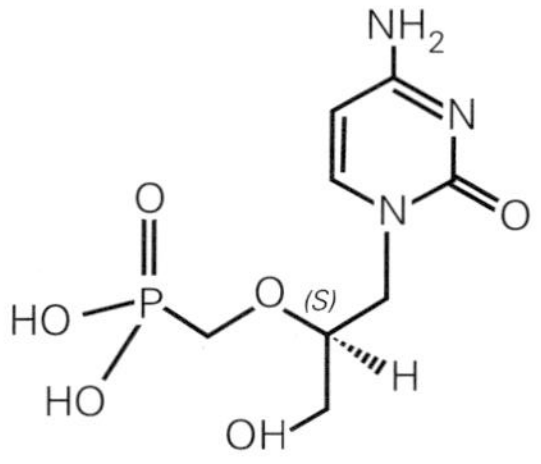
Darunavir (Prezista)	(1*R*,5*S*,6*R*)-2,8-dioxabicyclo[3.3.0] oct-6-yl]*N*-[(2*S*,3*R*)-4-[(4-aminophenyl) sulfonyl-(2-methylpropyl)amino]-3-hydroxy-1-phenyl-butan-2-yl]carbamate	Human immuno-deficiency virus (*Retroviridae*)	Protease inhibitor	Oral	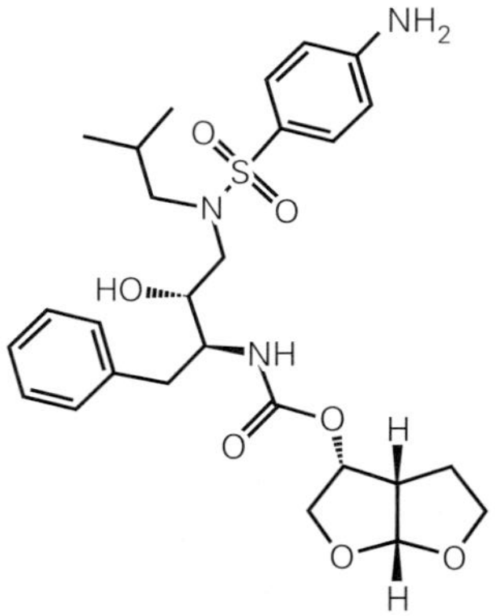

Drug name (Trade name**)	Chemical name*	Virus target(s)	Mode of action	Route of admini-stration	Structure
Delavirdine (Rescriptor)	*N*-[2-[4-[3-(1-methylethyl amino) pyridin-2-yl] piperazin-1-yl] carbonyl-1*H*-indol-5-yl] methanesulfonamide	Human immuno-deficiency virus (*Retroviridae*)	Non-nucleoside reverse transcriptase inhibitor	Oral	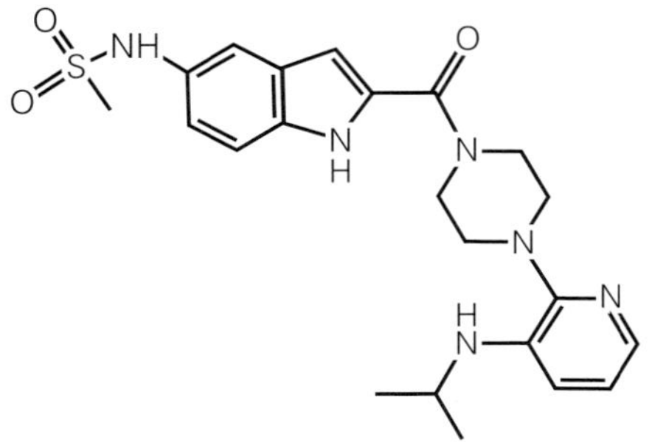
Didanosine (Videx)	9-[5-(hydroxymethyl) oxolan-2-yl]-3*H*-purin-6-one	Human immuno-deficiency virus (*Retroviridae*)	Nucleoside analog reverse transcriptase inhibitor (competitive and chain terminator)	Oral	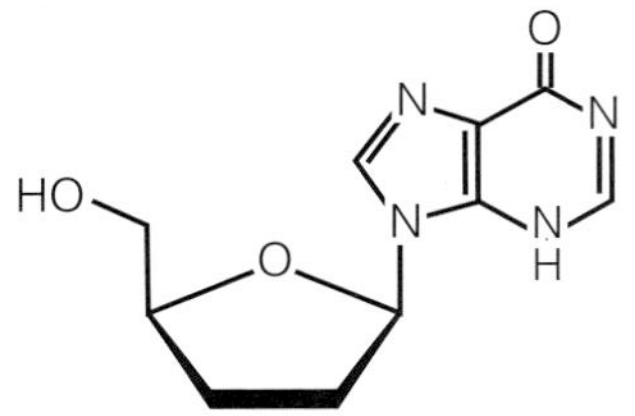
Docosanol (Abreva)	Docosan-1-ol	Herpes simplex virus (*Herpesviridae*) (and other enveloped viruses)	Fusion (entry) inhibitor	Topical	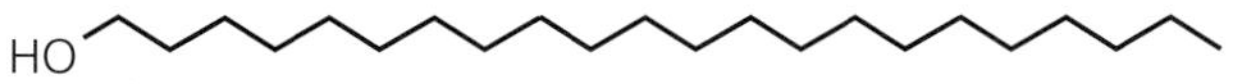
Efavirenz (Sustiva, Stocrin)	8-chloro-5-(2-cyclopropylethynyl)-5-(trifluoromethyl)-4-oxa-2-azabicyclo[4.4.0] deca-7,9,11-trien-3-one	Human immuno-deficiency virus (*Retroviridae*)	Non-nucleoside reverse transcriptase inhibitor	Oral	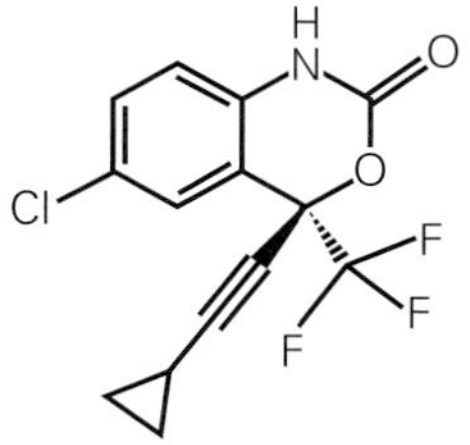

Emtricitabine (Emtriva, Coviracil)	4-amino-5-fluoro-1-[2-(hydroxymethyl)-1,3-oxathiolan-5-yl]-pyrimidin-2-one	Human immuno-deficiency virus (*Retroviridae*)	Nucleoside analog reverse transcriptase inhibitor (chain terminator)	Oral	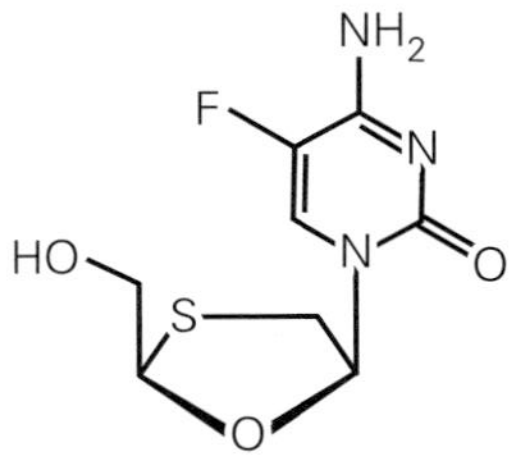
Enfuvirtide (Fuzeon)	Acetyl-YTSLIHSLIEESQ NQQEKNEQELLELDKW ASLWNWF-amide	Human immuno-deficiency virus (*Retroviridae*)	Fusion inhibitor (biomimetic peptide)	Subcutaneous	Structure too large to show in print. Please see http://www.garlandscience.com/viruses
Entecavir (Baraclude)	2-amino-9-[4-hydroxy-3-(hydroxymethyl)-2-methylidene-cyclopentyl]-3*H*-purin-6-one	Hepatitis B virus (*Hepadnaviridae*)	Reverse transcriptase inhibitor (nucleoside analog, multiple inhibitory stages)	Oral	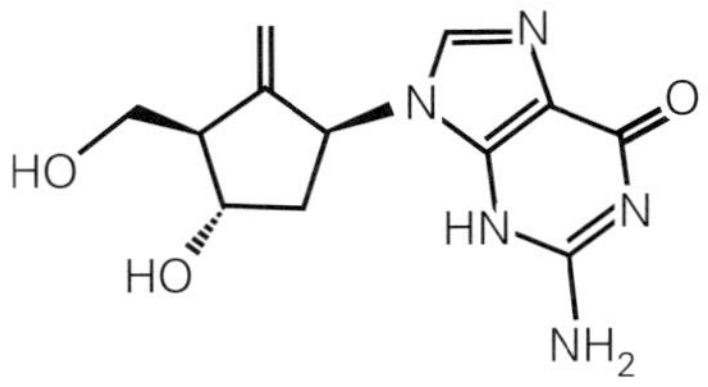
Etravirine (Intelence)	4-[6-amino-5-bromo-2-[(4-cyanophenyl) amino] pyrimidin-4-yl]oxy-3,5-dimethylbenzonitrile	Human immuno-deficiency virus (*Retroviridae*)	Non-nucleoside reverse transcriptase inhibitor	Oral	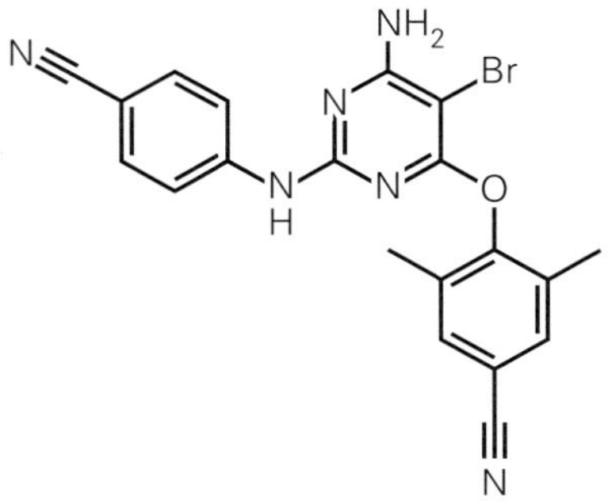
Famciclovir (Famvir)	[2-(acetyloxymethyl)-4-(2-aminopurin-9-yl)-butyl]acetate	Herpes simplex virus, varicella-zoster virus (*Herpesviridae*)	DNA polymerase inhibitor; prodrug of penciclovir	Oral	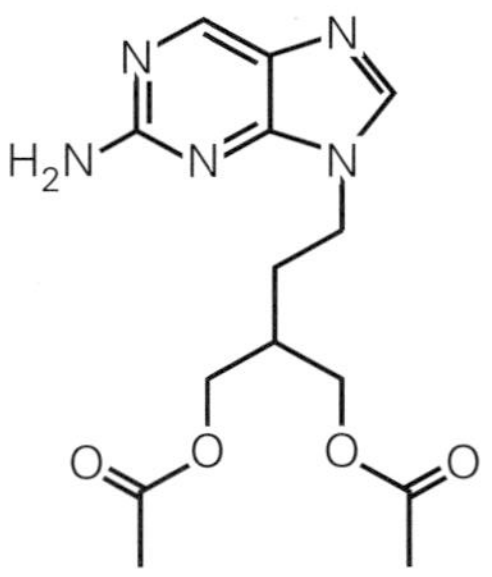

Drug name (Trade name**)	Chemical name*	Virus target(s)	Mode of action	Route of admini- stration	Structure
Fomivirsen (Vitravene)	(21 base antisense phosphorothioate oligonucleotide)	Cytomegalovirus (*Herpesviridae*)	Inhibits expression of the IE genes of cytomegalovirus (mechanism disputed)	Intravitreal	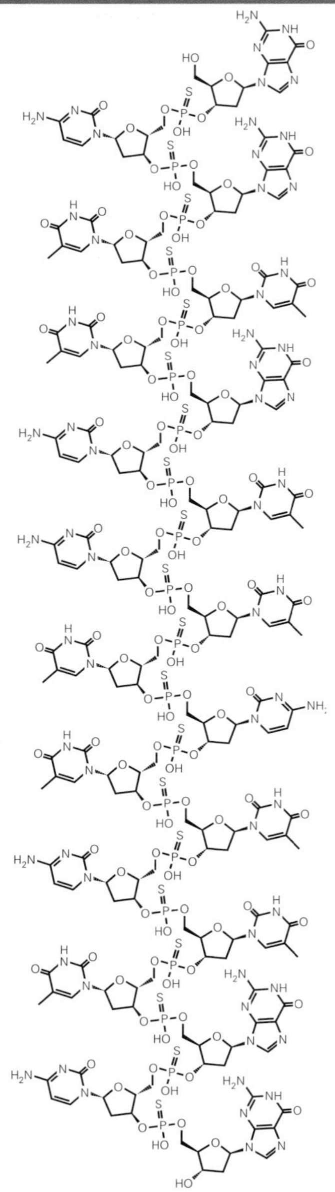

Fosamprenavir (Lexiva)	[(2*R*,3*S*)-1-[(4-aminophenyl) sulfonyl-(2-methylpropyl)amino]-3-{[(3*S*)-oxolan-]3-yl oxycarbonylamino}-4-phenyl-butan-2-yl] oxyphosphonic acid	Human immuno-deficiency virus (*Retroviridae*) (prodrug of and replacement for amprenavir)	Protease inhibitor	Oral	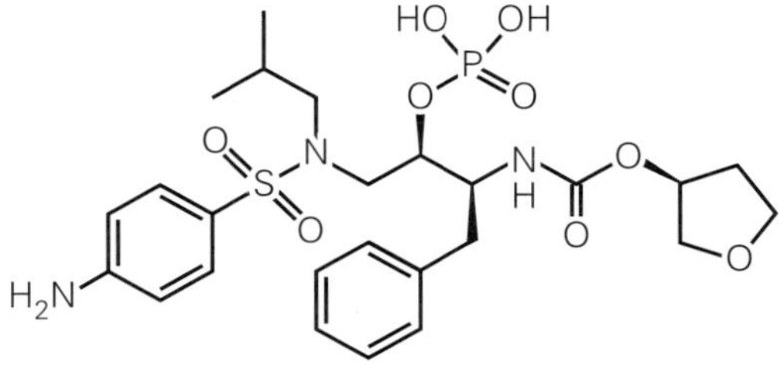
Foscarnet (Foscavir)	Trisodium phosphono-formate hexahydrate	Herpes simplex virus, other human herpesviruses (*Herpesviridae*)	DNA polymerase inhibitor (noncompetitive, at pyrophosphate binding site)	Intravenous	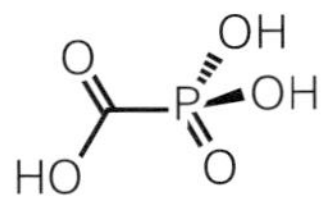
Ganciclovir (Cymevene, Cytovene, Vitrasert)	2-amino-9-(1,3-dihydroxypropan-2-yloxymethyl)-3*H*-purin-6-one propoxymethyl) guanine	Cytomegalovirus (*Herpesviridae*)	DNA polymerase inhibitor (competitive, nucleoside analog, chain terminator only after multiple incorporation)	Oral, intravenous, intravitreal	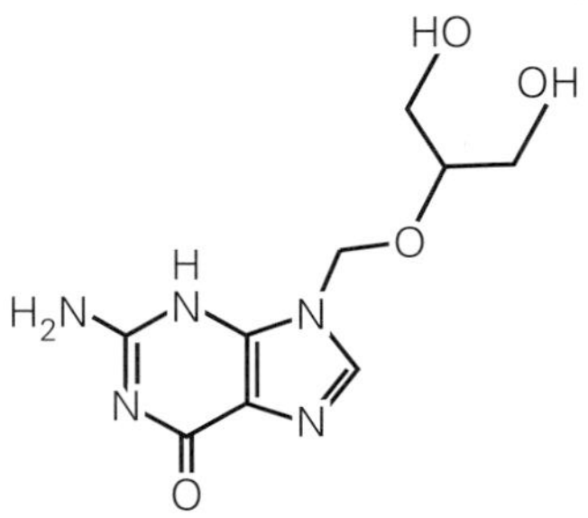
Idoxuridine (Herplex Liquifilm, Stoxil)	1-[(2*R*,4*S*,5*R*)-4-hydroxy-5-(hydroxymethyl) oxolan-2-yl]-5-iodo-pyrimidine-2,4-dione	Herpes simplex virus (*Herpesviridae*)	Incorporated into DNA, disrupts transcription and translation (nucleoside analog)	Ophthalmic	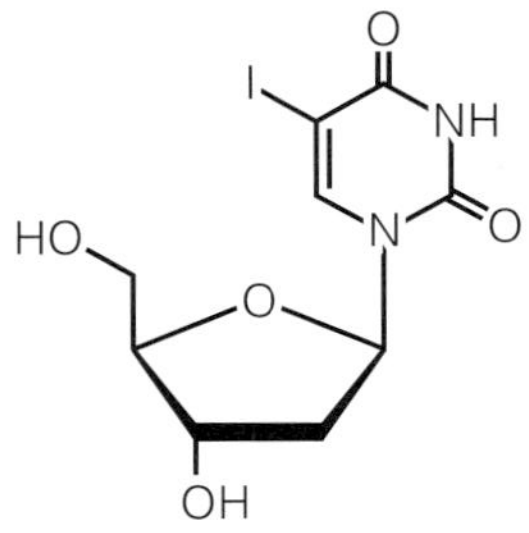
Imiquimod (Aldara)	1-isobutyl-1*H*-imidazo [4,5-c]quinolin-4-amine	Papillomaviruses (*Papovaviridae*)	Indirect action; stimulation of the innate immune system via Toll-like receptors	Topical	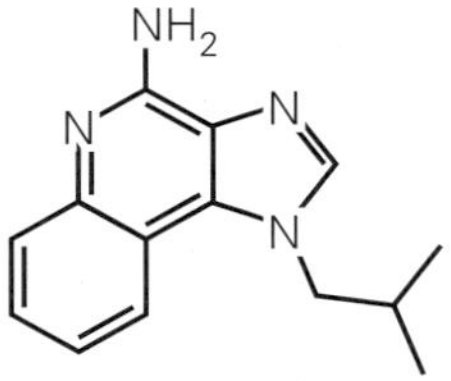

Drug name (Trade name**)	Chemical name*	Virus target(s)	Mode of action	Route of admini-stration	Structure
Indinavir (Crixivan)	1-[2-hydroxy-4-[(2-hydroxy-2,3-dihydro-1*H*-inden-1-yl) carbamoyl]-5-phenyl-pentyl]-4-(pyridin-3-ylmethyl)-*N*-tert-butyl-piperazine-2-carboxamide	Human immuno-deficiency virus (*Retroviridae*)	Protease inhibitor	Oral	
Inosine pranobex (Imunovir, isoprinosine)	9-[(2*R*,3*R*,4*S*,5*R*)-3,4-dihydroxy-5-(hydroxylmethyl) oxolan-2-yl]-3*H*-purin-6-one (inosine) plus acetamidobenzoic acid and dimethylamino-isopropanol	Herpes simplex virus (*Herpesviridae*), Papillomaviruses (*Papovaviridae*)	Indirect action; stimulation of immune system (T cells)	Oral	
Interferon alfa (IntronA, Roferon-A, Viraferon; pegylated as Pegasys, PegIntron, ViraferonPeg)	(Antiviral protein)	Hepatitis B virus (*Hepadnaviridae*), Hepatitis C virus (*Flaviviridae*), Papillomaviruses (*Papovaviridae*), Human herpesvirus 8 (*Herpesviridae*)	Stimulation of cell-mediated immunity and induction of cellular antiviral state	Injection (inhaled)	(Pegylated recombinant protein)
Lamivudine (Epivir)	L-2′,3′-dideoxy-3′-thiacytidine	Human immuno-deficiency virus (*Retroviridae*), Hepatitis B virus (*Hepadnaviridae*)	Nucleoside analog reverse transcriptase inhibitor (chain terminator)	Oral	

Laninamivir	(4*S*,5*R*,6*R*)-5-acetamido-4-carbamimidamido-6-[(1*R*,2*R*)-3-hydroxy-2-methoxypropyl]-5,6-dihydro-4*H*-pyran-2-carboxylic acid	Influenza A and B (*Orthomyxoviridae*)	Neuraminidase inhibitor	Inhaled	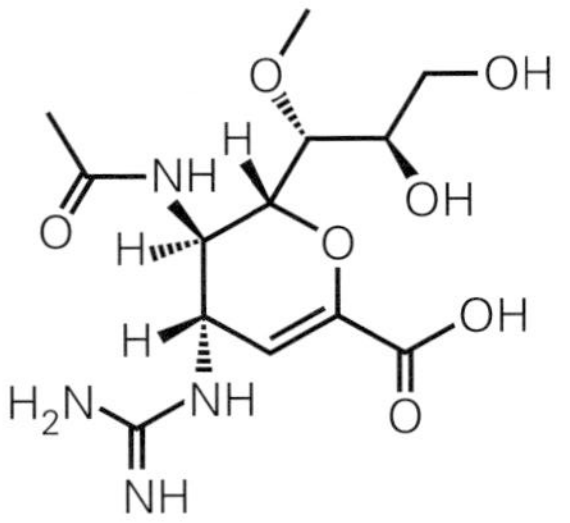
Lopinavir (with ritonavir as Kaletra)	(2*S*)-N-[(2*S*,4*S*,5*S*)-5-{[2-(2,6-dimethylphenoxy)acetyl]amino}-4-hydroxy-1,6-diphenylhexan-2-yl]-3-methyl-2-(2-oxo-1,3-diazinan-1-yl)butanamide	Human immunodeficiency virus (*Retroviridae*)	Protease inhibitor	Oral	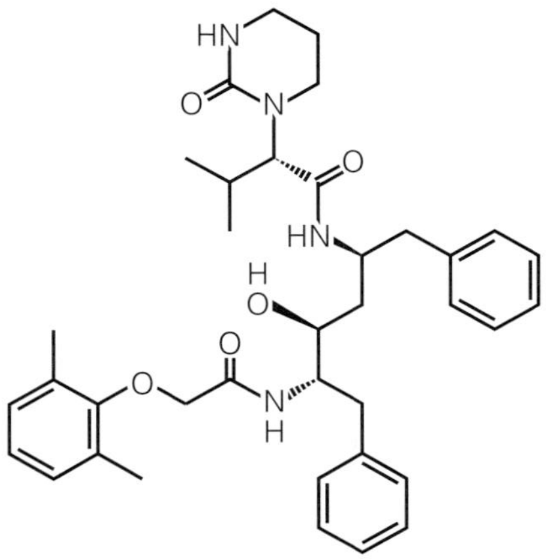
Maraviroc (Selzentry)	4,4-difluoro-*N*-{(1S)-3-[3-(3-isopropyl-5-methyl-4*H*-1,2,4-triazol-4-yl)-8-azabicyclo[3.2.1]oct-8-yl]-1-phenylpropyl}cyclohexanecarboxamide	Human immunodeficiency virus (*Retroviridae*)	Chemokine receptor antagonist (CCR5)	Oral	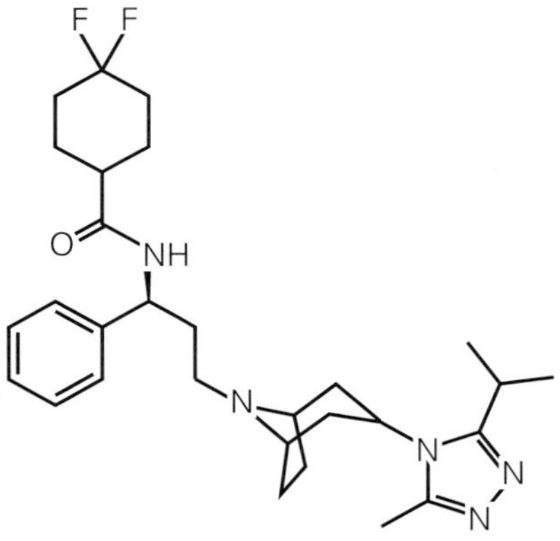

Drug name (Trade name**)	Chemical name*	Virus target(s)	Mode of action	Route of admini-stration	Structure
Nelfinavir (Viracept)	2-[2-hydroxy-3-(3-hydroxy-benzoyl)amino-4-phenylsulfanyl-butyl]-*N*-tertyl-1,2,3,4,4a,5,6,7,8,8a-hydroisoquinoline-3-oxamide	Human immuno-deficiency virus (*Retroviridae*)	Protease inhibitor	Oral	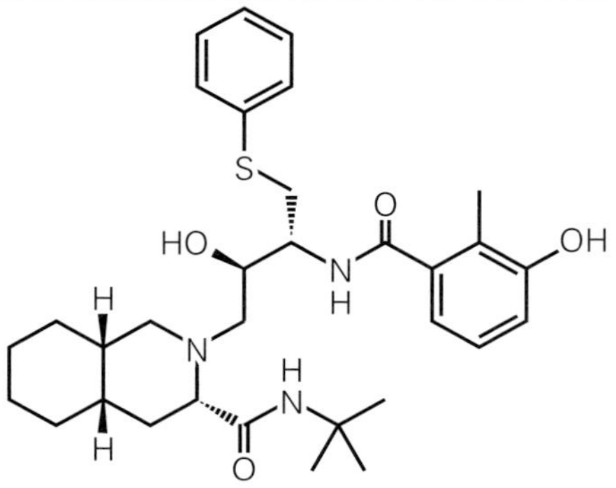
Nevirapine (Viramune)	1-cyclopropyl-5,11-dihydro-4-methyl-6H-dipyrido [3,2-b:2′,3′-e] [1,4] diazepin-6-one	Human immuno-deficiency virus (*Retroviridae*)	Non-nucleoside reverse transcriptase inhibitor	Oral	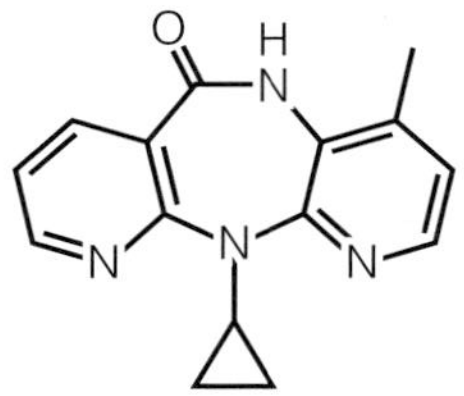
Oseltamivir (Tamiflu)	(3*R*,4*R*,5*S*)-4-acetylamino-5-amino-3-(1-ethylpropoxy)-1-cyclohexene-1-carboxylic acid ethyl ester	Influenza A and B (*Orthomyxo-viridae*)	Neuraminidase inhibitor	Oral	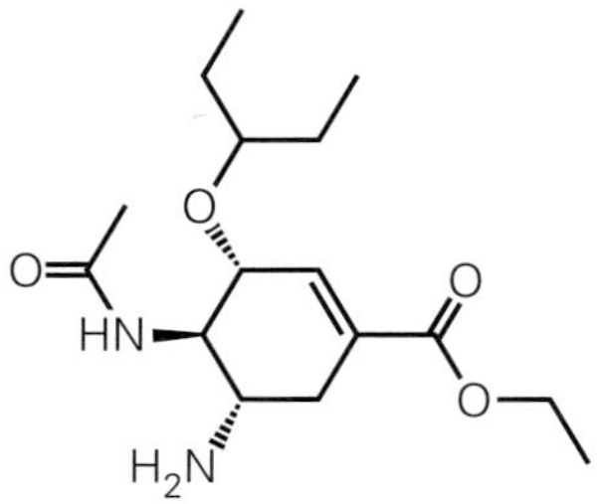
Palivizumab (Synagis)	(Monoclonal antibody)	Respiratory syncytial virus (*Paramyxo-viridae*)	Uptake inhibitor (neutralizing antibody)	Injection	(Humanized monoclonal antibody)
Penciclovir (Denavir, Fenistil)	2-amino-9-[4-hydroxy-3-(hydroxymethyl) butyl]-3*H*-purin-6-one	Herpes simplex virus (*Herpesviridae*)	DNA polymerase inhibitor (nucleoside analog)	Oral	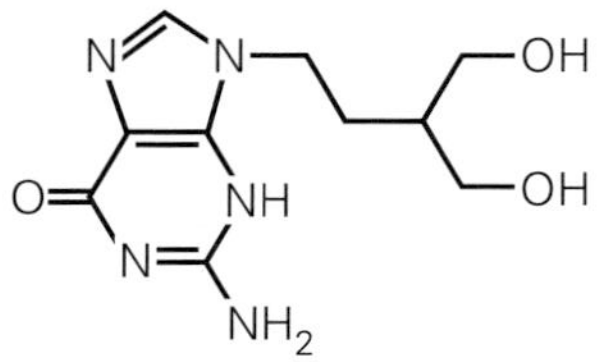

Peramivir	(1*S*,2*S*,3*S*,4*R*)-3-[(1S)-1-acetamido-2-ethyl-butyl]-4-(diaminomethyl-ideneamino)-2-hydroxy-cyclopentane-1-carboxylic acid	Influenza A and B (*Orthomyxo-viridae*)	Neuraminidase inhibitor	Intravenous	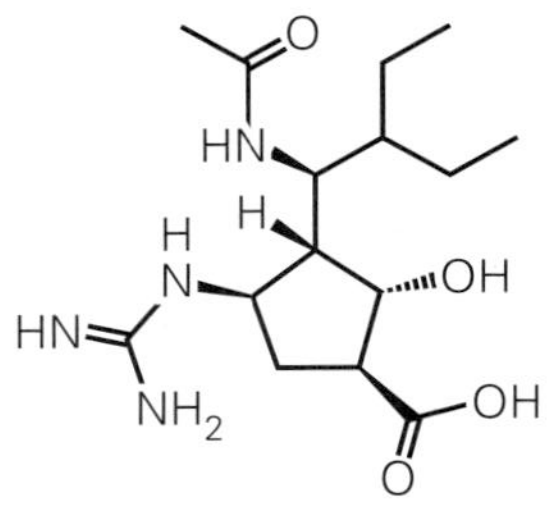
Raltegravir (Isentress)	*N*-(2-(4-(4-fluorobenzyl-carbamoyl)-5-hydroxy-1-methyl-6-oxo-1,6-dihydro-pyrimidin-2-yl)propan-2-yl)-5-methyl-1,3,4-oxadiazole-2-carboxamide	Human immuno-deficiency virus (*Retroviridae*)	Integrase inhibitor	Oral	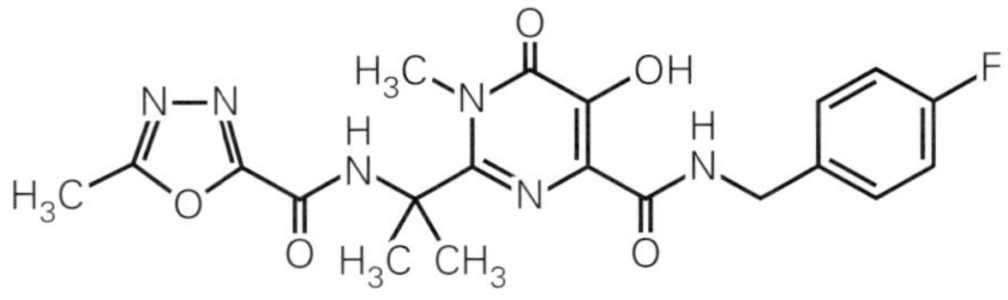
Ribavirin (Copegus, Rebetol, Ribasphere, Vilona, Virazole, and generics)	1-b-D-ribofuranosyl-1,2,4 triazole 3-carboxamide	Respiratory syncytial virus (*Paramyxo-viridae*), Lassa fever virus (*Arenaviridae*), yellow fever virus, hepatitis C virus (*Flaviviridae*) (in combination with interferon alfa), other viruses inhibited *in vitro*	Several proposed, none certain (ribonucleoside analog)	Oral, inhaled	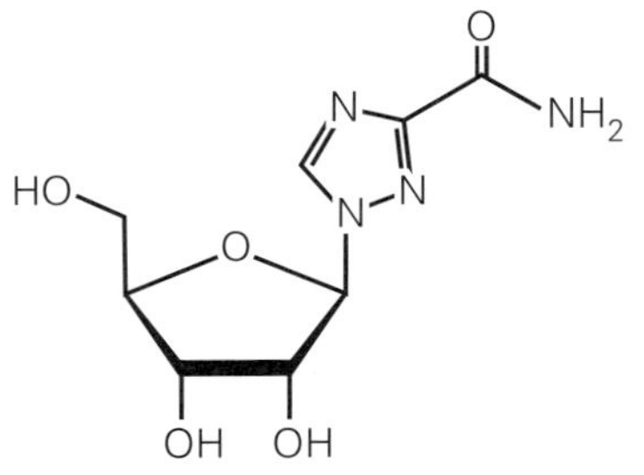
Rimantadine (Flumadine)	1-(1-adamantyl) ethanamine	Influenza A (*Orthomyxo-viridae*)	As amantadine	Oral	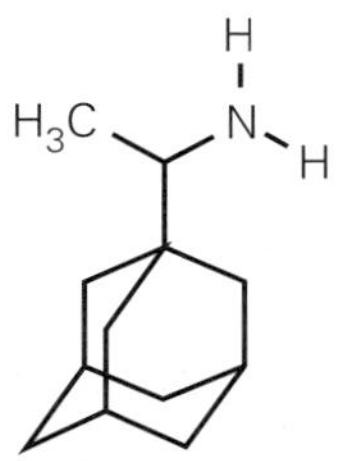

Drug name (Trade name**)	Chemical name*	Virus target(s)	Mode of action	Route of admini-stration	Structure
Ritonavir (Norvir)	1,3-thiazol-5-ylmethyl[3-hydroxy-5-[3-methyl-2-[methyl-[(2-propan-2-yl-1,3-thiazol- 4-yl)methyl] carbamoyl]amino-butanoyl]amino-1,6-diphenyl-hexan-2-yl] aminoformate	Human immuno-deficiency virus (*Retroviridae*)	Protease inhibitor	Oral	
Saquinavir (Fortovase, Invirase)	*N*-[1-benzyl-2-hydroxy-3-[3-(tert-butylcarbamoyl)-1,2,3,4,4a,5,6,7,8,8a-decahydroisoquinolin-2-yl]-propyl]-2-quinolin-2-ylcarbonylamino-butanediamide	Human immuno-deficiency virus (*Retroviridae*)	Protease inhibitor	Oral	
Stavudine (Zerit)	1-[5-(hydroxymethyl)-2,5-dihydrofuran-2-yl]-5-methyl-1*H*-pyrimidine-2,4-dione	Human immuno-deficiency virus (*Retroviridae*)	Nucleoside analog reverse transcriptase inhibitor (chain terminator)	Oral	

Tenofovir (as tenofovir disoproxil fumarate, Viread)	({[(2*R*)-1-(6-amino-9*H*-purin-9-yl)propan-2-yl]oxy}methyl) phosphonic acid	Human immuno-deficiency virus (*Retroviridae*)	Nucleotide analog reverse transcriptase inhibitor (chain terminator)	Oral	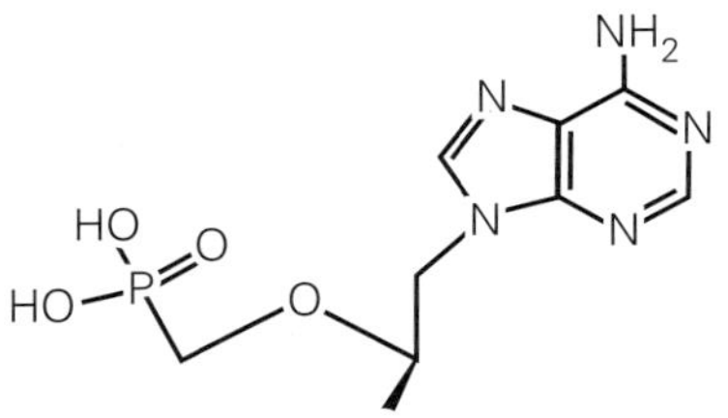
Tipranavir (Aptivus)	[*R*-(*R**,*R**)]-*N*-[3-[1-[5,6-dihydro-4-hydroxy-2-oxo-6- (2-phenylethyl)-6-propyl-2*H*-pyran-3-yl] propyl]phenyl]-5-(trifluoromethyl)-2-pyridinesulfonamide	Human immuno-deficiency virus (*Retroviridae*)	Protease inhibitor	Oral	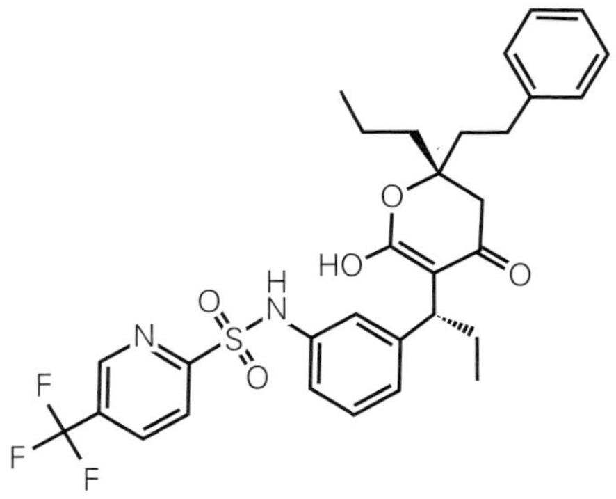
Trifluoro-thymidine (trifluridine, Viroptic)	1-[4-hydroxy-5-(hydroxymethyl) oxolan-2-yl]-5-(trifluoromethyl) pyrimidine-2,4-dione	Herpes simplex virus (*Herpesviridae*)	Incorporated into DNA, disrupts transcription and translation (nucleoside analog)	Ophthalmic	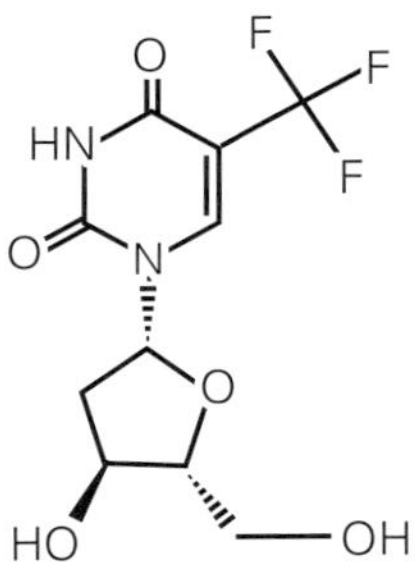
Tromantadine (Viru-Merz)	*N*-1-adamantyl-N-[2-(dimethylamino) ethoxy] acetamide	Herpes simplex virus (*Herpesviridae*)	Alters glycosylation, disrupts virus entry and uncoating	Topical	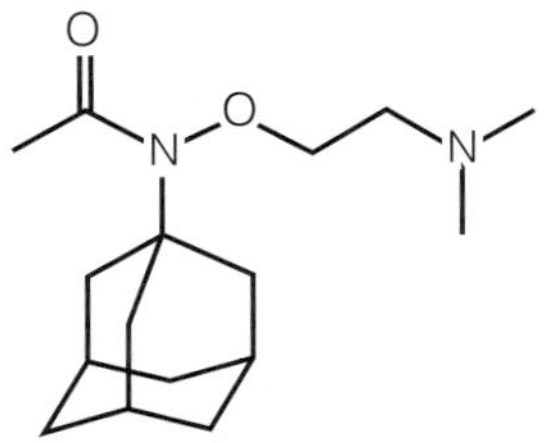

Drug name (Trade name**)	Chemical name*	Virus target(s)	Mode of action	Route of admini-stration	Structure
Valaciclovir (Valtrex)	2-[(2-amino-6-oxo-3,9-dihydropurin-9-yl) methoxy]ethyl-2-amino-3-methyl-butanoate	Herpes simplex virus, varicella-zoster virus (*Herpesviridae*)	DNA polymerase inhibitor (chain terminator; prodrug of aciclovir)	Oral	
Valganciclovir (Valcyte)	2-[(2-amino-6-oxo-3,6-dihydro-9*H*-purin-9-yl)methoxy]-3-hydroxypropyl (2*S*)-2-amino-3-methylbutanoate	Cytomegalovirus (*Herpesviridae*)	DNA polymerase inhibitor (competitive; prodrug of ganciclovir)	Oral	
Vidarabine, adenosine arabinoside (Ara-A, Vira-A)	2-(6-aminopurin-9-yl)-5-(hydroxymethyl) oxolane-3,4-diol hydrate	Herpes simplex virus, varicella-zoster virus (*Herpesviridae*). Other viruses *in vitro*	DNA synthesis inhibitor (unknown method, ribonucleoside analog)	Intravenous, ophthalmic	

Zalcitabine (Hivid)	4-amino-1-[(2*R*,5*S*)-5-(hydroxymethyl) oxolan-2-yl]-1,2-dihydropyrimidin-2-one	Human immuno-deficiency virus (*Retroviridae*)	Nucleoside analog reverse transcriptase inhibitor (competitive and chain terminator)	Oral	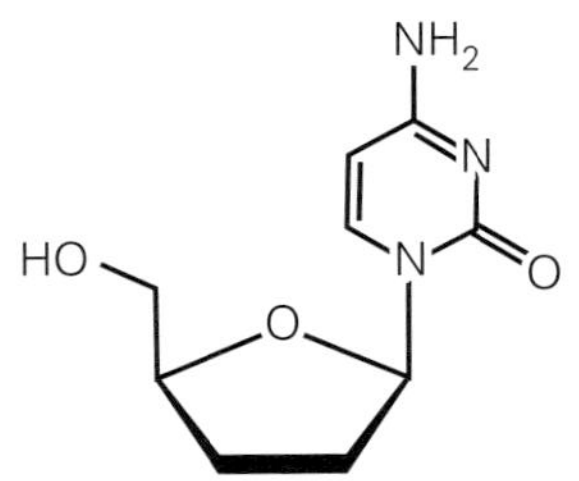
Zanamivir (Relenza)	5-acetamido-4-guanidino-6-(1,2,3-trihydroxypropyl)-5,6-dihydro-4*H*-pyran-2-carboxylic acid	Influenza A and B (*Orthomyxo-viridae*)	Neuraminidase inhibitor	Inhaled	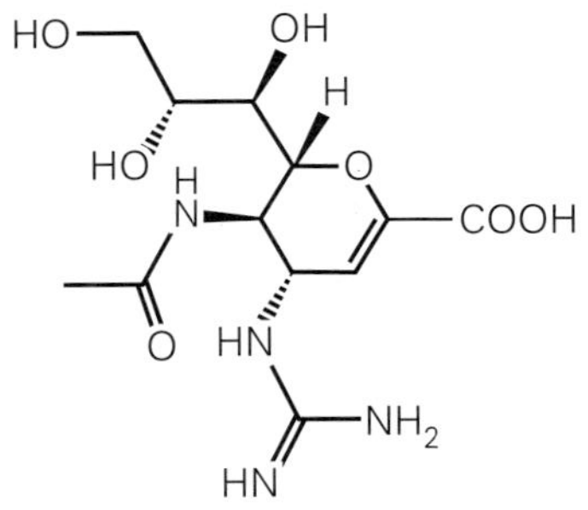
Zidovudine (Retrovir, Retrovis)	1-[(2*R*,4*S*,5*S*)-4-azido-5-(hydroxymethyl) oxolan-2-yl]-5-methyl-pyrimidine-2,4-dione	Human immuno-deficiency virus (Retroviridae)	Nucleoside analog reverse transcriptase inhibitor (chain terminator)	Oral	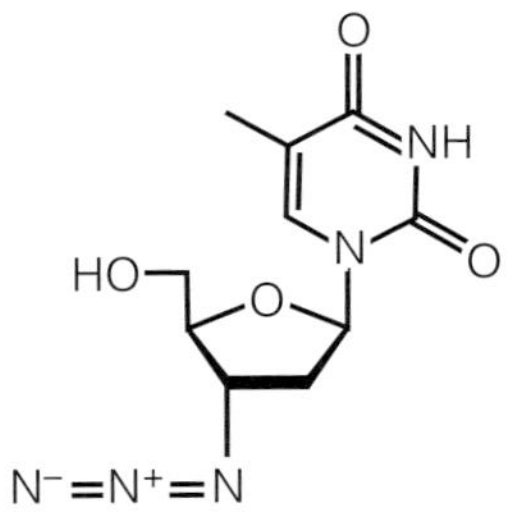

* The formal IUPAC naming system is used here for reasons of standardization, rather than the simpler and possibly more informative names that are sometimes used elsewhere (for example, in Figures 6.5 and 6.9).

** Trade names used here are trademarks (™) or registered trademarks (®).

Glossary

additive effect Effect produced when the combined effect of two drugs working together is the same as adding their individual effects together

adjuvants Substances added to increase the immunogenicity of antigens by a variety of effects. Used to increase the efficacy of vaccines

affinity The binding intensity of a protein for its ligand; more specifically, that of an immunoglobulin for its antigen

affinity maturation The increase in affinity produced by small, sequential changes in antibody structure after the production of the initial antibodies to a novel antigen

ambisense genomes Single-stranded viral genomes which contain both protein-coding (positive sense or sense) and complementary (negative sense or antisense) sequences. The latter need to be transcribed to produce mRNAs that can be used to make proteins

anergy An effect observed when T cells show a limited response to subsequent antigenic stimulation due to the lack of necessary co-stimulatory signals at their first encounter with that antigen

antagonistic effect Effect produced when the combined effect of two drugs working together is less than adding their individual effects together

antibody An immunoglobulin; a glycoprotein produced by B cells in response to antigenic stimulation which combines specific binding properties for its target antigen with immune signaling functions

antigen A protein or other molecule recognized by the immune system as a target

antigene inhibition Blocking of gene expression by a sequence complementary (of complementary base sequence) to the gene

antigenic drift Minor changes in the structure and immunogenicity of antigens, specifically of the surface proteins of influenza virus, caused by mutation

antigenic shift Major changes in the structure and immunogenicity of the surface proteins of influenza virus caused by gene exchange with related viruses

anti-idiotypic antibody Antibody molecules raised against antigen-binding sites of antibodies which thereby mimic the conformation of the antigen

antisense inhibition Blocking of gene expression by a nucleic acid of complementary base sequence (antisense) to the mRNA produced by the gene

antiviral state A complex state induced by interferons in which viral activity is suppressed by a variety of mechanisms

apoptosis Programmed cell death; a process by which a cell disintegrates in a controlled fashion (unlike the uncontrolled cell death referred to as necrosis)

arboviruses Viruses transmitted by arthropods (often insects), typically with a replication stage inside the cells of the insect host

attenuation Weakening of the pathogenic capability of a virus, usually by adaptation to other hosts or culture conditions

autoimmune disease Disease state where the body's own immune system targets self antigens (the body's own structures), causing destructive effects

bacteriophages Viruses that infect bacteria, named for their destructive effects on the bacterial host

Baltimore classification system A system of classification for viruses based on the nature of their genome and its replicative strategy. This system has the virtue of simplicity but has difficulty coping with complex viral systems

biological control The use of biological agents (including viruses) to control (usually by killing) a pest organism

biological weapons Weapons where the active agent is a biological agent (organism), typically a bacterium or virus. Classified as "weapons of mass destruction" and banned under international treaties

branched DNA system A reporter amplification system which produces an enhanced detection signal by the use of a highly branched DNA providing multiple binding sites

cap A 5′-5′-linked methylguanylate structure located at the 5′ end of a eukaryotic mRNA that is required for ribosome binding and subsequent translation. Some viruses can bypass this requirement

capsid The protein structure that contains the nucleic acid genome of a virus

capsomers The protein components of a capsid

caspases Cellular aspartic acid—specific cysteine proteinases involved in apoptosis

caveolae (singular caveola) Small (50–100 nm) pits in the cell membrane that are associated with the protein caveolin. May be used for entry by viruses

CD3 complex Formed of invariant transmembrane proteins, the CD3 complex activates cellular kinase enzymes which phosphorylate (and thus activate) a complex cascade of cellular signaling molecules, transmitting an activation signal to the T cell

CD4 A T cell surface marker protein, the co-receptor for the MHC-II–peptide complex. Found on helper T cells

CD8 A T cell surface marker protein, the co-receptor for the MHC-I–peptide complex. Found on cytotoxic T cells

cell tropism The targeting of specific cell types by viruses, often dependent on surface marker proteins expressed by target cells

clathrin-coated pits Pits at the cell surface associated with the protein clathrin. Mediate transport into the cell by pinching-off to form vesicles. May be used for entry by viruses

clinical trials A structured series of closely monitored and highly regulated tests of a novel medicine in human volunteers which are designed to provide proof of safety and efficacy

clonal selection Process whereby only those immune cells bearing receptors that encounter their specific antigen are selected for proliferation. Also contains the process of clonal deletion, whereby cells that recognize self molecules (those of the organism itself) are destroyed during maturation

complement A complex set of proteins that are present in normal blood and which are part of the innate immune response, as well as enhancers of phagocytosis and of cell killing by antibodies

complement fixation test One of the first tests to be developed for the detection of antibody. Measures the ability of antibodies to fix complement, preventing lysis of indicator red blood cells

complementarity-determining regions Three hypervariable regions formed of paired regions in the light and heavy chains of an immunoglobulin that combine to form the unique antigen-binding site

complementary DNA (cDNA) DNA that is transcribed from and thus of complementary sequence to an mRNA

concatamers Linked polymers of genomic nucleic acids

cytokines Small regulatory proteins that act as cell signaling molecules over short distances. They are produced by a range of cell types and control many aspects of both the innate and the adaptive immune responses

cytology The direct examination of cells within clinical specimens

cytotoxic Lethal to cells

cytotoxic T lymphocytes (CTLs) T cells with the CD8 surface marker protein that kill cells presenting MHC-I–bound antigens (indicating that they are infected). Mechanisms used involve cytokine synthesis, perforin release, and the induction of apoptosis

decatenation Unlinking of subunits from a larger structure

decoy oligonucleotides Nucleic acids corresponding to regulatory regions of the genome or transcripts from it, which bind to regulatory proteins and prevent their function

defective interfering (DI) particles Variant forms of viruses with incomplete (and thus faster replicating) genomes. Since they "steal" the missing functions from complete viruses they can interfere with their function

dendritic cells Highly branched cells of the immune system, involved in antigen presentation

DNA polymerase An enzyme that synthesizes DNA

dot blotting An assay technique where mixed DNA samples are bound as discrete zones or "dots" on a membrane. Following blocking of unused binding sites on the membrane, the presence of specific sequences is detected using probe nucleic acids

electrophoresis The resolution of biological macromolecules by their size, using movement through a gel matrix under the influence of an electric current

emerging disease Defined by the Centers for Disease Control and Prevention as a disease of infectious origin with an incidence that has increased within the last two decades, or threatens to increase in the near future

endogenous retroviruses DNA sequences within a host genome that appear to be retrovirus genomes (usually defective and/or incomplete). They are passed on to later generations with the host genome. A few may be able to reactivate and produce virus, but the vast majority are unable to do this

endosome A membrane-bound space within a eukaryotic cell. Involved in intracellular transport

envelope The outer lipid membrane of some viruses. Derived from host cell membranes

epidemic Major outbreak of an infectious disease

episome A large extrachromosomal plasmid-like DNA element with the ability to integrate into the host genome. Some viral genomes may be maintained as episomes

epitope A region of an antigen which reacts with elements of the immune system. The chemical nature of epitopes is highly variable

error catastrophe A situation where the high mutation rate of viral RNA genomes exceeds the level where viability of the population can be maintained

extinction Elimination of a specific species or strain

family A taxonomical classification below order and above genus. Virus family names end in "*viridae*"

fomite An inanimate object on which infectious organisms may be carried or transferred

frameshifting A shift during the translation of a messenger RNA which results in reading in a different frame. Since three RNA bases form a codon, specifying each amino acid, a shift of one or two bases will result in reading of two bases from one codon with one base from an adjacent codon, changing the amino acid specified. May terminate translation by the introduction of new stop codons

fusion inhibitors Molecules that inhibit the fusion of membranes. May be used as antiviral drugs, blocking viral fusion and entry at the cell surface

gene cloning The transfer of genetic material from one organism to another. Usually uses intermediate stages, which may be viral

gene expression Traditionally, the transcription of a gene to produce messenger RNAs and their subsequent translation to proteins. However, some genes may be expressed to produce non-protein elements such as regulatory RNAs

gene probing Use of nucleic acid probes to identify specific genetic sequences

gene therapy The introduction of a therapeutic gene into cells in order to correct a genetic defect

genetic engineering The application of nucleic acid-based techniques including cloning to produce novel organisms expressing genes from other sources

genetic manipulation; *see* **genetic engineering**

genome The complete hereditary information of an organism. For viruses, this may be DNA or RNA

genomics The study of genomes, usually by analysis of the nucleic acid sequence. Information on genome sequences requires extensive interpretation

genus A taxonomical classification below family and above species. For viruses, genus names end in "*virus*"

germ-line gene therapy Gene therapy where therapeutic genes are introduced into the reproductive system and passed on to the next generation. Due to the potential for introducing permanent changes, this remains highly contentious

glycoproteins Proteins with covalently attached sugar molecules

helper T cells T cells with the CD4 surface marker protein that promote the immune response via a complex system of signaling and cytokines, following antigenic stimulation via the MHC-II pathway

humanized monoclonal antibody A monoclonal antibody incorporating the complementarity-determining regions (and thus the antigen-binding site) from an antibody produced in a nonhuman species that have been transferred into the protein structure of a human antibody. Allows therapeutic use without stimulating a strong immune response

humoral immune response; *see* **serological immune response**

hybridization The sequence-specific binding of two nucleic acids with complementary sequences. Varying reaction conditions may vary the level of sequence-matching required

immunoassays Assays evaluating elements of the immune response, in particular the presence of specific antibodies

immunoglobulin; *see* **antibody**

***in situ* hybridization** Nucleic acid detection system used on biological specimens without the extraction of the nucleic acid from the original specimen. Can permit localization within cellular structures

inclusion bodies Bodies of viral material formed inside infected cells. Much larger than individual viruses, they can be detected by light microscopy, indicating infection

induration Localized hardening of a soft tissue or organ, often the result of inflammation

innate immune system The non-adaptive immune system, which produces a response to foreign antigens but which does not adapt to target them specifically

integrated pest management The use of multiple complementary systems of pest control to produce a stronger overall effect than can be achieved with a single approach

interferons A class of cytokines that mediate many responses, including the antiviral state within cells

International Committee for the Taxonomy of Viruses (ICTV) The international body responsible for the naming of viruses

Koch's postulates A set of four rules used to define an infectious organism as responsible for a particular disease. Since they date from the late-nineteenth century, they are no longer accurate in many cases and updated versions have been proposed

latency Of a virus, entry into an inactive state (that may involve integration into the host genome) permitting long-term survival within the host. Return to an active state may occur due to a variety of triggers

lead compound A candidate drug that has passed early stages of development and has been identified as promising enough to justify more expensive and complex later-stage testing

ligase chain reaction (LCR) A variant form of nucleic acid amplification assay. Relies on ligation (joining) rather than extension of primers. LCR primers carry molecules allowing their detection

lysis Bursting of a virus-infected cell, releasing the next generation of viruses

macrophages Phagocytic (myeloid) white blood cells with functions in innate immunity and antigen presentation. Bear multiple receptors for bacterial components

macropinocytosis Uptake of larger volumes of extracellular fluid by the cell. May be used for virus entry

major histocompatibility complex (MHC) The proteins of the MHC (known as human leukocyte antigen or HLA in humans) are expressed from a cluster of highly variable genes and present antigens to the immune system at the cell surface. The main components are the MHC-I proteins that present antigens processed within infected cells, and the MHC-II proteins that express proteins processed by specialized cells of the immune system. Other MHC proteins play a variety of roles in the function of the immune system. They play a key role in identifying cells as of "self" or foreign origin

membrane rafts Areas of the cell membrane with altered lipid content (in particular, enriched in cholesterol and sphingolipids) that have altered mechanical properties and are associated with a variety of functions, including viral budding from the cell

memory cells Cells of the immune system that provide immunological memory, enabling a more rapid and stronger response when re-exposed to an antigen

memory T cells T cells that provide immunological memory, enabling a more rapid and stronger response when re-exposed to an antigen

messenger RNA RNA that carries information from the genome to the translation machinery of the cell, where its sequence is used to produce proteins

metagenomics The simultaneous study of large numbers of genomes from environmental or clinical samples

microarrays A two-dimensional array of test materials (often nucleic acid probes) on a small area of a solid substrate that permits simultaneous assay for a wide variety of targets

micro-RNAs (miRNAs) Small RNAs (20–25 bases) found in all eukaryotic cells. They bind to complementary sequences on target mRNAs, limiting their translation, and can shut down the expression of target genes

mobile genetic elements DNA sequences able to move within the cell and/or within the genome

monoclonal antibodies Antibody derived from a single clone of a stimulated B cell, and thus targeting a single epitope. Their high specificity and flexibility has resulted in a wide variety of uses

mucosal immunity Immunity at the mucosal surface. The main antibody type in this location is secretory (dimeric) IgA, along with a limited repertoire of immune cells including neutrophils and innate-type T cells

negative sense Of a single-stranded nucleic acid, contains the complementary base sequence to messenger RNAs. Requires transcription to (positive sense) mRNAs to produce proteins. Viruses with negative-sense RNA genomes need to carry an enzyme capable of producing these mRNAs from the genome since this activity is lacking in cells

neuraminidase inhibitors Molecules that inhibit the function of neuraminidases (sialidases). These enzymes cleave sugar structures and are involved in the passage of the influenza virus both to and away from the host cell. Thus neuraminidase inhibitors are used as antiviral agents

neutralization Prevention of virus infectivity by antibodies. May be direct (by blocking of essential structures and activities) or complement-mediated (by signaling to and activating the complement system)

NK cells Natural killer cells; specialized cytotoxic lymphocytes lacking the T-cell receptor, they recognize and destroy cells lacking "self" MHC-I antigens. An important part of the system of innate immunity, they are activated by cytokines

non-nucleoside reverse transcriptase inhibitors (NNRTIs) A class of antiviral drugs that inhibits the activity of the reverse transcriptase enzyme of the *Retroviridae*. Unlike nucleoside analog-based inhibitors, they bind outside the active site and may generate very rapid resistance

non-self Lacking the cell surface marker proteins (MHC or HLA) that mark a cell as belonging to the host organism

notifiable diseases A list of diseases (which varies from country to country) where cases must be notified to the relevant authorities

nucleic acid sequence-based amplification (NASBA) An alternative system of amplification-based nucleic acid detection. The system is isothermal (does not require heating/cooling cycles), unlike PCR

nucleocapsid Nucleic acid plus capsid. A complex of proteins and the viral genomic nucleic acid

nucleoside analogs Molecules that resemble nucleosides. Used as antiviral drugs, where their incorporation into the forming nucleic acid disrupts its synthesis

occlusion bodies Bodies formed by *Baculoviridae*, formed of viral nucleocapsids embedded in a protein matrix. Released from lysing cells, they stabilize the virus and aid in the transmission of infection

off-label drug use Use of an approved drug for another indication, for which it has not yet been licensed for commercial use

oncogene A gene associated with the formation of cancers or, more broadly, with cancer-associated changes in cells

oncogenesis The formation of a cancer

opsonization Targeting of an antigen for an immune response, in particular for phagocytosis

order A level of taxonomical classification above family. The highest level of classification for a virus (though many levels exist above this for other organisms), ending "*virales*"

pandemic A worldwide outbreak of an infectious disease

passive immunity Use of transferred antibodies to provide immunity. "Passive" refers to bypassing the need for the recipient to make their own antibodies

phagocytosis The process of internalizing solid material (including other cells), carried out by specialized cells of the immune system

phosphorothioate modification Replacement of non-linking oxygen residues with sulfur in the phosphate backbone of a nucleic acid molecule, resulting in increased stability

polyadenylation A modification present at the 3′ terminus of eukaryotic mRNAs, consisting of a chain of adenosines

polymerase An enzyme that produces nucleic acids (RNA or DNA)

polymerase chain reaction (PCR) A system of nucleic acid detection based on the binding of primers to their target followed by extension. Multiple reaction cycles allow the production of very high levels of product from very low levels of target nucleic acid, since the product of each reaction can be used as a template for further reactions

positive sense Of a single-stranded nucleic acid, contains the base sequences of messenger RNAs. Positive-sense viral RNA genomes are usually able to be translated directly into proteins

primer Short, sequence-specific nucleic acid probes used in amplification-based assays

prions Protein-only infectious agents containing a modified isoform of a cellular protein. They appear to induce the production of new prions via a biocatalytic effect on normal cell proteins. Prions are highly resistant to inactivation

prodrugs A precursor form of an active drug that is converted to the active form by metabolism within the body. Typically used because of superior delivery of the prodrug form

proofreading Of a replicating nucleic acid, checking of the newly attached nucleotide against the template sequence, with removal if these do not match. This increases the accuracy of replication since such removal allows the subsequent attachment of a correct match to the template

prophylaxis The use of a treatment to prevent the occurrence of (rather than to resolve) a condition

protease inhibitors Molecules that inhibit the cleavage of a protein. May be used as antiviral drugs, as with inhibitors of the maturational protease (proteinase) of HIV

proviral integration Integration of a DNA copy of the RNA viral genome of the *Retroviridae* into the cellular genome

pseudovirion A synthetic virus-like structure. May be entirely lacking in nucleic acid, although some forms may be used for gene transfer. Unable to replicate

quantitative real-time PCR Due to the multiple rounds of amplification that occur, a simple PCR with quantitation of product only after the reaction is completed cannot be used to provide information on the amount of the target nucleic acid that is present. Quantitative real-time PCR is able to quantify the amount of nucleic acid target present by tracking production during the reaction process

quasispecies A cluster of related genomes arising from replication of a single species with a high mutation rate. Seen in particular with RNA viruses. The effect of such variation is that the variant forms needed to respond to a particular stimulus (e.g. an antiviral drug) are already present in the population

receptor A molecule on the cell surface (often but not always a protein) to which a virus binds prior to entering the cell. The presence of a specific receptor is often key to the capability of the virus to infect a particular type of cell

restriction enzymes Endonuclease enzymes that recognize specific sequences (usually palindromic sequences of four to six bases) within a target DNA molecule and then cut the DNA chain. Type II restriction enzymes cut at their recognition site and are used in a wide variety of molecular biological techniques

retrotransposons A form of mobile genetic element that replicates via an RNA stage that is copied back to DNA which can then integrate into the genome

reverse transcriptase An enzyme (first identified in retroviruses) that produces DNA from an RNA template

ribosomal RNA The RNA present (along with proteins) in the ribosomal structure, where it interacts with mRNAs and tRNAs

ribozyme From ribonucleic acid enzyme; a catalytic RNA which can mediate reactions (usually cleavage) in either itself or another RNA

RNA interference (RNAi) A system which controls the activity of genes through the actions of differing classes of small RNA molecules

RNA splicing The removal of internal sequences ("introns") from RNA transcripts to form a mature mRNA. Occurs with many eukaryotic mRNAs

RNA-induced silencing complex (**RISC**) A complex of proteins and RNA that uses an siRNA or an miRNA to mediate recognition of a complementary mRNA. Binding to a complementary strand (with variable levels of sequence specificity) triggers the degradation of that RNA

satellite virus A sub-virus that produces its own coat proteins but requires assistance from a co-infecting helper virus to replicate. More common in viruses infecting plants. Adeno-associated virus is a satellite virus infecting humans

scaffolding proteins Proteins that are required to form a viral capsid but which are removed during the final stages of assembly

segmented genome A viral genome formed of more than one segment of nucleic acid. May be carried in single or multiple virus particles

self antigens Antigens (often expressed on the cell surface) that mark a cell or protein as belonging to the host organism

self-sustained sequence replication (**3SR**); *see* **nucleic acid sequence-based amplification** (**NASBA**)

serological immune response That arm of the immune response that relies on antibodies

slow virus A virus infection that takes a very long time to produce a pathological effect, typically with very low levels of viral activity and a fatal outcome. Often affects the central nervous system

small interfering RNA (**siRNA**) Double-stranded small RNAs (20–25 base pairs) derived from larger dsRNAs that specifically inhibit gene expression

Southern blotting Technique for transferring nucleic acids (separated by size using electrophoresis) from a gel onto an adjacent membrane, to which they bind. After blocking the remaining binding sites on the membrane, probe nucleic acids may be used to detect specific sequences

species A taxonomical classification below genus. The lowest level of formal classification. When used for viruses many of the traditional definitions do not apply, and a viral species is a much looser definition than is the equivalent term used for more complex organisms

sterilizing immunity A level of immunity which completely removes the infectious agent which is targeted

stop codon A codon (three-base sequence) which, when read at the ribosome, stops the translation of an mRNA

subfamily An intermediate level of taxonomical classification between family and genus; not always used. For viruses, subfamily names end in "*virinae*"

supervirus A term loosely used to describe unusually large and complex viruses

surveillance In the context of virology, monitoring of the environment for instances of particular infections or for novel infections

sylvatic cycle Circulation of a virus in a host or system of hosts that is restricted to wild animals

synergistic effect Effect produced when the combined effect of two drugs working together is greater than adding their individual effects together

tegument A region of a herpesvirus lying between the nucleocapsid and the viral envelope; contains a large number of proteins that assist with infection of cells

Th1 A class of helper T lymphocytes primarily associated with the inflammatory and cytotoxic responses

Th2 A class of helper T lymphocytes primarily associated with B-cell activation and the serological response

therapeutic index The ratio of the level of a drug that produces toxic effects for the host organism to that required for the desired beneficial effect (e.g. antiviral effect)

tolerance Of the immune system, the process by which an immune response is not generated to a potentially immunogenic antigen

Toll-like receptors A family of at least ten transmembrane proteins, named for the Toll receptor first identified in the fruit fly. Toll-like receptors (TLRs) recognize specific molecular structures associated with pathogens and play an important role in inflammation and the innate immune response

transcription-based amplification (**TBA**); *see* **nucleic acid sequence-based amplification** (**NASBA**)

transduction The transfer of DNA (usually bacterial) between bacterial cells by a bacteriophage

transfection The introduction of DNA from an external source into a eukaryotic cell without the use of a virus

transfer RNA (**tRNA**) An RNA molecule that is attached to a specific amino acid and recognizes a specific codon in an mRNA during translation at the ribosome. The tRNA then mediates the incorporation of its attached amino acid into a forming protein

transformation The introduction of DNA from an external source into a prokaryotic cell without the use of a virus

transition state mimetics Drugs that mimic a transitional state of a substrate during an enzymic reaction, so that they can bind with high efficiency into the active site of the target enzyme

tumor suppressor genes Genes associated with controlling the development of cancers, often by effects on the cell cycle, apoptosis, or DNA repair

urban (**amplification and infection**) **cycle** Circulation of a virus in a host, or system of hosts, that includes humans (or domesticated animals)

vaccination Use of a weakened or partial form of a pathogenic organism to induce an immune response that protects against the virulent form

variolation Use of deliberate infection with a less pathogenic form of smallpox to protect against infection with more damaging forms

vector (1) A nucleic acid such as a plasmid or viral genome used to transfer genetic material; (2) A system used to carry the antigenic material (gene or protein) in a vaccine; (3) An additional host species involved in the transmission of an infectious agent, either passively (by simple transfer of the agent) or actively (by replication of the agent in the vector species)

viral vector A virus used as a vector for the transfer of genetic material

virion A complete virus particle

virocell concept The idea that a virus is simply a spore or a seed, and represents only the inactive form of virus. In this concept the "real" form of a virus is seen when it is replicating in an infected cell, in particular when viruses establish specific foci of infection in the cell, as with the large DNA viruses

viroids Sub-viral infectious agents consisting of small (220–375 bases) single-stranded RNAs with extensive internal base pairing. Other than the use of cellular metabolism they have no protein stage in their replication, and have no requirement for a helper virus. Found only in plants, though the hepatitis delta agent appears to be related to viroids

virotherapy The use of live viruses as therapeutic agents, for example in the destruction of tumors

virus-directed enzyme prodrug therapy The use of viruses to target and deliver precursor forms of drugs (prodrugs) to specific cells

virusoids Very small (200–400 bases) circular, single-stranded RNAs that replicate using the functions of a helper virus, and which take proteins from the helper. Only known to infect plants. The term is sometimes used more widely to include satellite viruses

xenotransplantation The transplantation of organs from one species to another

X-ray crystallography A means of analyzing the structure of crystals by the scattering of X-rays

yeast expression systems Yeast cells used for the expression of cloned genes and the production of proteins from them

zoonosis An infection transferred from an animal host to a human (or sometimes, from a human to an animal)

Index

Page numbers in **boldface** refer to major discussion of a topic. Page numbers followed by B refer to boxes, those followed by F refer to figures and those followed by T refer to tables.

B

N

O

Q

R

S

W

X

Y

Z